Sociobiology

THE ABRIDGED EDITION

By the Same Author

On Human Nature

Caste and Ecology in the Social Insects
with George F. Oster

Sociobiology: The New Synthesis

Life on Earth
with Thomas Eisner, Winslow R. Briggs, Richard E. Dickerson, Robert L. Metzenberg, Richard D. O'Brien, Millard Susman, William E. Boggs

The Insect Societies

A Primer of Population Biology
with William H. Bossert

The Theory of Island Biogeography
with Robert H. MacArthur

Sociobiology

THE ABRIDGED EDITION

Edward O. Wilson

THE BELKNAP PRESS
OF HARVARD UNIVERSITY PRESS
Cambridge, Massachusetts
and London, England · 1980

Printed in the United States of America

Library of Congress Cataloging in Publication Data

Wilson, Edward Osborne, 1929-
Sociobiology.

Bibliography: p.
Includes index.
1. Social behavior in animals. I. Title.
QL775.W539 591.5 79-17387
ISBN 0-674-81623-4
ISBN 0-674-81624-2 pbk.

Preface to the Abridged Edition

Modern sociobiology is being created by gifted investigators who work primarily in population biology, invertebrate zoology, including entomology especially, and vertebrate zoology. Because my training and research experience were fortuitously in the first two subjects and there was some momentum left from writing *The Insect Societies*, I decided to learn enough about vertebrates to attempt a general summary. The result was *Sociobiology: The New Synthesis*, published in 1975. The book met with substantial critical success. It has been widely used throughout the world as a source book in research, an advanced textbook, and a general reference work. However, its large size (697 double-column pages) and necessarily high cost prevented it from reaching much of the large audience of lay readers and students who have become interested in sociobiology and in the manner in which this particular work formulated it. General readers were further discouraged by the thickets of technical discussions and data summaries. In the present version, *Sociobiology: The Abridged Edition*, I have trimmed the text down to its essential introductory parts and most interesting case histories, while retaining the full basic structure of the original book. This shortened version is intended to serve both as a textbook and a semi-popular general account of sociobiology. Because of the unusual amount of interest and commentary it has generated, I have left the final chapter on human social behavior (Chapter 26 in the abridged version) virtually intact.

In writing on general sociobiology, I have had to venture into many disciplines in which I have no direct experience, and thus had to rely on the expertise and advice of colleagues at Harvard University and elsewhere. The generosity they displayed, guiding me patiently through films and publications, correcting my errors, and offering the kind of enthusiastic encouragement usually reserved for promising graduate students, is a testament to the communality of science.

These scientists also critically read most of the chapters in early draft. I am especially grateful to Robert L. Trivers for reading most of the book and discussing it with me from the time of its conception. Others who reviewed portions of the manuscript, with the chapter numbers listed after their names, are Ivan Chase (13), Irven DeVore (26), John F. Eisenberg (22, 23, 24, 25), Richard D. Estes (23), Robert Fagen (1–5, 7), Madhav Gadgil (1–5), Robert A. Hinde (7), Bert Hölldobler (8–13), F. Clark Howell (26), Sarah Blaffer Hrdy (1–13, 15–16, 26), Alison Jolly (25), A. Ross Kiester (7, 11–13), Bruce R. Levin (4, 5), Peter R. Marler (7), Ernst Mayr (11–13), Donald W. Pfaff (11), Katherine Ralls (15), Jon Seger (1–6, 8–13, 26), W. John Smith (8–10), Robert M. Woollacott (18), James Weinrich (1–5, 8–13), and Amotz Zahavi (5).

Illustrations, unpublished manuscripts, and technical advice were supplied by R. D. Alexander, S. A. Boorman, Jack Bradbury, F. H. Bronson, W. L. Brown, Noam Chomsky, P. A. Corning, Iain Douglas-Hamilton, Mary Jane West Eberhard, John F. Eisenberg, R. D. Estes, Charles Galt, Valerius Geist, Peter Haas, W. J. Hamilton III, Bert Hölldobler, Sarah Hrdy, Alison Jolly, J. H. Kaufmann, M. H. A. Keenleyside, A. R. Kiester, Hans Kummer, J. A. Kurland, M. R. Lein, B. R. Levin, P. R. Levitt, P. R. Marler, Ernst Mayr, G. M. McKay, D. B. Means, A. J. Meyerriecks, Martin Moynihan, R. A. Paynter, Jr., Katherine Ralls, Lynn Riddiford, P. S. Rodman, L. L. Rogers, Thelma E. Rowell, W. E. Schevill, N. G. Smith, Judy A. Stamps, R. L. Trivers, J. W. Truman, Jan Wind, E. N. Wilmsen, E. E. Williams, and D. S. Wilson.

Kathleen M. Horton assisted closely in bibliographic research, checked many technical details, and typed the manuscript through two intricate drafts. Nancy Clemente edited the manuscript, providing many helpful suggestions concerning organization and exposition.

Sarah Landry executed the drawings of animal societies presented in Chapters 19–26. In the case of the vertebrate species, her compositions are among the first to represent entire societies, in the correct demographic proportions, with as many social interactions displayed as can plausibly be included in one scene. In order to make the drawings as accurate as possible, we sought and were generously given the help of the following biologists who had conducted research on the sociobiology of the individual species: Robert T. Bakker (reconstruction of the appearance and possible social behavior of dinosaurs), Brian Bertram (lions), Iain Douglas-Hamilton (African elephants), Richard D. Estes (wild dogs, wildebeest), F. Clark Howell (reconstructions of primitive man and the Pleistocene mammal fauna), Alison Jolly (ring-tailed lemurs), James Malcolm (wild dogs), John H. Kaufmann (whip-tailed wallabies), Hans Kummer (hamadryas baboons), George B. Schaller (gorillas), and Glen E. Woolfenden (Florida scrub jays). Elso S. Barghoorn, Leslie A. Garay, and Rolla M. Tryon added advice on the depiction of the surrounding vegetation. Other drawings in this book were executed by Joshua B. Clark, and most of the graphs and diagrams by William G. Minty.

Certain passages have been taken with little or no change from *The Insect Societies*, by E. O. Wilson (Belknap Press of Harvard University Press, 1971); these include short portions of Chapters 1, 3, 6, 8, 9, 13, 14, and 16 in the present book as well as a substantial portion of Chapter 19, which presents a brief review of the social insects. Other excerpts have been taken from *A Primer of Population Biology*, by E. O. Wilson and W. H. Bossert (Sinauer Associates, 1971), and *Life on Earth*, by E. O. Wilson et al. (Sinauer Associates, 1973). Other portions come from my article "Group Selection and Its Significance for Ecology" (*BioScience*, vol. 23, pp. 631–638, 1973), copyright © 1973 by the President and Fellows of Harvard College. Other passages have been adapted from various of my articles in *Bulletin of the Entomological Society of America* (vol. 19, pp. 20–22, 1973); *Science* (vol. 179, p. 466, 1973; copyright © 1969, 1973, by the American Association for the Advancement of Science); *Scientific American* (vol. 227, pp. 53–54, 1972); *Chemical Ecology* (E. Sondheimer and J. B. Simeone, eds., Academic Press, 1970); *Man and Beast: Comparative Social Behavior* (J. F. Eisenberg and W. S. Dillon, eds., Smithsonian Institution Press, 1970). The quotations from the Bhagavad-Gita are taken from the Peter Pauper Press translation. The editors and publishers are thanked for their permission to reproduce these excerpts.

I wish further to thank the following agencies and individuals for permission to reproduce materials for which they hold the copyright: American Association for the Advancement of Science, representing *Science; American Zoologist;* Annual Reviews, Inc.; E. J. Brill Co.; Cambridge University Press; American Society of Ichthyologists and Herpetologists, representing *Copeia;* Dowden, Hutchinson and Ross, Inc.; Duke University Press and the Ecological Society of America, representing *Ecology;* Dr. Mary Jane West Eberhard; W. H. Freeman and Company, representing *Scientific American;* Dr. Charles S. Henry; Dr. J. A. R. A. M. van Hooff; Houghton Mifflin Company; Macmillan Publishing Company, Inc.; Professor Peter Marler; McGraw-Hill Book Company; Methuen and Co., Ltd.; Professor Daniel Otte; Dr. Katherine Ralls; Random House, Inc.; Professor Carl W. Rettenmeyer; the Royal Society, London; University of California Press; The University of Chicago Press, including representation of *The American Naturalist;* and Worth Publishers, Inc.

Finally, much of my personal research reported in the book has been supported continuously by the National Science Foundation. It is fair to say that I would not have reached the point from which a synthesis could be attempted if it had not been for this generous public support.

E. O. W.

Cambridge, Massachusetts
September 1979

Contents

Arjuna to Lord Krishna: *Although these are my enemies, whose wits are overthrown by greed, see not the guilt of destroying a family, see not the treason to friends, yet how, O Troubler of the Folk, shall we with clear sight not see the sin of destroying a family?*

Lord Krishna to Arjuna: *He who thinks this Self to be a slayer, and he who thinks this Self to be slain, are both without discernment; the Soul slays not, neither is it slain.*

PART I

Social Evolution

CHAPTER 1

The Morality of the Gene

Camus said that the only serious philosophical question is suicide. That is wrong even in the strict sense intended. The biologist, who is concerned with questions of physiology and evolutionary history, realizes that self-knowledge is constrained and shaped by the emotional control centers in the hypothalamus and limbic system of the brain. These centers flood our consciousness with all the emotions—hate, love, guilt, fear, and others—that are consulted by ethical philosophers who wish to intuit the standards of good and evil. What, we are then compelled to ask, made the hypothalamus and limbic system? They evolved by natural selection. That simple biological statement must be pursued to explain ethics and ethical philosophers, if not epistemology and epistemologists, at all depths. Self-existence, or the suicide that terminates it, is not the central question of philosophy. The hypothalamic-limbic complex automatically denies such logical reduction by countering it with feelings of guilt and altruism. In this one way the philosopher's own emotional control centers are wiser than his solipsist consciousness, "knowing" that in evolutionary time the individual organism counts for almost nothing. In a Darwinist sense the organism does not live for itself. Its primary function is not even to reproduce other organisms; it reproduces genes, and it serves as their temporary carrier. Each organism generated by sexual reproduction is a unique, accidental subset of all the genes constituting the species. Natural selection is the process whereby certain genes gain representation in the following generations superior to that of other genes located at the same chromosome positions. When new sex cells are manufactured in each generation, the winning genes are pulled apart and reassembled to manufacture new organisms that, on the average, contain a higher proportion of the same genes. But the individual organism is only their vehicle, part of an elaborate device to preserve and spread them with the least possible biochemical perturbation. Samuel Butler's famous aphorism that the chicken is only an egg's way of making another egg has been modernized: the organism is only DNA's way of making more DNA. More to the point, the hypothalamus and limbic system are engineered to perpetuate DNA.

In the process of natural selection, then, any device that can insert a higher proportion of certain genes into subsequent generations will come to characterize the species. One class of such devices promotes prolonged individual survival. Another promotes superior mating performance and care of the resulting offspring. As more complex social behavior by the organism is added to the genes' techniques for replicating themselves, altruism becomes increasingly prevalent and eventually appears in exaggerated forms. This brings us to the central theoretical problem of sociobiology: how can altruism, which by definition reduces personal fitness, possibly evolve by natural selection? The answer is kinship: if the genes causing the altruism are shared by two organisms because of common descent, and if the altruistic act by one organism increases the joint contribution of these genes to the next generation, the propensity to altruism will spread through the gene pool. This occurs even though the altruist makes less of a solitary contribution to the gene pool as the price of its altruistic act.

To his own question "Does the Absurd dictate death?" Camus replied that the struggle toward the heights is itself enough to fill a man's heart. This arid judgment is probably correct, but it makes little sense except when closely examined in the light of evolutionary theory. The hypothalamic-limbic complex of a highly social species, such as man, "knows," or more precisely it has been programmed to perform as if it knows, that its underlying genes will be proliferated maximally only if it orchestrates behavioral responses that bring into play an efficient mixture of personal survival, reproduction, and altruism. Consequently, the centers of the complex tax the conscious mind with ambivalences whenever the organisms encounter stressful situations. Love joins hate; aggression, fear; expansiveness, withdrawal; and so on—in blends designed not to promote the happiness and survival of the individual, but to favor the maximum transmission of the controlling genes.

The ambivalences stem from counteracting pressures on the units of natural selection. Their genetic consequences will be explored formally later in this book. For the moment suffice it to note that what is good for the individual can be destructive to the family; what preserves the family can be harsh on both the individual and the tribe to which its family belongs; what promotes the tribe can weaken the family and destroy the individual; and so on upward through the permutations of levels of organization. Counteracting selection on these different units will result in certain genes being multiplied and fixed, others lost, and combinations of still others held in static proportions. According to present theory, some of the genes will produce emotional states that reflect the balance of counteracting selection forces at the different levels.

I have raised a problem in ethical philosophy in order to characterize the essence of sociobiology. Sociobiology is defined as the systematic study of the biological basis of all social behavior. For the present it focuses on animal societies, their population structure, castes, and communication, together with all of the physiology underlying the social adaptations. But the discipline is also concerned with the social behavior of early man and the adaptive features of organization in the more primitive contemporary human societies. Sociology *sensu stricto,* the study of human societies at all levels of complexity, still stands apart from sociobiology because of its largely structuralist and nongenetic approach. It attempts to explain human behavior primarily by empirical description of the outermost phenotypes and by unaided intuition, without reference to evolutionary explanations in the true genetic sense. It is most successful, in the way descriptive taxonomy and ecology have been most successful, when it provides a detailed description of particular phenomena and demonstrates first-order correlations with features of the environment. Taxonomy and ecology, however, have been reshaped entirely during the past forty years by integration into neo-Darwinist evolutionary theory—the "Modern Synthesis," as it is often called—in which each phenomenon is weighed for its adaptive significance and then related to the basic principles of population genetics. It may not be too much to say that sociology and the other social sciences, as well as the humanities, are the last branches of biology waiting to be included in the Modern Synthesis. One of the functions of sociobiology, then, is to reformulate the foundations of the social sciences in a way that draws these subjects into the Modern Synthesis. Whether the social sciences can be truly biologicized in this fashion remains to be seen.

This book is a condensation of *Sociobiology: The New Synthesis,* which made an attempt to codify sociobiology into a branch of evolutionary biology and particularly of modern population biology. I believe that the subject has an adequate richness of detail and aggregate of self-sufficient concepts to be ranked as coordinate with such disciplines as molecular biology and developmental biology. In the past its development has been slowed by too close an identification with ethology and behavioral physiology. In the view presented here, the new sociobiology should be compounded of roughly equal parts of invertebrate zoology, vertebrate zoology, and population biology. Figure 1-1 shows the schema with which I closed *The Insect Societies,* suggesting how the amalgam can be achieved. Biologists have always been intrigued by comparisons between societies of invertebrates, especially insect societies, and those of vertebrates. They have dreamed of identifying the common properties of such disparate units in a way that would provide insight into all aspects of social

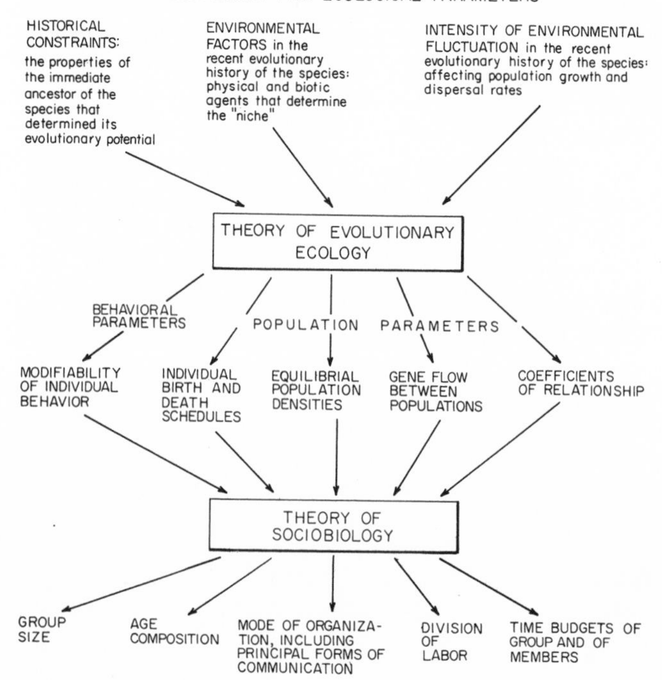

evolution, including that of man. The goal can be expressed in modern terms as follows: when the same parameters and quantitative theory are used to analyze both termite colonies and troops of rhesus macaques, we will have a unified science of sociobiology. This may seem an impossibly difficult task. But as my own studies have advanced, I have been increasingly impressed with the functional similarities between invertebrate and vertebrate societies and less so with the structural differences that seem, at first glance, to constitute such an immense gulf between them. Consider for a moment termites and monkeys. Both are formed into cooperative groups that occupy territories. The group members communicate hunger, alarm, hostility, caste status or rank, and reproductive status among themselves by means of something on the order of 10 to 100 nonsyntactical signals. Individuals are intensely aware of the distinction between groupmates and nonmembers. Kinship plays an important role in group structure and probably served as a chief generative force of sociality in the first place. In both kinds of society there is a well-marked division of labor, although in the insect society there is a much stronger reproductive component. The details of organization have been evolved by an evolutionary optimization process of unknown precision, during which some measure of added fitness was given to individuals with cooperative tendencies—at least toward relatives. The fruits of cooperativeness depend upon the particular conditions of the environment and are available to only a minority of animal species during the course of their evolution.

This comparison may seem facile, but it is out of such deliberate oversimplification that the beginnings of a general theory are made. The formulation of a theory of sociobiology constitutes, in my opinion, one of the great manageable problems of biology for the next twenty or thirty years. The prolegomenon of Figure 1-1 guesses part of its future outline and some of the directions in which it is most likely to lead animal behavior research. Its central precept is that the evolution of social behavior can be fully comprehended only through an understanding, first, of demography, which yields the vital information concerning population growth and age structure, and, second, of the genetic structure of the populations, which tells us what we need to know about effective population size in the genetic sense, the coefficients of relationship within the societies, and the amounts of gene flow between them. The principal goal of a general theory of sociobiology should be an ability to predict features of social organization from a knowledge of these population parameters combined with information on the behavioral constraints imposed by the genetic constitution of the species. It will be a chief task of evolutionary ecology, in turn, to derive the population parameters from a knowledge of the evolutionary history of the species and of the environment in which the most recent segment of that history unfolded. The most important feature of the prolegomenon, then, is the sequential relation between evolutionary studies, ecology, population biology, and sociobiology.

In stressing the tightness of this sequence, however, I do not wish to underrate the filial relationship that sociobiology has had in the past with the remainder of behavioral biology. Although behavioral biology is traditionally spoken of as if it were a unified subject, it is now emerging as two distinct disciplines centered on neurophysiology and on sociobiology, respectively. The conventional wisdom also speaks of ethology, which is the naturalistic study of whole patterns of animal behavior, and its

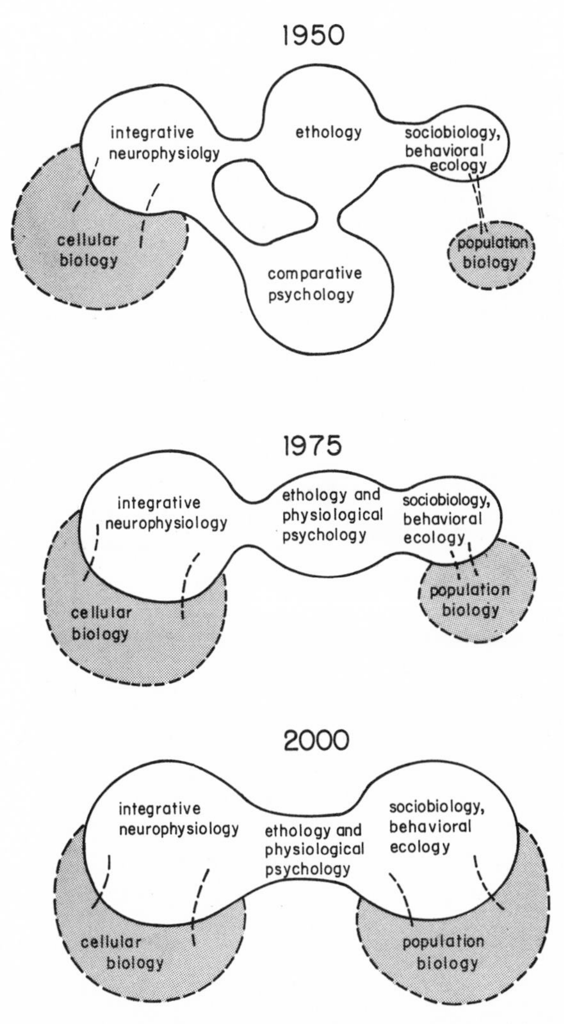

Figure 1-2 A subjective conception of the relative number of ideas in various disciplines in and adjacent to behavioral biology to the present time and as it might be in the future.

companion enterprise, comparative psychology, as the central, unifying fields of behavioral biology. They are not; both are destined to be cannibalized by neurophysiology and sensory physiology from one end and sociobiology and behavioral ecology from the other (see Figure 1-2).

I hope not too many scholars in ethology and psychology will be offended by this vision of the future of behavioral biology. It seems to be indicated both by the extrapolation of current events and by consideration of the logical relationship behavioral biology holds with the remainder of science. The future, it seems clear, cannot lie with the ad hoc terminology, crude models, and curve fitting that characterize most of contemporary ethology and comparative psychology. Whole patterns of animal behavior will inevitably be explained within the framework, first, of integrative neurophysiology, which classifies neurons and reconstructs their circuitry, and, second, of sensory physiology, which seeks to characterize the cellular transducers at the molecular level. Endocrinology will continue to play a peripheral role, since it is concerned with the cruder tuning devices of nervous activity. To pass from this level and reach the next really distinct discipline, we must travel all the way up to the society and the population. Not only are the phenomena best described by families of models different from those of cellular and molecular biology, but the explanations become largely evolutionary. There should be nothing surprising in this distinction. It is only a reflection of the larger division that separates the two greater domains of evolutionary biology and functional biology. As Lewontin (1972) has truly said: "Natural selection of the character states themselves is the essence of Darwinism. All else is molecular biology."

CHAPTER 2

Elementary Concepts of Sociobiology

The higher properties of life are emergent. To specify an entire cell, we are compelled to provide not only the nucleotide sequences but also the identity and configuration of other kinds of molecules placed in and around the cell. To specify an organism requires still more information about both the properties of the cells and their spatial positions. A society can be described only as a set of particular organisms, and even then it is difficult to extrapolate the joint activity of this ensemble from the instant of specification, that is, to predict social behavior. To cite one concrete example, Maslow (1936) found that the dominance relations of a group of rhesus monkeys cannot be predicted from the interactions of its members matched in pairs. Rhesus monkeys, like other higher primates, are intensely affected by their social environment—an isolated individual will repeatedly pull a lever with no reward other than the glimpse of another monkey. Moreover, this behavior is subject to higher-order interactions. The monkeys form coalitions in the struggle for dominance, so that an individual falls in rank if deprived of its allies. A second-ranking male, for example, may owe its position to protection given it by the alpha male or support from one or more close peers. Such coalitions cannot be predicted from the outcome of pairwise encounters, let alone from the behavior of an isolated monkey.

The recognition and study of emergent properties is holism, once a burning subject for philosophical discussion by such scientists as Lloyd Morgan (1922) and W. M. Wheeler (1927a), but later, in the 1940's and 1950's, temporarily eclipsed by the triumphant reductionism of molecular biology. The new holism is much more quantitative in nature, supplanting the unaided intuition of the old theories with mathematical models. Unlike the old, it does not stop at philosophical retrospection but states assumptions explicitly and extends them in mathematical models that can be used to test their validity. In the sections to follow we will examine several of the properties of societies that are emergent and hence deserving of a special language and treatment. We begin with a straightforward, didactic review of a set of the most basic definitions, some general for biology, others peculiar to sociobiology.

Basic Definitions

Society: a group of individuals belonging to the same species and organized in a cooperative manner. The terms *society* and *social* need to be defined broadly in order to prevent the exclusion of many interesting phenomena. Such exclusion would cause confusion in all further comparative discussions of sociobiology. Reciprocal communication of a cooperative nature, transcending mere sexual activity, is the essential intuitive criterion of a society. Thus it is difficult to think of a bird egg, or even a honeybee larva sealed in its brood cell, as a member of the society that produced it, even though it may function as a true member at other stages of its development. It is also not satisfying to view the simplest aggregations of organisms, such as swarms of courting males, as true societies. They are often drawn together by mutually attractive stimuli, but if they interact in no other way it seems excessive to refer to them by a term stronger than aggregation. By the same token a pair of animals engaged in simple courtship or a group of males in territorial contention can be called a society in the broadest sense, but only at the price of diluting the expression to the point of uselessness. Yet aggregation, sexual behavior, and territoriality are important *properties* of true societies, and they are correctly referred to as social behavior. Bird flocks, wolf packs, and locust swarms are good examples of true elementary societies. So are parents and offspring if they communicate reciprocally. Although this last, extreme example may seem at first trivial, parent-offspring interactions are in fact often complex and serve multiple functions.

Furthermore, in many groups of organisms, from the social insects to the primates, the most advanced societies appear to have evolved directly from family units.

Another way of defining societies is by delimiting particular groups. Since the bond of the society is simply and solely communication, its boundaries can be defined in terms of the curtailment of communication. Altmann (1965) has expressed this aspect: "A society . . . is an aggregation of socially intercommunicating, conspecific individuals that is bounded by frontiers of far less frequent communication."

Aggregation: a group of individuals of the same species, comprised of more than just a mated pair or a family, gathered in the same place but not internally organized or engaged in cooperative behavior. Winter congregations of rattlesnakes and ladybird beetles, for example, may provide superior protection for their members, but unless they are organized by some behavior other than mutual attraction they are better classified as aggregations than as true societies.

Colony: in strict biological usage, a society of organisms which are highly integrated, either by physical union of the bodies or by division into specialized zooids or castes, or by both. In the vernacular and even in some technical descriptions, a colony can mean almost any group of organisms, especially if they are fixed in one locality. In sociobiology, however, the word is best restricted to the societies of social insects, together with the tightly integrated masses of sponges, siphonophores, bryozoans, and other "colonial" invertebrates.

Individual: any physically distinct organism.

Group: a set of organisms belonging to the same species that remain together for any period of time while interacting with one another to a much greater degree than with other conspecific organisms. The word *group* is thus used with the greatest flexibility to designate any aggregation or kind of society or subset of a society. The expression is especially useful in accommodating descriptions of certain primate societies in which there exists a hierarchy of levels of organization constructed of nested subsets of individuals belonging to a single large congregation.

Population: a set of organisms belonging to the same species and occupying a clearly delimited area at the same time. This unit—the most basic but also one of the most loosely employed in evolutionary biology—is defined in terms of genetic continuity. In the case of sexually reproducing organisms, the population is a geographically delimited set of organisms capable of freely interbreeding with one another under natural conditions. The special population used by model builders is the *deme,* the smallest local set of organisms within which interbreeding occurs freely. The idealized deme is panmictic, that is, its members breed completely at random. Put another way, panmixia means that each reproductively mature male is equally likely to mate with each reproductively mature female, regardless of their location within the range of the deme. Although not likely to be attained in absolute form in nature, especially in social organisms, panmixia is an important simplifying assumption made in much of elementary quantitative theory.

In sexually reproducing forms, including the vast majority of social organisms, a *species* is a population or set of populations within which the individuals are capable of freely interbreeding under natural conditions. By definition the members of the species do not interbreed freely with those of other species, however closely related they may be genetically. The existence of natural conditions is a basic part of the definition of the species. In establishing the limits of a species it is not enough merely to prove that genes of two or more populations can be exchanged under experimental conditions. The populations must be demonstrated to interbreed fully in the free state.

A population that differs significantly from other populations belonging to the same species is referred to as a *geographic race* or *subspecies.* Subspecies are separated from other subspecies by distance and geographic barriers that prevent the exchange of individuals, as opposed to the genetically based "intrinsic isolating mechanisms" that hold species apart. Subspecies, insofar as they can be distinguished with any objectivity at all, show every conceivable degree of differentiation from other subspecies. At one extreme are the populations that fall along a cline —a simple gradient in the geographic variation of a given character. In other words, a character that varies in a clinal pattern is one that changes gradually over a substantial portion of the entire range of the species. At the other extreme are subspecies consisting of easily distinguished populations that are differentiated from one another by numerous genetic traits and exchange genes across a narrow zone of intergradation.

What is the relation between the population and the society? Here we arrive unexpectedly at the crux of theoretical sociobiology. The distinction between the two categories is essentially as follows: the population is bounded by a zone of sharply reduced gene flow, while the society is bounded by a zone of sharply reduced communication. Often the two zones are the same, since social bonds tend to promote gene flow among the members of the society to the exclusion of outsiders. For example, detailed field studies by Stuart and Jeanne Altmann (1970) on the yellow baboons (*Papio cynocephalus*) of Amboseli show that in this species the society and the deme are essentially the same thing. The baboons are internally organized by dominance hierarchies and are usually hostile toward outsiders. Gene exchange occurs between troops by the emigration from one to another of subordinate males, who typically leave their home troop after the loss of a fight or during competition for estrous females. Using the Altmanns' data, Cohen (1969b) estimated the immigration rate into one large troop to be 8.043×10^{-3} individuals per group per day, a degree of flow that is many orders of magnitude below that which occurred between subgroups belonging to the same troop.

Communication: action on the part of one organism (or cell) that alters the probability pattern of behavior in another organism (or cell) in an adaptive fashion. This definition conforms well both to our intuitive understanding of communication and to the procedure by which the process is mathematically analyzed.

Coordination: interaction among units of a group such that the overall effort of the group is divided among the units without leadership being assumed by any one of them. Coordination may be influenced by a unit in a higher level of the social hierarchy, but such outside control is not essential. The formation of a fish school, the exchange of liquid food back and forth by worker ants, and the encirclement of prey by a pride of lions are all examples of coordination among organisms at the same organizational level.

Hierarchy: in ordinary sociobiological usage, the dominance of one member of a group over another, as measured by superiority in aggressive encounters and order of access to food, mates, resting sites, and other objects promoting survivorship and reproductive fitness. Technically, there need be only two individuals to make such a hierarchy, but chains of many individuals in descending order of dominance are more frequent. More generally, a hierarchy can be defined without reference to dominance as a system of two or more levels of units, the higher levels controlling at least to some extent the activities of the lower levels in order to achieve the goal of the group as a whole (Mesarović et al., 1970). Hierarchies without dominance are common in social insect colonies and occur in certain facets of the behavior of such highly coordinated mammals as higher primates and social canids. The more advanced animal societies are in general organized at one or at most two hierarchical levels and consist of individuals tightly connected by relatively few kinds of social bonds and communicative signals. Human societies, in contrast, are typically organized through many hierarchical levels and are comprised of numerous individuals loosely joined by very many kinds of social bonds and an extremely rich language.

Regulation: in biology, the coordination of units to achieve the maintenance of one or more physical or biological variables at a constant level. The result of regulation is termed *homeostasis*. The most familiar form of homeostasis is physiological: a properly tuned organism maintains constant values in pH, in concentrations of dissolved nutrients and salts, in proportions of active enzymes and organelles, and so forth, which fall close to the optimal values for survival and reproduction. Like a man-designated machine system, physiological homeostasis is self-regulated by internal feedback loops that increase the values of important variables when they fall below certain levels and decrease them when they exceed other, higher values. At a higher level, social insects display marked homeostasis in the regulation of their own colony populations, caste proportions, and nest environment. This form of steady-state maintenance has aptly been termed social homeostasis by Emerson (1956).

The Multiplier Effect

Social organization is the class of phenotypes furthest removed from the genes. It is derived jointly from the behavior of individuals and the demographic properties of the population, both of which are themselves highly synthetic properties. A small evolutionary change in the behavior pattern of individuals can be amplified into a major social effect by the expanding upward distribution of the change into multiple facets of social life. This phenomenon can be referred to as the multiplier effect. Consider, for example, the differing social organizations of the related olive baboon (*Papio anubis*) and hamadryas baboon (*P. hamadryas*). These two species are so close genetically that they interbreed extensively where their ranges overlap and could reasonably be classified as no more than subspecies. The hamadryas male is distinguished by its proprietary attitude toward females, which is total and permanent, whereas the olive male attempts to appropriate females only around the time of their estrus. This apparently genetic difference is only one of degree, and would scarcely be noticeable if one's interest were restricted in each species to the activities of a single dominant male and one consort female. Yet this trait alone is enough to account for profound differences in social structure, affecting the size of the troops, the relationship of troops to one another, and the relationship of males within each troop (Kummer, 1971).

Multiplier effects can speed social evolution still more when an individual's behavior is strongly influenced by the particularities of its social experience. This process, called socialization, becomes increasingly prominent as one moves upward phylogenetically into more intelligent species, and it reaches its maximum influence in the higher primates. Although the evidence is still largely inferential, socialization appears to amplify phenotypic differences among primate species. As an example, take the diverging pathways of development of social behavior observed in young olive baboons (*Papio anubis*) as opposed to Nilgiri langurs (*Presbytis johnii*) (Eimerl and DeVore, 1965; Poirier, 1972b). The baboon infant stays close to its mother during the first month of its life, and the mother discourages the approach of other females. But afterward the growing infant associates freely with adults. It is even approached by the males, who often draw close to the mother, smacking their lips in the typical conciliatory signal in order to be near the youngster. From the age of nine months, male baboons progressively lose the protection of their mothers, who reject them with increasing severity. As a result they come to mingle with other members of the troop even more quickly and freely. The social structure of the olive baboon is consistent with this program of socialization. Adult males and females mingle freely, and peripheral groups of males and solitary individuals are rare or absent. Langur social development is far more sex oriented than that of baboons. The infant is relinquished readily to other adult females, who pass it around. But it has little contact with adult males, who are chased away whenever they disturb the youngster. Juvenile males begin to associate with adult males only after eight months, while young females do not permit contact until the onset of sexual activity at the age of three years. The young males spend most of their free time playing. As the play-fighting becomes rougher and requires more room, they tend to drift to the periphery of the group, well apart from the infants and adults. The langur society reflects this form of segregated rearing. Adult males and females tend to remain apart. Groups of peripheral males are common,. and they often interact aggressively with the dominant males within the troops in an attempt to penetrate and gain ascendancy.

The Evolutionary Pacemaker and Social Drift

The multiplier effect, whether purely genetic in basis or reinforced by socialization and other forms of learning, makes behavior the part of the phenotype most likely to change in response to long-term changes in the environment. It follows that when evolution involves both structure and behavior, behavior should change first and then structure. In other words, behavior should be the evolutionary pacemaker. This is an old idea, with roots extending back at least as far as the sixth edition of Darwin's *Origin of Species* (1872) and the principle of *Funktionswechsel* expressed by Anton Dohrn (1875). Dohrn postulated that the function of an organ, which in retrospect we can view to be most clearly expressed in its behavior, is continually changing and dichotomizing over many generations according to the experience of the organism. Changes in the structure of the organ represent accommodations to these functional shifts. Among recent zoologists, Wickler (1967a,b) has most explicity argued the same point of view with reference to behavior, citing many examples from birds and fishes. Among the tetraodontiform fishes, to take one of the simpler and clearer cases, a number of species are able to inflate themselves tremendously with water or air as a protective device against predators. In young porcupine fishes of the genus *Diodon*, the median fins disappear into pouches of the skin that fold inward during inflation. The inflated stage has become irreversible in the diodontid genus *Hyosphaera*, while the tetraodontid globe fish, *Kanduka michiei*, not only is permanently inflated but also has lost the dorsal fin and reduced the anal fin to vestigial form.

Social behavior also frequently serves as an evolutionary pacemaker. The entire process of ritualization, during which a behavior is transformed by evolution into a more efficient signaling device, typically involves a behavioral change followed by morphological alterations that enhance the visibility and distinctiveness of the behavior.

The relative lability of behavior leads inevitably to *social drift*, the random divergence in the behavior and mode of organization of societies or groups of societies. The term *random* means that the behavioral differences are not the result of adaptation to the particular conditions by which the habitats of one society differ from those of other societies. If the divergence has a genetic basis, the hereditary component of social drift is simply the same as genetic drift, an evolutionary phenomenon discussed in Chapter 4. The component of divergence based purely on differences in experience can be referred to as tradition drift (Burton, 1972). The amount of variance within a population of societies is the sum of the variances due to genetic drift, tradition drift, and their interaction.

The Concept of Adaptive Demography

All true societies are differentiated populations. When cooperative behavior evolves, it is put to service by one kind of individual on behalf of another, either unilaterally or mutually. A male and a female cooperate to hold a territory, a parent feeds its young, two nurse workers groom a honeybee queen, and so forth. This being the case, the behavior of the society as a whole can be said to be defined by its demography. The breeding females of a bird flock, the helpless infants of a baboon troop, and the middle-aged soldiers of a termite colony are examples of demographic classes whose relative proportions help determine the mass behavior of the group to which they belong.

The proportions of the demographic classes also affect the fitness of the group and, ultimately, of each individual member. A group comprised wholly of infants or aging males will perish—obviously. Another, less deviant, group has a higher fitness that can be defined as a higher probability of survival, which can be translated as a longer waiting time to extinction. Either measure has meaning only over periods of time on the order of a generation in length, because a deviant population allowed to reproduce for one to several generations will go far to restore the age distribution of populations normal for the species. Unless the species is highly opportunistic, that is, unless it follows a strategy of colonizing empty habitats and holding on to them only for a relatively short time, the age distribution will tend to approach a steady state. In species with seasonal natality and mortality, which is to say nearly all animal species, the age distribution will undergo annual fluctuation. But even then the age distribution can be said to approach stability, in the sense that the fluctuation is periodic and predictable when corrected for season.

A population with a stable age distribution is not ipso facto well adjusted to the environment. It can be in a state of gradual decline, destined ultimately for extinction; or it can be increasing, in which case it may still be on its way to a population crash that leads to a decimation of numbers, strong deviation in the age distribution, and possibly even extinction. Only if its growth is zero when averaged out over many generations can the population have a chance of long life. There is one remaining way to be a success. A population headed for extinction can still possess a high degree of fitness if it succeeds in sending out propagules and creates new populations elsewhere. This is the basis of the opportunistic strategy, to be described in greater detail in Chapter 4.

We can therefore speak of a "normal" demographic distribution as the age distribution of the sexes and castes that occurs in populations with a high degree of fitness. But to what extent is the demographic distribution itself really adaptive? This is a semantic distinction that depends on the level at which natural selection acts to sustain the distribution. If selection operates to favor individuals but not groups, the demographic distribution will be an incidental effect of the selection. Suppose, for example, that a species is opportunistic, and females are strongly selected for their capacity to produce the largest number of offspring in the shortest possible time. Theory teaches that evolution will probably proceed to reduce the maturation time, increase the reproductive effort and progeny size, and shorten the natural life span. The demographic consequence will be a flattening of the age pyramid. A squashed age distribution is a statistical property of the population. It is a secondary effect of the selection that occurred at the individual level, contributes nothing of itself to the fitness of either the individual or the population, and therefore cannot be said to be adaptive in the usual sense of the word.

Now consider a colony of social insects. The demographic distribution, expressed in part by the age pyramid, is vital to the fitness of the colony as a whole and particularly to that of the progenitrix queen, with reference to whom the nonreproductive members can be regarded as a somatic extension. If too few soldiers are present at the right moment, the colony may be demolished by a predator; or if too few nurse workers of the appropriate age are not always available, the larvae may starve to death. Thus the demographic distribution is adaptive, in the sense that it is tested directly by natural selection. It can be shaped by altering growth thresholds, so that a lower or higher proportion of nymphs or larvae reaching a certain weight, or detecting a sufficient amount of certain odorous secretion, is able to metamorphose into a given caste. It can also be shaped by changing the periods of time an individual spends at a certain task. For example, if each worker has a shortened tenure as a nurse, the percentage of colony members who are active nurses at any moment will be less. Finally, the demographic distribution can be changed by altering longevity: if soldiers die sooner, their caste will be less well represented numerically in each moment of time.

With reference to social behavior, the two most important components of a demographic distribution are age and size. In Figure 2-1 I have represented age-size frequency distributions as they might appear in two societies (*A* and *B*) subjected to little selection at the level of the society, as opposed to the distribution in a society (*C*) in which such group selection has been a major force. All can agree that demography is more interesting when it is adaptive. The patterns are likely to be not only more complex but more meaningful. Nonadaptive demography follows from a study of the behavior and life cycles of individuals; but adaptive demography must be analyzed holistically before the behavior and life cycles of the individuals take on meaning.

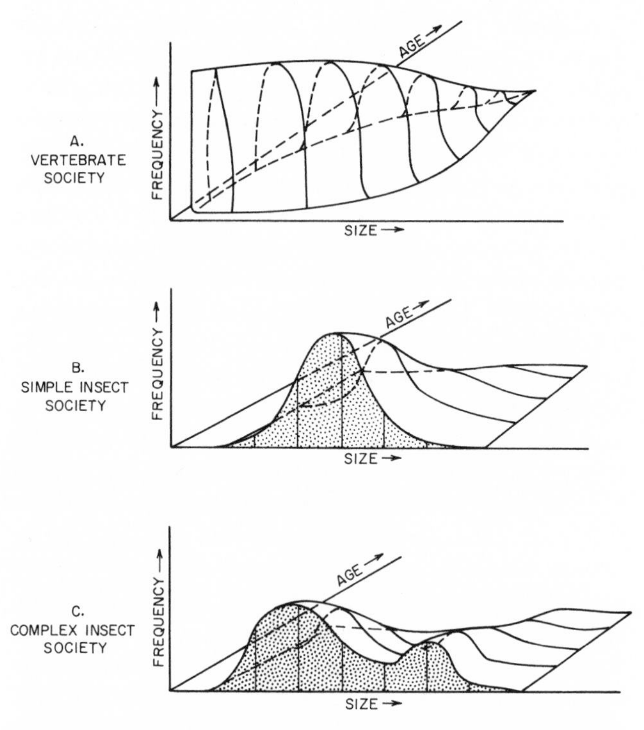

Figure 2-1. The age-size frequency distributions of three kinds of animal societies. These examples are based on the known general properties of real species, but their details are imaginary. *A*: The distribution of the "vertebrate society" is nonadaptive at the group level and therefore is essentially the same as that found in local populations of otherwise similar, nonsocial species. In this particular case the individuals are shown to be growing continuously throughout their lives, and mortality rates change only slightly with age. *B*: The "simple insect society" may be subject to selection at the group level, but its age-size distribution does not yet show the effect and is therefore still close to the distribution of an otherwise similar but nonsocial population. The age shown is that of the imago, or adult instar, during which most or all of the labor is performed for the colony, and no further increase in size occurs. *C*: The "complex insect society" has a strongly adaptive demography, reflected in its complex age-size curve: there are two distinct size classes, and the larger is longer lived.

Qualities of Sociality

The following set of ten qualities of sociality can be both measured and ultimately incorporated into models of particular social systems (see also Figure 2-2):

1. *Group size.* Joel Cohen (1969,a,b, 1971) has shown the existence of orderly patterns in the frequency of distribution of group size among primate troops. In the case of closed, relatively stable groups, much (but not all) of the information can be accounted for with stochastic models that assume constant gain rates through birth and immigration and constant loss rates through death and emigration. Orderliness also occurs in the frequency of distributions of casual subgroups in monkeys and man, and can be predicted in good part by reference to the variation in the attractiveness of groups of different

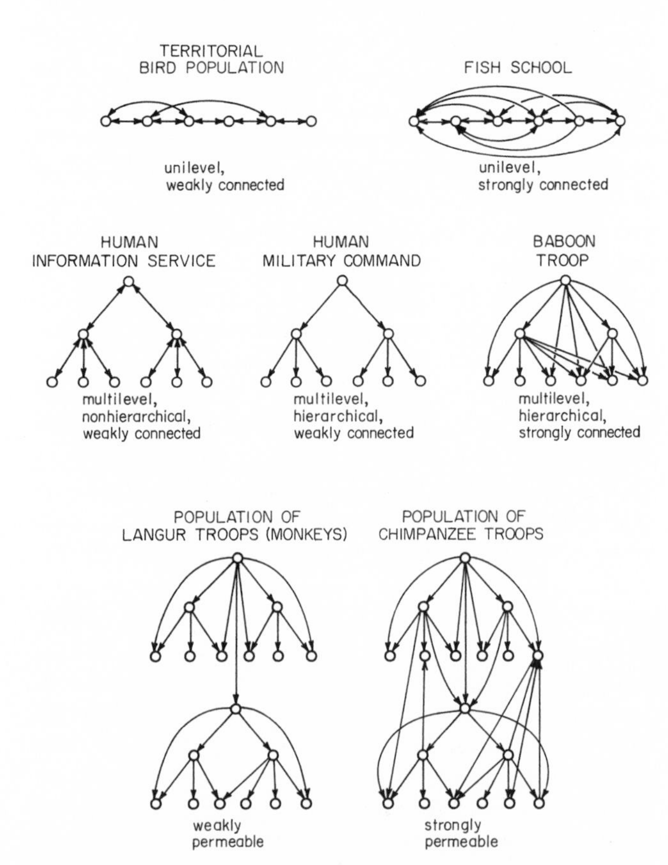

Figure 2-2 Seven social groups depicted as networks in order to illustrate variation in several of the qualities of sociality. The traits of the social groups are abstracted, and the details are imaginary.

sizes and in the attractiveness and joining tendency of individual group members.

2. *Demographic distribution.* The significance of these frequency distributions and the degree of their stability were discussed in the previous section on adaptive demography.

3. *Cohesiveness.* Intuitively, we expect that the closeness of group members to one another is an index of the sociality of the species. This is true, first, because the effectiveness of group defense and group feeding is enhanced by tight formations and, second, because the widest range of communication channels can be brought into play at close range. There is indeed a correlation between physical cohesiveness and the magnitude of the other nine social parameters listed here, but it is only loose. Honeybee colonies, for example, are more cohesive than nesting aggregations of solitary halictid bees. But chimpanzee troops and human societies are much less cohesive than fish schools and herds of cattle.

4. *Amount and pattern of connectedness.* The network of communication within a group can be patterned or not. That is, different kinds of signals can be directed preferentially at particular individuals or classes of individuals; or else, in the unpatterned case, all signals can be directed randomly for periods of time at any individuals close enough to receive them. In unpatterned networks, such as fish schools and temporary roosting flocks of birds, the number of arcs per node in the network, meaning the number of individuals contacted by the average member per unit of time, provides a straightforward measure of the sociality. This is a number that increases with the cohesiveness of the group or, in the case of animals that communicate over distances exceeding the diameter of the aggregation, the size of the group. In the case of patterned networks, the situation is radically different. Hierarchies with multiple levels can be constructed with relatively few arcs (see Figure 2-2). Provided the members are also performing separate functions, the degree of coordination and efficiency of the group as a whole can be vastly increased over an unpatterned network containing a comparable number of members, even if the degree of connectedness (the number of arcs per member) is much lower. All higher forms of societies, those recognized to possess a strong development of the other nine social qualities, are characterized by an advanced degree of patterning in connectedness. They are not always characterized by a large amount of connectedness, however.

5. *Permeability.* To say that a society is closed means that it communicates relatively little with nearby societies of the same species and seldom if ever accepts immigrants. A troop of langurs (*Presbytis entellus*) is an example of a society with low permeability. Exchanges between troops consist mostly of aggressive encounters over territory, and, at least in the dense populations of southern India, immigration is mostly limited to the intrusion of males who usurp the position of the dominant male (Ripley, 1967; Sugiyama, 1967). At the opposite extreme are the very permeable troops of chimpanzees, in which groups temporarily fuse and exchange members freely. All other things being equal, an increase in permeability should result in an increase in gene flow through entire populations and a reduced degree of genetic relationship between any two members chosen at random from within a single society.

6. *Compartmentalization.* The extent to which the subgroups of a society operate as discrete units is another measure of the complexity of the society. When confronted with danger, a herd of wildebeest flees as a disorganized mob, with the mothers turning individually to defend themselves and their calves only if overtaken. Zebra herds, in contrast, sort out into family groups, with each dominant stallion maneuvering to place himself between the predator and his harem. When the danger passes, the families merge again into a single formation.

7. *Differentiation of roles.* Specialization of members of a group is a hallmark of advance in social evolution. One of the theorems of ergonomic theory is that for each species (or genotype) in a particular environment there exists an optimum mix of coordinated specialists that performs more efficiently than groups of equal size consisting wholly of generalists (Wilson, 1968a; see also Chapter 14). It is also true that under many circumstances mixes of specialists can perform qualitatively different tasks not easily managed by otherwise equivalent groups of generalists, whereas the reverse is not true. Packs of African wild dogs, to cite one case, break into two "castes" during hunts: the adult pack that pursues, and the adults that remain behind at the den with the young. Without this division of labor, the pack could not subdue a sufficient number of the large ungulates that constitute its chief prey (Estes and Goddard, 1967).

8. *Integration of behavior.* The inverse of differentiation is integration: a set of specialists cannot be expected to function as well as a group of generalists unless they are in the correct proportions and their behaviors coordinated. The following example is among the most striking known in the social insects. Minor workers of the American ant *Pheidole dentata* forage singly for food outside the nest. When they discover a food particle too large to carry home, they lay an odor trail back to the nest. The trail chemical is carried in the poison gland and released through the

sting when the tip of the abdomen is dragged over the ground. The trail attracts and guides both the other minor workers and members of the soldier caste, all of whom then assist in the cutting up and transport of the food. The soldiers, who cannot lay odor trails, are specialized for yet another function: they defend the food from intruders, especially members of other ant colonies. Together the two castes perform the same task of foraging, perhaps with greater efficiency, as do the workers of other ant species that contain a single caste.

9. *Information flow.* Norbert Wiener said that sociology, including animal sociobiology, is fundamentally the study of the means of communication. Indeed, many of the social qualities I am listing here could, with varying degrees of effort, be subsumed under communication. The magnitude of a communication system can by measured in three ways: the total number of signals, the amount of information in bits per signal, and the rate of information flow in bits per second per individual and in bits per second for the entire society.

10. *Fraction of time devoted to social behavior.* The allocation of individual effort to the affairs of the society is one fair measure of the degree of sociality. This is the case whether effort is measured by the percentage of the entire day devoted to it, by the fraction of time devoted to it out of all the time spent engaged in any activity, or by the fraction of energy expended. Social effort reflects, but is not an elementary function of, cohesiveness, differentiation, specialization, and rate of information flow. R. T. Davis and his coworkers (1968) detected a rough correlation with these several traits within the primates. Lemurs (*Lemur catta*), generally regarded to have a somewhat simple social organization, devote approximately 20 percent of their time to social behavior, while pig-tailed macaques (*Macaca nemestrina*) and stump-tailed macaques (*M. speciosa*), which by other criteria are relatively sophisticated social animals, invest about 80 and 90 percent of their time, respectively, in social acts.

The Concept of Behavioral Scaling

In the early years of vertebrate sociobiology it was customary for observers to assume that social structures, no less than ethological fixed action patterns, are among the invariant traits by which species can be diagnosed. If a short-term field study revealed no evidence of territoriality, dominance hierarchies, or some other looked-for social behavior, then the species as a whole was characterized as lacking the behavior. Even so skillful a field zoologist as George Schaller could state confidently on the basis of relatively few data that the gorilla "shares its range and its abundant food resources with others of its kind, disdaining all claims to a plot of land of its own."

Experience has begun to stiffen our caution about generalizing beyond single populations of particular species and at times other than the period of observation. Largely because of the multiplier effect, social organization is among the most labile of traits. The following case involving Old World monkeys is typical. Troops of vervets (*Cercopithecus aethiops*) observed by Struhsaker (1967a,b) at the Amboseli Masai Game Reserve, Kenya, are strictly territorial and maintain rigid dominance hierarchies by frequent bouts of fighting. In contrast, those studied by J. S. Gartlan (cited by Thelma Rowell, 1969) in Uganda had no visible dominance structure at the time of observation, exchange of males occurred frequently between troops, and fighting was rare.

In some cases differences of this sort are probably due to geographic variation of a genetic nature, originating in the adaptation of local populations to peculiarities in their immediate environment. Some fraction can undoubtedly also be credited to tradition drift. (The entire pattern of variation based on all contributing factors is referred to by geneticists as the norm of reaction.)

But a substantial percentage of cases do not represent permanent differences between populations at all: the societies are just temporarily at different points on the same *behavioral scale*. Behavioral scaling is variation in the magnitude or in the qualitative state of a behavior which is correlated with stages of the life cycle, population density, or certain parameters of the environment. It is a useful working hypothesis to suppose that in each case the scaling is adaptive, meaning that it is genetically programmed to provide the individual with the particular response more or less precisely appropriate to its situation at any moment in time. In other words, the entire scale, not isolated points on it, is the genetically based trait that has been fixed by natural selection (Wilson, 1971b). To make this notion clearer, and before I take up concrete examples, consider the following imaginary case of aggressive behavior programmed to cope with varying degrees of population density and crowding. At low population densities, all aggressive behavior is suspended. At moderate densities, it takes a mild form, such as intermittent territorial defense. At high densities, territorial defense is sharp, while some joint occupancy of land is also permitted under the regime of dominance hierarchies. Finally, at extremely high densities, the system may break down almost completely, transforming the pattern of aggressive encounters into homosexuality, cannibalism, and other symptoms of "social pathology." Whatever the specific program that slides individual responses up and down the aggression scale, each of the various degrees of aggressiveness is adaptive at an appropriate population density—short of the rarely recurring pathological level. In sum, it is the total *pattern* of aggressive responses that is adaptive and has been fixed in the course of evolution.

In the published cases of scaled social response, the most frequently reported governing parameter is indeed population density. True threshold effects suggested in the imaginary example do exist in nature. Aggressive encounters among adult hippopotami, for example, are rare where populations are low to moderate. However, when populations in the Upper Semliki near Lake Edward became so dense that there was an average of one animal to every 5 meters of riverbank, males began to fight viciously, sometimes even to the death (Verheyen, 1954). When snowy owls (*Nyctea scandiaca*) live at normal population densities, each bird maintains a territory about 5,000 hectares in extent, and it does not engage in territorial defense. But when the owls are crowded together, particularly during the times of lemming highs in the Arctic, they are forced to occupy areas covering as few as 120 hectares. Under these conditions they defend the territories overtly, with characteristic sounds and postures (Frank A. Pitelka, in Schoener, 1968a).

The Dualities of Evolutionary Biology

The theories of behavioral biology are riddled with semantic ambiguity. Like buildings constructed hastily on unknown ground, they sink, crack, and fall to pieces at a distressing rate for reasons seldom understood by their architects. In the special case of sociobiology, the unknown substratum is usually evolutionary theory. We should therefore set out to map the soft areas in the relevant parts of evolutionary biology. With remarkable consistency the most troublesome evolutionary concepts can be segregated into a series of dualities. Some are simple two-part classifications, but others reflect more profound differences in levels of selection and between genetic and physiological processes.

Adaptive versus nonadaptive traits. A trait can be said to be adaptive if it is maintained in a population by selection. We can put the matter more precisely by saying that another trait is nonadaptive, or "abnormal," if it reduces the fitness of individuals that consistently manifest it under environmental circumstances that are usual for the species. In other words, deviant responses in abnormal environments may not be nonadaptive—they may simply reflect flexibility in a response that is quite adaptive in the environments ordinarily encountered by the species. A trait can be switched from an adaptive to a nonadaptive status by a simple change in the environment. For example, the sickle-cell trait of human beings, determined by the heterozygous state of a single gene, is adaptive under living conditions in Africa, where it confers some degree of resistance to falciparum malaria. In Americans of African descent, it is nonadaptive, for the simple reason that its bearers are no longer confronted by malaria.

The pervasive role of natural selection in shaping all classes of traits in organisms can be fairly called the central dogma of evolutionary biology. When relentlessly pressed, this proposition may not produce an absolute truth, but it is, as G. C. Williams disarmingly put the matter, the light and the way. A large part of the contribution of Konrad Lorenz and his fellow ethologists can be framed in the same metaphor. They convinced us that behavior and social structure, like all other biological phenomena, can be studied as "organs," extensions of the genes that exist because of their superior adaptive value.

How can we test the adaptation dogma in particular instances? There exist situations in which social behavior temporarily manifested by animals seems clearly to be abnormal, because it is possible to diagnose the causes of the deviation and to identify the response as destructive or at least ineffectual. When groups of hamadryas baboons were first introduced into a large enclosure in the London Zoo, social relationships were highly unstable and males fought viciously over possession of the females, sometimes to the death (Zuckerman, 1932). But these animals had been thrown together as strangers, and the ratio of males to females was higher than in the wild. Kummer's later studies in Africa showed that under natural conditions hamadryas societies are stable, with the basic unit composed of several adult females and their offspring dominated by one or two males.

Furthermore, what is adaptive social behavior for one member of the family may be nonadaptive for another. The Indian langur males who invade troops, overthrow the leaders, and destroy their offspring are clearly improving their own fitness, but at a severe cost to the females they take over as mates. When male elephant seals fight for possession of harems, they are being very adaptive with respect to their own genes, but they reduce the fitness of the females whose pups they trample underfoot.

Monadaptive versus polyadaptive traits. Social evolution is marked by repeated strong convergence of widely separate phylogenetic groups. The confusion inherent in this circumstance is worsened by the still coarse and shifting nature of our nomenclatural systems. Ideally, we should try to have a term for each major functional category of social behavior. This semantic refinement would result in most kinds of social behavior being recognized as monadaptive, that is, possessing only one function. In our far from perfect language, however, most behaviors are artificially construed as polyadaptive. Consider the polyadaptive nature of "aggression," or "agonistic behavior," in monkeys. Males of langurs, patas, and many other species use aggression to maintain troop distance. Similar behavior is also employed by a diversity of species, including langurs, to establish and to sustain dominance hierarchies. Male hamadryas baboons use aggression to herd females and discourage them from leaving the harems. Aggression, in short, is a vague term used to designate an array of behaviors, with various functions, that we intuitively feel resemble human aggression.

Some social behavior patterns nevertheless remain truly polyadaptive even after they have been semantically purified. Allogrooming in rhesus monkeys, for examples, serves the typically higher primate function of conciliation and bond maintenance. Yet it retains a second, apparently more primitive, cleansing function, because monkeys kept in isolation often develop severe infestations of lice. In some bird species, flocking behavior undoubtedly serves the dual function of predator evasion and improvement in foraging efficiency.

Reinforcing versus counteracting selection. A single force in natural selection acts on one or more levels in an ascending hierarchy of units: the individual, the family, the group, and possibly even the entire population or species. If affected genes are uniformly favored or disfavored at more than one level, the selection is said to be reinforcing. Evolution, meaning changes in gene frequency, will be accelerated by the additive effects inserted at multiple levels. This process should offer no greater problem to mathematicians. By contrast, the selection might be counteracting in nature: genes favored by a selective process at the individual level could be opposed by the same process at the family level, only to be favored again at the population level, and so on in various combinations. The compromise gene frequency is of general importance to the theory of social evolution, but it is mathematically difficult to predict. It will be considered formally in Chapters 5 and 14.

Ultimate versus proximate causation. The division between functional and evolutionary biology is never more clearly defined than when the proponents of each try to make a pithy statement about causation. Consider the problem of aging and senescence. Contemporary functional biologists are preoccupied with four competing theories of aging, all strictly physiological: rate-of-living, collagen wear, autoimmunity, and somatic mutation (Curtis, 1971). If one or more of these factors can be firmly implicated in a way that accounts for the whole process in the life of an individual, the more narrowly trained biochemist will consider the problem of causation solved. However, only the proximate causation will have been demonstrated. Meanwhile, as though dwelling in another land, the theoretical population geneticist works on senescence as a process that is molded in time so as to maximize the reproductive fitness in particular environments (Hamilton, 1966). These specialists are aware of the existence of physiological processes but regard them abstractly as elements to be jiggered to obtain the optimum time of senescence according to the schedules of survivorship and fertility that prevail in their theoretical populations. This approach attempts to solve the problem of ultimate causation.

How is ultimate causation linked to proximate causation? Ultimate causation consists of the necessities created by the environment: the pressures imposed by weather, predators, and other stressors, and such opportunities as are presented by unfilled living space, new food sources, and accessible mates. The species responds to environmental exigencies by genetic evolution through natural selection, inadvertently shaping the anatomy, physiology, and behavior of the individual organisms. In the process of evolution, the species is constrained not only by the slowness of evolutionary time, which by definition covers generations, but also by the presence or absence of preadapted traits and certain deep-lying genetic qualities that affect the rate at which selection can proceed. These prime movers of evolution (see Chapter 3) are the ultimate biological causes, but they operate only over long spans of time. The anatomical, physiological, and behavorial machinery they create constitutes the proximate causation of the functional biologist. Operating within the lifetimes of organisms, and sometimes even within milliseconds, this machinery carries out the commands of the genes on a time scale so remote from that of ultimate causation that the two processes sometimes seem to be wholly decoupled.

Ideal versus optimum permissible traits. When organisms are thought of as machine analogs, their evolution can be viewed as a gradual perfecting of design. In this conception there exist ideal traits for survival in particular environments. There would be the ideal hammer bill and extrusible tongue for woodpeckers, the ideal caste system for army ants, and so forth. But we know that such traits vary greatly from species to species, even those belonging to the same phyletic group and occupying the same narrowly defined niche. In particular, it is disconcerting to find frequent cases of species with an advanced state or intermediate states of the same character.

Take the theoretical problem created by the primitively social insect species. Why have they progressed no further? Two extreme possibilities can be envisioned (Wilson, 1971a). First, there is what might be termed the "disequilibrium case." This means that the species is still actively evolving toward a higher social level. The situation can arise if social evolution is so slow that the species is embarked on a particular adaptive route but is still in transit (see Figure 2-3A).

Implicit in a disequilibrium hypothesis is the assumption that the advanced social state, or some particular advanced social state, is the *summum bonum*, the solitary peak defined by the ideal trait toward which the species and its relatives are climbing.

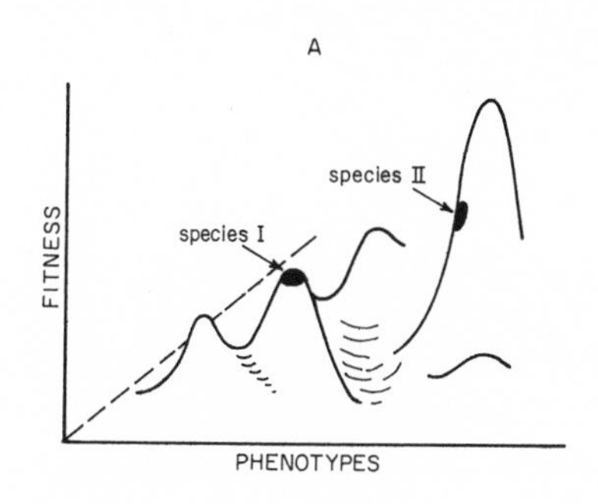

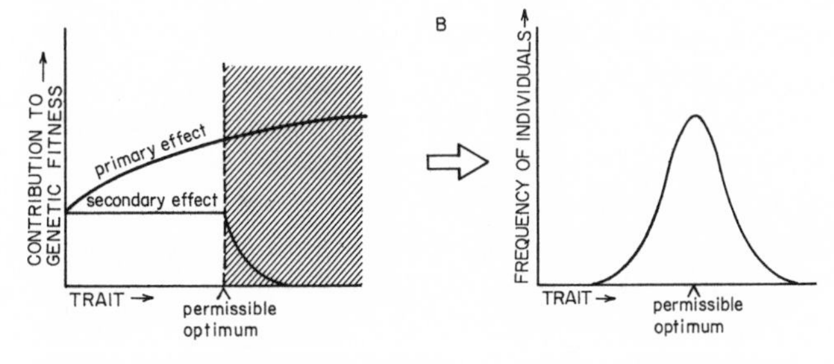

Figure 2-3 Concepts of optimization in evolutionary theory. *A*: An adaptive landscape: a surface of phenotypes (imaginary in this case) in which the similarity of the underlying genotypes is indicated by the nearness of the points on the surface and their relative fitness by their elevation. Species I is at equilibrium on a lower adaptive peak; it is characterized by an optimal permissible trait which is less perfect than the ideal conceivable trait. Species II is in disequilibrium because it is still evolving toward another permissible optimum. *B*: The species shown here is at equilibrium at the permissible optimum of a particular trait. Although the primary function of the trait would be ever more improved by an indefinite intensification of the trait, its secondary effect begins to reduce the fitness conferred on the organism when the trait exceeds a certain value. The threshold value is by definition the permissible optimum.

The opposite extreme is the "equilibrium case," in which species at different levels of social evolution are more or less equally well adapted. There can be multiple adaptive peaks, corresponding to "primitive," "intermediate," and "advanced" stages of sociality. In somewhat more concrete terms, the equilibrium hypothesis envisions lower levels of sociality as compromises struck by species under the influence of opposing selection pressures. The imaginary species represented in Figure 2-3B is favored by an indefinite intensification of the primary effect of the trait. But the evolution of the trait cannot continue forever, because a secondary effect begins to reduce the fitness of the organism when the trait exceeds a certain value. The species equilibrates at this, the permissible, optimum. For example, among males of mountain sheep and other harem-forming ungulates dominance rank is strongly correlated with the size of the horns. The upper limit of horn size must therefore be set by other effects, presumably mechanical stress and loss of maneuverability caused by excessive horn size, together with the energetic cost of growing and maintaining the horns.

Deep versus shallow convergence. At this stage of our knowledge it is desirable to begin an analysis of evolutionary convergence per se, for the reason that an analogy recognized between two behaviors in one case may be a much more profound and significant phenomenon than an analogy recognized in another case. It will be useful to make a rough distinction between instances of evolutionary convergence that are deep and those that are shallow. The primary defining qualities of deep convergence are two: the complexity of the adaptation and the extent to which the species has organized its way of life around it. The eye of the vertebrate and the eye of the cephalopod mollusk constitute a familiar example of a very deep convergence. Other characteristics associated with deep convergence, but not primarily defining the phenomenon, are degree of remoteness in phylogenetic origin, which helps determine the summed amount of evolution the two phyletic lines must travel to the point of convergence, and stability. Very shallow convergence is often marked by genetic lability. Related species, and sometimes populations within the same species, differ in the degree to which they show the trait, and some do not possess it at all.

Among the deepest and therefore most interesting cases of convergence in social behavior is the development of sterile workers castes in the social wasps, most of which belong to the family Vespidae, and in the social bees, which have evolved through nonsocial ancestors ultimately from the wasp family Sphecidae. The convergence of worker castes of ants and termites is even more profound, in that the adult forms have become flightless and reduced their vision as adaptations to a subterranean existence. Also, their phylogenetic bases are considerably farther apart: the ants originated from tiphiid wasps, and the termites from primitive social cockroaches. At the opposite extreme, numerous examples of shallow convergence can be listed from the evolution of territoriality and dominance hierarchies, an aspect of the subject that will be explored in detail in Chapter 13.

Grades versus clades. Evolution consists of two simultaneously occurring processes: while all species are evolving vertically through time, some of them split into two or more independently evolving lines. In the course of vertical evolution a species, or a group of species, ultimately passes through series of stages in certain morphological, physiological, or behavioral traits. If the stages are distinct enough, they are referred to as evolutionary grades. Phylogenetically remote lines can reach and pass through the same grades, in which case we speak of the species making up these lines as being convergent with respect to the trait. Different species often reach the same grade at different times. A separate evolving line is referred to as a clade, and a branching diagram that shows how species split and form new species is called a cladogram (Simpson, 1961; Mayr, 1969). The full phylogenetic tree contains the information of the cladogram, plus some measure of the amount of divergence between the branches, plotted against a time scale. Sociobiologists are interested in both the evolutionary grades of social behavior and the phylogenetic relationships of the species within them. An excellent paradigm from the literature of social wasps is provided in Figure 2-4.

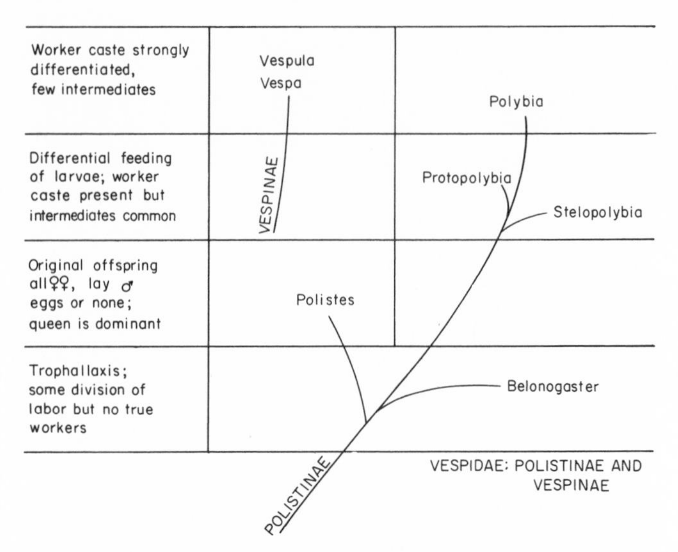

Figure 2-4 Cladograms of two groups of social wasps, the subfamilies Polistinae and Vespinae of the family Vespidae, are projected against the evolutionary grades of social behavior. The grades, which ascend from less advanced to more advanced states, are labeled on the left. The clades, or separate branches, are genera of the wasps. (Redrawn from Evans, 1958.)

Instinct versus learned behavior. In the history of biology no distinction has produced a greater semantic morass than the one between instinct and learning. Some recent writers have attempted to skirt the issue altogether by declaring it a nonproblem and refusing to continue the instinct-learning dichotomy as part of modern language. Actually, the distinction remains a useful one, and the semantic difficulty can be cleared up rather easily.

The key to the problem is the recognition that instinct, or innate behavior, as it is often called, has been intuitively defined in two very different ways:

1. An innate behavioral difference between two individuals or two species is one that is based at least in part on a genetic difference. We then speak of differences in the hereditary component of the behavior pattern, or of innate differences in behavior, or, most loosely, of differences in instinct.

2. An instinct, or innate behavior pattern, is a behavior pattern that either is subject to relatively little modification in the lifetime of the organism, or varies very little throughout the population, or (preferably) both.

The first definition can be made precise, since it is just a special case of the usual distinction made by geneticists between inherited and environmentally imposed variation. It requires, however, that we identify a difference between two or more individuals. Thus, by the first definition, blue eye color in human beings can be proved to be genetically different from brown eye color. But it is meaningless to ask whether blue eye color alone is determined by heredity or environment. Obviously, both the genes for blue eye color and the environment contributed to the final product. The only useful question with reference to the first definition is whether human beings that develop blue eye color instead of brown eye color do so at least in part because they have genes different from those that control brown eye color. The same reasoning can be extended without change to different patterns of social behavior.

The second intuitive definition of instinct can be most readily grasped by considering one of the extreme examples that fits it. The males of moth species are characteristically attracted only to the sex pheromones emitted by the females of their own species.

In some cases they may be "fooled" by the pheromones of other, closely related species, but rarely to the extent of completing copulation. The sex pheromone of the silkworm moth (*Bombyx mori*) is 10, 12-hexadecadienol. The male responds only to this substance, and he is more sensitive by several orders of magnitude to one particular geometric isomer (*trans*-10-*cis*-12-hexadecadienol) than to the other isomers. Moreover, the discrimination takes place at the level of the sensilla trichodea, the hairlike olfactory receptors distributed over the antennae. Only when these organs encounter the correct pheromone do they send nervous impulses to the brain, triggering the efferent flow of commands that initiate the sexual response. In not the remotest sense is learning involved in such a machinelike response, which is typical of much of the behavior of arthropods and other invertebrates. Very few invertebrate zoologists feel self-conscious about alluding to this behavior as innate or instinctive, and they have in mind both the first and second definitions.

At the opposite extreme, we have the plastic qualities of human speech and vertebrate social organization, and no one feels correct in labeling these traits instinctive by the second definition. A moment's reflection on the intermediate cases reveal that they cannot be classified by a strict criterion comparable to the presence or absence of a genetic component used in the first definition. Therefore, the second definition can never be precise, and it really has informational content only when applied to the extreme cases.

CHAPTER 3

The Prime Movers of Social Evolution

This chapter consists of an excursion into what can be termed the natural history of sociobiology, as opposed to its basic theory. Its organizing theme is the following argument. The major determinants of social organization are the demographic parameters (birth rates, death rates, and equilibrium population size), the rates of gene flow, and the coefficients of relationship. In both an evolutionary and functional sense these factors, to be analyzed more formally in Chapter 4, orchestrate the joint behaviors of group members. But as population biologists come to understand them better, they see that the chain of causation has been traced only one link down. What, we then ask, determines the determinants? These prime movers of social evolution can be divided into two broad categories of very diverse phenomena: *phylogenetic inertia* and *ecological pressure*.

Phylogenetic inertia, similar to inertia in physics, consists of the basic properties of the population that determine the extent to which its evolution can be deflected in one direction or another, as well as the amount by which its rate of evolution can be speeded or slowed. Environmental pressure is simply the set of all the environmental influences, both physical conditions such as temperature and humidity and the living part of the environment, including prey, predators, and competitors, that constitute the agents of natural selection and set the direction in which a species evolves.

Social evolution is the outcome of the genetic response of populations to ecological pressure within the constraints imposed by phylogenetic inertia. Typically, the adaptation defined by the pressure is narrow in extent. It may be the exploitation of a new kind of food, or the fuller use of an old one, superior competitive ability against perhaps one formidable species, a stronger defense against a particularly effective predator, the ability to penetrate a new, difficult habitat, and so on. Such a unitary adaptation is manifest in the choice and interplay of the behaviors that make up the social life of the species. As a consequence, social behavior tends to be idiosyncratic. That is why any current discussion of the prime movers must take the form of natural history. The remainder of this chapter, then, consists of a survey of the many kinds of phylogenetic inertia and ecological pressure, together with a first attempt to assess their relative importance.

Phylogenetic Inertia

High inertia implies resistance to evolutionary change, and low inertia a relatively high degree of lability. Inertia includes a great deal of what evolutionists have always called preadaptation—the fortuitous predisposition of a trait to acquire functions other than the ones it originally served—but there are aspects of the process involved that fall outside the ordinary narrow usage of that term. Furthermore, as I hope to establish here, there is an advantage to continuing the physical analogy into at least the initial stages of evolutionary behavioral analysis.

Sociobiologists have found examples of phylogenetic diversity that are the outcome of inertial differences between evolving lines. One of the most striking is the restricted appearance of higher social behavior within the insects. Of the 12 or more times that true colonial life (eusociality) has originated in the insects, only once—in the termites—is this event known to have occurred outside the single order Hymenoptera, that is, in insects other than ants, bees, and wasps. W. D. Hamilton (1964) has argued with substantial logic and documentation that this peculiarity stems from the haplodiploid mode of sex determination used by the Hymenoptera and a few other groups of organisms, in which fertilized eggs produce females and unfertilized eggs produce males. One consequence of haploidiploidy is that females are more closely related to their sisters than they are to their own daughters. Therefore, all other things being equal, a female will contribute more genes to the next generation by rearing a sister than by rearing a daughter. The likely result in evolution is the origin of sterile female castes and of a tight colonial organization centered on a single fertile female. This in fact is the typical condition of the hymenopterous societies. (For a full critique of the advantages and difficulties of this idea, see Wilson, 1971a, and Lin and Michener, 1972, as well as Chapter 19).

An important component of inertia is the genetic variability of a population, or, more precisely, the amount of phenotype variability referable to genetic variation. The rate at which a population responds to selection depends exactly on the amount of this variability. Inertia in this case is measured by the rate of change of relative frequencies of genes that already exist in the population. If an environmental change renders old features of social organization inferior to new ones, the population can evolve relatively quickly to the new mode provided the appropriate genotypes can be assembled from within the existing gene pool. The population will proceed to the new mode at a rate that is a function of the product of the degree of superiority of the new mode, referred to as the intensity of selection, and the amount of phenotypic variability that has a genetic basis. Imagine some nonterritorial population faced with an environmental change that makes territoriality strongly advantageous. Suppose that a small fraction of the individuals occasionally display the rudiments of territorial behavior, and that this tendency has a genetic basis. We can expect the population to evolve relatively quickly, say, over the order of 10 to at most 100 generations, to arrive at a primarily territorial mode of organization. Now consider a second population in identical circumstances, but with the occasional display of territorial behavior having no genetic basis—any genotype in the population is equally likely to develop it. In other words, genetic variability in the trait is zero. In this second case, the species will not evolve in the direction of territorial behavior.

There are some intriguing cases in which populations have failed to alter their social behavior to what seems to be a more adaptive form. The gray seal (*Halichoerus grypus*) has extended its range in recent years from the North Atlantic ice floes, where it breeds in pairs or in small groups, southward to localities where it breeds in large, crowded rookeries along rocky shores. Under the new circumstances the females might be expected to adopt the habit, characteristic of other colonial pinnipeds, of limiting their attention strictly to their own pups. But this has not occurred. Instead, mothers fail to discriminate between pups during mammary feeding, and many of the weaker young die of starvation (E. A. Smith, 1968). The spotted hyenas of the Serengeti, unlike their relatives in the Ngorongoro Crater, subsist on game that is migratory during large parts of the year. Yet this population still behaves as though it were dealing with fixed ungulate populations, in ways that seem adapted to an environment like that of Ngorongoro Crater rather than to the unstable conditions of the Serengeti. The cubs are immobile and dependent on the mother over long periods of time, and they are not whelped at the most favorable season of the year. Several specific behavior patterns of the hyenas are clearly connected with an obsolete territorial system. They include stereotyped forms of scent marking, "border patrols," and direct aggression toward intruders (Kruuk, 1972).

We are led to ask whether the gray seal and hyena populations have failed to adapt because the required social alterations are not within their immediate genetic grasp. Or do they have the capacity and are evolving, but have not yet had sufficient time? A third possibility is that the requisite genetic variability is present but the populations cannot evolve further because of gene flow from nearby populations adapted to other circumstances. The last hypothesis, that of genetic "swamping," is basically the explanation offered by Kummer (1971) to account for maladaptive features in the social organization of baboon populations living just beyond the limits of the species' preferred habitats.

Success or failure in evolving a particular social mechanism often depends simply on the presence or absence of a particular *preadaptation*—a previously existing structure, physiological process, or behavior pattern which is already functional in another context and available as a stepping stone to the attainment of a new adaptation. Avicularia and vibracula, two of the more bizarre forms of specialized individuals found in bryozoan colonies, occur only within the ectoproct order Cheilostomata. The reason is simple: only the cheilostomes possess the operculum, a lidlike cover that protects the mouth of the organism. The essential structures of the specialized castes, the beak of the avicularium, which is used to fight off enemies, and the seta of the vibraculum, were both derived in evolution from the operculum (Ryland, 1970). Passerine birds accommodate the increased demands of territorial defense and reproduction in the breeding season by raising their total energy expenditure. But the same option is closed to the hummingbirds, whose hovering flight is already energetically very costly. Instead, hummingbirds maintain a nearly constant energy expenditure and simply devote less time during the breeding period to their nonsocial activities (Stiles, 1971). Social parasitism is rampant in the ants but virtually absent in the bees and termites. The reason appears to be simply that ant queens often return to nests of their own species after nuptial flights, predisposing them to enter nests of other species as well, while queens of bees and termites do not (Wilson, 1971a).

The kind of food on which the species feeds can also guide the evolution of social behavior. First, dispersed, predictable food sources tend to lead to territorial behavior, while patchily distributed sources unpredictable through time favor colonial existence. Second, large, dangerous prey promote high degrees of cooperative and reciprocally altruistic behavior. Still another very general relation concerns the position on the trophic ladder: herbivores maintain the highest population densities and smallest home ranges, while top carnivores such as wolves and tigers are scarcest and utilize the largest home ranges. The reason is the substantial leakage of energy through respiration as the energy is passed up the food chains from plants to herbivores and thence to carnivores and top carnivores. In fact, only about 10 percent of the energy is transferred successfully from one trophic level to the next. This figure is imprecise but still accurate enough to account for an important general feature of the organization of ecosystems: food chains seldom have more than four or five links. The explanation is that a 90 percent reduction (approximately) in productivity results in only $(1/10)^4 = 0.0001$ of the energy removed from the green plants being available to the fifth trophic level. In fact, the top carnivore that is utilizing only 0.0001 as many calories as are produced by the plants on which it ultimately depends must be both sparsely distributed and far-ranging in its activities. Wolves must travel many kilometers each day to find enough energy. The ranges of tigers and other big cats often cover hundreds of square kilometers, while polar bears and killer whales travel back and forth over even greater distances. This demanding existence has exerted a strong evolutionary influence on the details of social behavior.

The components of phylogenetic inertia also include many *antisocial factors,* the selection pressures that tend to move the population to a less social state (Wilson, 1972a). Social insects and probably other highly colonial organisms have to contend with the "reproductivity effect": the larger the colony, the lower its rate of production of new individuals per colony member (Michener, 1964a; Wilson, 1971a). Large colonies, in other words, usually produce a higher total of new individuals in a given season, but the number of such individuals divided by the number already present in the colony is less. Ultimately, this means that social behavior can evolve only if large colonies survive at a significantly higher rate than small colonies and if individuals protected by colonies survive better than those left unprotected. Otherwise, the lower reproductivity of larger colonies will cause natural selection to reduce colony size and perhaps to eliminate social life altogether.

In mammals the principal antisocial factor appears to be chronic food shortage. Adult male coatis (*Nasua narica*) of Central American forests join the bands of females and juveniles only while large quantities of food are ripening on the trees, at which time mating takes place. In other seasons, when food is scarcer, the males are actively repulsed by the bands. The females and young begin to forage cooperatively for invertebrates on the forest floor, while the solitary males concentrate on somewhat larger prey (Smythe, 1970). The moose (*Alces americana*), unlike many of the other great horned ungulates, is essentially solitary in its habits. Not only do the bulls stay apart outside the rutting season, but the cows drive away the yearling calves at just about the age when these young animals are able to fend off wolves, their principal predators. Geist (1971a) has argued persuasively that this curtailment of social behavior, which would otherwise confer added protection against wolves, has been forced in evolution by the species' opportunistic feeding strategy. Moose depend to a great extent on second-growth forage, particularly that emerging after fires. This food source is patchy in distribution and subject to periodic shortages, especially when the winter snow is high.

Another potentially antisocial force is sexual selection. When circumstances favor the evolution of polygamy (see Chapter 15), sexual dimorphism increases. Typically, the males become larger, more aggressive, and conspicuous by virtue of their exaggerated display behavior and secondary anatomical characteristics. The result is that adult males are less likely to be closely integrated into the society formed by the females and juveniles. This is the apparent explanation for the female-centered societies that characterize deer, African plains antelopes, mountain sheep, and certain other ungulates whose males fight to establish harems during the rutting season. In the elephant seals, sea lions, and other strongly dimorphic pinnipeds, the large size and aggressive behavior of the males occasionally results in accidental injury or death to the young. Size dimorphism can also lead to different energetic requirements and sleeping sites, which have an even more disruptive effect.

Another, possibly widespread antisocial factor is the loss of efficiency and individual fitness through inbreeding. Social organization, by closing groups off from one another, tightening the association of kin, and reducing individual movement, tends to restrict gene flow within the population as a whole. The result is increasing inbreeding and homozygosity (see Chapter 4). We have little information on the importance of this factor in real populations. If significant at all, it almost certainity varies greatly in effect from case to case, because of the idiosyncratic qualities of social organization and gene flow that characterize the biology of individual species.

Different categories of behavior vary enormously in the amount of phylogenetic inertia they display. Among those characterized by relatively low degrees of inertia are dominance, territoriality, courtship behavior, nest building, and oriented movements. Behaviors possessing high inertia include complex learning, feeding responses, oviposition, and parental care. In the case of low inertial systems, large components of the behavior can be added or discarded, or even the entire category evolved or discarded, in the course of evolution from one species to another. At least four aspects of a behavioral category, or any particular evolving morphological or physiological system underlying behavior, determine the inertia:

1. *Genetic variability*. This property of populations can be expected to cause differences between populations in low inertial social categories.
2. *Antisocial factors*. The processes are idiosyncratic in their occurrence and can be expected to generate inertia at various levels.
3. *The complexity of the social behavior*. The more numerous the components constituting the behavior, and the more elaborate the physiological machinery required to produce each component, the greater the inertia.
4. *The effect of the evolution on other traits*. To the extent that efficiency of other traits is impaired by alterations in the social system, the inertia is increased. For instance, if installment of territorial behavior cuts too far into feeding time or exposes individuals to too much predation, the evolution of the territorial behavior will be slowed or stopped.

Ecological Pressure

The natural history of sociobiology has begun to yield a very interesting series of ecological correlations. Some environmental factors tend to induce social evolution, others do not. Moreover, the form of social organization and the degree of complexity of the society are strongly influenced by only one or a very few of the principal adaptations of the species: the food on which it specializes, the degree to which seasonal change of its habitat forces it to migrate, its most dangerous predator, and so forth. To examine this generalization properly, let us next review the factors that have been identified as principal selective forces in field studies of particular social species.

Defense against Predators

An Ethiopian proverb says, "When spider webs unite, they can halt a lion." Defensive superiority is the adaptive advantage of cooperative behavior reported most frequently in field studies, and it is the one that occurs in the greatest diversity of organisms. It is easy to imagine the steps by which social integration of populations can be made increasingly complex by the force of sustained predation. The mere concentration of members of the same species in one place makes it more difficult for a predator to approach any one member without detection. Flying foxes (*Pteropus*), which are really large fruit bats, form dense sleeping aggregations in trees. Each male has his own resting position determined by dominance interactions with other males. The lower, more perilous branches of the trees serve as warning stations for the colony as a whole. Any predator attempting to climb the tree launches the entire colony into the air and out of reach (Neuweiler, 1969). In his study of arctic ground squirrels (*Spermophilus undulatus*), Ernest Carl (1971) was personally able to stalk isolated individuals to within 3 meters—close enough, in all probability, for a predator such as the red fox to make the rush and kill. But he found it impossible to close in on groups. From distances as great as 300 meters the *Spermophilus* set up waves of alarm calls, which increased in intensity and duration as the intruder came closer. By noting the quality and source of the alarm calls, Carl was even able to judge the shifting positions of predators as they passed through the *Spermophilus* colonies. Individual ground squirrels can probably make the same judgment. Similar observations were made by King (1955) on the black-tail prairie dog (*Cynomys ludovicianus*). These rodents live in particularly dense, well-organized communities, the so-called towns, and it is probably one reward of their population structure that they suffer only to a minor degree from predation.

Another social way of avoiding predators is to use marginal individuals of the group as a shield. Since predators tend to seize the first individual they encounter, there is a great advantage for each individual to press toward the center of its group. The result in evolution would be a "herd instinct" that centripetally collapses populations into local aggregations. Francis Galton was the first to comprehend the effects of such an elementary natural selection for geometric pattern. In 1871 he described the behavior of cattle exposed to lions in the Damara country of South Africa:

> Yet although the ox had so little affection for, or individual interest in, his fellows, he cannot endure even a momentary severance from his herd. If he be separated from it by stratagem or force, he exhibits every sign of mental agony; he strives with all his might to get back again and when he succeeds, he plunges into its middle, to bathe his whole body with the comfort of close companionship.

The result of centripetal movement is some of the most visually impressive but least organized of all forms of social behavior. Centripetal movement generates not only herds of cattle but also fish and squid schools, bird flocks, heronries, gulleries, terneries, locust swarms, and many other kinds of elementary motion groups and nesting associations. In more recent years this idea of the "selfish herd" has been developed persuasively, principally by means of circumstantial evidence and plausibility arguments, by G. C. Williams (1964, 1966) and W. D. Hamilton (1971a).

Nearly equivalent in effect to herding and schooling is synchronized breeding. When colonial birds "crowd" their reproductive effort into a short time span, each mated pair confronts the waiting predators at the time the predators are most probably well fed and hence likely to ignore any particular chick. The pairs that start too early or too late will produce chicks that are the equivalent of the cattle living dangerously on the margin of a herd. The absence of synchrony among the pairs is the equivalent of what happens when the members of the herd scatter and expose themselves to increased predation. This inference is supported by the independent studies of Patterson (1965) on the population of the black-headed gull *Larus ridibundus* at Ravenglass in England. In 1962 most eggs were laid between the sixth and fifteenth days after the first eggs appeared, and of these, 11 percent gave rise to fledged young. But of the smaller number of eggs laid five days before and five days following this period, only 3.5 percent produced fledged young. Comparable results were obtained in 1963. The chief predators of the chicks, carrion crows and herring gulls, were simply saturated by the brief superabundance of their small prey.

Synchronized breeding, of unknown physiological origin, also occurs in social ungulates. The reproductive cycle of the wildebeest (*Connochaetes taurinus*) is characterized by sharp peaks of mating and birth. Mating occurs during a short interval in the middle of the long rainy season. Calving begins abruptly about eight months later and continues at a fairly constant rate for two to three weeks, during which 80 percent of the births occur. The remaining 20 percent occur at a slowly declining rate over the following four to five months. The synchronization of birth is even more precise than these data suggest: the majority of births occur in the forenoon, in large aggregations on calving grounds usually located on short grass (Estes, 1966). When a cow is thrown slightly out of phase, she is able to interrupt delivery at any stage prior to the emergence of the calf's head, thus giving her another chance to join the mass parturition. The synchronization almost certainly has among its results the saturation of local predators and the increased survival rate of the newborn calves. To this benefit is added an extraordinary precocity on the part of the calves: they are able to stand and to run within an average of seven minutes after their birth. And they *must* be able to do this because the cows will defend them only if both are overtaken in flight.

Simply moving together in a group can reduce the individual's risk of encountering a hungry predator, because aggregation makes it difficult for a particular predator to find any prey at all. Suppose that a large fish has no way of tracking smaller fish and feeds only when it encounters the prey in the course of random searching. Brock and Riffenburgh (1960) have pursued a basic geometric and probability model to prove formally what intuition suggests, that as a prey population coalesces into larger and larger schools, the average distance between the schools increases, and there is a corresponding decrease in the frequency of the detection of schools by a randomly moving predator. Since one predator consumes no more than a fixed average number of prey at each encounter, the school size need only exceed this number in order for some of its members to escape. Thus above a certain level, increase in school size confers a mounting degree of protection on its members. The same conclusion applies to herds, flocks, and other constantly moving groups. It loses force to the degree that the hunted group settles down, follows predictable migratory paths, can be tracked from place to place by the predators, or is easier to detect in the first place.

Substantial evidence also exists of the greater effectiveness of group defense. In experiments on two European butterflies, the small tortoiseshell *Aglais urticae* and the peacock butterfly *Inachis io*, Erna Mosebach-Pukowski (1937) found that caterpillars in crowds were eaten less frequently than solitary ones. A study of ascalaphid neuropterans by Charles Henry (1972) has revealed what is virtually a controlled evolutionary experiment on the efficiency of group defense. The adults of these insects superficially resemble dragonflies and are sometimes popularly called owlflies. The female of *Ululodes mexicana* lays eggs in packets on the sides of twigs, then deposits a set of highly modified eggs called repagula ("barriers") farther down the stem. The repagula form a sticky barrier that prevents ants and other crawling predatory insects from reaching the nearby hatching larvae. Thus protected, the larvae quickly scatter from the oviposition site. A second ascalaphid species, *Ascaloptynx furciger,* employs a very different strategy. The modified eggs are used as food by the young owlfly larvae. They are not sticky and do not prevent predators from attacking the larvae. Unlike *Ululodes,* however, the *Ascaloptynx* larvae strongly aggregate and present potential enemies with a bristling mass of sharp, snapping jaws (Figure 3-1). The response is seen only when the *Ascaloptynx* are threatened by larger insects. Smaller insects such as fruit flies are treated as prey and captured by larvae that approach them singly. Henry's experiment demonstrated that larvae can be subdued by predators such as ants if they are caught alone, but that when defending en masse they are relatively safe.

Cooperative behavior within the group, the essential ingredient that turns an aggregation into a society, can improve defensive capability still further.

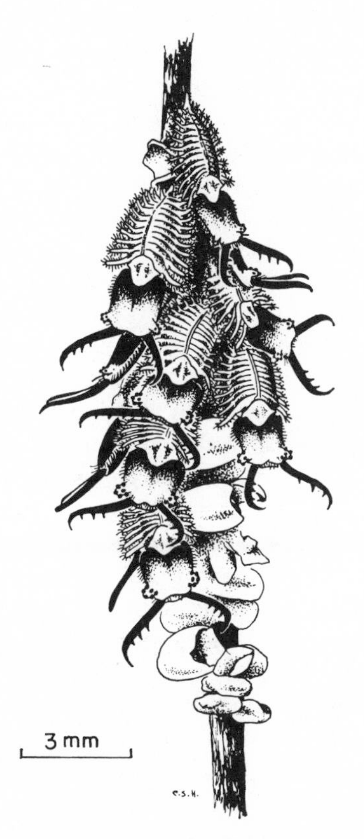

Figure 3-1 The mass defensive response of newly hatched owlfly larvae (*Ascaloptynx furciger*). When confronted by insect predators who crawl up the stem toward them, the larvae bunch together, turn to face the enemy, raise their heads, and rapidly and repeatedly snap their jaws. (From Henry, 1972.)

Among the bees, cooperative defense seems also to have been a principal element in the evolution to complex sociality. Bees are influenced by the reproductivity effect—the decline of individual reproduction with increase in group size—which is a component of phylogenetic inertia that slows or reverses social evolution in primitively social insects. The effect has been overcome in halictid bees, according to Michener (1958), by the improved defense against parasitic and predatory arthropods that associations of little groups of nestmates provide. Several observers besides Michener have witnessed guard bees protecting their nest entrances against ants and mutillid wasps.

Social ungulates that move in large amorphous herds, such as the wildebeest and Thomson's gazelle, do not cooperate in active defense against lions and other predators (Kruuk, 1972; Schaller, 1972). They depend chiefly on flight to escape. But ungulates that form small discrete units, comprised of one or more harems and other kinship groups, are more aggressive toward predators and mutually assist one another. Sometimes they move in complex patterns resembling military maneuvers. One of the most striking is the celebrated perimeter defense thrown up by musk oxen (*Ovibos moschatus*) against wolves. The adults form a circle with their heads facing outward, while the calves retreat to the center of the circle.

Even more impressive are the defensive maneuvers of large terrestrial primates. This is particularly true of the response called *mobbing:* the joint assault on a predator too formidable to be handled by a single individual in an attempt to disable it or at least drive it from the vicinity, even though the predator is not engaged in an attack on the group (Hartley, 1950). When presented with a stuffed leopard, for example, a troop of baboons goes into an aggressive frenzy. The dominant males dash forward, screaming and charging and retreating repeatedly in short rushes. When the "predator" does not react, the males grow more confident, slashing at the hind portions of the dummy with their long canines and dragging it for short distances. After a while other members of the troop join in the attack. Finally, the troop calms down and continues on its way (DeVore, 1972). Chimpanzees show a similar response to leopard models.

Mobbing in birds is a well-defined behavioral pattern that occurs irregularly in a wide diversity of taxonomic groups, from certain hummingbirds, vireos, and sparrows to jays, thrushes, vireos, warblers, blackbirds, sparrows, finches, towhees, and still others (S. A. Altmann, 1956). It is apparently absent in other species of hummingbirds, vireos, and sparrows, and at least some doves. The attacks are normally directed at predatory birds, particularly hawks and owls, when they passively intrude into the territorial or roosting areas of the smaller birds. The mobbing calls are high-pitched, loud, and easy for human observers to localize. As Marler (1959) pointed out, the mobbing calls of different bird species are strongly convergent. In the majority of cases they are loud clicks, 0.1 second or less in duration and spread over at least 2 or 3 kiloherz of frequencies in the 0–8 kiloherz range. These two properties combine to provide a biaural receptor system, which birds possess as well as human beings, with an instant fixation on the sound source. Thus, alerted birds are able to fly toward the predator being harassed, and sizable mobs are quickly assembled. Furthermore, different species respond to one another's calls, since all make nearly the same sound, and mobbing becomes a cooperative venture.

Increased Competitive Ability

The same social devices used to rebuff predators can be used to defeat competitors. Gangs of elk approaching salt licks are able to drive out other animals, including porcupines, mule deer, and even moose, simply by the intimidating appearance of the massed approach of the group (Margaret Altmann, 1956). Observers of the African wild dog (*Lycaon pictus*) have noted that coordinated pack behavior is required not only to capture game but also to protect the prey from hyenas immediately after the kill. The wild dogs and hyenas in turn each compete with lion prides.

Elsewhere (Wilson, 1971a) I have characterized as "bonanza strategists" a class of subsocial beetle species adapted to exploit food sources that are very rich but at the same time scattered and ephemeral: dung (*Platystethus* among the Staphylinidae; and Scarabaeidae), dead wood (Passalidae, Platypodidae, Scolytidae), and carrion (*Nicrophorus* among the Silphidae). When individuals discover such a food source, they are assured of a supply more than sufficient to rear their brood. They must, however, exclude others who are seeking to utilize the same bonanza. Territorial behavior is commonplace in all of these groups. Sometimes, as in *Nicrophorus*, fighting leads to complete domination of the food site by a single pair. It is probably no coincidence that the males, and to a lesser extent the females, of so many of the species are equipped with horns and heavy mandibles—a generalization that extends to other bonanza strategists that are not subsocial, for example, the Lucanidae, the Ciidae, and many of the solitary Scarabaeidae. By the same token there is an obvious advantage to remaining in the vicinity of the food site to protect the young.

Within the higher social insects, group action is the decisive factor in aggressive encounters between colonies. It is a common observation that ant queens in the act of founding colonies as well as young colonies containing workers—the weaker units—are destroyed in large numbers by other, larger colonies belonging to the same species. Newly mated queens of *Formica fusca*, for example, are captured and killed as they run past the nest entrances (Donisthorpe. 1915); the same fate befalls a large percentage of the colony-founding queens of the Australian meat ant *Iridomyrmex detectus* and red imported fire ant *Solenopsis invicta*. Queens of *Myrmica* and *Lasius* are harried by ant colonies, including those belonging to their own species, and finally they are either driven from the area or killed (Brian, 1955, 1956a,b; Wilson, 1971a).

Territorial fighting among mature colonies is common but not universal in ants. The most dramatic battles known within species are those conducted by the common pavement ant *Tetramorium caespitum*. First described by the Reverend Henry C. McCook (1879) from observations in Penn Square, Philadelphia, these "wars" can be witnessed in abundance on sidewalks and lawns in towns and cities of the eastern United States throughout the summer. Masses of hundreds or thousands of the small dark brown workers lock in combat for hours at a time, tumbling, biting, and pulling one another, while new recruits are guided to the melee along freshly laid odor trails. Although no careful study of this phenomenon has been undertaken, it appears superficially to be a contest between adjacent colonies in the vicinity of their territorial boundaries. Curiously, only a minute fraction of the workers are injured or killed.

In the case of competition within the same species, we should expect to find that groups generally prevail over individuals, and larger groups over small ones. Consequently, competition, when it comes into play, should be a powerful selective force favoring not only social behavior but also large group size. Lindburg (1971) demonstrated a straightforward case of this relationship in a local population of free-ranging rhesus monkeys (*Macaca mulatta*) he studied in northern India. The population was divided into five troops, most of which had overlapping home ranges and therefore came into occasional contact. In the pairwise aggressive encounters that occurred, one group usually retreated, and this was almost invariably the smaller one. The same selective pressures should operate to favor coalitions or cliques within societies. The phenomenon does occur commonly in wolves and those primate species, such as baboons and rhesus monkeys, in which dominance hierarchies play an important role in social organization. In other words, coalitions are known in aggressive animals that have a sufficiently high degree of intelligence to remember and exploit cooperative relationships.

Increased Feeding Efficiency

In addition to preventing individual organisms from being turned into energy by predators, social behavior can assist in converting other organisms into energy. There are two major categories of social feeding: *imitative foraging* and *cooperative foraging*. In imitative foraging the animal simply goes where the group goes, and eats what it eats, as in schools of fish or flocks of birds. The pooled knowledge and efficiency of such a feeding assemblage exceeds that of an otherwise similar but independently acting group of individuals, but the outcome is a by-product of essentially selfish actions on the part of each member of the assemblage. In cooperative foraging there is some measure of at least temporarily altruistic restraint, the behaviors of the group members are often diversified, and the modes of communication are typically complex. Some of the most advanced of all societies, possibly including those of primitive man, are based upon a strategy of cooperative hunting. Indeed, the qualities we intuitively associate with higher social behavior—altruism, differentiation of group members, and integration of group members by communication—are the same ones that evolve in a straightforward way to implement cooperative foraging.

Some of the most striking cases of imitative foraging are provided by communally feeding birds. In the Central Valley of California, enormous flocks of starlings leave their roosts and fly in straight lines to food sources as distant as 80 kilometers. The lengths of the flights are greatest in winter, when food is in shortest supply (W. J. Hamilton III and Gilbert, 1969). By following a flock the individual starling has the greatest chance of locating adequate amounts of food on a given day, since it is utilizing the knowledge of the most experienced birds in the group. Also, it will expend the least amount of energy reaching the food. Theoretically, the prime factor for colonial roosting and nesting, as Horn (1968) has shown in an elegant geometric analysis, is that the food supply be considerably variable in space and time. That is, food must appear in unpredictable, irregular patches in the environment. If it occurs in patches but is available in certain spots permanently or at predictable intervals, individuals will simply roost or nest as closely as possible to those spots, and fly singly to them. But if the food is evenly distributed through the environment and concentrated enough to more than repay the energy expended in its defense, the individuals will stake out separate territories from which they exclude other birds (see Figure 3-2). The important feature of such colonial life is that the group be concentrated enough to forage more or less as a unit. Horn's principle applies to many kinds of colonial birds, from blackbirds and swallows to herons, ibises, spoonbills, and various seabirds. Terns, for example, are an extreme example of seabirds that nest in aggregations and forage in

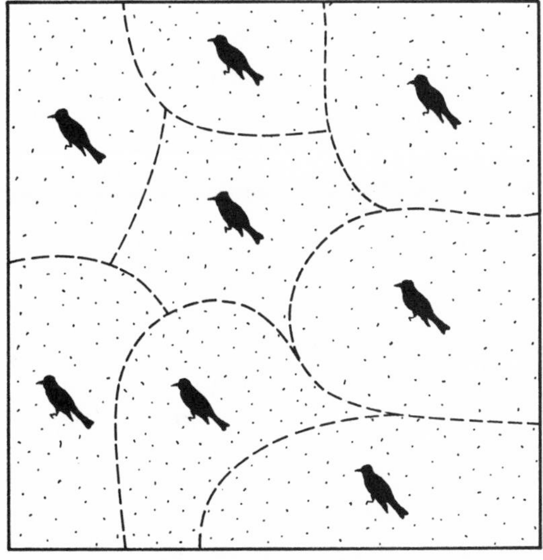

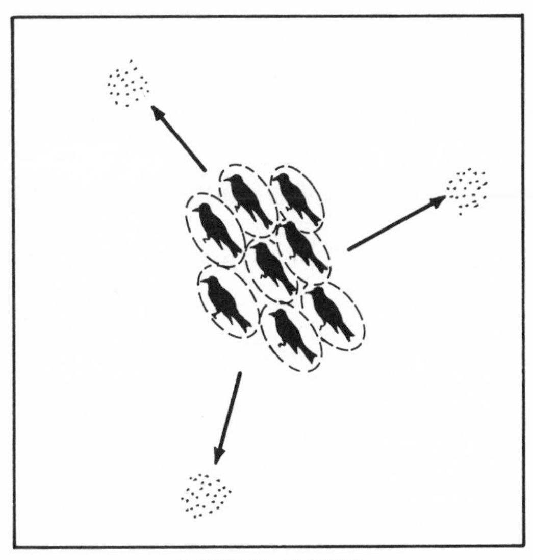

Figure 3-2 Horn's principle of group foraging. If food is more or less evenly distributed through the environment and can be defended economically, it is energetically most efficient to occupy exclusive territories (above). But if food occurs in unpredictable patches, the individuals should collapse their territories to roosting spots or nest sites, and forage as a group (below).

groups for highly unpredictable food patches: schools of small fish that move near the surface of the ocean.

Among the vertebrates, pack-hunting mammals have evolved the strongest forms of cooperative foraging. The behavior allows these animals to take prey that are otherwise difficult if not impossible to catch. In his pioneering study of the wolves of Mount McKinley National Park, for example, Murie (1944) found that these carnivores could capture their principal large prey, Dall's sheep, only with difficulty. On a typical day a pack trots from one herd to another in search of a weak or sick individual, or a stray surprised on terrain in which it is at a disadvantage. A lone wolf can trap a healthy sheep only with great difficulty if it is on a slope; the sheep outdistances the wolf easily by racing it up the slope. Two or more wolves are able to hunt with greater success because they spread out and often are able to maneuver the sheep into a downhill race or force it onto flat land. Under both circumstances they hold the advantage. Where wolves hunt moose, as they do for example in the Isle Royale National Park of Michigan, cooperative hunting is required both to trap and to disable the prey (Mech, 1970).

The most social canids of all are the African wild dogs, *Lycaon pictus*. These relatively small animals are superbly specialized for hunting the large ungulates of the African plains, including gazelles, zebras, and wildebeest. The packs, often under the guidance of a lead dog, take aim on a single animal and chase it at a dead run. They pursue the target relentlessly, sometimes through crowds of other ungulates who either stand and watch or scatter away for short distances. The *Lyacaon* do not ordinarily stalk their prey while in the open, although they sometimes use cover to approach animals more closely. Estes and Goddard (1967) watched a pack race blindly over a low crest in the apparent hope of surprising animals on the other side—in this one instance no quarry was there. Fleeing prey frequently circle back, a tactic that can help shake off a solitary pursuer. This maneuver, however, tends to be fatal when employed against a wild dog pack: the dogs lagging behind the leader simply swerve toward the turning animal and cut the loop. Once they have caught up to the prey, the dogs seize it on all sides and swiftly tear it to pieces.

A far more elaborate mode of cooperative foraging is practiced by ants whose workers use odor trail systems to guide followers to food. The well-analyzed case of the red fire ants of the genus *Solenopsis* can serve as a paradigm. When a worker of *S. invicta* finds a particle of food too large to carry, it heads home slowly, at frequent intervals extruding its sting so that the tip is drawn lightly over the ground surface. As the sting touches the ground, a pheromone flows down from Dufour's gland, a tiny organ located next to the hind gut. Each worker possesses only a small fraction of a nanogram of the trail substance at any given time, so the pheromone must be a very potent attractant. In 1959 I showed that it is possible to induce the complete recruitment process in the fire ant with artificial trails made from extracts of single Dufour's glands. Such trails induce following by dozens of individuals over a meter or more.

The waggle dance of the honeybee is in a sense the *ne plus ultra* of foraging communication, since it utilizes symbolic messages to direct workers to targets prior to leaving for the trip. It also operates over exceptionally long distances, exceeding the reach of any other known animal communication with the possible exception of the songs of whales. The waggle dance will be described in more detail in another context, in Chapter 8.

Penetration of New Adaptive Zones

Occasionally a social device permits a species to enter a novel habitat or even a whole new way of life. One case is provided by the staphylinid beetle *Bledius spectabilis,* which has evolved a complexity of maternal care rarely attained in the Coleoptera. The change has permitted the species to penetrate one of the harshest of all environments available to any insect: the intertidal mud of the European coast, where the beetle must subsist on algae and face extreme hazards from both the high salinity and periodic shortages of oxygen. The female constructs unusually wide tunnels in her brood nest, which are kept ventilated by tidal water movements and by renewed burrowing activity on the part of the female. If the mother is taken away, her brood soon perishes from lack of oxygen. The female also protects the eggs and larvae from intruders, and from time to time forages outside the nest for a supply of algae (Bro Larsen, 1952.)

Increased Reproductive Efficiency

Mating swarms, which rank with the most dramatic visual phenomena of the insect world, are formed by a diversity of species belonging to such groups as the mayflies, cicadas, coniopterygid neuropterans, mosquitoes and other nematoceran flies, empidid dance flies, braconid wasps, termites, and ants. They normally occur only during a short period of time at a certain hour of the day or night. Their primary function is to bring the sexes together for nuptial displays and mating. Termites and some ants fill the air with diffuse clouds of individuals that mate either while traveling through the air or after falling to the ground. Nematoceran flies, dance flies, and some ant species typically gather in concentrated masses over prominent landmarks such as a bush, tree, or patch of bare earth. It is plausible (but unproved) that swarming is most advantageous to members of rare species and to those living in environments where the optimal time for mating is unpredictable. Newly mated ant queens and royal termite couples, for example, require soft, moist earth in which to excavate their first nest cells and to rear the first brood of workers. In drier climates their nuptial swarms usually occur immediately after heavy rains first break a prolonged dry spell. A second potential function of the swarms is to promote outcrossing. If mature individuals of scarce species began sexual activity immediately after emerging, or in response to very local microclimatic events, rather than traveling relatively long distances to join swarms, the amount of inbreeding would be much greater.

Increased Survival at Birth

Evolving animal species are faced with two broad options in designing their birth process. First, they can invest time after the formation of zygotes by incubating the eggs, by bearing live young, or by otherwise assisting the embryos through the birth process. Failing one of these relatively involved procedures, they can simply deposit the eggs and gamble that the young will hatch and survive. In both alternatives the major risk comes from predators. Animals taking the second option, the simple ovipositors, also generally make an effort to conceal the eggs. Techniques include burying the eggs deep in the soil, inserting them into crevices, placing them on specially constructed stalks, and encrusting them with secretions that harden into an extra shell. These procedures improve the chances of survival of the embryos, but they make it more difficult for the newly hatched young to reach the outer world.

The green turtle (*Chelonia mydas*) is one species in which group behavior by the newly born increases the survival of individuals. The female journeys every second or third year to the beach of her birth to lay between 500 and 1,000 eggs. The entire lot is parceled out at up to 15 intervals in clutches of about 100. Each clutch is deposited in a deep, flask-shaped hole excavated by the mother turtle, who then pulls sand in to bury it. In watching this process, Archie Carr and his coworkers gained the impression that mass effort on the part of the hatchlings is required to escape from the nests. They tested the idea by digging up clutches and reburying the eggs in lots of 1 to 10. Of 22 hatchlings reburied singly, only 6, or 27 percent, made it to the surface. Those that came out were too unmotivated or poorly oriented to crawl down to the sea. When allowed to hatch in groups of 2, the little turtles emerged at a strikingly higher rate —84 percent—and journeyed to the water in a normal manner. Groups of 4 or more achieved virtually perfect emergence. Observations of the process through glass-sided nests revealed that emergence does depend on group activity. The first young to hatch do not start digging at once but lie still until others have appeared. Each hatching adds to the working space, because the young turtles and crumpled egg shells take up less room between them than the unhatched, spherical eggs. The excavation then proceeds by a division of labor. By relatively uncoordinated digging and squirming, the hatchlings in the top layer scratch down the ceiling, while those around the side undercut the walls, and those on the bottom trample and compact the sand that falls down from above. Gradually the whole mass of individuals moves upward to the surface.

Improved Population Stability

Under a variety of special circumstances, social behavior increases the stability of populations. Specifically, it acts either as a buffer to absorb stress from the environment and to slow population decline, or as a control preventing excessive population increase, or both. The primary result is the damping of amplitude in the fluctuation of population numbers around a consistent, predictable level. One secondary result of such regulation is that in a fixed period of time the population has a lesser probability of extinction than another, otherwise comparable population lacking regulation. In other words, the regulated population persists longer. Does a longer population survival time really benefit the individual belonging to it, whose own life span may be many orders of magnitude shorter than that of the population? Or has the regulation originated solely by selection at the level of the population, without reference to individual fitness? The third possibility is that the population stability is an epiphenomenon—an accidental by-product of individual selection with no direct adaptive value of its own.

These alternative explanations of the relation between social organization and population regulation will be explored further in Chapters 4 and 5. Suffice it for the moment simply to note what the relation is. Territories are areas controlled by animals who exclude strangers. Members of a population who cannot obtain a territory wander singly or in groups through less desirable habitats, consequently suffering a relatively high rate of mortality. They constitute an excess that drains off quickly. Since the number of possible territories is relatively constant from year to year, the population remains correspondingly stable.

Modification of the Environment

Manipulation of the physical environment is the ultimate adaptation. If it were somehow brought to perfection, environmental control would insure the indefinite survival of the species, because the genetic structure could at last be matched precisely to favorable conditions and freed from the capricious emergencies that endanger its survival. No species has approached full environmental control, not even man. Yet in a lesser sense all adaptations modify the environment in ways favorable to the individual. Social adaptations, by virtue of their great power and sophistication, have achieved the highest degree of modification.

At a primitive level, animal aggregations alter their own physical environment to an extent disproportionately greater than the extent achieved by isolated individuals, and sometimes even in qualitatively novel ways. This general effect was documented in detail by G. Bohn, A. Drzewina, W. C. Allee, and others in the 1920's and 1930's (Allee, 1931, 1938). Consider the following two examples from the flatworms. *Planaria dorotocephala,* like most protistans and small invertebrates, is very vulnerable to colloidal suspensions of heavy metals. Kept at a certain marginal concentration of colloidal silver in 10 cubic centimeters of water, a single planarian shows the beginning of head degeneration within 10 hours. But

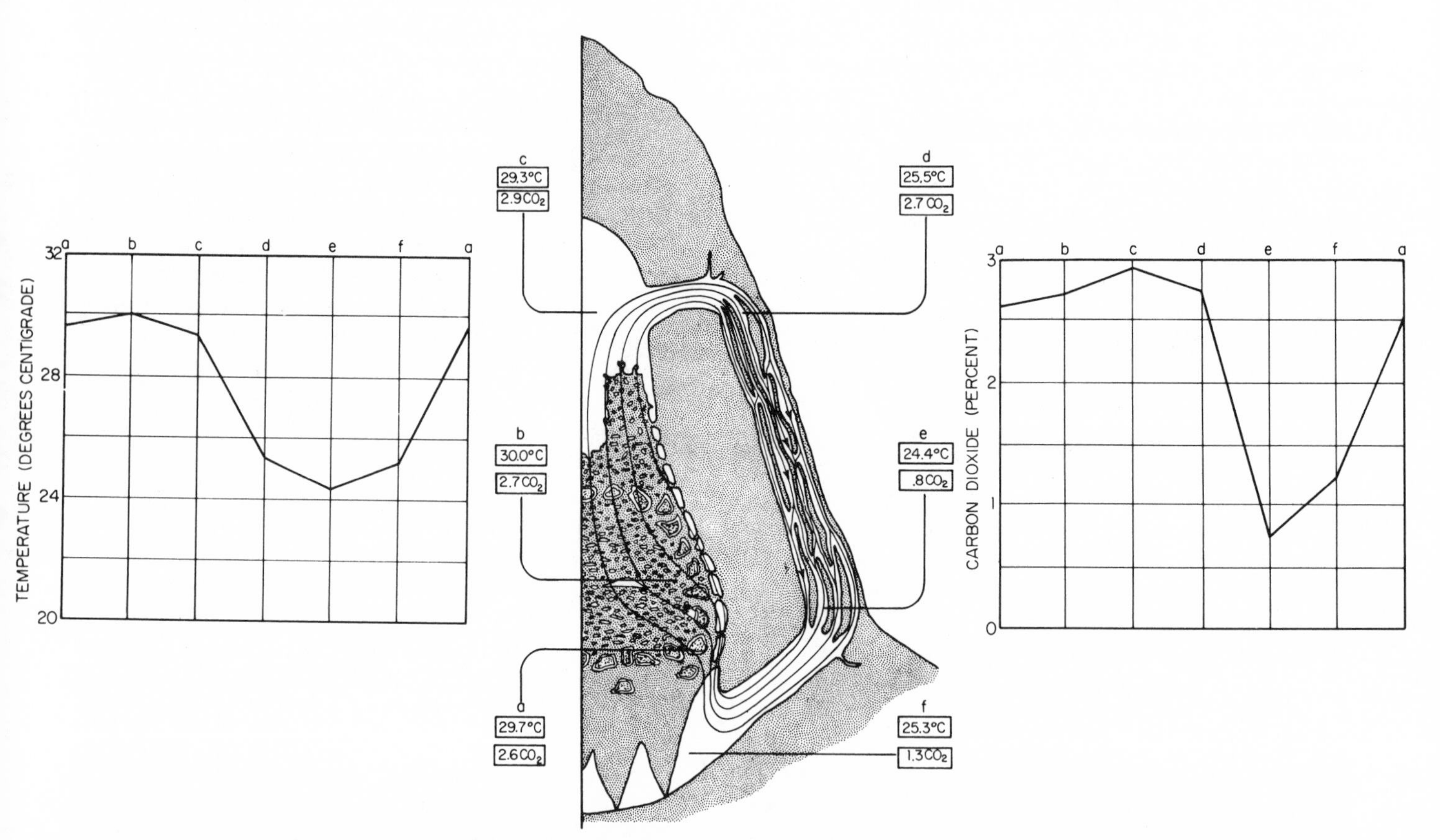

Figure 3-3 Air flow and microclimatic regulation in a nest of the African fungus-growing termite *Macrotermes bellicosus.* Half of a longitudinal section of the nest is shown here. At each of the positions indicated, the temperature (in degrees C) is shown in the upper rectangle and the percentage of carbon dioxide appears in the lower rectangle. As air warms in the central core of the nest (*a, b*) from the metabolic heat of the huge colony inside, it rises by convection to the large upper chamber (*c*) and then out to a flat, capillarylike network of chambers (*d*) next to the outer nest wall. In the outer chambers the air is cooled and refreshed. As this occurs, it sinks to the lower passages of the nest beneath the central core (*e, f*). The graphs at the sides show how the temperature and carbon dioxide change during circulation. These changes are brought about by the diffusion of gases and the radiation of heat through the thin, dry walls of the ridge. (Modified from Lüscher, 1961. From "Air-conditioned Termite Nests," by M. Lüscher ® 1961 by Scientific American, Inc. All rights reserved.)

lots of 10 worms or more maintained in the same concentration and volume survive for at least 36 hours with no externally obvious effects. The greater resistance of the group is due to the smaller amount of the toxic substance that each worm has to remove from its immediate vicinity in order to lower the concentration of the substance to a level beneath the lethal threshold. When single marine turbellarians of the species *Procerodes wheatlandi* are placed in small quantities of fresh water, they soon die and disintegrate. Groups survive longer and sometimes indefinitely. The effect is due to the higher rate at which calcium is emitted in groups, either by secretion from healthy individuals or by the disintegration of those unfortunate enough to succumb first. The group is therefore exposed to a dangerously hypotonic condition for a shorter period of time.

Adaptive design in environmental control attains its clearest expression in the higher social insects. To take one of the most extreme cases yet discovered, the complex architecture of the great nests of fungus-growing termites functions as an air-conditioning machine, the basic principles of which are illustrated in Figure 3-3.

The Reversibility of Social Evolution

Two broad generalizations have begun to emerge that will be reinforced in subsequent chapters: the ultimate dependence of particular cases of social evolution on one or a relatively few idiosyncratic environmental factors; and the existence of antisocial factors that also occur in a limited, unpredictable manner. If the antiosocial pressures come to prevail at some time after social evolution has been initiated, it is theoretically possible for social species to be returned to a lower social state or even to the solitary condition. At least two such cases have been suggested. Michener (1964b, 1965) observed that allodapine bees of the genus *Exoneurella* are a little less than fully social, since the females disperse from the nest before being joined by their daughters. This condition appears to have been derived from the behavior still displayed by the closely related genus *Exoneura,* in which the mother and daughters remain in association. Michener (1969) also noted that reversals may have occurred in the primitively social species of the halictine sweat bees. The most likely selective force, inferred from field studies on the halictines, is the relaxation of pressure from such nest parasites as mutillid wasps. The second case is from the vertebrates. In the ploceine weaverbirds, as in most other passerine groups, the species that nest in forests and feed primarily on insects are solitary in habit, or at most territorial. According to Crook (1964), these species have evolved from other ploceines that live in savannas, eat seeds—and, like many other passerine groups similarly specialized—nest in colonial groups, in a few cases of very large size.

CHAPTER 4

The Relevant Principles of Population Biology

In 1886 August Weismann expressed metaphorically the central dogma of evolutionary biology:

> It is true that this country is not entirely unknown, and if I am not mistaken, Charles Darwin, who in our time has been the first to revive the long-dormant theory of descent, has already given a sketch, which may well serve as a basis for the complete map of the domain; although perhaps many details will be added, and many others taken away. In the principle of natural selection, Darwin has indicated the route by which we must enter this unknown land.

Sociobiology will perhaps be regarded by history as the last of the disciplines to have remained in the "unknown land" beyond the route charted by Darwin's *Origin of Species.* In the first three chapters of this book we reviewed the elementary substance and mode of reasoning in sociobiology. Now let us proceed to a deeper level of analysis based at last on the principle of natural selection. The ultimate goal is a stoichiometry of social evolution. When perfected, the stoichiometry will consist of an interlocking set of models that permit the quantitative prediction of the qualities of social organization—group size, age composition, and mode of organization, including communication, division of labor, and time budgets—from a knowledge of the prime movers of social evolution discussed in Chapter 3. This chapter provides a brief review of important concepts in current theoretical population biology, arranged and exemplified in a way that stresses applications to sociobiology. The synopsis assumes a knowledge of elementary evolutionary theory and genetics at the level usually provided by beginning courses in biology. It also requires familiarity with mathematics through elementary probability theory and calculus.

Microevolution

The process of sexual reproduction creates new genotypes each generation but does not in itself cause evolution. More precisely, it creates new combinations of genes but does not change gene frequencies. If, in the simplest possible case, the frequencies of two alleles a_1 and a_2 on the same locus are p and q, respectively, and they occur in a Mendelian population within which sexual breeding occurs at random, $p + q = 1$ by definition; and the frequencies of the diploid genotypes can be written as the binomial expansion

$$(p + q)^2 = 1$$
$$p^2 + 2pq + q^2 = 1$$

where p^2 is the frequency of a_1a_1 individuals (a_1 homozygotes), $2pq$ is the frequency of a_1a_2 individuals (heterozygotes), and q^2 is the frequency of a_2a_2 individuals (a_2 homozygotes). The same result, usually called the *Hardy-Weinberg Law,* can be obtained in an intuitively clearer manner by noting that where breeding is random, the chance of getting an a_1a_1 individual is the product of the frequencies of the a_1 sperm and a_1 eggs, or $p \times p = p^2$. Likewise, a_2a_2 individuals must occur with frequency $q \times q = q^2$; and heterozygotes are generated by p sperm mating with q eggs (yielding a_1a_2 individuals) plus q sperm mating with p eggs (yielding a_2a_1 individuals), for a total of $2pq$. This result holds generation after generation. Thus sexual reproduction allows an individual to produce offspring with a diversity of genotypes, all similar to but different from its own. Yet the process does not alter the frequencies of the genes; it does not cause evolution.

Microevolution, which is evolution in its slightest, most elemental form, consists of changes in gene frequency. It is caused by one or a combination of the following five agents: mutation pressure, segregation distortion (meiotic drive), genetic drift, gene flow, and selection. Each is briefly described below.

1. *Mutation pressure:* the increase of allele a_1 at the expense of a_2 due to the fact that a_2 mutates to a_1 at a higher rate than a_1 mutates to a_2. Because mutation rates are mostly 10^{-4}/organism (or cell)/generation or

less, mutation pressure is not likely to compete with the other evolutionary forces, which commonly alter gene frequencies at rates that are orders of magnitude higher.

2. *Segregation distortion:* the unequal representation of a_1 and a_2 in the initial production of gametes by heterozygous individuals. Segregation distortion, also known as meiotic drive, can be due to mechanical effects in the cell divisions of gametogenesis, in which one allele or the other is favored in the production of the fully formed gametes. This process, however, is difficult to distinguish from gamete selection, a true form of natural selection due to the differential mortality of cells during the period between the reductional division of meiosis and zygote formation. True segregation distortion appears to be sufficiently rare to be of minor general importance.

3. *Genetic drift:* the alteration of gene frequencies through sampling error. To gain an immediate intuitive understanding of what this means, consider the following simple experiment in probability theory. Suppose we take a random sample of 10 marbles from a very large bag containing exactly half black and half white marbles. Despite the 1:1 ratio in the bag, we cannot expect to draw exactly 5 white and 5 black marbles each time. In fact, we expect from the binomial probability distribution that the probability of obtaining a perfect ratio is only

$$\frac{10!}{5!5!}\left(\frac{1}{2}\right)^{10} = 0.246$$

There is, however, a small probability—$2(\frac{1}{2})^{10} = 0.002$—of drawing a sample of either all white or all black marbles. This thought experiment is analogous to sampling in a small population of sexually reproducing organisms. In a 2-allele Mendelian system, a stable population of N parental individuals produces a large number of gametes whose allelic frequencies closely reflect those of the parents; this gamete pool is comparable to the bag of marbles. From the pool, approximately $2N$ gametes are drawn to form the next generation of N individuals. If $2N$ is small enough, and if the sampling is not overly biased by the operation of other forces such as selection, the proportions of a_1 and a_2 alleles (comparable to the black and white marbles) can change considerably from generation to generation by sampling error alone. Thus, at one locus or at many loci, a small population can literally drift in a random, unpredictable direction during evolution.

4. *Gene flow:* the immigration of groups of genetically different individuals into the population. Two categories of gene flow can be usefully distinguished: intraspecific flow between geographically separate populations or societies of the same species; and interspecific hybridization. The former occurs almost universally within sexually reproducing plant and animal species and is a major determinant of the patterns of geographic variation. Interspecific hybridization occurs during breakdowns of normal species-isolating barriers. Ordinarily it is temporary, or at least rapidly shifting in nature. Although much less common than gene flow within species, it has a greater per generation effect because of the larger number of gene differences that normally separates species. Aside from selection, gene flow is the quickest way by which gene frequencies can be altered.

5. *Selection:* the change in relative frequency in genotypes due to differences in the ability of their phenotypes to obtain representation in the next generation. Selection, whether artificial selection as deliberately practiced on populations by man or natural selection as it occurs everywhere beyond the conscious intervention of man, is overwhelmingly the most important force in evolution and the only one that assembles and holds together particular ensembles of genes over long periods of time. Variation in the relative frequency of a genotype can stem from many causes: different abilities in direct competition with other genotypes; differential survival under the onslaught of parasites, predators, and changes in the physical environment; variable reproductive competence; variable ability to penetrate new habitats; and so forth. The production of a superior variant in any or all of such categories represents *adaptation*. The devices of adaptation, together with genetic stability in constant environments and the ability to generate new genotypes to cope with fluctuating environments, constitute the *components of fitness* (Thoday, 1953). Natural selection means only that one genotype is increasing at a greater rate than another. The absolute growth rate is meaningless in this regard. All of the tested genotypes may be increasing or decreasing in absolute terms while nonetheless differing in their relative increase or decrease. Acting upon genetic novelties created by mutation, natural selection is the agent that molds virtually all of the characteristics of species.

A selective force may act on the variation of a population in several radically different ways. The principal ensuring patterns are illustrated in Figure 4-1. In the diagrams, the phenotypic variation, measured along the horizontal axis, is given as normally distributed, with the frequencies of individuals measured along the vertical axis. Normal distributions are common but not universal among continuously varying characters, such as size, maturation time, and mental qualities. Stabilizing selection, sometimes also called optimizing selection, consists of a disproportionate elimination of extremes, with a

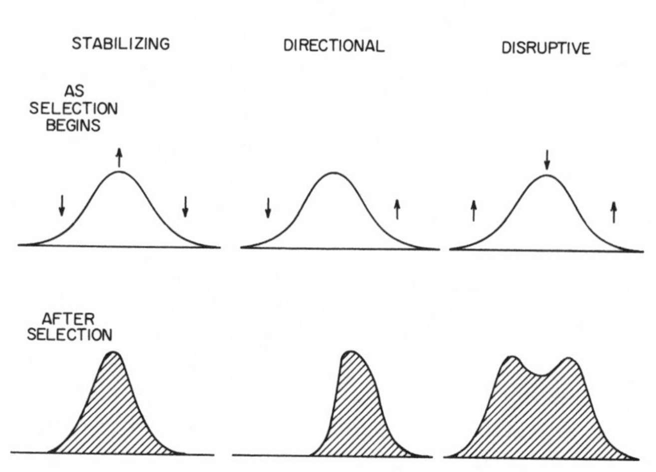

Figure 4-1 Results of adverse (↓) and favorable (↑) selection on various parts of the population frequency distribution of a phenotypic character. The heights of different points on the frequency distribution curves represent the frequencies of individuals in the populations, and the horizontal axes the phenotypic variation. The top figure of each pair shows the pattern as the selection begins; the bottom figure shows the pattern after selection.

consequent reduction of variance; the distribution "pulls in its skirts," as shown in the left-hand pair. This pattern of selection occurs in all populations. Variance is enlarged each generation by mutation pressure, recombination, and possibly also by immigrant gene flow; stabilizing selection constantly reduces the variance about the optimum "norm" best adapted to the local environment. Balanced genetic polymorphism (as opposed to social caste polymorphism) is sometimes effected by a special, very simple kind of stabilizing selection. In a simple two-allele system, the heterozygote a_1a_2 is favored over the homozygotes a_1a_1 and a_2a_2, and each generation sees a reduction in homozygotes. But the gene frequencies remain constant, and as a result the same diploid frequencies recur in each following generation, in a Hardy-Weinberg equilibrium, prior to the action of selection. True disruptive selection (often called diversifying selection) is a rarer phenomenon, or at least one less well known. It is caused by the existence of two or more accessible adaptive norms along the phenotypic scale, perhaps combined with preferential mating between individuals of the same genotype. Recent experimental evidence suggests that it might occasionally result in the creation of new species. Directional selection (or dynamic selection, as it is sometimes called) operates against one end of the range of variation and hence tends to shift the entire population toward the opposite end. It is the principal pattern through which progressive evolution is achieved.

Phenodeviants and Genetic Assimilation

The student of social evolution is especially concerned with rare events that give small segments of populations unusual opportunities to innovate and thus, perhaps, to increase their fitness and affect the future of the species to a disproportionate degree. One such phenomenon found by geneticists is the appearance of *phenodeviants,* scarce aberrant individuals that appear regularly in populations because of the segregation of certain unusual combinations of individually common genes (Lerner, 1954; Milkman, 1970). Examples include pseudotumors and missing or defective crossveins in the wings of *Drosophila,* crooked toes in chickens, and diabetes in mammals. The traits often appear in larger numbers when stocks are being intensively selected for some other trait or are being inbred (the two processes usually amount to the same thing). They are often highly variable, and deliberate selection can further modify their penetrance and expressivity. The appearance of phenodeviants is generally part of the genetic load that slows evolution in other traits. Yet, clearly, they also represent potential points of departure for new pathways in evolution.

Closely related to phenodeviation is the special sequence of events referred to by C. H. Waddington as *genetic assimilation.* An extreme theoretical example is presented in Figure 4-2. Suppose that in each generation a few individuals possess unusual combinations of genes that give them the potential to develop a trait in certain environments, but under ordinary circumstances the species does not encounter conditions that favor the development of the trait. When finally the environment changes long enough to permit the manifestation of the trait in some members of the species, the trait confers superior fitness. In the new circumstances, the genes that provide the potential also increase in proportion. In time they may become so common that most individuals contain a sufficient number to develop the trait *even in the old environment.* If the environment now returns to that original state, all or a substantial number of the individuals will still develop the trait spontaneously.

Because behavior, and especially social behavior, has the greatest developmental plasticity of any category of phenotypic traits, it is also theoretically the most subject to evolution by genetic assimilation. Behavioral scales, such as those that range within one species from territorial behavior to dominance hierarchies, could be created by the appearance of a few individuals capable of shifting their behavior in one direction or another when the environment is altered for the first time. If the environment remains

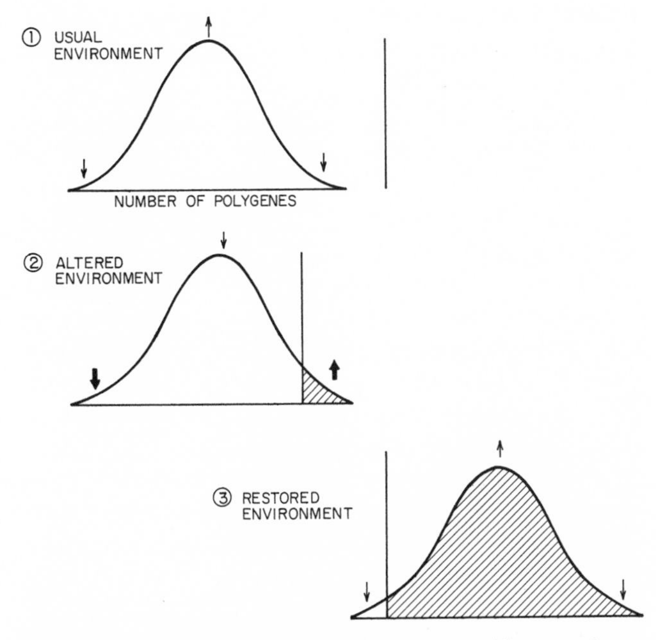

Figure 4-2 Genetic assimilation can occur if an environmental change causes the previously hidden genetic potential of some extreme individual to be exposed. (1) The ordinary environment never permits the development of the potential, but (2) a few individuals attain it when the environment changes. If the trait thus "unveiled" provides increased fitness, those genotypes with the potential will increase in the altered environment; and the population may then evolve to a point where most individuals develop the trait spontaneously even if the environment returns to its orginal state, as indicated in the bottom diagram (3).

changed in a way that strongly favors these genotypes, the species as a whole may shift further by dropping one end of the scale previously occupied. Most species of chaetodontid butterfly fishes, for example, are exclusively territorial in their habits. *Chelmon rostratus* and *Heniochus acuminatus*, however, form schools organized into dominance hierarchies (Zumpe, 1965). Other species in related groups display scales connecting the two behaviors. As different as the two extremes appear superficially, it is not difficult to imagine how one could evolve into the other, especially if differing developmental capacities were subjected to natural selection with the aid of genetic assimilation. The process would be further intensified if members of the same species mimicked one another to any appreciable extent. Cultural innovation of the sort recorded in birds and primates could be the first step, provided that the creativity has a genetic basis. Finally, it is even possible for the most plastic species, including man, to pass through repeated assimilative episodes in the development of higher mental faculties.

Inbreeding and Kinship

Most kinds of social behavior, including perhaps all of the most complex forms, are based in one way or another on kinship. As a rule, the closer the genetic relationship of the members of a group, the more stable and intricate the social bonds of its members. Reciprocally, the more stable and closed the group, and the smaller its size, the greater its degree of inbreeding, which by definition produces closer genetic relationships. Inbreeding thus promotes social evolution, but it also decreases heterozygosity in the population and the greater adaptability and performance generally associated with heterozygosity. It is thus important in the analysis of any society to take as precise a measure as possible of the degrees of inbreeding and relationship.

Three measures of relationship, originally devised by Sewall Wright, are used routinely in population genetics:

Inbreeding coefficient: Symbolized by f or F, the inbreeding coefficient is the probability that both alleles on one locus in a given individual are identical by virtue of identical descent. Any value of f above zero implies that the individual is inbred to some degree, in the sense that both of its parents share an ancestor in the relatively recent past. (In defining "recent," we must recognize that virtually all members of a Mendelian population share a common ancestor if their pedigree is traced far enough back.) If the two alleles in question are identical (because they are descended from a single allele possessed by an ancestor), they are said to be *autozygous;* if not identical, they are called *allozygous.*

Coefficient of kinship. Also called the *coefficient of consanguinity,* the coefficient of kinship is the probability that a pair of alleles drawn at random from the same locus in two individuals will be autozygous. The coefficient of kinship is numerically the same as the inbreeding coefficient; it refers to two alleles drawn from the parents in one generation, whereas the inbreeding coefficient refers to the alleles after they have been combined in an offspring. The coefficient of kinship is ordinarily symbolized as f_{IJ} (or F_{IJ}), where I and J (or any other subscripts) refer to the two individuals compared.

Coefficient of relationship. Designated by r, the coefficient of relationship is the fraction of genes in two individuals that are identical by descent, averaged over all of the loci. It can be derived from the previous two coefficients in a straightforward way that will be explained shortly.

Let us next examine the intuitive basis of the first two measures. Figure 4-3 presents a derivation of the

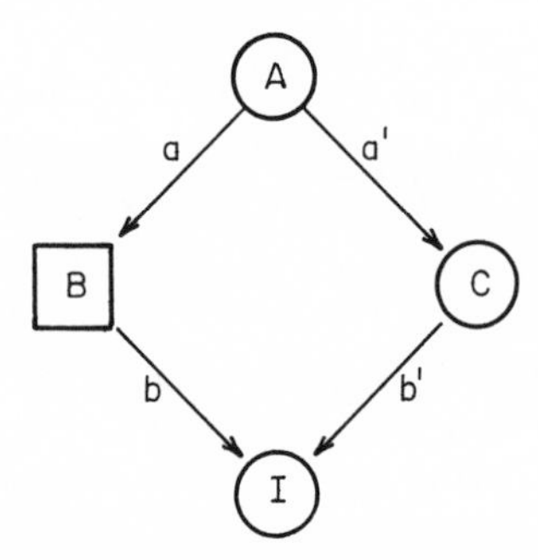

Figure 4-3 Pedigree of an organism (I) produced by the mating of two half sibs (B and C). The computation of the inbreeding coefficient of I is explained in the text.

inbreeding coefficient of an offspring (I) produced by a mating between half sibs (B and C), individuals related to each other by the joint possession of one parent. Females are enclosed in circles and males in squares, while the alleles are symbolized by lowercase letters (a, a', b, b'). The inbreeding coefficient of I is computed as follows. Only the individuals descended from the common ancestor (A) are shown. The probability that a and b are the same is ½, since a makes up half the alleles in B at that locus and therefore half the gametes that B might contribute to I. The probability that a and a' are the same is also ½, because once one allele is chosen at random (label it a), the chance that the second allele chosen at random (label this a') is the same as the first is ½, provided that A itself is not inbred and therefore is unlikely to have two identical alleles to start with. The probability that a' and b' are identical is ½, since a' makes up half the alleles at the locus and therefore half the gametes that C might contribute to I. The probability that b and b' are identical is the coefficient of kinship of B and C, as well as the inbreeding coefficient of I. Because $b = b'$ if and only if $b = a = a' = b'$, the coefficient is the product of the three probabilities just indicated.

$$f_{BC} = f_I = \frac{1}{2} \times \frac{1}{2} \times \frac{1}{2} = \frac{1}{8}$$

Notice that if we count the steps in the path leading from one parent to the common ancestor back to the second parent ($B\underline{A}C$, where the common ancestor is underlined), we obtain the number of times (three) by which the probability ½ must be multiplied against itself. This simple procedure is the basis of *path analysis*, by which coefficients in even more complex pedigrees can be readily computed. Each possible path leading to every common ancestor is traced separately. The inbreeding coefficient is the sum of the probabilities obtained from every separate path. The technique is shown in the three somewhat more involved cases analyzed in Figure 4-4.

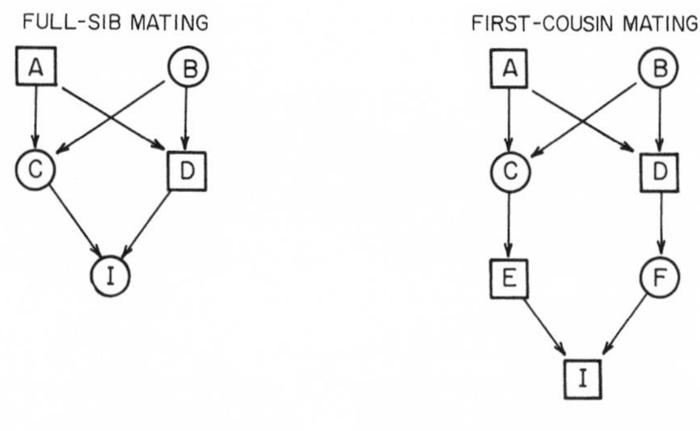

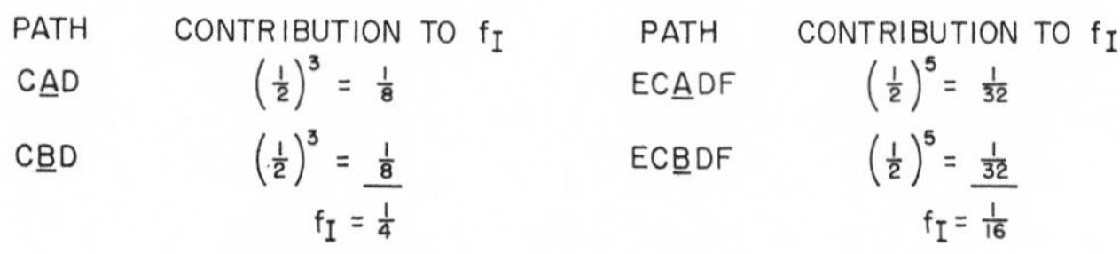

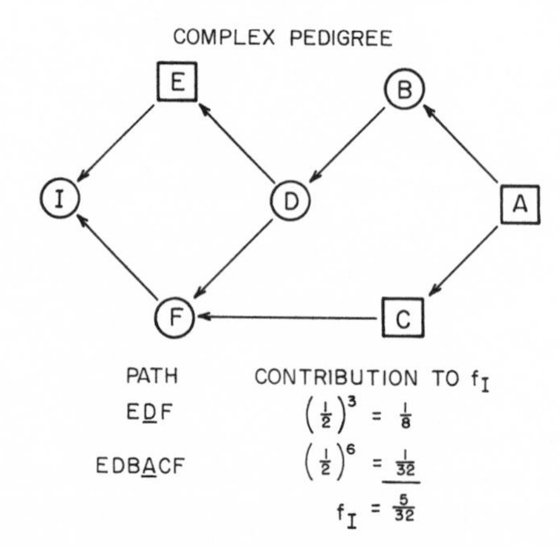

Figure 4-4 Path analysis and calculation of inbreeding coefficients in three pedigrees. The procedure is explained in the text.

The analysis must be modified if the common ancestor is itself inbred. If its inbreeding coefficient is indicated as f_A, then the probability that two alleles drawn randomly from it will be autozygous is ½ $(1 + f_A)$, and the inbreeding coefficient of the ultimate descendant (or at least of one separate path contributing to its coefficient) is

$$f_I = \left(\frac{1}{2}\right)^n (1 + f_A)$$

where n is the number of individuals in the path, as before.

The meaning of the coefficient of relationship can now be made clearer. It is related in the following way to the coefficient of kinship (f_{IJ}) and the inbreeding coefficients (f_I and f_J) of the two organisms compared:

$$r_{IJ} = \frac{2f_{IJ}}{\sqrt{(f_I + 1)(f_J + 1)}}$$

in the absence of dominance or epistasis. If neither individual is inbred to any extent, that is, $f_I = f_J = 0$, the fraction of shared genes (r_{IJ}) is twice their coefficient of kinship. But if each individual is completely inbred, that is, $f_I = f_J = 1$, r_{IJ} is the same as the coefficient of kinship. Suppose that r_{IJ} of two outbred individuals is known to be 0.5. This means that ½ of the genes in I are identical by common descent with ½ of the genes in J, when all the loci (or at least a large sample of them) are considered. Then if we consider one locus, the probability of drawing an allele from I and one from J which are identical by common descent (this probability is f_{IJ}, the coefficient of kinship) is the following: the probability of drawing the correct allele from I (½) times the probability of drawing the correct one from J (½), or ¼. In other words, $r_{IJ} = 2f_{IJ}$. Suppose, in contrast, that I and J were totally inbred. In this unlikely circumstance all allele pairs in I and J are autozygous. As a result the fraction of alleles shared by I and J is the same as the fraction of loci shared by them. If 50 percent of the alleles in I are identical to 50 percent of the alleles in J, 50 percent of the loci are also shared in toto and 50 percent are not shared at all, because all the loci are autozygous.

The coefficients of kinship and relationship can be estimated indirectly, in the absence of pedigree information, by recourse to data on the similarity of blood types and other phenotypic traits among individuals, as well as information on migration (Morton, 1969; Morton et al., 1971; Cavalli-Sforza and Bodmer, 1971). In 1948 G. Malécot showed that in systems of populations with uniform rates of gene flow, the mean coefficient of kinship between individuals selected from different populations can be expected to decline exponentially as the distance (d) separating them increases:

$$f(d) = ae^{-bd}$$

where a and b are fitted constants.

As populations are fragmented and their viscosity, or slowness of dispersal, increases the degree of kinship among immediately adjacent individuals grows larger. Consequently, the prospect for social evolution increases, since cooperative and even altruistic acts will pay off more in terms of the perpetuation of genes shared by common descent. Yet sides effects also arise that can progressively reduce the fitnesses of both individuals and local groups when viscosity is increased, and hence bring social evolution to a standstill. As inbreeding increases, homozygous recombinants increase in frequency more than heterozygous ones, spreading the variation more evenly over the possible diploid types.

Thus a crucial parameter in short-term social evolution is the size and degree of closure of the group. It is conventional to speak of N, the number of individuals in the group (or in the populations embracing it), as though it were composed of equal numbers of each sex all equally likely to contribute progeny. This ideal state is seldom realized. Instead it is necessary to define the *effective population number:* the number of individuals in an ideal, randomly breeding population with 1:1 sex ratio which would have the same rate of heterozygosity decrease as the real population under consideration. Typically, the effective population number, abbreviated N_e, is well below the real population number. By measuring it, we obtain a truer picture of the likely course of microevolutionary events within the population.

The effective population numbers of the few real populations measured so far have generally turned out to be low. In the house mouse *Mus musculus* they are on the order of 10 or less, with male dominance exerting a strong depressing influence. Deer mice (*Peromyscus maniculatus*) form relatively stable territorial populations, in spite of the ebb and flow of emigrating juveniles, and the effective numbers range from 10 to 75. Leopard frogs (*Rana pipiens*) studied by Merrell (1968) had N_e values ranging from 48 to 102, which because of the strongly unequal sex ratios favoring males are well below the actual numbers of adult frogs inhabiting natural habitats. Tinkle (1965) studied side-blotched lizards (*Uta stansburiana*) with unusual care by marking and tracking young individuals until they reached reproductive age. He found that N_e ranged from 16 to 90 in six local populations, with a mean of 30; these figures did not depart far from the actual census numbers. From uncorrected census data, social vertebrates in general appear to have effective population sizes on the order of 100 or less.

The general occurrence of small effective deme sizes in social animals bring them into the range envisaged in Sewall Wright's original "island model" (1943): a population divided into many very small demes and affected by genetic drift that restricts genetic variation within individual demes but increases it between them. Such a population would conceivably be more adaptable than an undivided population of equal size because of its greater overall genetic variation. Where the genotypes of one deme fail, those of the next might succeed, with the end result of preserving the species. As a corollary result, such a population will also evolve more quickly.

We now ask, specifically what is the risk encountered by increased inbreeding and decreased heterozygosity by these social populations? Heterozygosity per se generally raises the viability and reproductive performance of organisms. The extreme case of the relation is heterosis, the temporary improvement in fitness that results from a massive increase in the frequency of heterozygotes over many loci from the outcrossing of two inbred strains. Wallace (1958, 1968) obtained essentially the same effect by irradiating populations of *Drosophila melanogaster* continuously. Instead of the expected decline in the population from accumulated lethal and subvital mutations, he got the opposite trend as sufficient numbers of these mutations expressed beneficial effects in the heterozygous state. Of course if a heterotic stock is then inbred, its performance declines precipitously because of the quick reversal from heterozygous to homozygous states created in a large fraction of the population through elementary Mendelian recombination. Even so, ordinary populations sustain high levels of heterozygous loci, and any increase in inbreeding will result in a decrease in average population performance, part of which will be due to a raising of average mortality by the production of more lethal homozygotes. The formal theory of this decline has been considered at length by Crow and Kimura (1970) and Cavalli-Sforza and Bodmer (1971). The essential relation can be stated as follows. If some trait, such as size, intelligence, motor skill, sociability, or whatever, possesses a degree of heritability, and if some of the loci display either dominance or superior heterozygote performance, or both, inbreeding will cause a decline of the trait within the population. The decline will affect not only the trait averaged over the population as a whole, but also the performance of an increasing number of individuals.

A case of inbreeding depression of a human trait—chest circumference in males—is given in Figure 4-5. Further studies by Schull and Neel (1965) and others have demonstrated depression effects in overall size, neuromuscular ability, and academic performance. A recent study of children of incest in Czechoslovakia confirms the dangers of extreme inbreeding in human beings. A sample of 161 children born to women who had had sexual relations with their fathers, brothers, or sons were afflicted to an unusual degree: 15 were stillborn or died within the first year of life, and more than 40 percent suffered from various physical and mental defects, including severe mental retardation, dwarfism, heart and brain deformities, deaf-mutism, enlargement of the colon, and urinary-tract abnormalities. In contrast, a group of 95 children born to the same women through nonincestuous relations conformed closely to the population at large. Five died during the first year of life, none had serious mental deficiencies, and only 4.5 percent had physical abnormalities (Seemanova, 1971).

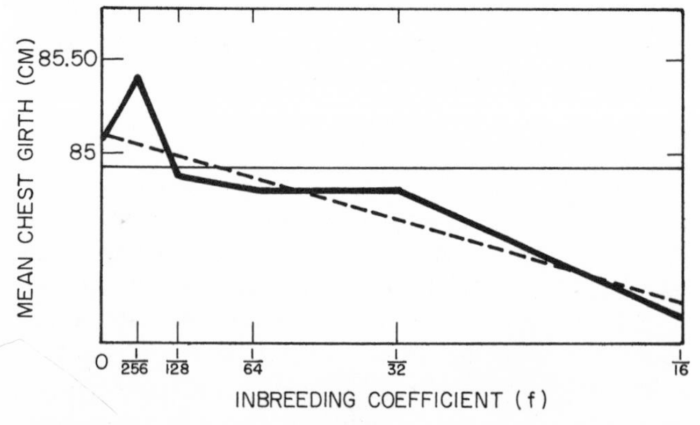

Figure 4-5 Inbreeding depression of chest size in men born in the Parma Province of northern Italy between 1892 and 1911. (From Cavalli-Sforza and Bodmer, 1971; after Barrai, Cavalli-Sforza, and Mainardi, 1964.)

In addition to a straightforward decline in competence, the loss of heterozygosity reduces the ability to buffer the development of structures against fluctuations in the environment. Hence less heterozygosity increases the chance of producing less adaptive variants such as phenodeviants. It further reduces the genetic diversity of offspring, a loss that can result in the loss of entire blood lines, or even social groups, when the environment changes.

In view of the clear dangers of excessive homozygosity, we should not be surprised to find social groups displaying behavioral mechanisms that avoid incest. These strictures should be most marked in small, relatively closed societies. Incest is in fact generally avoided in such cases. Virtually all young lions, for example, leave the pride of their birth and wander as nomads before joining the lionesses of another pride. A few of the young lionesses also transfer in this fashion (Schaller, 1972). A closely similar pattern is followed by many Old World monkeys and apes (Itani, 1972). Even when the young males remain with their troops they seldom mate with their mothers, possibly because of the lower rank they occupy with respect to both their mothers and older males for long periods of time. In the small territorial family groups of the white-handed gibbon *Hylobates lar*, the father drives sons from the group when they attain sexual maturity, and the mother drives away her daughters (Carpenter, 1940). Young female mice (*Mus musculus*) reared with both female and male parents later prefer to mate with males of a different strain, thus rejecting males most similar to the father.

Despite such strong anecdotal evidence, however, we are not yet able to say whether incest avoidance in these animals is a primary adaptation in response to inbreeding depression or merely a felicitous byproduct of dominance behavior that confers other advantages on the individual conforming to it. It is necessary to turn to human beings to find behavior patterns uniquely associated with incest taboos. The most basic process appears to be what Tiger and Fox (1971) have called the precluding of bonds. Teachers and students find it difficult to become equal colleagues even after the students equal or surpass their mentors; mothers and daughters seldom change the tone of their original relationship. More to the point, fathers and daughters, mothers and sons, and brothers and sisters find their primary bonds to be all-exclusive, and incest taboos are virtually universal in human cultures. Studies in Israeli kibbutzim, the latest by Joseph Shepher (1971), have shown that bond exclusion among age peers is not dependent on sibship. Among 2,769 marriages recorded, none was between members of the same kibbutz peer group who had been together since birth. There was not even a single recorded instance of heterosexual activity, despite the fact that no formal or informal pressures were exerted to prevent it.

In summary, small group size and the inbreeding that accompanies it favor social evolution, because they ally the group members by kinship and make altruism profitable through the promotion of autozygous genes (hence, one's own genes) among the recipients of the altruism. But inbreeding lowers individual fitness and imperils group survival by the depression of performance and loss of genetic adaptability. Presumably, then, the degree of sociality is to some extent the evolutionary outcome of these two opposed selection tendencies. How are the forces to be translated into components of fitness and then traded off in the same selection models? This logical next step does not seem feasible at the present time, and it stands as one of the more important challenges of theoretical population genetics. A few of the elements necessary for the solution will be given in the analysis of group selection in Chapter 5.

Assortative and Disassortative Mating

Assortative mating, or *homogamy*, is the nonrandom pairing of individuals who resemble each other in one or more phenotypic traits. Human couples, for example, tend to pair off according to similarity in size and intelligence. Sternopleural bristle number, which may simply reflect total size, and certain combinations of chromosome inversions have been found to be associated with assortative mating in *Drosophila* fruit flies (Parsons 1967; Wallace, 1968). In domestic chickens and deer mice (*Peromyscus maniculatus*), color varieties prefer their own kind (Blair and Howard, 1944; Lill, 1968). Assortative mating can be based upon kin recognition, in which case its consequences are identical to those of inbreeding. Or it can be based strictly on the matching of like phenotypes, either without reference to kinship or in conjunction with the avoidance of incest, as in the case of human beings. "Pure" assortative mating of the latter type has effects similar to those of inbreeding, but it results in a less rapid passage to homozygosity, affects only those loci concerned with the homogamous trait or closely linked to it (whereas inbreeding affects all loci), and, in the case of polygenic inheritance, causes an increase in variance.

Disassortative mating has been documented fewer times in nature than assortative mating, and in a disproportionate number of instances it has involved chromosomal and genic polymorphs in insects (Wallace, 1968). The effects of disassortative mating are of course generally the reverse of those caused by assortative mating. In additive polygenic systems there is a tendency to "collapse" variation toward the mean. However, in the case of genetic polymorphism, diversity is preserved and even stabilized, since scarcer phenotypes are the beneficiaries of preferential mating and the underlying genotypes will therefore tend to increase until the advantage of scarcity is lost.

Population Growth

Natural selection can be viewed simply as the differential increase of alleles within a population. It does not matter whether the population as a whole is increasing, decreasing, or holding steady. So long as one allele is increasing relative to another, the population is evolving. In fact, a population can be evolving rapidly, responding to natural selection and hence "adapting," at the same time that it is going extinct. The conceptualization and measurement of growth, then, is the meeting place of population genetics and ecology.

The rate of increase of a population is the difference between the rate of addition of individuals due to birth and immigration and the rate of subtraction due to death and emigration:

$$\frac{dN}{dt} = B + I - D - E$$

where N is the population size, and B, I, D, E are the rates at which individuals are born, immigrate, die, and emigrate. A society, even if nearly closed, comprises a population in which all four of these rates are significant. In larger populations, however, including the set of all conspecific societies that make up a given population, a realistic modeling effort can be started by setting $I = E = 0$ (no individuals enter the population or leave it) and varying B and D, the birth and death rates. In the simplest model of exponential growth, it is assumed that there exist some average fertilities and probabilities of death over all the individuals in the population. This means that B and D are each proportional to the number of individuals (N). In other words, $B = bN$ and $D = dN$, where b and d are the average birth and death rates per individual per unit time. Then

$$\frac{dN}{dt} = bN - dN$$

$$= (b - d)N = rN$$

where r ($= b - d$) is called the intrinsic rate of increase (or "Malthusian parameter") of the population for that place and time. The solution of the equation is

$$N = N_0 e^{rt}$$

where N_0 is the number of organisms in the population at the moment we begin our observations, and t is the amount of time elapsed after the observations begin. The units of time chosen (hours, days, years, or whatever) determine the value of r. (The symbol r is not to be confused with the same symbol used to denote the coefficient of relationship. The fact that the same letter has been used for two major parameters is one of the inconveniences resulting from the largely independent histories of ecology and genetics.)

Theoretically, each population has an optimum environment—physically ideal, with abundant space and resources, free of predators and competitors, and so forth—where its r would reach the maximum possible value. This value is sometimes referred to formally as r_{max}, the maximum intrinsic rate of increase. Obviously, the rates of increase actually achieved in the great majority of the less-than-perfect environments are well below r_{max}. For example, although the realized values of r of most human populations are very high, enough to create the current population explosion, they are still several times smaller than r_{max}, the value of r that would be obtained if human beings made a maximum reproductive effort in a very favorable environment. The values of r vary enormously among species. Almost all human populations increase at a rate of 3 percent or less per year ($r = 0.03$ per year). The value of r in unrestricted rhesus populations is about 0.16 per year, while in the prolific Norway rat it is 0.015 *per day*.

Since any value of r above zero will eventually produce more individuals of the species than there are atoms in the visible universe, the exponential growth model is obviously incomplete. The problem lies in the implicit assumption that b and d are constants, with values independent of N. A new and more realistic postulate is that b and d are functions of N, say, linear functions for simplicity:

$$b = b_o - k_b N$$
$$d = d_o + k_d N$$

In this case, b_o and d_o are the values approached as the population size becomes very small, k_b is the slope of the decrease for the birth rate, and k_d is the slope of the increase for the death rate. The equations state that the birth rate decreases and the death rate increases as the population increases, both of which are plausible assertions that have been documented in some species in nature. We substitute the new values of b and d into the model to find:

$$\frac{dN}{dt} = [(b_o - k_b N) - (d_o + k_d N)]N$$

This is one form of the basic equation for logistic population growth. Note that when b becomes equal to d, the population reaches a stable size. That is, the population can maintain itself at the value of N such that

$$b_o - k_b N = d_o + k_d N$$

$$N = \frac{b_o - d_o}{k_b + k_d}$$

This particular value of N is called the carrying capacity of the environment and usually is given the shorthand symbol K. For any value of N less than K the population will grow, and for any value greater than K it will decline; and the change will occur until K is reached (Figure 4-6). Taking the shorthand notations

$$K = \frac{(b_o - d_o)}{(k_o + k_d)}$$

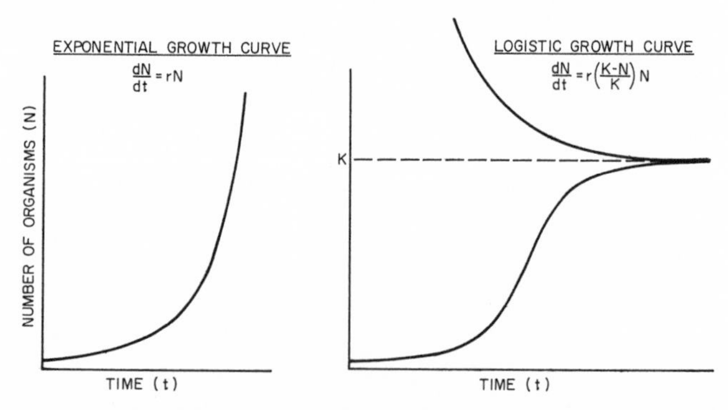

Figure 4-6 Two basic equations for the growth and regulation of populations (written as differential equations) and the solutions to the equations (drawn as curves). Two logistic curves are shown, one starting above K and descending toward this asymptote and the other starting from near zero and ascending toward the same asymptote.

and

$$r = b_o - d_o$$

and substituting them into the logistic differential equation just derived, we obtain

$$\frac{dN}{dt} = rN\left(\frac{K-N}{K}\right)$$

This is the familiar form of the logistic equation for the growth and regulation of animal populations. Usually the equation is stated flatly in this way, then the constants are defined and discussed with reference to their possible biological meaning. The derivation given here reveals the intuitive basis for the model.

For all values of N less than or equal to K, the solution of the logistic equation gives a symmetric S-shaped curve of rising N as time passes, with the maximum population growth rate (the "optimum yield") occurring at $K/2$. Some laboratory populations conform well to the pure logistic, while a few natural populations can at least be fitted to it empirically. Schoener (1973) has shown that in at least one circumstance—the limitation of population growth by competition of individuals scrambling for resources as opposed to competition by direct interference—the growth curve cannot be expected to be S-shaped. Instead it will turn evenly upward and over to approach the asymptote.

Density Dependence

Why should populations be expected to attain particular values of the carrying capacity K, asymptotically or otherwise, and remain there? Ecologists often distinguish density-independent effects from density-dependent effects in the environment. A density-independent effect alters birth, death, or migration rates, or all three, without having its impact influenced by population density. As a result it does not regulate population size in the sense of tending to hold it close to K. Imagine an island whose southern half is suddenly blanketed by ash from a volcanic eruption. All of the organisms on this part of the island, roughly 50 percent of the total from each population, are destroyed. Beyond doubt the volcanic eruption was a potent controlling factor, but its effect was density-independent. It reduced all of the populations by 50 percent no matter what their densities at the time of the eruption, and hence could not serve in a regulatory capacity. Most density-independent depressions in population size may be due to sudden, severe changes in weather. Journals devoted to birdlore, natural history, and wildlife management are filled with anecdotes of hail storms killing most of the young of local wading bird populations, late hard freezes causing a crash in the small mammal populations, fire destroying most of a saw grass prairie, and so forth. An important theoretical consideration is that populations whose growth is governed exclusively by density-independent effects probably are destined for relatively early extinction. The reason is that unless there are density-dependent controls always acting to guide the population size toward K, the population size will randomly drift up and down. It may reach very high levels for a while, but eventually it will head down again. And if it has no density-dependent controls to speed up its growth at lower levels while it is down, it will eventually hit zero. The density-independent population is like a gambler playing against an infinitely powerful opponent, which in this case is the environment. The environment can never be beaten, at least not in such a way that the population insures its own immortality. But the population, being composed of a finite number of organisms, will itself eventually be beaten, that is, reduced to extinction. For this reason biologists believe that most existing populations have some form of density-dependent controls that ward off extinction.

An astonishing diversity of biological responses have been identified as density-dependent controls. Most of them are implicated in one way or another in social behavior and, indeed, much social behavior is comprehensible only by reference to the role it plays in population control. This generalization will be borne out in the following briefly annotated catalog of the principal classes of controls.

Emigration

The single most widespread response to increased population density throughout the animal kingdom is restlessness and emigration. Hydras produce a bubble beneath their pedal disk and float away (Łomnicki and Slobodkin, 1966). Pharaoh's ants (*Monomorium pharaonis*) remove their brood from the nest cells, swarm feverishly over the nest surface, and depart for other sites located by worker scouts (Peacock and Baxter, 1950). Mice (*Mus, Peromyscus*) sharply increase their level of locomotor activity and begin to explore away from their accustomed retreats. Every overpopulated range of songbirds and rodents contains floater populations, consisting of individuals without territories who live a perilous vagabond's existence along the margins of the preferred habitats. Sometimes the movements become directed and unusually persistent, a trend that reaches an evolutionary extreme in the "marches" of the lemmings. The wanderers are the juveniles, the subordinates, and the sickly—the "losers" in the territorial contests for the optimum living places. However, these meek of the earth are not necessarily doomed; their circumstances have simply forced them into the next available strategy, which is to "get out while the getting is good," to search with the possibility of finding a less crowded environment. In fact, many individuals do succeed in this endeavor, and as a result they play a key role in enlarging the total population size, in extending the range of the species, and possibly even in pioneering in the genetic adaptation to new habitats (Lidicker, 1962; Christian, 1970; G. C. Clough, in Archer, 1970).

Stress and Endocrine Exhaustion

In 1939 R. G. Green, C. L. Larson, and J. F. Bell observed a population crash of the snowshoe hare (*Lepus americanus*) in Minnesota and drew a remarkable conclusion about it. They deduced the primary cause to be shock disease, a hormone-mediated idiopathic hypoglycemia that can be identified by liver damage and disturbances in several aspects of carbohydrate metabolism. The implication was that when conditions are persistently crowded the hares suffer an excessive endocrine response from which they cannot recover. Even when individuals collected during the population decline were placed in favorable laboratory conditions, they lived for only a short time. Many vertebrate physiologists and ecologists have subsequently explored the effect of crowding and aggressive interaction on the endocrine system. And conversely, they have speculated on the multifarious ways in which endocrine-mediated physiological responses can serve as density-dependent controls by increasing mortality and emigration, diminishing natality, and slowing growth. In general, raising the population density increases the rate of individual interactions, and this effect triggers a complex sequence of physiological changes: increased adrenocortical activity, depression of reproductive function, inhibition of growth, inhibition of sexual maturation, decreased resistance to disease, and inhibition of growth of nursing young apparently caused by deficient lactation.

Reduced Fertility

An inverse relation between population density and birth rate has been demonstrated in laboratory and free-living populations of many species of insects, birds, and mammals. Fission rates of protistans and lower invertebrates invariably decline in laboratory cultures if other restraining factors are removed and the organisms are allowed to multiply at will. Best et al. (1969) traced the control in the planarian *Dugesia dorotocephala* to a secretion released by the animals themselves into the surrounding water. In the house mouse (*Mus musculus*), a species probably typical of rodents in its population dynamics, the decline in birth rate in laboratory populations was found to be due to decreased fertility in mature females, inhibition of maturation and increased intrauterine mortality (Christian, 1961). In fact, almost unlimited means exist by which crowding can reduce fertility. To take one of numerous examples, pigeon fanciers are aware that when the birds are too crowded, the males interfere with one another's attempts to copulate, and female fertility declines.

Infanticide and Cannibalism

Guppies (*Lebistes reticulatus*) are well known for the stabilization of their populations in aquaria by the consumption of their excess young. In one experiment Breder and Coates (1932) started two colonies, one below and one above the carrying capacity, by introducing a single gravid female in one aquarium and 50 mixed individuals in a second, similar aquarium. Both populations converged to 9 individuals and stabilized there, because all excess young were eaten by the residents. Cannibalism is commonplace in the social insects, where it serves as a means of conserving nutrients as well as a precise mechanism for regulating colony size. The colonies of all termite species so far investigated promptly eat their own dead and injured. Cannibalism is in fact so pervasive in termites that it can be said to be a way of life in these insects.

Nomadic male lions of the Serengeti plains frequently invade the territories of prides and drive away or kill the resident males. The cubs are also sometimes killed and eaten during territorial disputes (Schaller, 1972). High-density populations of langurs (*Presbytis entellus, P. senex*) display a closely similar pattern of male aggression. The single males and their harems are subject to harassment by peripheral male groups, who sometimes succeed in putting one of their own in the resident male's position. Infant mortality is much higher as a direct result of the disturbances. In the case of *P. entellus*, the young are actually murdered by the usurper (Hrdy, 1977).

Competition

Competition is defined by ecologists as the active demand by two or more organisms for a common resource. When the resource is not sufficient to meet the requirements of all the organisms seeking it, it becomes a limiting factor in population growth. When, in addition, the shortage of the resource limits growth with increasing severity as the organisms become more numerous, then competition is by definition one of the density-dependent factors. Competition can occur between members of the same species (intraspecific competition) or between individuals belonging to different species (interspecific competition). Either process can serve as a density-dependent control for a given species, although the more precise regulation of population size is likely to occur when the competition is primarily intraspecific. The techniques of competition are extremely diverse, and will be explored more fully in later chapters on territory and aggression. An animal that aggressively challenges another over a piece of food is obviously competing. So is another animal that marks its territory with a scent, even when other animals avoid the territory solely because of the odor and without ever seeing the territory owner. Competition also includes the using up of resources to the detriment of other organisms, whether or not any aggressive behavioral interaction also occurs. A plant, to take an extreme case, may absorb phosphates through its root system at the expense of its neighbors, or cut off its neighbors from sunlight by shading them with its leaves.

For the moment, it is useful to classify competition into two broad modes, scramble and contest (Nicholson, 1954). Scramble competition is exploitative. The winner is the one who uses up the resource first, without specific behavioral responses to other competitors who may be in the same area. It is the struggle of small boys scrambling for coins tossed on the ground before them. If the boys stood up and fought, with the winner appropriating all the coins within a certain radius, the process would be contest competition. Examples of this latter, more fully animallike behavior are territoriality and dominance hierarchies. Competition theory is a relatively advanced field in ecological research; important recent reviews include those by Schoener (1973), May (1973), and May et al. (1976).

Predation and Disease

Because their numbers can be counted, predators and parasites exert the most easily quantifiable density-dependent effects. As local populations of the host species increase in numbers, its enemies are able to encounter and to strike individuals at a higher frequency. This "functional response," as it is called by ecologists (Holling, 1959), is enhanced in cases where the parasites and predators migrate to the foci of greatest density. Alternatively or concurrently, the parasites and predators can exert their influence on their victims by a long-term "numerical response," in which their own populations build up over two or more generations because of the increased survivorship and fecundity afforded by the improved food supply.

Like competition, predator-prey interaction lies at the heart of community ecology and has been the object of intensive theoretical and experimental research. Among the most significant recent reviews are those by Krebs (1972), MacArthur (1972), May (1973), and May et al. (1976). A brief elementary introduction to the basic theory is provided by Wilson and Bossert (1971).

Social Convention and Epideictic Displays

Suppose that animals *voluntarily* agreed to curtail reproduction when they became aware of rising population density. For instance, males could compete with other males in a narrowly restricted manner for access to females, as in a territorial display, with the loser simply withdrawing from the contest short of bloodshed or exhaustion on either side. This technique of slowing population growth by ritualized means has been called conventional behavior by V. C. Wynne-Edwards (1962). Its most refined form might be the epideictic display, a conspicuous message "to whom it may concern" by which members of a population reveal themselves and allow all to assess the density of the population. The correct response to evidence of an overly dense population would be voluntary birth control or removal of one's self from the area. This idea, with strong roots going back to W. C. Allee (in Allee et al., 1949), was developed in full by Olavi Kalela (1954) and Wynne-Edwards. It is fundamentally different from the remainder of the conception of density dependence, because it implies altruism of individuals. And altruism of individuals directed at entire groups can evolve only by natural selection at the group level.

Few ecologists believe that social conventions play a significant role in population control, and many doubt if such a role exists at all. The reason for scepticism is twofold. First, the intensity of group extinction required to fix an altruistic gene must be high, and the problem becomes acute when the altruism is directed at entire Mendelian populations. Because the formal theory of group selection is complex and has many ramifications in sociobiology, it will be left to a chapter by itself (Chapter 5). At that time the feasibility of population control by social conventions will be examined. The second reason for doubt is the difficulty of demonstrating the phenomenon in nature. To prove a functional social convention, and hence population-level selection, is to accomplish the onerous feat of proving (as opposed to disproving) a null hypothesis: the other density-dependent controls, based upon individual opposed to group selection, must all be eliminated one by one.

Intercompensation

A great deal of variation has been observed in density-dependent controls between species, between laboratory and free-ranging populations of the same species, and even among free-ranging populations of the same species. Much of this is due to the property of intercompensation. This means that if the environment changes to relieve the population of pressure from a previously sovereign effect, the population will increase until it reaches a second equilibrium level where another effect halts it. For example, if the predators that normally keep a certain herbivore population in balance are removed, the population may increase to a point where food becomes critically short. If a superabundance of food is then supplied, the population may increase still further—until intense overcrowding triggers an epizootic disease or a severe stress syndrome. The rodent experiments of Calhoun, Christian, Krebs, Lidicker, and others have been instructive in revealing the sequences of intercompensating controls in a variety of species. Calhoun's "behavioral sink"—in which most individuals behaved abnormally and failed to reproduce—can be viewed as a rat population that was allowed to rise above nearly all the controls the species encounters in nature. Sociopathology, if caused by crowding, can be viewed as a control that is nonadaptive in the sense that it lies beyond the limit of a species' repertory and therefore does not contribute to either individual or group fitness.

Population Cycles of Mammals

The population cycles of mammals, and especially of rodents, have loomed large—too large—in the central literature of sociobiology. This is a doubly unfortunate circumstance because of the confusing, often bitter controversies that have risen around the cycles. The real problem, aside from the practical difficulties in obtaining data, is the fact that population cycles have traditionally been subjected to the advocacy method of doing science. Each of several density-dependent controls has had its own theory, school of thought, and set of champions: emigration, stress and endocrine exhaustion, cyclical selection for aggressive genotypes, predation, and nutrient depletion. A plausible model and supporting data have been marshaled behind each process to advance it as the premier factor in nature. To express the matter in such a way is not to denigrate the work of the researchers, which is of high quality and imaginative. And paradoxically, all could be at least partly correct. But inconsistencies have arisen from the tendency to generalize from restricted labortory experiments and field observations of only one to several populations, together with a failure by a few key authors to perceive the possible role of intercompensation. It does seem plausible that intercompensation could be responsible for much of the great variation in operating controls from population to population and from one environment to another. If any rule can be drawn from the existing data, it is perhaps that in free-living rodent populations the principal density-dependent control is most often territoriality combined with emigration, followed by depletion of food supply and predation, in that order. Endocrine-induced changes are difficult to evaluate, but they appear to fall in the secondary ranks of the controls. When they occur they may affect female fertility primarily. Endocrine exhaustion, as easy as it is to induce in laboratory populations by the lifting of other controls, is perhaps rare or absent in most free-living populations. Genetic changes in aggressive behavior are also hard to evaluate. It seems probable that they amplify cycles but are nevertheless subordinate to territorial aggression and emigration as density-dependent controls.

Life Tables

The vital demographic information of a closed population is summarized in two separate schedules: the *survivorship schedule*, which gives the number of individuals surviving to each particular age, and the *fertility schedule*, which gives the average number of daughters that will be produced by one female at each particular age. First consider survivorship. Let age be represented by x. The number surviving to a particular age x is recorded as the proportion or frequency (l_x) of organisms that survive from birth to age x, where the frequency ranges from 1.0 to 0. Thus, if we measure time in years, and find that only 50 percent of the members of a certain population survive to the age of one year, then $l_1 = 0.5$. If only 10 percent survive to an age of 7 years, $l_7 = 0.1$; and so on. The process can be conveniently represented in survivorship curves. Figure 4-7A shows the three basic forms such curves can take. The curve for type I, which is approached by human beings in advanced civilizations and by carefully nurtured populations of plants and animals in the garden and laboratory, is generated when accidental mortality is kept to a minimum. Death comes to most members only when they reach the age of senescence. In survivorship of type II, the probability of death remains the same at every age. That is, a fixed fraction of each age group is removed—by predators, or accidents, or whatever—in each unit of time. The annual adult mortality of the white stork, for example, is steady around 21 percent, while that of the yellow-eyed penguin is 13 percent. Type II survivorship, therefore, takes the form of negative exponential decay. When plotted on a semilog scale (l_x on logarithmic

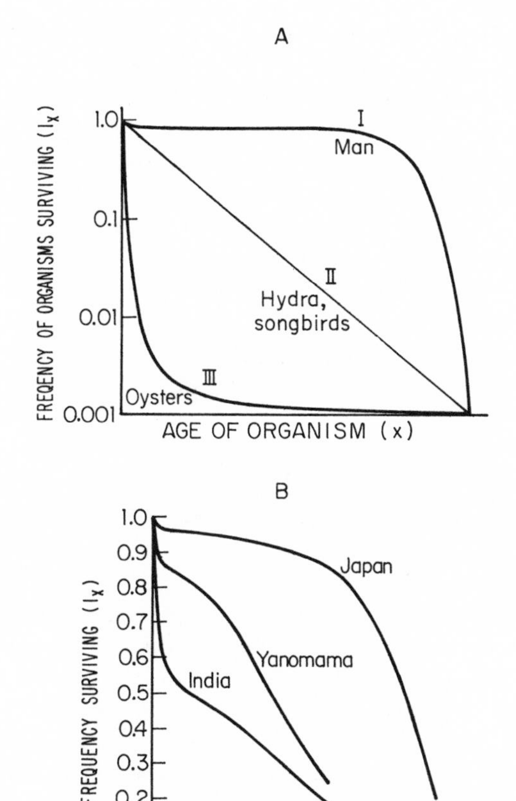

Figure 4-7 Survivorship curves. *A*: the three basic types. *B*: variation in the survivorship curves among human populations, from type I to type III (modified from Neel, 1970). The vertical axis of *A* is on a logarithmic scale.

scale, x on normal scale), the curve is a straight line. Type III is the most common of all in nature. It occurs when large numbers of offspring, usually in the form of spores, seeds, or eggs, are produced and broadcast into the environment. The vast majority quickly perish; in other words, the survivorship curve plummets at an early age. Those organisms that do survive by taking root or by finding a safe place to colonize have a good chance of reaching maturity. The shape of the survivorship curve depends on the condition of the environment, with the result that it can vary widely from one population to another within the same species. In man himself, the variation ranges all the way from type I to type III (see Figure 4-7B).

The fertility schedule consists of the age-specific birth rates; during each period of life the average number of female offspring born to each female is specified. To see how such a schedule is recorded, consider the following imaginary example: at birth no female has yet given birth ($m_0 = 0$); during the first year of her life still no birth occurs ($m_1 = 0$); during the second year of her life the female gives birth on the average to 2 female offspring ($m_2 = 2$); during the third year of her life she gives birth on the average to 4.5 female offspring ($m_3 = 4.5$); and so on through the entire life span.

From the survivorship and fertility schedules we can obtain the *net reproductive rate*, symbolized by R_0, and defined as the average number of female offspring produced by each female during her entire lifetime. It is a useful figure for computing population growth rates. In the case of species with discrete, nonoverlapping generations, R_0 is in fact the exact amount by which the population increases each generation. The formula for the net reproductive rate is

$$R_0 = \sum_{x=0}^{\infty} l_x m_x$$

To see more explicitly how R_0 is computed, consider the following simple imaginary example. At birth all females survive ($l_0 = 1.0$) but of course have no offspring ($m_0 = 0$); hence $l_0 m_0 = 1 \times 0 = 0$. At the end of the first year 50 percent of the females still survive ($l_1 = 0.5$) and each gives birth on the average to 2 female offspring ($m_1 = 2$); hence $l_1 m_1 = 0.5 \times 2 = 1.0$. At the end of the second year 20 percent of the original females still survive ($l_2 = 0.2$), and each gives birth on the average at that time to 4 female offspring ($m_2 = 4$); hence $l_2 m_2 = 0.2 \times 4 = 0.8$. No female lives into the third year ($l_3 = 0$; $l_3 m_3 = 0$). The net reproductive rate is the sum of all the $l_x m_x$ values just obtained:

$$R_0 = \sum_{x=0}^{\infty} l_x m_x$$

	$l_x m_x$ at birth ($x = 0$)		$l_x m_x$ 1st year ($x = 1$)		$l_x m_x$ 2nd year ($x = 2$)		$l_x m_x$ 3rd year ($x = 3$)
=	0	+	1.0	+	0.8	+	0

$= 1.8$

We can proceed to the method whereby r, the intrinsic rate of increase, can be computed precisely from the survivorship and fertility schedules. We start with the solution of the exponential growth equation

$$N_t = N_0 e^{rt}$$

Let t = maximum age that a female can reach, and N_0 be only one female. Thus we have set out to find the number of descendants a single female will produce, including her own offspring, the offspring they produce, et seq., during the maximum life span one female can enjoy. Since $N_0 = 1$

$$N_{\text{max age}} = e^{r(\text{max age})}$$

$$= \sum_{x=0}^{\text{max age}} l_x m_x e^{r(\text{max age}-x)}$$

In words, the total number of individuals stemming from a single female is the sum of the expected number of offspring produced by that female at each age x of the female ($l_x m_x$) times the number of offspring that each of these sets of offspring will produce from the time of their birth to the maximum age of the original female (max age $- x$). Substituting and rearranging, we obtain

$$e^{r(\text{max age})} = e^{r(\text{max age})} \sum_{x=0}^{\text{max age}} l_x m_x e^{-rx}$$

$$\sum_{x=0}^{\text{max age}} l_x m_x e^{-rx} = 1$$

Or, in continuous distributions of l_x and m_x,

$$\int_0^{\text{max age}} l_x m_x e^{-rx} dx = 1$$

For "max age" we can further substitute ∞, since the two are biologically equivalent. This formulation can be referred to as the Euler equation or the Euler-Lotka equation, after the eighteenth-century mathematician Leonhard Euler, who first derived it, and A. J. Lotka, who first applied it to modern ecology. Since we know the values of m_x and l_x, the Euler-Lotka equation permits us to solve the intrinsic rate of increase, r.

The Euler-Lotka equation has potentially powerful applications throughout sociobiology. Each l_x and m_x value can have underlying social components. Conversely, the adaptive value, r, of each genotype is determined in part by the way its social responses affect each l_x and m_x. Heritability in the $l_x m_x$ schedules has been documented in *Drosophila, Aedes* mosquitoes, lizards, and human beings; and it is surely a universal quality of organisms. Therefore, the fine details of life history, meaning the survivorship-fertility schedules and their determinants, can be expected to respond to natural selection. In fact, the entire evolutionary strategy of a species can be described abstractly by these schedules.

To cite one relatively simple example, longevity and low fertility are compensatory traits favored by natural selection under either one of two opposite environmental conditions. If the environment is very stable and predictable, survivorship and hence longevity are improved for species that can appropriate part of the habitat, key their activities to its rhythms, or otherwise take advantage of the stability. Such organisms will not find it a good strategy to seed their homes with large numbers of offspring, who become potential competitors. At the other extreme, a harsh, unpredictable environment will cause some (but not all) species to evolve a tough, durable mature stage that utilizes its energies more successfully for survival than in reproductive effort. It can be shown that the best strategy for such organisms is to engage in highly irregular reproduction keyed to the occasional good times (Holgate, 1967). Longevity is further improved when the survival of progeny is not only low but unpredictable in time (Murphy, 1968).

The Stable Age Distribution

An important principle of ecology is that any population allowed to reproduce itself in a constant environment will attain a stable age distribution. (The only exception occurs in those species that reproduce synchronously at a single age). This means that the proportions of individuals belonging to different age groups will maintain constant values for generation after generation. Suppose that upon making a census of a certain population, we found 60 percent of the individuals to be 0–1 year old, 30 percent to be 1–2 years old, and 10 percent to be 2 years or older. If the population had existed for a long time previously in a steady environment, this is likely to be a stable age distribution. Future censuses will therefore yield about the same proportions. Stable age distributions are approached by any population in a constant environment, regardless of whether the population is increasing in size, decreasing, or holding steady. Each population has its own particular distribution for a given set of environmental conditions.

Reproductive Effort

In the fundamental equations of population biology, effort expended on reproduction is not to be measured directly in time or calories. What matters is benefit and cost in future fitness. Suppose that the female of a certain kind of fish spawns heavily in the first year of her maturity, with the result that enough eggs are released to produce 20 surviving fry. However, the expenditure of effort and energy invariably costs the female her life. Imagine next a second kind of fish, the female of which makes a lesser effort, resulting in only 5 surviving fry but entailing a negligible risk to life, with the result that she can expect to make five or ten such efforts in one breeding season. The reproductive effort of the second fish, measured in units of future fitness sacrificed at each spawning, is far less than that of the first fish, but in this particular case we can expect populations of the second fish to increase faster. The general question is: In order to attain a given m_i at age i, what will be the reduction in future l_i and m_i? The problem has been the object of a series of theoretical investigations by G. C. Williams (1966), Tinkle (1969), Gadgil and Bossert (1970), and Fagen (1972), who have used variations on the Euler-Lotka equation (or its intuitive equivalent) to investigate the effect on fitness of various relations between l_i and m_i over all ages. It makes sense to describe reproductive effort in terms of its physiological and behavioral enabling devices, such as proportion of somatic tissue converted to gonads and the amount of time spent in courtship and parental care. However, the performance of these devices must be converted into units in the life tables before their effects on genetic evolution can be computed.

Only fragmentary data exist that can be related to the reproductive effort models. The wildlife literature contains many anecdotes of male animals that lose their lives because of a momentary preoccupation with territorial contest or courtship. Schaller (1972), for example, observed that "when two warthog boars fought, a lioness immediately tried to catch one; a courting reedbuck lost his life because he ignored some lions nearby." When barnacles spawn, their growth rate is substantially reduced (Barnes, 1962), with the result that they are able to produce fewer gametes in the next breeding season and are more subject to elimination by other barnacles growing next to them. Murdoch (1966) demonstrated that the survival of females of the carabid beetle *Agonum fuliginosum* from one breeding season to the next is inversely proportional to the amount of reproduction in the first. In general, the smaller and shorter-lived the organism, the greater its reproductive effort as measured by the amount of fertility per season. A striking example from the lizards is given in Figure 4-8. The expected negative correlation between life span and fertility is based on the assumption, probably true for many kinds of organisms in addition to lizards, that there exists an inverse relation between the time an animal puts into reproduction and its chance of survival. However, in social animals this simple trade-off is easily averted. A dominant male, for example, may invest large amounts of time in activities related more or less directly to reproduction, and still enjoy higher survivorship by virtue of its secure position within a territory or at the head of a social group.

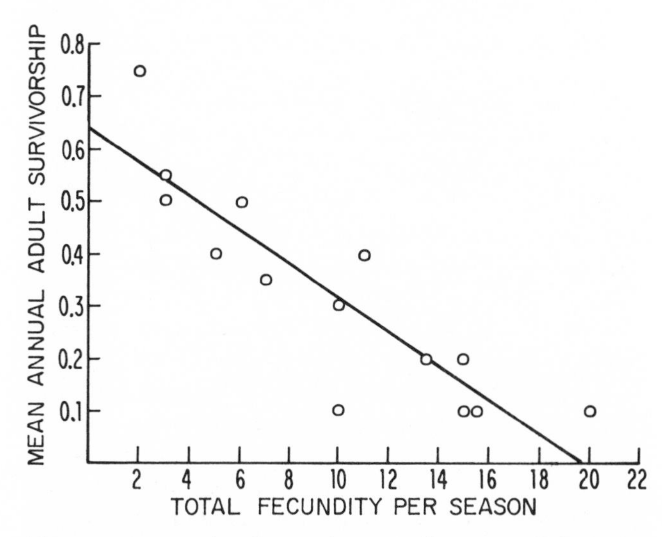

Figure 4-8 A rule of reproductive effort exemplified: the inverse relationship between the rate at which individual lizard females reproduce and the length of their lives, as measured by annual survivorship. Each point represents a different species. (From Tinkle, 1969.)

r and *K* Selection

The demographic parameters r and K are determined ultimately by the genetic composition of the population. As a consequence, they are subject to evolution, in ways that have only recently begun to be carefully examined by biologists. Suppose that a species is adapted for life in a short-lived, unpredictable habitat, such as the weedy cover of new clearings in forests, the mud surfaces of new river bars, or the bottoms of nutrient-rich rain pools. Such a species will succeed best if it can do three things well: (1) discover the habitat quickly; (2) reproduce rapidly to use up the resources before other, competing species exploit the habitat, or the habitat disappears altogether; and (3) disperse in search of other new habitats as the existing one becomes inhospitable. Such a species, relying upon a high r to make use of a fluctuating environment and ephemeral resources, is known as an "r strategist," or "opportunistic species" (MacArthur and Wilson, 1967). One extreme case of r strategist is the fugitive species, which is consistently wiped out of the places it colonizes, and survives only by its ability to disperse and fill new places at a high rate (Hutchinson, 1951). The r strategy is to make full use of habitats that, because of their temporary nature, keep many of the populations at any given moment on the lower, ascending parts of the growth curve. Under such extreme circumstances, genotypes in the population with high r will be consistently favored. Less advantage will accrue to genotypes that substitute an ability to compete in crowded circumstances (when $N = K$ or close to it) for the precious high r. The process is referred to as r selection.

A "K strategist," or "stable species," characteristically lives in a longer-lived habitat—an old climax forest, for example, a cave wall, or the interior of a coral reef. Its populations, and those of the species with which it interacts, are consequently at or near their saturation level K. No longer is it very advantageous for a species to have a high r. It is more important for genotypes to confer competitive ability, in particular the capacity to seize and to hold a piece of the environment and to extract the energy produced by it. In higher plants this K selection may result in larger individuals, such as shrubs or trees, with a capacity to crowd out the root systems of and to deny sunlight to other plants that germinate close by. In animals K selection could result in increased specialization (to avoid interference with competitors) or an increased tendency to stake out and to defend territories against members of the same species. All else being equal, those genotypes of K strategists will be favored that are able to maintain the densest populations at equilibrium. Genotypes less able to survive and to reproduce under these long-term conditions of crowding will be eliminated.

Of course the two forms of selection cannot be mutually exclusive. In all cases, r is subject to at least some evolutionary modification, upward or downward, while few species are so consistently prevented from approaching K that they are not subject to some degree of K selection. King and Anderson (1971) and Roughgarden (1971) have, in fact, independently defined sets of conditions in which competing r and K alleles can coexist in balanced polymorphism. But in many instances where extreme K selection occurs, resulting in a stable population of long-lived individuals, the result must be an evolutionary decrease in r. For a genotype or a species that lives in a stable habitat, there is no Darwinian advantage to making a heavy commitment to reproduction if the effort reduces the change of individual survival. At the opposite extreme, it does pay to make a heavy reproductive effort, even at the cost of life, if the temporary availability of empty habitats guarantees that at least a few of one's offspring will find the resources they need in order to survive and to reproduce. Most of the r strategists' offspring will perish during the dispersal phase, but a few are likely to find an empty habitat in which to renew the life cycle.

The expected correlates of r and K selection in ecology and behavior are numerous and complex (Table 4-1). In general, higher forms of social evolution should be favored by K selection. The reason is that population stability tends to reduce gene flow and thus to increase inbreeding, while at the same time promoting land tenure and the multifarious social bonds that require longer life in more predictable environments.

Table 4-1 Some of the correlates of r selection and K selection (modified from Pianka, 1970).

Correlate	r selection	K selection
Climate	Variable and/or unpredictable: uncertain	Fairly constant and/or predictable: more certain
Mortality	Often catastrophic, nondirected, density-independent	More directed, density-dependent
Suvivorship	Often type III	Usually types I and II
Population size	Variable in time, nonequilibrium; usually well below carrying capacity of environment; unsaturated communities or portions thereof; ecological vacuums; recolonization each year	Fairly constant in time, equilibrium; at or near carrying capacity of the environment; saturated communities; no recolonization necessary
Intraspecific and interspecific competition	Variable, often lax	Usually keen
Attributes favored by selection	1. Rapid development 2. High r_{max} 3. Early reproduction 4. Small body size 5. Semelparity: single reproduction	1. Slower development, greater competitive ability 2. Lower resource thresholds 3. Delayed reproduction 4. Larger body size 5. Iteroparity: repeated reproductions
Length of life	Short, usually less than 1 year	Longer, usually more than 1 year
Emphasis in energy utilization	Productivity	Efficiency
Colonizing ability	Large	Small
Social behavior	Weak, mostly schools, herds, aggregations	Frequently well developed

The rodents are one of many groups of animals containing both r-selected and K-selected species. Judging from the account by Christian (1970), *Microtus pennsylvanicus* stands at the r extreme of the spectrum. In pre-Columbian times this abundant vole species may have been restricted to temporary wet grasslands, such as "beaver meadows" created by the abandonment of beaver dams. These temporary meadows give way rapidly to seral stages of reforestation, so that species dependent on them must adopt a strategy of rapid population growth and efficient dispersal. *M. pennsylvanicus* goes through marked population fluctuations that produce large numbers of "floaters," nonterritorial animals that emigrate long distances. Christian observed the invasion of one beaver meadow by these voles in less than a week after its creation, during a year when the *M. pennsylvanicus* population was very high. The voles had to cross inhospitable forest tracts to reach the newly opened habitat. The r strategy preadapted *M. pennsylvanicus* to life in the rapidly changing, meadowlike environments of agricultural land, where today it is a dominant species over a large part of North America. Other North American microtine rodents, particularly the deer mice of the genus *Peromyscus*, are closer to the K end of the scale. They originally inhabited the continuous habitats of North America, particularly the eastern deciduous forests and the central plains. Their populations are more stable, and they seldom irrupt in the spectacular fashion of the voles and lemmings. The beaver (*Castor canadensis*) is close to what we can designate as a true K selectionist. To a large extent this mammal designs and stabilizes its own habitats with the dams and ponds it creates. Protected from predators by its large size and secure aquatic lodges, and provided with a rich food source, the beaver has both low mortality and low birth rates. The young disperse away from the parental lodge only after a couple of years of residence. As a result, beaver populations are much more stable than those of microtine rodents.

With the discussion of the concept of r and K selection, our brief review of fundamental population biology is complete, and we are now prepared, in Chapter 5, to explore the more complex but pivotal subject of group selection in the evolution of social behavior.

CHAPTER 5

Group Selection and Altruism

Reporter: *When you ran Finland onto the map of the world, did you feel you were doing it to bring fame to a nation unknown by others?*
Nurmi: *No. I ran for myself, not for Finland.*
Reporter: *Not even in the Olympics?*
Nurmi: *Not even then. Above all, not then. At the Olympics, Paavo Nurmi mattered more than ever.*

Who does not feel at least a tinge of admiration for Paavo Nurmi, the ultimate individual selectionist? At the opposite extreme, we shared a different form of approval, warmer in tone but uneasily loose in texture, for the Apollo 11 astronauts who left their message on the moon, "We came in peace for all mankind." This chapter is about natural selection at the levels of selection in between the individual and the species. Its pivot will be the question of altruism, the surrender of personal genetic fitness for the enhancement of personal genetic fitness in others.

Group Selection

Selection can be said to operate at the group level, and deserves to be called group selection, when it affects two or more members of a lineage group as a unit. Just above the level of the individual we can delimit various of these lineage groups: a set of sibs, parents, and their offspring; a close-knit tribe of families related by at least the degree of third cousin; and so on. If selection operates on any of the groups as a unit, or operates on an individual in any way that affects the frequency of genes shared by common descent in relatives, the process is referred to as *kin selection*. At a higher level, an entire breeding population may be the unit, so that populations (that is, demes) possessing different genotypes are extinguished differentially, or disseminate different numbers of colonists, in which case we speak of *interdemic (or interpopulation) selection*. The ascending levels of selection are visualized in Figure 5-1. The classification adopted here is approximately that recommended by J. L. Brown (1966).

Selection can also operate at the level of species or entire clusters of related species. The process, well known to paleontologists and biogeographers, is responsible for the familiar patterns of dynastic succession in major groups such as ammonites, sharks, graptolites, and dinosaurs through geologic time (Simpson, 1953; P. J. Darlington, 1971). It is even possible to conceive of the differential extinction of entire ecosystems, involving all trophic levels (Dunbar, 1960, 1972). However, selection at these highest levels is not likely to be important in the evolution of altruism, for the following simple reason. In order to counteract individual selection, it is necessary to have population extinction rates of comparable magnitude. New species are not created at a sufficiently fast pace to be tested in this manner, at least not when the species are so genetically divergent as those ordinarily studied by the biogeographers. The same restriction applies a fortiori to ecosystems.

Pure kin and pure interdemic selection are the two poles at the ends of a gradient of selection on ever enlarging nested sets of related individuals. They are sufficiently different to require different forms of mathematical models, and their outcomes are qualitatively different. Depending on the behavior of the individual organisms and their rate of dispersal between societies, the transition zone between kin selection and interdemic selection for most species probably occurs when the group is large enough to contain somewhere on the order of 10 to 100 individuals. At that range, one reaches the upper limit of family size and passes to groups of families. One also finds the upper bound in the number of group members one animal can remember and with whom it can therefore establish personal bonds. Finally, 10 to 100 is the range in which the effective population numbers (N_e) of a great many vertebrate species fall. Thus, aggregations of more than 100 are genetically

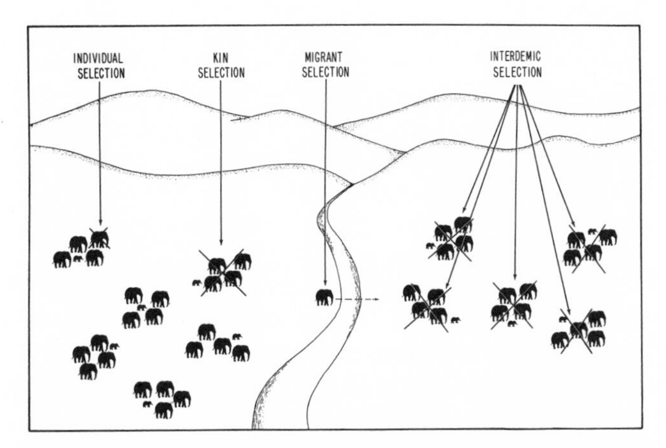

Figure 5-1 Ascending levels of selection. Group selection in the broadest sense can consist of either kin selection, in which the actions of individuals differentially favor relatives, or interdemic selection (also called interpopulation selection), in which entire populations are diminished or extinguished at different rates. Every gradation between these two extremes is possible. The differential tendency to disperse is referred to as migrant selection.

fragmented, and the geometry of their distribution is important to their microevolution.

Interdemic (Interpopulation) Selection

A cluster of populations belonging to the same species may be called a metapopulation. The metapopulation is most fruitfully conceived as an amebalike entity spread over a fixed number of patches (Levins, 1970). At any moment of time a given patch may contain a population or not; empty patches are occasionally colonized by immigrants that form new populations, while old populations occasionally become extinct, leaving an empty patch. If $P(t)$ is the proportion of patches which support populations at time t, m is the proportion of patches (whether already occupied or not) receiving migrants in an instant of time, and $\bar{E}$ is the proportion of populations becoming extinct in an instant of time,

$$\frac{dP}{dt} = mg(P) - \bar{E}P$$

The function $g(P)$ must decrease with the proportion of sites already occupied, a relation that can exist in the simple logistic form

$$\frac{dP}{dt} = mP(1 - P) - \bar{E}P$$

At equilibrium the proportion of occupied patches is

$$P = 1 - \frac{\bar{E}}{m}$$

where the metapopualtion as a whole can persist only if $\bar{E} < m$. Thus the system is metaphorically viewed through evolutionary time as a nexus of patches, each patch winking into life as a population colonizes it and winking out again as extinction occurs. At equilibrium the rate of winking and the number of occupied sites are constant, despite the fact that the pattern of occupancy is constantly shifting. The imagery can be translated into reality only when the observer is able to delimit real Mendelian populations in the system. The complications that arise from this problem are illustrated in Figure 5-2.

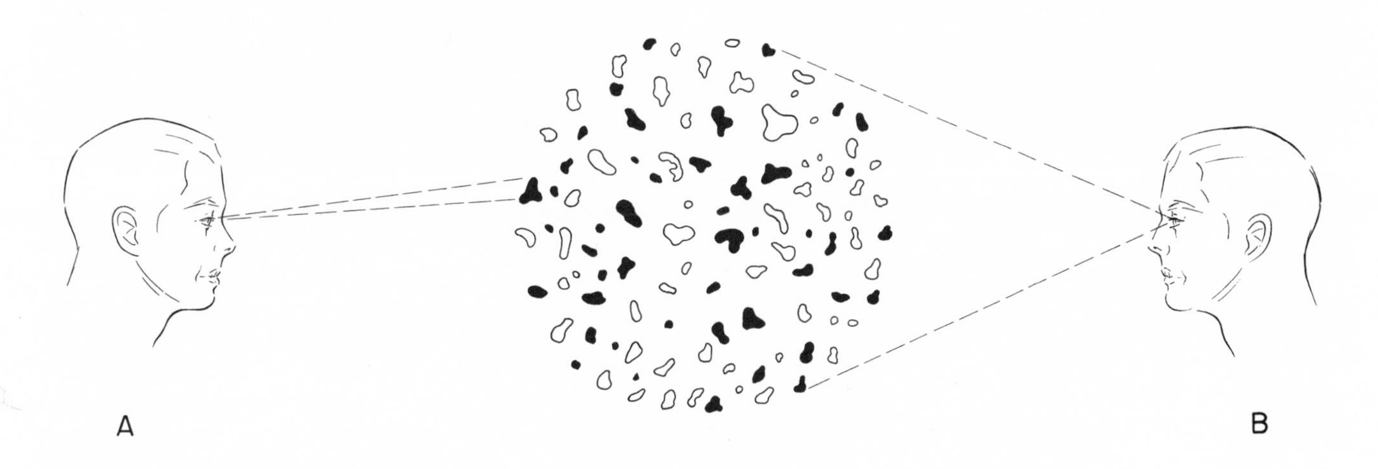

Figure 5-2 The metapopulation is a set of populations occupying a cluster of habitable sites. Because of constantly recurring extinction not totally canceled by new immigration, some percentage of the sites is always unfilled, although different ones are empty at different times.

In considering interdemic selection, it is important to distinguish the timing of the extinction event in the history of the population (Figure 5-3). There are two moments at which extinction is most likely: at the very beginning, when the colonists are struggling to establish a hold on the site, and soon after the population has reached (or exceeded) the carrying capacity of the site, and is in most danger of crashing from starvation or destruction of the habitat. The former event can be called *r* extinction and the latter *K* extinction, in appreciation of the close parallel this dichotomy makes with *r* and *K* selection. When populations are more subject to *r* extinction, altruist traits favored by group selection are likely to be of the "pioneer" variety. They will lead to clustering of the little population, mutual defense against enemies, and cooperative foraging and nest building. The ruling principle will be the maximum *average* survival and fertility of the group as a whole; in other words, the maximization of *r*. In *K* extinction the opposite is true. The premium is now on "urban qualities" that keep population size below dangerous levels. Extreme pressure from density-dependent controls of an external nature is avoided. Mutual aid is minimized, and personal restraint in the forms of underutilization of the habitat and birth control comes to the fore.

These two levels of extinction can be distinguished in the populations of the aphid *Pterocomma populifoliae* as described by Sanders and Knight (1968). The species is highly opportunistic, colonizing sucker stands of bigtooth aspen and multiplying rapidly to create small, isolated populations. Extinction rates are very high. The earliest colonies, composed of first-generation colonists, are wiped out by errant predators, including spiders and adult ladybird beetles. Older, more established colonies acquire resident predators, such as syrphid and chamaemyid flies and some ladybird beetles, who breed along with them. These predators, aided by the emigration of many of the surviving aphids themselves, often eliminate entire colonies.

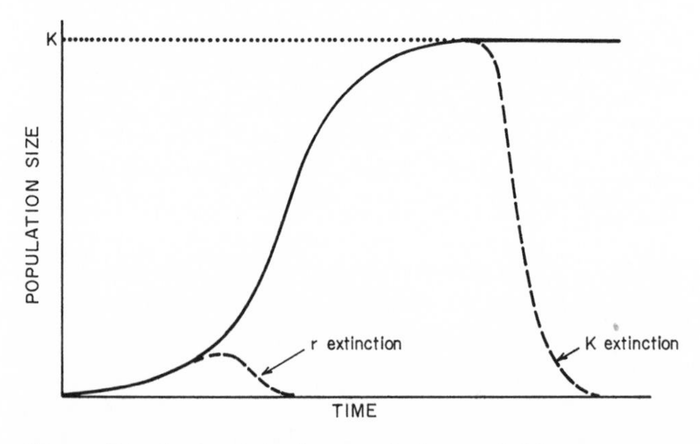

Figure 5-3 Extinction of a population probably most commonly occurs at an early stage of its growth, particularly when the first colonists are trying to establish a foothold (*r* extinction), or after the capacity of the environment has been reached or exceeded and a crash occurs (*K* extinction). The consequences in evolution are potentially radically different. (From Wilson, 1973.)

Very young, growing populations are likely to consist of individuals who are closely related. Interdemic selection by *r* extinction is therefore intrinsically difficult to separate from kin selection, and in extreme cases it is probably identical with it. A second feature that makes the process difficult to analyze is the change in gene frequencies due to genetic drift. In populations of ten or so individuals drift can completely swamp out the overall effect of differential extinction within the metapopulation. For these reasons, analysis has been concentrated on larger populations, and the most general results obtained are more easily applicable to interdemic selection by *K* extinction.

Our current understanding of interdemic selection can be most clearly understood if approached through its historical development. In 1932 Haldane constructed a few elements of a general theory that are equally applicable to kin and interdemic selection. He thought he could dimly see how altruistic traits increase in populations. "A study of these traits involves the consideration of small groups. For a character of this type can only spread through the population if the genes determining it are borne by a group of related individuals whose chances of leaving offspring are increased by the presence of these genes in an individual member of the group whose private viability they lower." Haldane went on to prove that the process is feasible if the groups are small enough for altruists to confer a quick advantage. He saw that the altruism could be stable in a metapopulation if the genes were fixed in individual groups by drift, made possible by the small size of the groups or at least of the new populations founded by some of their emigrants. For some reason Haldane overlooked the role of differential population extinction, which might have led him to the next logical step in developing a full theory.

A separate thread of thought led to the development, in the 1930's and 1940's, of the "island model" of population genetics by Sewall Wright. The island model was related explicitly to the evolution of altruistic behavior by Wright in his 1945 essay review of G. G. Simpson's *Tempo and Mode in Evolution*. The formulation was nearly identical to that of Haldane, although made independently of it. Wright conceived of a set of populations diverging by genetic drift and adaptation to local environments but exchanging genes with one another. The pattern is that which Wright has persistently argued to be "the greatest creative factor of all" in evolution. In the special case considered here, the disadvantageous (for example, altruistic) genes can prevail over all the metapopulation if the populations they aid are small enough to allow them to drift to high values, and if the aided populations thereby send out a disproportionate number of emigrants. Like Haldane, Wright did not consider the effect of differential extinction on the equilibrial metapopulation. Nor did the model come any closer to a full theory of altruistic evolution. It is a curious twist that when W. D. Hamilton reinitiated group selection theory twenty years later, he was inspired not by the island model but by Wright's studies of relationship and inbreeding, which led to the topic of kin selection.

The next step in the study of interdemic selection was taken by ecologists largely unaware of genetic theory. Kalela (1954, 1957) postulated group selection as the mechanism responsible for reproductive restraint in subarctic vole populations. He saw food shortages as the ultimate controlling factor but believed that self-control of the populations during times of food plenty prevented starvation during food shortages. Kalela correctly deduced that self-control in matters of individual fitness can only be evolved if the groups not possessing the genes for self-control are periodically decimated or extinguished as a direct consequence of their lack of self-control. Kalela added one more feature to his scheme that substantially increased its plausibility. He suggested that rodent populations in many cases really consist of expanded family groups, so that self-restraint is the way for genetically allied tribes to hold their ground while other tribes of the same species eat themselves into extinction. In other words, the most forceful mode of interdemic selection is one that approaches a special form of kin selection. Kalela believed that the same kind of population structure and group selection might characterize many other rodents, ungulates, and primates.

It remained for Wynne-Edwards to bring the subject to the attention of a wide biological audience. In his book *Animal Dispersion in Relation to Social Behaviour* (1962), he carried the theory of self-control by group selection to its extreme, thereby forcing an evaluation of its strengths and weaknesses. The principal concept in Wynne-Edwards' theory is density-dependent population control by means of social conventions, which are devices by which individual animals curtail their own individual fitness—that is, their survivorship, or fertility, or both—for the good of group survival. The density-dependent effects cited by Wynne-Edwards as involving social conventions run virtually the entire gamut: lowered fertility, reduced status in hierarchies, abandonment or direct killing of offspring, endocrine stress, deferment of growth and maturity. Sacrifice in each of these categories he viewed as an individual contribution to maintain populations below crash levels. Much of social behavior was reinterpreted by Wynne-Edwards to be epideictic displays, modes of communication by which members of populations inform each other of the density of the population as a whole and therefore the degree to which each member should decrease its own individual fitness. Examples of epideictic displays (which are distinguished from the epigamic displays that function purely in courtship) include the formation of mating swarms by insects, flocking in birds, and even vertical migration of zooplankton. The displays, then, are the most evolved communicative part of social conventions.

Specifically, Wynne-Edwards proposed the hypothesis that animals voluntarily sacrifice personal survival and fertility in order to help control population growth. He also postulated that this is a very widespread phenomenon among all kinds of animals. Furthermore, he did not stop at kin groups, as had Kalela, but suggested that the mechanism operates in Mendelian populations of all sizes, representing all breeding structures. Alternative hypotheses explaining social phenomena, such as nuptial synchronization, antipredation, and increased feeding efficiency, were either dismissed or minimized.

Wynne-Edwards' book prompted large numbers of biologists, including theoreticians, to address themselves to the serious issues of group selection and genetic social evolution. In the long series of reviews and fresh studies that followed, culminating in G. C. Williams' *Adaptation and Natural Selection* (1966), one after another of Wynne-Edwards' propositions about specific "conventions" and epideictic displays were knocked down on evidential grounds or at least matched with competing hypotheses of equal plausibility drawn from models of individual selection. But for a long time neither critics nor sympathizers could answer the main theoretical question raised by this controversy: What are the deme sizes, interdemic migration rates, and differential deme survival probabilities necessary to counter the effects of individual selection? Only when population genetics was extended this far could biologists hope to evaluate the significance of extinction rates and to rule out one or the other of the various competing hypotheses in particular cases.

The first quantitative models derived from population genetics, formulated by Richard Levins (1970) and S. A. Boorman and P. R. Levitt (1972, 1973), agree in one important respect: the spread of an altruist gene by means of pure interdemic selection is an improbable event. The metapopulation must pass through a very narrow "window" framed by strict parameter values: steeply descending extinction functions, preferably approaching a step function with a threshold value of the frequency of the altruist gene; high extinction rates comparable in magnitude (in populations per generation) to the opposing individual selection (in individuals per population per generation); and the existence of moderately large metapopulations broken into many semiisolated populations. Even after achieving all these conditions, the metapopulation is likely to be infiltrated only partway by the gene. (A critical review of the models is presented in Wilson, 1975).

What this means in practice is that most of the wide array of "social conventions" hypothesized by Wynne-Edwards and other authors are probably not true. Moreover, self-restraint on behalf of the entire population is *least* likely in the largest, most stable populations, where social behavior is the most highly developed. Examples include the breeding colonies of seabirds, the communal roosts of starlings, the leks of grouse, the warrens of rabbits, and many of the other societal forms cited by Wynne-Edwards as the best examples of altruistic population control. In these cases one must favor alternative hypotheses that involve either kin selection or individual selection. Even so, a mechanism for the evolution of population-wide cooperation has been validated, and the hypothesis of social conventions must either be excluded or kept alive for each species considered in turn.

One should also bear in mind that the real population is the unit whose members are freely interbreeding. Such a unit can exist firmly circumscribed in the midst of a seemingly vast population—which is really a metapopulation in evolutionary time. Consider a population of rodents in which tens of thousands of adults hold small territories over a continuous habitat of hundreds of square kilometers. The aggregation seems vast, yet each ridge of earth, each row of trees, and each streamlet could cut migration sufficiently to delimit a true population. The effective population size might be 10 or 100, despite the fact that a hawk's-eye view of the entire metapopulation makes it seem continuous. Not even the little habitat barriers are required. If the rodents move about very little, or return faithfully to the site of their birth to breed, the population neighborhood will be small, and the effective population size low. The delimitation of such local populations could be sharpened by the development of cultural idiosyncrasies such as the learned dialects of birds or the inherited burrow systems of social rodents. With increasing delimitation and reduction in population size, the selection involved also slides toward the kin-selection end of the scale. To evaluate definitively the potential intensity of interdemic selection it is necessary to estimate the size of the neighborhood, the effective size of the populations, and the rate at which the true populations become extinct (see Figure 5-1).

The chief role of interdemic selection may turn out to lie not in forcing the evolution of altruism per se, but rather in serving as a springboard from which other forms of altruistic evolution are launched. Suppose that the altruists also have a tendency to cooperate with one another in a way that ultimately benefits each altruist at the expense of the nonaltruists. Cliques and communes may require personal sacrifice, but if they are bonded by possession of one inherited trait, the trait can evolve as the groups triumph over otherwise comparable units of noncooperating groups. The bonding need not even require prolonged sacrifice, only the trade-offs of reciprocal altruism. The formation of such networks requires either a forbiddingly high starting gene frequency or a large number of random contacts with other individuals in which the opportunity for trade-offs exists (Trivers, 1971). These frequency thresholds might be reached by interdemic selection that initially favors other aspects of the behavior that are not altruistic.

There do exist special conditions under which interdemic selection can spread altruist genes rapidly in a population. Maynard Smith (1964) suggested a model in which local populations are first segregated and allowed to grow or to decline for a while in ways influenced by their genetic composition. Then individuals from different populations mix and interbreed to some extent before going on to form new populations. Suppose that the populations were mice in haystacks, with each haystack being colonized by a single fertilized female. If a/a are altruist individuals, and A/A and A/a selfish individuals, the a allele would be eliminated in all haystacks where A-bearing individuals were present. But if pure a/a populations contributed more progeny in the mixing and colonizing phases, and if there were also a considerable amount of inbreeding (so that pure a/a populations were more numerous than expected by chance alone), the altruist gene would spread through the population. D. S. Wilson (1975) argues that many species in nature go through regular cycles of segregation and mixing and that altruist genes can be spread under a wide range of realistic conditions beyond the narrow one conceived by Maynard Smith. All that is required is that the absolute rate of increase of the altruists be greater. It does not matter that their rate of increase relative to the nonaltruists in the same population is (by definition) less during the period of isolation. Provided the rate of increase of the population as a whole is enhanced enough by the presence of altruists, they will increase in frequency through the entire metapopulation.

What specific traits would interdemic selection be expected to produce? Under some circumstances the altruism would oppose *r* selection. There is a fundamental tendency for genotypes that have the highest *r* to win in individual selection, and their advantage is enhanced in species that are opportunistic or otherwise undergo regular fluctuations in population size. But the greater the fluctuation, the higher the extinction rate. Thus interdemic selection would tend to damp population cycles by a lower fertility and an early, altruistic sensitivity to density-dependent controls. There is also a fundamental tendency for genotypes that can sustain the highest density to prevail (*K* selection; see Chapter 4). But high density contaminates the environment, attracts predators, and promotes the spread of disease, all of which increase the extinction rates of entire populations. Altruism promoted by these effects might include a higher physiological sensitivity to crowding and a greater tendency to disperse even at the cost of lowered fitness. Levins (1970) has pointed out that mixtures of genotypes in populations of fruit flies and crop plants often attain a higher equilibrium density than pure strains, but under a variety of conditions one strain excludes the others competitively. If higher densities result in the production of more propagules without incurring a greater risk of extinction to the mother population, an antagonism between group and individual selection will result. Also, genetic resistance to disease or predation often results in lowered fitness in another component, as exemplified by sickle-cell anemia. In the temporary absence of this pressure, individual selection "softens" the population as a whole, which will be disfavored in interdemic selection when the pressure is exerted again.

It is also true, as Madhav Gadgil (1975) has pointed out, that pure interdemic selection, acting apart from kin selection, can lead to exceptionally selfish and even spiteful behavior. Suppose, for example, that the particular circumstances of interdemic selection within a given species dictate a reduction in population growth. Then the "altruist" who curtails its personal reproduction might just as well spend its spare time cannibalizing other members of the population—also to the benefit of the deme as a whole. Another seemingly spiteful behavior that could be favored by *K* extinction is the maintenance of excessively large territories.

Kin Selection

Imagine a network of individuals linked by kinship within a population. These blood relatives cooperate or bestow altruistic favors on one another in a way that increases the average genetic fitness of the members of the network as a whole, even when this behavior reduces the individual fitness of certain members of the group. The members may live together or be scattered throughout the population. The essential condition is that they jointly behave in a way that benefits the group as a whole, while remaining in relatively close contact with the remainder of the population. This enhancement of kin-network welfare in the midst of a population is *kin selection*.

Kin selection can merge into interdemic selection by an appropriate spatial rearrangement. As the kin network settles into one physical location and becomes physically more isolated from the rest of the species, it approaches the status of a true population. A closed society, or one so nearly closed that it exchanges only a small fraction of its members with other societies each generation, is a true Mendelian population. If in addition the members all treat one another without reference to genetic relationship, kin selection and interdemic selection are the same process. If the closed society is small, say, with 10 members or less, we can analyze group selection by the theory of kin selection. If it is large, containing an effective breeding size of 100 or more, or if the selection proceeds by the extinction of entire demes of any size, the theory of interdemic selection is probably more appropriate.

The personal actions of one member toward another can be conveniently classified into three categories in a way that makes the analysis of kin selection more feasible. When a person (or animal) increases the fitness of another at the expense of his own fitness, he can be said to have performed an act of *altruism*. Self-sacrifice for the benefit of offspring is altruism in the conventional but not in the strict genetic sense, because individual fitness is measured by the number of surviving offspring. But self-sacrifice on behalf of second cousins is true altruism at both levels; and when directed at total strangers such abnegating behavior is so surprising (that is, "noble") as to demand some kind of theoretical explanation. In contrast, a person who raises his own fitness by lowering that of others is engaged in *selfishness*. While we cannot publicly approve the selfish act, we do understand it thoroughly and may even sympathize. Finally, a person who gains nothing or even reduces his own fitness in order to diminish that of another has committed an act of *spite*. The action may be sane, and the perpetrator may seem gratified, but we find it difficult to imagine his rational motivation. We refer to the commitment of a spiteful act as "all too human"—and then wonder what we meant.

The concept of kin selection to explain such behavior was originated by Charles Darwin in *On The Origin of Species.* Darwin had encountered in the social insects the "one special difficulty, which at first appeared to me insuperable, and actually fatal to my whole theory." How, he asked, could the worker castes of insect societies have evolved if they are sterile and leave no offspring? This paradox proved truly fatal to Lamarck's theory of evolution by the inheritance of acquired characters, for Darwin was quick to point out that the Lamarckian hypothesis requires characters to be developed by use or disuse of the organs of individual organisms and then to be passed directly to the next generation, an impossibility when the organisms are sterile. To save his own theory, Darwin introduced the idea of natural selection operating at the level of the family rather than of the single organism. In retrospect, his logic seems impeccable. If some of the individuals of the family are sterile and yet important to the welfare of fertile relatives, as in the case of insect colonies, selection at the family level is inevitable. With the entire family serving as the unit of selection, it is the capacity to generate sterile but altruistic relatives that becomes subject to genetic evolution. To quote Darwin, "Thus, a well-flavoured vegetable is cooked, and the individual is destroyed; but the horticulturist sows seeds of the same stock, and confidently expects to get nearly the same variety; breeders of cattle wish the flesh and fat to be well marbled together; the animal has been slaughtered, but the breeder goes with confidence to the same family" (*Origin of Species,* 1859:237). Employing his familiar style of argumentation, Darwin noted that intermediate stages found in some living species of social insects connect at least some of the extreme sterile castes, making it possible to trace the route along which they evolved. As he wrote, "With these facts before me, I believe that natural selection, by acting on the fertile parents, could form a species which regularly produce neuters, either all of a large size with one form of jaw, or all of small size with jaws having a widely different structure; or lastly, and this is the climax of our difficulty, one set of workers of one size and structure, and simultaneously another set of workers of a different size and structure" (*Origin of Species,* 1859:24). Darwin was speaking here about the soldiers and minor workers of ants.

The modern genetic theory of altruism, selfishness, and spite was launched by William D. Hamilton in a series of important articles (1964, 1970, 1971a,b, 1972). Hamilton's pivotal concept is *inclusive fitness:* the sum of an individual's own fitness plus the sum of all the effects it causes to the related parts of the fitnesses of all its relatives. When an animal performs an altruistic act toward a brother, for example, the inclusive fitness is the animal's fitness (which has been lowered by performance of the act) plus the increment in fitness enjoyed by that portion of the brother's hereditary constitution that is shared with the altruistic animal. The portion of shared heredity is the fraction of genes held by common descent by the two animals and is measured by the coefficient of relationship, r (see Chapter 4). Thus, in the absence of inbreeding, the animal and its brother have $r = 1/2$ of their genes identical by common descent. Hamilton's key result can be stated very simply as follows. A genetically based act of altruism, selfishness, or spite will evolve if the average inclusive fitness of individuals within networks displaying it is greater than the inclusive fitness of individuals in otherwise comparable networks that do not display it.

Consider, for example, a simplified network consisting solely of an individual and his brother (Figure 5-4). If the individual is altruistic he will perform some sacrifice for the benefit of the brother. He may surrender needed food or shelter, or defer in the choice of a mate, or place himself between his brother and danger. The important result, from a purely evolutionary point of view, is loss of genetic fitness—a reduced mean life span, or fewer offspring, or both—which leads to less representation of the altruist's personal genes in the next generation. But at least half of the brother's genes are identical to those of the altruist by virtue of common descent. Suppose, in the extreme case, that the altruist leaves no offspring. If his altruistic act more than doubles the brother's personal representation in the next generation, it will ipso facto increase the one-half of the genes identical to those in the altruist, and the altruist will actually have gained representation in the next generation. Many of the genes shared by such brothers will be the ones that encode the tendency toward altruistic behavior. The inclusive fitness, in this case determined solely by the brother's contribution, will be great enough to cause the spread of the altruistic genes through the population, and hence the evolution of altruistic behavior.

The model can now be extended to include all relatives affected by the altruism. If only first cousins were benefited ($r = 1/8$), the altruist who leaves no offspring would have to multiply a cousin's fitness eightfold; an uncle ($r = 1/4$) would have to be advanced fourfold; and so on. If combinations of relatives are benefited, the genetic effect of the altruism

Figure 5-4 The basic conditions required for the evolution of altruism, selfishness, and spite by means of kin selection. The family has been reduced to an individual and his brother; the fraction of genes in the brother shared by common descent ($r = ½$) is indicated by the shaded half of the body. A requisite of the environment (food, shelter, access to mate, and so on) is indicated by a vessel, and harmful behavior to another by an axe. *Altruism:* the altruist diminishes his own genetic fitness but raised his brother's fitness to the extent that the shared genes are actually increased in the next generation. *Selfishness:* the selfish individual reduces his brother's fitness but enlarges his own to an extent that more than compensates. *Spite:* the spiteful individual lowers the fitness of an unrelated competitor (the unshaded figure) while reducing that of his own or at least not improving it; however, the act increases the fitness of the brother to a degree that more than compensates.

is simply weighted by the number of relatives of each kind who are affected and their coefficients of relationship. In general, k, the ratio of gain in fitness to loss in fitness, must exceed the reciprocal of the average coefficient of relationship ($\bar{r}$) to the ensemble of relatives:

$$k > \frac{1}{\bar{r}}$$

Thus in the extreme brother-to-brother case, $1/r = 2$; and the loss in fitness for the altruist who leaves no offspring is said to be total (that is $= 1.0$). Therefore in order for the shared altruistic genes to increase, k, the gain-to-loss ratio, must exceed 2. In other words, the brother's fitness must be more than doubled.

The evolution of selfishness can be treated by the same model. Intuitively, it might seem that selfishness in any degree pays off so long as the result is the increase of one's personal genes in the next generation. But this is not the case if relatives are being harmed to the extent of losing too many of their genes shared with the selfish individual by common descent. Again, the inclusive fitness must exceed 1, but this time the result of exceeding that threshold is the spread of the selfish genes.

Finally, the evolution of spite is possible if it, too, raises inclusive fitness. The perpetrator must be able to discriminate relatives from nonrelatives, or close relatives from distant ones. If the spiteful behavior causes a relative to prosper to a compensatory degree, the genes favoring spite will increase in the population at large. True spite is a commonplace in human societies, undoubtedly because human beings are keenly aware of their own blood lines and have the intelligence to plot intrigue. Human beings are unique in the degree of their capacity to lie to other members of their own species. They typically do so in a way that deliberately diminishes outsiders while promoting relatives, even at the risk of their own personal welfare (Wallace, 1973). Examples of spite in animals may be rare and difficult to distinguish from purely selfish behavior. This is particularly true in the realm of false communication. As Hamilton drily put it, "By our lofty standards, animals are poor liars." Chimpanzees and gorillas, the brightest of the nonhuman primates, sometimes lie to one another (and to zookeepers) to obtain food or to attract company (Hediger, 1955:150; van Lawick-Goodall, 1971). The mental capacity exists for spite, but if these animals lie for spiteful reasons this fact has not yet been established. Even the simplest physical techniques of spite are ambiguous in animals. Male bowerbirds sometimes wreck the bowers of the neighbors, an act that appears spiteful at first (Marshall, 1954). But bowerbirds are polygynous, and the probability exists that the destructive bird is able to attract more females to his own bower. Hamilton (1970) has cited cannibalism in the corn ear worm (*Heliothis zea*) as a possible example of spite. The first caterpillar that penetrates an ear of corn eats all subsequent rivals, even though enough food exists to see two or more of the caterpillars through to maturity. Yet even here, as Hamilton concedes, the trait might have evolved as pure selfishness at a time when the *Heliothis* fed on smaller flowerheads or small corn ears of the ancestral type. Many other examples of the killing of conspecifics have been demonstrated in insects, but almost invariably in circumstances where the food supply is limited and the aggressiveness is clearly selfish as opposed to spiteful (Wilson, 1971b).

The Hamilton models are beguiling in part because of their transparency and heuristic value. The coefficient of relationship, r, translates easily into "blood," and the human mind, already sophisticated in the intuitive calculus of blood ties and proportionate altruism, races to apply the concept of inclusive fitness to a reevaluation of its own social impulses. But the Hamilton viewpoint is also unstructured. The conventional parameters of population genetics, allele frequencies, mutation rates, migration, group size, and so forth, are mostly omitted from the equations. As a result, Hamilton's mode of reasoning can be only loosely coupled with the remainder of genetic theory, and the number of predictions it can make is unnecessarily limited.

Reciprocal Altruism

The theory of group selection has taken most of the good will out of altruism. When altruism is conceived as the mechanism by which DNA multiplies itself through a network of relatives, spirituality becomes just one more Darwinian enabling device. The theory of natural selection can be extended still further into the complex set of relationships that Robert L. Trivers (1971) has called *reciprocal altruism.* The paradigm offered by Trivers is Good Samaritan behavior in human beings. A man is drowning, let us say, and another man jumps in to save him, even though the two are not related and may not even have met previously. The reaction is typical of what human beings regard as "pure" altruism. However, upon reflection one can see that the Good Samaritan has much to gain by his act. Suppose that the drowning man has a one-half chance of drowning if he is not assisted, whereas the rescuer has a one-in-twenty chance of dying. Imagine further that when the rescuer drowns the victim also drowns, but when the rescuer lives the victim is always saved. If such episodes were extremely rare, the Darwinist calculus would predict little or no gain to the fitness of the rescuer for his attempt. But if the drowning man reciprocates at a future time, and the risks of drowning stay the same, it will have benefited both individuals to have played the role of rescuer. Each man will have traded a one-half chance of dying for about a one-tenth chance. A population at large that enters into a series of such moral obligations, that is, reciprocally altruistic acts, will be a population of individuals with generally increased genetic fitness. The trade-off actually enhances personal fitness and is less purely altruistic than acts evolving out of interdemic and kin selection.

In its elementary form the Good Samaritan model still contains an inconsistency. Why should the rescued individual bother to reciprocate? Why not cheat? The answer is that in an advanced, personalized society, where individuals are identified and the record of their acts is weighed by others, it does not pay to cheat even in the purely Darwinist sense. Selection will discriminate against the individual if cheating has later adverse affects on his life and reproduction that outweigh the momentary advantage gained. Iago stated the essence in *Othello:* "Good name in man and woman, dear my lord, is the immediate jewel of their souls."

Trivers has skillfully related his genetic model to a wide range of the most subtle human behaviors. Aggressively moralistic behavior, for example, keeps would-be cheaters in line—no less than hortatory sermons to the believers. Self-righteousness, gratitude, and sympathy enhance the probability of receiving an altruistic act by virtue of implying reciprocation. The all-important quality of sincerity is a metacommunication about the significance of these messages. The emotion of guilt may be favored in natural selection because it motivates the cheater to compensate for his misdeed and to provide convincing evidence that he does not plan to cheat again. So strong is the impulse to behave altruistically that persons in experimental psychological tests will learn an instrumental conditioned response without advance explanation and when the only reward is to see another person relieved of discomfort (Weiss et al., 1971).

Human behavior abounds with reciprocal altruism consistent with genetic theory, but animal behavior seems to be almost devoid of it. Perhaps the reason is that in animals relationships are not sufficiently enduring, or memories of personal behavior reliable enough, to permit the highly personal contracts associated with the more human forms of reciprocal altruism. Almost the only exceptions I know occur just where one would most expect to find them—in the more intelligent monkeys, such as rhesus macaques and baboons, and in the anthropoid apes. Members of troops are known to form coalitions or cliques and to aid one another reciprocally in disputes with other troop members. Chimpanzees, gibbons, African wild dogs, and wolves also beg food from one another in a reciprocal manner.

Altruistic Behavior

Armed with existing theory, let us evaluate some reported cases of altruism among animals. In the following review each class of behavior will insofar as possible be examined in the light of competing hypotheses that counterpoise altruism and selfishness.

Thwarting Predators

The social insects contain many striking examples of altruistic behavior evolved by family-level selection. The altruistic responses are directed not only at offspring and parents but also at sibs and even nieces, nephews, and cousins (Wilson, 1971a). The soldier caste of most species of termites and ants is largely limited in function to colony defense. Soldiers are often slow to respond to stimuli that arouse the rest of the colony, but when they do react, they normally place themselves in the position of maximum danger. When nest walls of higher termites such as *Nasutitermes* are broken open, for example, the white, defenseless nymphs and workers rush inward toward the concealed depths of the nests, while the soldiers press outward and mill aggressively on the outside of the nest. W. L. Nutting (personal communication) witnessed soliders of *Amitermes emersoni* in Arizona emerge well in advance of the nuptial flights, wander widely around the nest vicinity, and effectively engage in combat all foraging ants that might have endangered the emerging winged reproductives. I have observed that injured workers of the fire ant *Solenopsis invicta* leave the nest more readily and are more aggressive on the average than their uninjured sisters. Dying workers of the harvesting ant *Pogonomyrmex badius* tend to leave the nest altogether. Both effects may be no more than nonadaptive epiphenomena, but it is also likely that the responses are altruistic. To be specific, injured workers are useless for functions other than defense, while dying workers pose a sanitary problem. Honeybee workers possess barbed stings that tend to remain embedded in their victims when the insects pull away, causing part of their viscera to be torn out and the bees to be fatally injured (Sakagami and Akahira, 1960). The suicide seems to be a device specifically adapted to repel human beings and other vertebrates, since the workers can sting intruding bees from other hives without suffering the effect (Butler and Free, 1952). A similar defensive maneuver occurs in the ant *P. badius* and in many polybiine wasps, including *Synoeca surinama* and at least some species of *Polybia* and *Stelopolybia* (Rau, 1933; W. D. Hamilton, personal communication). The fearsome reputation of social bees and wasps is due to their general readiness to throw their lives away upon slight provocation.

Although vertebrates are seldom suicidal in the manner of the social insects, many place themselves in harm's way to defend relatives. The dominant males of chacma baboon troops (*Papio ursinus*) position themselves in exposed locations in order to scan the environment while the other troop members forage. If predators or rival troops approach, the dominant males warn the others by barking and may move toward the intruders in a threatening manner, perhaps accompanied by other males. As the troop retreats, the dominant males cover the rear (Hall, 1960). Essentially the same behavior has been observed in the yellow baboon (*P. cynocephalus*) by the Altmanns (1970). When troops of hamadryas baboons, rhesus macaques, or vervets meet and fight, the adult males lead the combat (Struhsaker, 1967a,b; Kummer, 1968). The adults of many ungulates living in family groups, such as musk oxen, moose, zebras, and kudus, interpose themselves between predators and the young. When males are in charge of harems, they usually assume the role; otherwise the females are the defenders. This behavior can be rather easily explained by kin selection. Dominant males are likely to be the fathers or at least close relatives of the weaker individuals they defend. Something of a control experiment exists in the large migratory herds of ungulates such as wildebeest and bachelor herds of gelada monkeys. In these loose societies the males will threaten sexual rivals but will not defend other members of their species against predators.

Parental sacrifice in the face of predators attains its clearest expression in the distraction displays of birds (Armstrong, 1947; R. G. B. Brown, 1962; Gramza, 1967). A distraction display is any distinctive behavior used to attract the attention of an enemy and to draw it away from an object that the animal is trying to protect. In the great majority of instances the display directs a predator away from the eggs or young. Bird species belonging to many different families have evolved their own particular bag of tricks, including display flights and pretending to be young birds. The commonest, however, is injury feigning, which varies according to the species from simple interruptions of normal movements to the exact imitation of injury or illness. The female nighthawk (*Chordeiles minor*), for instance, deserts her nest when approached, flies conspicuously at low levels, and finally settles on the ground in front of the intruder (and away from her nest) with wings drooping or outstretched (Figure 5-5).

There are other ways a defender can risk its life besides simply confronting the enemy. If the defender

Figure 5-5 Distraction display of the female nighthawk. In order to draw intruders away from her nest, the bird often alights and either droops her wings (*A*) or holds them outstretched (*B*). (Original drawing by J. B. Clark; based on Gramza, 1967.)

just attempts to alarm other members of its species, it attracts attention to itself and runs a greater risk. Alarm communication in social insects is altruistic in a very straightforward way. In most species it is closely coupled with suicidal attack behavior. Even when the insect flees while releasing an alarm pheromone or stridulating, the signal cannot help but attract the intruder to it. Alarm communication in vertebrates is much more ambiguous. When small birds of many species discover a hawk, owl, or other potential enemy resting in the vicinity of their territories, they mob it while uttering characteristic clicking sounds that attract other birds to the vicinity. This behavior is relatively safe, because the predator is not in a position to attack. The aggression of the small birds often drives the predator from the neighborhood. Thus inciting or joining a mob would appear to increase personal fitness. However, the warning calls of the same small birds are very different in content and significance from mobbing calls. They are uttered by such diverse species as blackbirds, robins, thrushes, reed buntings, and titmice. When a hawk is seen flying overhead, the birds crouch low and emit a thin, reedy whistle. In contrast to the mobbing call, the warning call is acoustically designed to make it difficult to locate in space. The continuity of the sound, extending over a half-second or more, eliminates time cues that reveal the direction. A pure tone of about 7 kiloherz is used, just above the frequency required for phase difference location but below the optimum for generating biaural intensity differences (Marler, 1957; Marler and Hamilton, 1966). The bird is evidently "trying" to avoid the great danger posed by the hawk. Then why does it bother at all? Why warn others if it has already perceived the danger itself? Warning calls seem prima facie to be altruistic. Maynard Smith (1965) hypothesized that they originate by kin selection; not just the mate and offspring but also more distant relatives are benefited. He devised a model proving that genes for such an altruistic trait can be maintained in a balanced polymorphic state.

The same considerations hold for warning behavior in rodents, and here the kin-selection hypothesis is more tractable for testing. Colonies of black-tail prairie dogs and arctic ground squirrels set up waves of alarm calls when they sight a predator (King, 1955; Carl, 1971). Since the calling animals remain at or above the burrow entrance when they could scurry to safety, the action could well be altruistic. In a careful analysis of Belding's ground squirrel (*Spermophilus beldingi*), Sherman (1977) showed that protection of kin is the explanation for alarm calling that best fits all the circumstances of the behavior.

Cooperative Breeding

The reduction of personal reproduction in order to favor the reproduction of others is widespread among organisms and offers some of the strongest indirect evidence of kin selection. The social insects, as usual, are clear-cut in this respect. The very definition of higher sociality ("eusociality") in termites, ants, bees, and wasps entails the existence of sterile castes whose basic functions are to increase the oviposition rate of the queen, ordinarily their mothers, and to rear the queen's offspring, ordinarily their brothers and sisters. The case of "helpers" among birds is also strongly suggestive (Skutch, 1961; Lack,

1968; Woolfenden, 1974a,b). Among the many cases of helpers assisting other birds to rear their young, including moorhens, Australian blue wrens, thornbills, anis, and others, the assistance is typically rendered by young adults to their parents. Consequently, just as in the social insects, the cooperators are rearing their own brothers and sisters (see Chapter 21).

In some respects "aunt" and "uncle" behavior in monkeys and apes superficially resembles the cooperative brood care of social insects and birds. Childless adults take over the infants of others for short periods during which they carry the young about, groom them, and play with them. The baby-sitting may seem to be altruistic, but there are other explanations. Adult males of the Barbary macaque use infants in ritual presentations to conciliate other adult males. The "aunts" of rhesus and Japanese macaques also use baby-sitting to form alliances with mothers of superior rank. Furthermore, the possibility cannot be excluded that aunting behavior provides training in the manipulation of infants that improves the performance of young females when they bear their first young (see Chapter 16).

Outright adoption of infants and juveniles has also been recorded in a few mammal species. Jane van Lawick-Goodall (1971) recorded three cases of adult chimpanzees adopting young orphaned siblings at the Gombe Stream National Park in Tanzania. As she noted, it is strange (but significant for the theory of kin selection) that the infants were adopted by siblings rather than by an experienced female with a child of her own, who could supply the orphan with milk as well as with more adequate social protection. During studies of African wild dogs in the Ngorongoro Crater conducted by Estes and Goddard (1967), a mother died when her nine pups were only five weeks old. The adult males of the pack continued to care for them, returning to the den each day with food until the pups were able to join the pack on hunting trips. The small size of wild dog packs makes it probable that the males were fathers, uncles, cousins, or other similarly close relatives. Males of the hamadryas baboon normally adopt juvenile females (Kummer, 1968). This unusual adaptation is clearly selfish in nature, since in hamadryas society adoption is useful for the accumulation of a harem.

Assistance in the reproductive effort of others can take even stranger forms. In the southeastern Texas population of the wild turkey (*Meleagris gallopavo*), brothers assist each other in the fierce competition for mates (Watts and Stokes, 1971). The union of brothers begins in the late fall, when the birds are six to seven months old. At that time the young males break away from their brood flock together. They maintain their bond as a sibling group for the rest of their lives, so that even when all its brothers die, a male does not attempt to join another sibling group. In the winter the brotherhood joins flocks of other juvenile birds. At this time their status is determined by a series of combats, in which the young males wrestle in fighting-cock style for as long as two hours, pecking at each other's head and neck and striking with the wings. The winner of the match becomes the dominant member of the pair for life. Such contests are conducted at three levels. First, the brothers struggle with each other until one emerges as the unchallenged dominant. Next, brotherhoods meet in contest until one group, usually the largest, achieves ascendancy over all the others in the winter flock. Finally, different flocks contend with one another whenever they meet, again settling dominance at the group level. The final result of this elaborate tournament is that one male comes to hold the dominant position in the entire local population of turkeys. When the males and females gather on the mating ground in February, each brotherhood struts and fantails in competition with the others. The brothers display in synchrony with each other in the direction of the watching females. When a female is ready to mate, the subordinate brothers yield to their dominant sibling, and the subordinate brotherhoods yield to the dominant one. As a result only a small fraction of the mature males inseminate the females. Of 170 males belonging to four display groups watched by Watts and Stokes, no more than 6 cocks accounted for all the mating.

Food Sharing

Other than suicide, no behavior is more clearly altruistic than the surrender of food. The social insects have carried food sharing to a high art. In the higher ants, the "communal stomach," or distensible crop, together with a specially modified gizzard, forms a complex storage and pumping system that functions in the exchange of liquid food between members of the same colony (Eisner, 1957). In both ants and honeybees, newly fed workers often press offerings of regurgitated food on nestmates without being begged, and they may go so far as to expend their supply to a level below the colony average (Lindauer, 1961). The regurgitation results in at least two consequences of importance to social organization beyond the mere feeding of the hungry. First, because workers tend to hold a uniform quantity and quality of food in their crops at any given moment, each individual is continuously apprised of the condition of the colony as a whole. Its personal hunger and thirst are approximately those of the entire colony, and in a literal sense what is good for one worker is good for the colony. Second, the regurgitated food contains pheromones, as well as special nutriments manufactured by exocrine glands and other substances of social importance.

Altruistic food sharing among adults is known among African wild dogs, where it permits some individuals to remain at the dens with the cubs while others hunt (Kühme, 1965; H. and Jane van Lawick-Goodall, 1971). The donors carry fresh meat directly to the recipients or else regurgitate it in front of them. Occasionally a mother dog will allow other adults to suckle milk. Altruistic food sharing has also been reported on several occasions in the higher anthropoids. In captive gibbons the exchange is initiated by one animal trying to take food from another, either by grasping the food or by holding the partner's hand while taking the food. The partner usually lets some of the food go without protest. Under some circumstances it will resist by keeping it out of reach or, rarely, by threatening or fighting. The offering of food without solicitation does not appear to occur (Berkson and Schusterman, 1964). Chimpanzees also successfully beg food from one another, especially part of the small mammals that the apes occasionally kill as prey. This benevolent behavior is in sharp contrast to that of baboons: when they kill and eat small antelopes, the dominant males appropriate the meat, and fighting is frequent (Kummer, 1971).

Ritualized Combat, Surrender, Amnesty

The mere forbearance of an enemy can be a form of altruism. Fighting between animals of the same species is typically ritualized. By precise signaling, a beaten combatant can immediately disclose when it is ready to leave the field, and the winner will normally permit it to do so without harm. African wild dogs display submission by an open-mouth grimace, a lowering and turning of the head and neck, and a belly-up groveling motion of the body. The loser thus exposes itself even more to the bites of its needle-toothed opponent. But at this point the attack either moderates or stops altogether. Male mantis shrimps fight with explosive extensions of their second maxillipeds. One strike from these hammer-shaped appendages is enough to tear another animal apart. But fatalities seldom occur, because each shrimp is careful to aim at the heavily armored tail segment of its opponent (Dingle and Caldwell, 1969). Other examples of ritualized aggression can be multiplied almost endlessly from the literature, and indeed they form a principal theme of Konrad Lorenz's celebrated book *On Aggression*. They also pose a considerable theoretical difficulty: Why not always try to kill or maim the enemy outright? And when an opponent is beaten in a ritual encounter, why not go ahead and kill him then? Allowed to run away, to paraphrase the childhood rhyme, the opponent may live to fight another day—and win next time. So in a sense the kindness shown an enemy seems altruistic, an unnecessary risk of personal fitness. One explanation is that mercy is "good for the species," since it allows the greatest number of individuals to remain healthy and uninjured. That hypothesis requires interdemic selection of a high intensity, because at the level of individual selection the greatest fitness in such encounters would always seem to accrue to the genotype that "plays dirty." A second hypothesis is that ritualization arises from kin selection: the need to win fights without eliminating the genes shared with others by common descent. The explanation could well hold in many particular cases, for example the wrestling matches between the brother turkeys in Texas. But in other species the highly ritualized encounters are held between individuals that are at best distantly related. A third hypothesis, suggested by Maynard Smith and G. R. Price (1973; see also Price in Maynard Smith and Ridpath, 1972), explains ritualized fighting as the outcome of purely individual selection. It recognizes that a great many animal species actually display two forms of combat, ritualized fighting and escalated fighting. The escalated form is invoked when an animal is hurt by an opponent. This particular form of behavioral scaling will be stabilized in evolution because it is disadvantageous either to engage in escalated fighting too readily or never to use it all.

The Field of Righteousness

In conclusion, although the theory of group selection is still rudimentary, it has already provided insights into some of the least understood and most disturbing qualities of social behavior. Above all, it predicts ambivalence as a way of life in social creatures. Like Arjuna faltering on the Field of Righteousness, the individual is forced to make imperfect choices based on irreconcilable loyalties—between the "rights" and "duties" of self and those of family, tribe, and other units of selection, each of which evolves its own code of honor. No wonder the human spirit is in constant turmoil. Arjuna agonized, "Restless is the mind, O Krishna, turbulent, forceful, and stubborn; I think it no more easily to be controlled than is the wind." And Krishna replied, "For one who is uncontrolled, I agree the Rule is hard to attain; but by the obedient spirits who will strive for it, it may be won by following the proper way." In the opening chapter of this book, I suggested that a science of sociobiology, if coupled with neurophysiology, might transform the insights of ancient religions into a precise account of the evolutionary origin of ethics and hence explain the reasons why we make certain moral choices instead of others at particular times. Whether such understanding will then produce the Rule remains to be seen. For the moment, perhaps it is enough to establish that a single strong thread does indeed run from the conduct of termite colonies and turkey brotherhoods to the social behavior of man.

PART II

Social Mechanisms

CHAPTER 6

Group Size and Reproduction

Natural selection extended long enough always leads to compromise. Each selection pressure guiding genetic change in a population is opposed by other selection pressures. As the population evolves, the strongest pressure eventually weakens while opposing ones intensify. When these forces finally strike a balance, the population phenotypes can be said to be at their evolutionary optimum; and evolution has passed from the dynamic to the stabilizing state. A convenient way of visualizing the process with special reference to social evolution is shown in Figure 6-1. The axes of the graphs measure variation of two social traits in some quality, say, degree of complexity or intensity. The organisms in the population are represented by points on the plane, the position of each being determined by the phenotype it possesses in the two social traits. The cluster is densest near its center. This by definition constitutes the statistical mode of the population. For each environment there exists only one or a set of very few positions at which the statistical mode is favored by selection over less common phenotypes. If the population is not centered on one of these positions, the resulting dynamic selection will tend to move it there. Thus selection superimposes a kind of force field upon the plane of phenotypes. The position at which the population comes temporarily to rest is the ensemble of phenotypes around which selection forces are balanced, and it constitutes the evolutionary compromise. This steady state has been labeled the evolutionary stable strategy (ESS) by Maynard Smith and Price (1973), and its analysis has been developed further by Maynard Smith (1976, 1978).

If this equilibrium case is generally true in social systems, weak and intermediate stages in phylogenetic successions among living species represent earlier compromises rather than evolution in progress. The population phenotypes have simply been balanced by selection forces at some early point such as the lower left-hand area in Figure 6-1, rather than continuing to move away, as shown in the example depicted. It is reasonable to postulate, as a working hypothesis, that most social species are at least temporarily stabilized. Some have halted well down on the scale, to remain "primitive" species, while others have moved farther along before stabilizing (as seen in the right-hand graph of Figure 6-1), to become the "advanced" species.

Examples of counteracting selection forces are easy to find in nature. The intensity of aggressive behavior is undoubtedly limited by a destructiveness to self and relatives that causes a loss in genetic fitness comparable to the gains accrued from the defeat of the enemy. Destructive behavior is easy to document in nature. Male hamadryas baboons, for example, sometimes injure the females over which they are fighting, and bull elephant seals trample pups to death while flopping around in their spectacular territorial battles. Similarly, evidence of evolutionary compromise that limits destructiveness is readily found. In general, fighting among animals of the same species seldom passes from the ritual to the escalated stage, that is, to the point where serious injury is mutually inflicted (see Chapter 11). An inverse form of compromise is fashioned during the evolution of submissive behavior. Animals belonging to dominance systems submit to their superiors, signaling their state of mind with displays that are sometimes very specialized and elaborate, but they cannot be pushed beyond certain clear limits. At some level of harassment, the persecuted animal turns on its attacker with escalated fighting or deserts the group altogether. A more precise measure of the level of compromise can be obtained from the amount of time spent grooming other animals. In many dominance systems the subordinate individual grooms its superiors as a conciliation device. Rhesus monkeys are so punctilious in this matter that the rank of the animal can be ascertained simply by observing which group members it grooms and

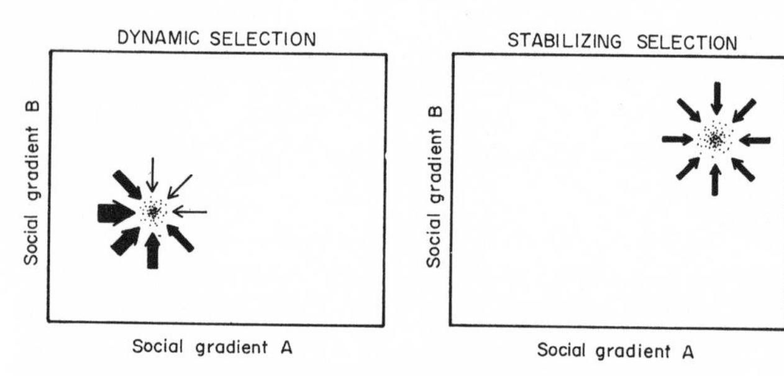

Figure 6-1 Evolution in two social traits is viewed here as the movement of an entire population of organisms on a plane of phenotypes. The rate and direction of movement is determined by the force field of opposing selection pressures (left figure). The stable states of social traits are reached when the selection pressures balance, the condition called stabilizing selection (right figure).

by whom it is groomed. How much time, from the animal's point of view, should be devoted to grooming others? Just enough to consolidate and advance its position. This cynical hypothesis is at least consistent with direct observations of the shifting dominance relations within rhesus troops.

Compromise is also manifest throughout the evolution of sexual behavior. The males of polygamous birds tend to evolve greater size, brighter plumage, and more conspicuous displays as devices for acquiring multiple mates. The trend is opposed by the greater ease with which predators are able to locate and capture the more dramatic males. As a result the sex ratio is progressively unbalanced with advancing age. The sex ratio of newly hatched great-tailed grackles (*Cassidix mexicanus*) is balanced, but within two months after the breeding season the ratio among first-year and adult birds is 1 male to 1.34 females, while five months later, in the following spring, the ratio has fallen to 1:2.42. Selander (1965), who discovered this case, believes that the higher mortality of males is partly a result of their greater vulnerability to predators, which in turn is due to their conspicuous coloration and the loss of flight maneuverability caused by the long tails used in display. A second handicap appears to be their larger size, which reduces efficiency at foraging.

The Determinants of Group Size

The number of members in a society can be wholly understood only by recourse to the concept of evolutionary compromise. We will approach this subject by considering first the purely functional parameters that influence group size, or more precisely those that determine the frequency distributions of groups of variable size. Then we will proceed to a consideration of the selection pressures that have led to particular values of the functional parameters. The total analysis must answer questions at two levels. First, what forces add individuals to groups and subtract them, and what magnitude of these forces must operate to create the observed frequency distributions? Second, to what extent has natural selection shaped responses to the forces, or even moderated the forces themselves? In proceeding from the first level to the next, the analysis will shift from phenomenological to fundamental theory.

The phenomenological theory has been largely developed by Joel E. Cohen (1969a, 1971). Cohen took his inspiration from earlier, partially successful attempts by sociologists to fit the size-frequency distributions of human groups to a Poisson distribution with the zero term (that is, frequency of groups with no members) eliminated. Groups were defined as clusters of people laughing, smiling, talking, working, or engaging in other activities indicating face-to-face interaction. The populations studied included pedestrians on city streets, masses of shoppers in department stores, and elementary school children at play. When the numbers of little clusters containing one person, two persons, and upward are counted, they fit in most instances a Poisson distribution with a truncated zero term. Cohen extended this approach to Old World monkeys. But he went deeper into the basic problem by deriving frequency distributions from a stochastic model in which groups of varying size and composition possess varying abilities to attract and to hold temporary members. Three parameters were entered: a is the rate (per unit of time) at which a single individual in a system of freely forming groups joins a group solely because of the attractions of group membership, independently of the size of any particular group; b is the rate at which the lone individual joins a group because of the attractiveness of individuals in the group, where the degree of attractiveness of the group can therefore be expected to change with the number of members in it; and d is the rate at which an individual member already in a group departs because of some personal decision of its own that is independent of the size of the group. Consider a closed population of individuals that are freely forming into casual groups containing variable numbers of individuals. The number of groups containing a certain number of members is designated as n_i, where $i(= 1, 2, 3 \ldots)$ represents the number of members. In the simplest kind of social system the rate of change in the number of casual groups of a certain size is conjectured to be

$$\frac{dn_i(t)}{dt} = an_{i-1}(t) + b(i-1)n_{i-1}(t) - an_i(t) - b(i)n_i(t) - d(i)n_i(t) + d(i+1)n_{i+1}(t)$$

This formula states that in a short interval of time the number of groups of size i is being *increased* at time t by:

1. The number of groups one member less in size $(i - 1)$ times the rate (a) at which individuals join groups independently of their membership; this increases a given group of size $i - 1$ to size i. *Plus*
2. The number of groups one member less in size $(i - 1)$ times the rate $[b(i - 1)]$ at which individuals join groups of that size owing to the attractiveness of individuals in it. *Plus*
3. The number of groups one member more in size $(i + 1)$ times the rate $[d(i + 1)]$ at which individuals spontaneously leave groups of that size; this decreases a given group of size $i + 1$ to size i.

The number of groups of size i is being simultaneously *decreased* by:

1. The number of groups already at size i times the rate (a) at which individuals are attracted to groups regardless of membership; this increases the number of members from i to $i + 1$. *Plus*
2. The number of groups already at size i times the rate $[b(i)]$ at which individuals are attracted to groups of size i owing to membership. *Plus*
3. The number of groups already at size i times the rate of spontaneous departure from groups of that size $[d(i)$.]

A second equation is simultaneously entered for the rate of increase of solitary individuals in the system. The basic model can be made more complex, as Cohen (1971) has shown, by adding terms that include other increments of attraction and expulsion as functions of group size and composition.

The single most important result of Cohen's basic model is the demonstration that at equilibrium ($dn_i = 0$ for all i), the frequency distribution of casual group size in a closed population should be a zero-truncated Poisson distribution when $b = 0$—in other words, when the specific membership or size of a group does not influence its attractiveness—and a zero-truncated binomial distribution when b is a positive number. Existing data from several primate species, including man, conform reasonably well to one or the other of the two distributions. Cohen has further demonstrated that estimated values of the ratios a/d and b/d are a species characteristic. As shown in Table 6-1, a general decrease in the role of individual attractiveness is apparent when one passes from the more elementary to more advanced social groups. Whether this rule will hold in larger samples remains to be seen. The important point is that a great many previously jumbled numerical data have been put into preliminary order in a surprisingly simple way. Hope has thus been engendered

Table 6-1 Values of the ratios of attraction rates (a, b) to the spontaneous departure rate (d) in groups of two monkey species and man, estimated from the basic Cohen model of casual group formation.

Group	a/d	b/d
Vervets (*Cercopithecus aethiops*)	1.15	0.66
Baboons (*Papio cynocephalus*)	0.12	0.16
Nursery children	0.33	0.10
Miscellaneous human groups	0.86	0

that at least one of the coarser qualities of sociality, group size, can be fully derived from models that specify as their first principles the forms and magnitude of individual interactions.

In addition to the *casual societies*, or casual groups, just considered, there exist *demographic societies*. The difference between the two is only a matter of duration in time, but its consequences are fundamental. The casual group forms and dissipates too quickly for birth and death rates to affect its statistical properties; immigration and emigration into and out of the population as a whole are also insignificant. The demographic society, in contrast, is far more nearly closed than the casual group, and it persists for long enough periods of time for birth, death, and migration between demes to play leading roles. One way in which a population can exist at both levels is for a more or less closed society to exist demographically while the membership of casual groups within it changes kaleidoscopically on a shorter time scale. In a separate modeling effort, Cohen (1969b) showed that when members of demographic societies are born, die, and migrate from one society to another at positive rates not dependent on the size of the group to which they belong, the frequency distribution of societies with varying numbers of members can be expected to approximate the negative binomial distribution with the zero term truncated. If, on the other hand, the individual birth rate is temporarily zero, or the number of offspring born in each group per unit of time is constant regardless of the size of the group, the frequency distribution should approach a zero-truncated Poisson distribution. These predictions appear to be well borne out by existing data from primate field studies. Langur and baboon troops, in which the demographic parameters are more or less independent of group size, conform to a negative binomial distribution. Gibbon troops are societies in which only one infant is born at a time regardless of group size, a fact which causes the individual birth rate to be a decreasing function of the number of members; group size in this case is Poisson distributed. During healthy periods, howler monkey troops fit negative binomial distributions, but following an epidemic in which young were temporarily eliminated, their size-frequency distribution shifted to the Poisson—as anticipated. It is a curious fact that although the form of the frequency distributions is correctly predicted by the most elementary stochastic model that incorporates demography, the dynamics of the model are not faithful to the single detailed set of demographic data (from yellow baboons) that were available to Cohen. In other words, the internal structure of the model must be made more complex in some way that cannot yet be guessed.

We can now turn to the evolutionary origins of group size by treating the entire subject in terms of the following argument. The immediately determining parameters are themselves adaptations on the part of individual organisms. The attractiveness of a group to a solitary animal is ultimately determined by the relative advantage of joining the group, measured by the gain in inclusive genetic fitness. Whether the organism attempts to migrate from one semiclosed demographic society to another is also under the direct sovereignty of natural selection. The birth rate, as shown in Chapter 4, is another parameter very sensitive to selection, because not only is it a key component of reproductive fitness, but it also contributes—negatively—to the survival rate of the parents. Of all the parameters determining group size, only the death rate can be said to escape classification as a direct adaptation to the environment.

We can postulate that the modal size of groups will be simply the outcome of the interaction of the parameter values that confer maximum inclusive fitness. In all social species the modal group size will therefore represent a compromise. The size must be greater than one because of the advantages of group foraging, or group defense, or any one of the combination of the "prime movers" of social evolution reviewed in Chapter 3. But it cannot be indefinitely large, since beyond a certain number the food runs out, or the defense can no longer be coordinated effectively, and so forth. The upper limits of group size are unfortunately much more difficult to discern in field studies than are the initial advantages favoring sociality at lower numbers. We can only speculate, for example, about the disadvantages of excessive size in fish schools. The food supply must ultimately be limiting, of course. As the schools grow larger their energy demands increase directly with the volume occupied by the fish, but the rate of energy acquisition increases with the outer surface of the fish school. Energy requirements, in other words, increase with the cube of the school's diameter, and energy input with its square—a disparity analogous to the weight-surface law of organismic growth. There are other potential disadvantages of large size. By clumping into schools, fish are found less often by predators. If schools become very large, however, there is a strong incentive for predators to track them continuously and to develop special orientation and other behavioral devices for staying close.

Because group size appears to be ultimately the result of evolutionary compromise, it is most efficiently analyzed by optimization theory. In Figures 6-2 and 6-3 are shown two graphical models to illustrate this approach; their curves hypothesize the general form of the functions. The first, inspired by a proposition about group territoriality by Crook (1972), assumes an exclusive or at least overwhelming role for the energy budget. In this extreme case, foraging by small groups is more effective in energy yielded per individual animal per unit of time than is solitary foraging within populations of equal density. Crook argued, correctly I believe, that although the energy requirements of the society increase linearly with the number of its members, the amount of territory that the group can effectively defend decelerates after a certain point. If defensible territory is translated into energy yield, it becomes clear that a maximum group size exists, above which demand exceeds the yield and either mortality or emigration must redress the balance. When the population as a whole is limited by energy, as opposed to some other density-dependent factor such as predation, group

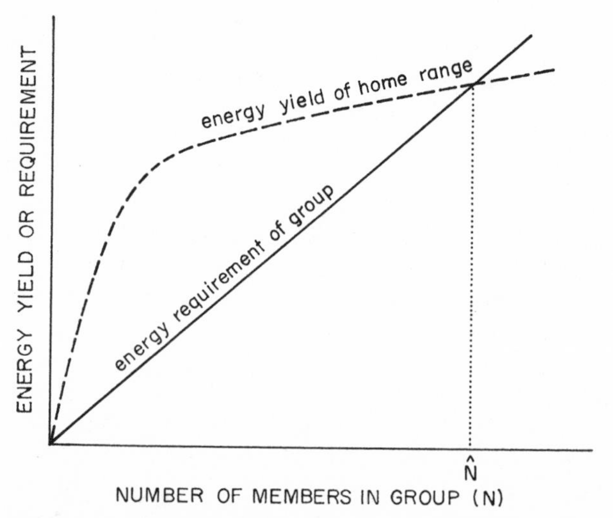

Figure 6-2 The optimization of group size is represented in this extreme model as an exclusive function of the energy budget. As group size increases, the energy requirement increases at the same rate, but the energy yield decelerates after an initial rapid rise. If unopposed by other selection pressures, the modal group size N should change in evolution to a point where the energy yield of the home range is fully utilized.

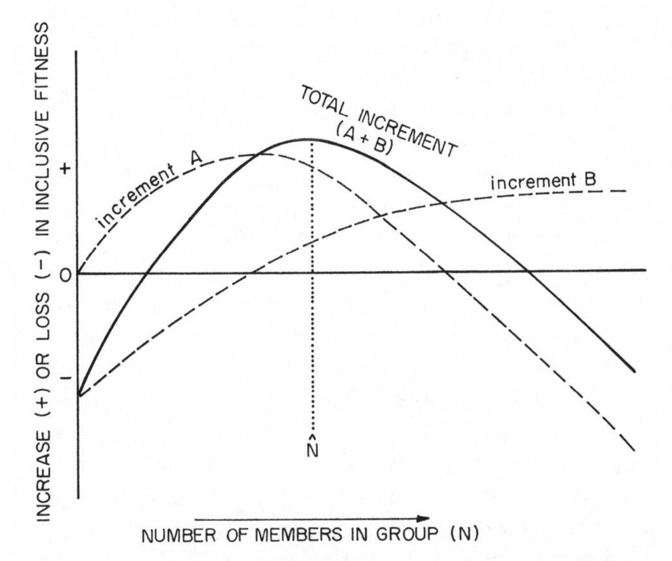

Figure 6-3 In this more general model, the optimum group size is given as a function of the maximum summed components of genetic fitness. Two social contributors to fitness are indicated as *A* and *B*; they could be, for example, increments due to superior group foraging and superior group defense against predators.

size will tend to evolve toward the maximum. There will also be a tendency for the group to become territorial in behavior. The more stable the environment, and the more evenly distributed the food in time and space, the more nearly the group size will approach the theoretical maximum. In a capricious environment, however, the optimal group size will normally be less than the theoretical maximum. The reason is that the energy yield of a home range measured over a long period of time is based on intervals of both superabundance and scarcity. The group must be small enough to survive the more prolonged periods of scarcity.

The energy-budget model takes into account only the components of fitness added by group superiority in territorial defense and foraging techniques. A more general modeling effort, which can encompass all of the components of genetic fitness, is presented in elementary form in Figure 6-3. Three ideas are incorporated into this more complicated graph: all components of fitness enhanced by group activity must inevitably decline beyond a certain group size; the increment curves, that is, contributions to fitness as a function of group size, usually differ from one component to another; and the optimum group size is that at which the sum of group-related increments in fitness is largest. This figure is no more than a representation of postulates; the data for drawing the fitness curves represented in it do not yet exist.

Adjustable Group Size

The significance of the casual society, as opposed to the demographic, lies in its adjustable size. The number of animals can be fitted to the needs and opportunities of the moment. The total breeding population consists of a constantly moving set of nuclear units—the individual animals, the family, the colony, or whatever—that band together temporarily to form larger aggregates of variable size. The aggregates can be passive, consisting of nuclear units that relax their mutual aversion temporarily to utilize a common resource, or they can actively cooperate to achieve some common goal. The goal of the casual society may be any activity that promotes inclusive fitness, from feeding to defense or hibernation.

Excellent examples of casual groups are provided by what Kummer (1971) has called the fusion-fission societies of the higher primates. The nuclear unit of the hamadryas baboon society is the harem, consisting of a dominant male, his females and offspring, and often an apprentice male with a developing harem. In the evening the harems aggregate at the sleeping cliffs, where they are relatively well protected from leopards and other predators and where, in fact, the baboons are in a position to cooperate by constituting a more efficient alarm and defense system than individual harems could provide. In the mornings the sleeping groups separate into smaller foraging bands or individual harems that proceed separately to the feeding grounds. There exists a clear relation between the amount of resource and the size of such foraging groups. At Danakil, Ethiopia, isolated acacia trees, whose flowers and beans are the baboons' main food source, are gleaned by individual harems. The density of baboons in each tree is consequently such that each harem member is able to keep the usual feeding distance of several meters between it and other members. As a result the movements of subordinate animals are unimpeded by the dominants. In groves of ten or more acacias, the feeding group is the entire band, consisting of at least several harems whose mutual tolerance and attraction are unusually high. Again, density is low enough to preserve individual distance, and the feeding efficiency of each animal remains adequate. During the dry season water becomes the critical resource. River ponds are kilometers apart, and each is visited by aggregations of as many as a hundred or more baboons.

Chimpanzees are organized into still more flexible societies. Casual groups of very variable size form, break up, and re-form with ease, apparently in direct response to the availability of food (Reynolds, 1965). Chimpanzees must search for locally abundant food that is irregular in space and time. The ability to disperse and to assemble rapidly is clearly adaptive; chimps even use special calls to recruit others to rich food finds. This strategy may be contrasted with that of gorillas. These great apes feed to a large extent on the leaves and shoots of plants, a relatively evenly distributed food resource. Gorilla societies are semiclosed and demographic in composition, and they patrol regular but broadly overlapping home ranges.

Some primitive human societies that depend on hunting and gathering of patchily distributed resources form casual societies not unlike the chimpanzee model (Lee, 1968; Turnbull, 1968; Sugiyama, 1972). The nuclear unit of the !Kung people of the Kalahari is the family or a small number of families. Units band together in camps for periods of two to several weeks, during which time most of the game captured by the men and the crop of nuts and other vegetable food gathered by the women and children are shared equally. The Mbuti pygmies of the Congo Forest have a much less organized society. The nuclear unit is more nearly the individual, rather than the family. Pygmies move about the forest according to the distribution of game, honey, fruit, and other kinds of patchily distributed foods. Groups form and divide in a very loose manner to make the fullest use of food discoveries.

The Multiplication and Reconstitution of Societies

Relatively few observations have been made of the division and internal changes in animal societies through demographic time. The taxa which are best understood are the mammals, particularly the primates, and the social insects. A variety of multiplication procedures is used by both, some of which are similar and represent convergent evolution. In general, the societies of both taxa are matrifocal, and as a consequence societal division depends on the willingness of breeding females to form fresh associations with males and to move to new locations. At the same time, mammalian societies differ from insect societies in three basic details with respect to their multiplication and internal construction: they are genetically less uniform, they usually if not invariably divide as a result of aggressive interactions among the members or with invading outsiders, and the timing and behavioral responses of their emigrations are far less rigidly programmed.

In Old World monkey societies, which have been studied with exceptional care by Japanese and Americans during the past 20 years, male aggression is the initial impetus that leads to group reorganization. Disruption is caused by one or the other of three forms of interactions: the rise of young males within the hierarchy, the attractive power of solitary bachelors outside the troop, or invasion by bands of bachelors. During Sugiyama's (1967) field study of the langur *Presbytis entellus* at Dharwar, India, troops underwent an important reorganization on the average of once every 27 months. *P. entellus* troops consist of either groups of adult females and juveniles ruled over by a single male, or bands of bachelors. In one instance a group of 7 males attacked and displaced the resident male. Fighting then erupted among the usurpers, until 6 were ousted and only one remained in control. Two other changes directly observed by Sugiyama ended in troop division. Once a solitary male attacked and defeated the resident male, then decamped with all the members of the troop except one adult female, who remained behind with her old consort. In another instance, a large band of 60 or more bachelor males repeatedly attacked several bisexual troops, forcing the resident males to retreat temporarily. During the fighting, small groups of females joined the bachelor band, which eventually moved into a new territory. Finally, true to the despotic nature of langur society, all of the males except the dominant individual deserted, leaving him in control of the females. A common feature of the various divisions and reorganizations was the intolerance of the new leaders toward the offspring of the former resident males, leading in some cases to infanticide by biting. In general, the juvenile populations soon came to consist entirely of the offspring of the new tyrants.

Macaques have more stable communities. Changes occur less frequently than in the langur troops, and they are normally precipitated by events within the societies rather than by the invasion of outsiders. Troops of Japanese monkeys (*Macaca fuscata*) divide when subgroups of females and their offspring gradually drift away from the main troop, visiting the feeding areas at different periods and staying outside the influence of the dominant male. Under such circumstances they become associated with subordinate adult males who have also left the troop and live in solitude or in association with other expatriate males from the same original troop. The new troops then organize themselves into the mild dominance system that characterizes the Japanese species (Sugiyama, 1960; Furuya, 1963; Mizuhara, 1964). Warfare does not appear to occur between the bachelor groups and the established bands.

The details of group fission differ strikingly from one mammalian species to the next. Mech (1970) has marshaled persuasive indirect evidence from his own data and those of Adolf Murie to suggest that new wolf packs are founded by an adult breeding pair who mate and leave the mother group. The new pack is soon enlarged by the first litter of about six pups. The young wolves then remain with the parents through at least the following winter while growing in size and acquiring competence in hunting. The black-tail prairie dog, a colonial rodent, has a radically different process of group fission (King, 1955). Burrows are occupied communally by coteries consisting of as many as two males and five females with their offspring. During the breeding season the females possessing young pups close off part of the burrow and defend it from other coterie members. The other adults, together with the yearlings, then construct new burrow systems nearby. The prairie dog "towns" are also extended by adults, who appear to be repelled by the incessant grooming demands of the juveniles. This pattern is the opposite of that of other mammals, including other rodents, in which it is the juveniles who form the bulk of the emigrants.

Insect societies are overall as diverse in their modes of reproduction as the mammalian societies. One of their most generally distinctive features is nuptial flights. The males and virgin queens depart from the nests at certain hours of the day set by circadian rhythms. The timing varies among the species: mid-morning for some species, late afternoon for others, midnight or the predawn hours for still others, and so on around the clock. The queens are able to attract the males quickly. The sexual forms of the more abundant species mingle in nuptial swarms that form over conspicuous environmental features such as treetops and open fields. After being inseminated by one to several males, each queen drops to the ground, sheds her wings, and runs about in search of a suitable nest site. If she is one of the tiny minority that escapes death from predators and hostile colonies of her own species, she constructs a brood chamber and lays a batch of eggs. The first adults to emerge in the brood are small workers, all sterile females, who immediately assist their mother in rearing sisters and putting the colony on a firmer footing. The males play no part in this effort. Whether they have participated in mating or not, they soon separate from the nuptial groups and wander about alone, destined to die within hours from accidents or the attacks of predators. During the early stages of their growth, the new colonies produce only workers. After a period normally requiring one to several years, males and virgin queens begin to appear just before the season of nuptial flights. Colonies that generate new queens are said to be "mature," in the sense that they are now able to reproduce themselves directly.

In general, the study of group size and multiplication is in a very early and incomplete stage. As more information becomes available from field studies of societal life cycles, we can expect to see the data related in novel and exciting ways to the basic theory of population biology.

CHAPTER 7

The Development and Modification of Social Behavior

Social behavior, like all other forms of biological response, is a set of devices for tracking changes in the environment. No organism is ever perfectly adapted. Nearly all the relevant parameters of its environment shift constantly. Some of the changes are periodic and predictable, such as the light-dark cycles and the seasons. But most are episodic and capricious, including fluctuations in the number of food items, nest sites, and predators, random alterations of temperature and rainfall within seasons, and others. The organism must track these parts of its environment with some precision, yet it can never hope to respond correctly to every one of the multifactorial twists and turns—only to come close enough to survive for a little while and to reproduce as well as most. The difficulty is exacerbated by the fact that the parameters change at different rates and often according to independent patterns. In each season, for example, a plant contends with irregularities in humidity on a daily basis, while over decades or centuries its species as a whole must adapt to a steadily increasing or decreasing average annual rainfall. An aphid has to thwart predators that vary widely in abundance from day to day, while over many years, the aphid species faces changes not only in the abundance but also in the species composition of its enemies.

Organisms solve the problem with an immensely complex multiple-level tracking system. At the cellular level, perturbations are damped and homeostasis maintained by biochemical reactions that commonly take place in less than a second. Processes of cell growth and division, some of them developmental and some merely stabilizing in effect, require up to several orders of magnitude more time. Higher organismic tracking devices, including social behavior, require anywhere from a fraction of a second to a generation or slightly more for completion. Figure 7-1 suggests how organismic responses can be classified according to the time required. All the responses together form an ascending hierarchy. That is, slower changes reset the schedules of the faster responses. For instance, a shift into a more advanced stage of the life cycle brings with it new programs of behavior and physiological responses, and the release of a hormone alters the readiness to react to a given stimulus with learned or instinctive behavior. In both cases, the slower response alters the potential of the faster one.

Even more profound changes occur at the level of entire populations during periods longer than a generation. In ecological time populations wax or wane, and their age structures shift, in reaction to environmental conditions. These are the demographic responses indicated by the middle curve of Figure 7-1. Ecological time is so slow that large sequences of organismic responses occur within it, few of which affect the outcome single-handedly. But ecological time is also generally too quick to bracket extensive evolutionary change. When the observation period is prolonged still further to about ten or more generations, the population begins to respond perceptibly by evolution. Long-term shifts in the environment permit certain genotypes to prevail over others, and the genetic composition of the population moves perceptibly to a better adapted statistical mode. The hierarchical nature of the tracking system is preserved, since the newly prevailing genotypes are likely to have different demographic parameters from those prevailing earlier, as well as different physiological and behavioral response curves. The time intervals are now spoken of as being evolutionary in scale—long enough to encompass many demographic episodes, so long, in fact, that separate events at the organismic level are reduced to insignificance.

The concept of the multiple-level, hierarchical tracking system will be expanded in the remainder of this chapter to provide a clearer picture of social behavior as a form of adaptation. The account begins at

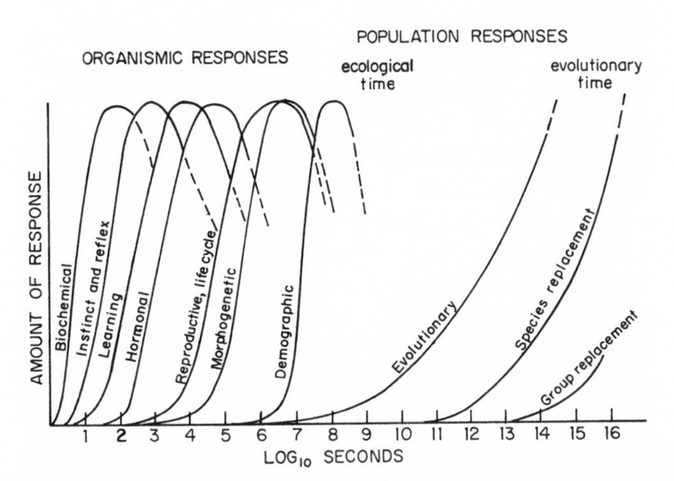

Figure 7-1 The full hierarchy of biological responses. Organismic responses are evoked by changes in the environment detectable within a life span, population responses to long-term trends. The hierarchy ascends with an increase in the response time; that is, any given response tends to alter the pattern of the faster responses. Beyond evolutionary responses are replacements of one species by another or even entire groups of related species by other such groups. The particular response curves shown here are imaginary.

the evolutionary time scale and works downward through the hierarchy to learning, play, and socialization. The important point to keep in mind is that such phenomena as the hormonal mediation of behavior, the ontogenetic development of behavior, and motivation, although sometimes treated in virtual isolation as the proper objects of entire disciplines, or else loosely connected under the rubric of "developmental aspects of behavior," are really only sets of adaptations keyed to environmental changes of different durations. They are not fundamental properties or organisms around which the species must shape its biology, in the sense that the chemistry of histone or the geometry of the cell membrane can be so described. The phenomena cannot be generally explained by searching for limiting features in the adrenal cortex, vertebrate midbrain, or other controlling organs, for the reason that these organs have themselves evolved to serve the requirements of special multiple tracking systems possessed by particular species.

Tracking the Environment with Evolutionary Change

All social traits of all species are capable of a significant amount of rapid evolution beginning at any time. This statement may seem exaggerated at first, but as a tentative generalization it is fully justified by the facts. All that the potential for immediate evolution requires is *heritability*—inherited variability—within populations. Moderate degrees of heritability have been demonstrated in the widest conceivable array of characteristics, including crowding and dominance ability in chickens, visual courtship displays in doves, size and dispersion of mouse groups, degree of closure of dog packs, dispersal tendency in milkweed bugs, and many other parameters in vertebrates and insects. One extensive research program devoted to the genetics of social behavior of dogs uncovered significant amounts of heritability in virtually every trait subjected to analysis (Scott and Fuller, 1965).

It is possible to proceed from information on heritability to a prediction of evolutionary rate. One of the remarkable results of the elementary theory of population genetics is that with only a moderate intensity of natural selection—say, a 50 percent advantage of one genotype over another at the same locus—one gene can be largely substituted for another in a population within ten generations or less. And in fact, geneticists, starting with behaviorally neutral stocks, have created lines of *Drosophila* whose adult flies orient toward or away from gravity and light. In only ten generations Ayala (1968) was able to achieve a doubling of the equilibrial adult population size within *Drosophila serrata* confined to overcrowded bottles. He then discovered that the result had come about at least in part because of a quick shift to strains in which the adults are quieter in disposition and thus less easily knocked over and trapped in the sticky culture medium (Ayala, personal communication).

Equally rapid behavioral evolution has been achieved in rodents through artificial selection, although the genetic basis is still unknown owing to the greater technical difficulties involved in mammalian genetics. Examples of the traits affected included running behavior in mazes, defecation and urination rates under stress, fighting ability, tendency of rats to kill mice, and tameness toward human observers (Parsons, 1967). Behavior can also evolve in laboratory rodent populations in the absence of artificial selection. When Harris (1952) provided laboratory-raised prairie deer mice (*Peromyscus maniculatus bairdii*), a form whose natural habitat is grassland, with both simulated grassland and forested habitats in the laboratory, the mice chose the grassland. This response indicated the presence of a genetic component of habitat choice inherited from their immediate ancestors. Ten years and 12–20 generations later, however, the laboratory descendants of Harris' mice had lost this unaided tendency to choose the habitat (Wecker, 1963). If first exposed to grassland habitat, they later chose it, as expected. But if they were exposed to woodland first, they later failed to show preference for either habitat. These results indicate that a predisposition remained to select the ancestral habitat, although the short period of evolution had weakened it considerably.

Clearly, there is every justification from both genetic theory and experiments on animal species to postulate that rapid behavioral evolution is at least a possibility, and that it can ramify to transform every aspect of social organization. The crucial independent parameters nevertheless remain the intensity and persistence of natural selection. If one or the other is very low, or if the selection is stabilizing, significant evolutionary change will consume far more time than the theoretical minimum of ten generations. Conceivably, millions of generations might be needed. Thus while theory and laboratory experimentation have established the maximum possible rate of behavioral evolution, it will be necessary to study free-living populations to find out how far reality falls below that rate.

The Hierarchy of Organismic Responses

Scanning downward through the hierarchy of tracking devices, from evolutionary and morphogenetic change to the increasingly sophisticated degrees of learning, one encounters a steady rise in the specificity and precision of the response. A genetic or pervasive anatomical alteration is scarcely perfectible at all. It is final in the sense that the organism has made its choice and must live or die by it, with further change being left to later generations. Short-term learning, in contrast, can be shaped to very fine particularities in the environment—the direction of a flash of light or the strength of a gust of wind—and augmented or discarded quickly as new circumstances dictate. At this level of maximum precision in behavioral adaptation, the organism can remake itself many times over during its lifetime.

A clear trend in the evolution of the organismic hierarchies is the increasingly fine adjustments made by larger organisms. Above a certain size, multicellular animals can assemble enough neurons to program a complex repertory of instinctive responses. They can also engage in more advanced forms of learning and add an endocrine system of sufficient complexity to regulate the onset and intensity of many of the behavioral acts.

Species from amebas to man can be classified into evolutionary grades according to the length of the response hierarchy and the degree of concentration of power in the lower, more finely tuned responses. It would be both premature and out of place to attempt a formal classification here. Yet to support my main argument, let me at least suggest three roughly defined grades, one at the very bottom, one at the top, and one approximately midway between. The paradigms used are species, although the particular traits characterizing them define the grades in accordance with the usual practice of phylogenetic analysis:

Lowest grade: the complete instinct-reflex machine. The representative organism is so simply constructed that it must depend largely or wholly on token stimuli from the environment to guide it. Perhaps a negative phototaxis keeps it always in the darkness, a circadian rhythm makes it most active just before dawn, a shrinking photoperiod causes it to encyst in the fall, the odor of a certain polypeptide attracts it to prey and induces engorgement, an epoxy terpenoid identifies the presence of a mate and causes it to shed gametes, and so forth—in fact, with this short list we have come close to exhausting its repertory. Endowed with no more than a nerve net or simple central nerve cord containing on the order of hundreds or thousands of neurons, our organism has virtually no leeway in the responses it can make. It is like a cheaply constructed servomechanism: all its components are committed to the performance of the minimal set of essential responses. Possibly no real species exactly fits such a description, but the type is at least approached by sponges, coelenterates, acoel flatworms, and many other of the most simply constructed lower invertebrates.

Middle grade: the directed learner. The organism has a fully elaborated central nervous system with a brain of moderate size, containing on the order of 10^5 to 10^8 neurons. As in the organisms belonging to the lowest grade, some of the behavior is stereotyped, wholly programmed, dependent on unconditioned sign stimuli, and species-specific. A moderate amount of learning occurs, but it is typically narrow in scope and limited in responsiveness to a narrow range of stimuli. It results in behavior as stereotyped as the most neurally programmed "instinct." The level of responsiveness may be strongly influenced by the hormone titers, which themselves are adjusted to a sparse set of cues received from the environment. The true advance that defines this intermediate evolutionary grade is the capacity to handle *particularity* in the environment. Depending on the species, the organism may be able to identify not just a female of the species, but also its mother; not only can it gravitate to the kind of habitat for which its species is adapted, but it can remember particular places as well, and will regard one area as its personal home range; it not only can hide but may retreat to a refugium, the location of which it has memorized; and so forth. Examples of this evolutionary grade are found among the more intelligent arthropods, such as lobsters and honeybee workers, the cephalopods, the cold-blooded vertebrates, and the birds.

Highest grade: the generalized learner. The organism has a brain large enough to carry a wide range of memories, some of which possess only a low probability of ever proving useful. Insight learning may be performed, yielding the capacity to generalize from one pattern to another and to juxtapose patterns in ways that are adaptively useful. Few if any complex behaviors are wholly programmed morphogenetically at the neural level. Among vertebrates at least, the endocrine system still affects response thresholds, but since most behaviors have been shaped by complex episodes of learning and are strongly dependent on the context in which stimuli are received, the role of hormones varies greatly from moment to moment and from individual to individual. The process of socialization in this highest grade of organism is prolonged and complex. Its details vary greatly among individuals. The key social feature of the grade, which is represented by man, the chimpanzee, baboons, macaques, and perhaps some other Old World primates and social Canidae (see Chapters 24 and 25), is a *perception of history.* The organism's knowledge is not limited to particular individuals and places with attractive or aversive associations. It also remembers relationships and incidents through time, and it can engineer improvements in its social status by relatively sophisticated choices of threat, conciliation, and alliances. It seems to be able to project mentally into the future, and in a few, extreme cases deliberate deception is practiced.

The remainder of this chapter will complete the examination of the hierarchy of environmental tracking devices, commencing with morphogenesis and caste formation and proceeding downward to the most precise forms of learning and cultural transmission.

Tracking the Environment with Morphogenetic Change

The most drastic response to fluctuations in the environment short of genetic change itself is the modification of body form. Many phyletic lines of invertebrates have adopted this strategy. In principle, the genome is altered to increase its plasticity of expression. Two or more morphological types, which also normally differ in physiological and behavioral traits, are available to the developing organism. Acting on token stimuli that indicate the overall condition of the environment, the organism "chooses" the type into which it will transform itself. Thus, developing *Brachionus* rotifers grow long spines when they detect the odor of predaceous rotifers belonging to the genus *Asplancha.* The new armament prevents them from being consumed. For their part, *Asplancha* (specifically, *A. sieboldi*) can respond to the stimulus of cannibalism and supplementary vitamin E by growing into a gigantic form capable of consuming larger prey. The giant is only one of three distinct morphological types in which the species exists (Gilbert, 1966, 1973). Aphids of many species develop wings when the onset of crowded conditions is signaled by increased tactile stimulation from neighbors. Given the power of flight, these insects are free to depart in search of uncrowded host plants. As populations of plague locusts grow dense, making contacts among the individual hoppers more frequent, they pass from the solitary to the gregarious phase. The transformation takes place over three generations.

The most elaborate forms of morphogenetic response are the caste systems of the social insects and the colonial invertebrates. With rare exceptions the caste into which an immature animal develops is based not on possession of a different set of genes but solely on receipt of such environmental stimuli as the presence or absence of pheromones from other colony members, the amount and quality of food received at critical growth periods, the ambient temperature, and the photoperiod prevailing during critical growth periods. The proportions of individuals shunted into the various castes are adaptive with respect to the survival and reproduction of the colony as a whole. Caste systems will be discussed in greater detail in Chapter 14.

Hormones and Behavior

Elaborate endocrine systems have evolved in two principal groups of animals, the phylum Arthropoda, including particularly the insects, and the phylum Chordata, including particularly the vertebrates. Since these two taxonomic groups also represent end-points in the two great branches of animal phylogeny, namely, the arthropod and echinoderm-chordate superphyla, their endocrine systems can safely be said to have evolved wholly independently. There are basic differences not only in structure and biochemistry but also in function. Arthropod hormones serve to mediate the events of growth, metamorphosis, and ovarian development. Their role in behavior appears to be limited to the stimulation of the production of pheromones and the indirect regulation of reproductive behavior through their influence on gonadal development. Vertebrate hormones have a much wider repertory. They help to regulate numerous purely physiological events, including growth, development, metabolism, and ionic balance. They also exercise profound effects on sexual and aggressive behavior, subjects that will be considered later in Chapters 11 and 15.

At this point there is a need only to draw two broad generalizations about the relation between hormones and behavior in vertebrates. The first is that the function of hormones is to "prime" the animal. Hormones affect the intensity of its drives, or to use a more neutral and professionally approved expression, the level of its motivational states. In addition, they directly alter other physiological processes and large sectors of the behavioral repertory of animals. However, they are relatively crude as controls. Their effects cannot be quickly turned on or off. They track medium-range fluctuations in the environment, such as the seasonal changes made predictable by steady increases or decreases in the daily photoperiod, the stress of extreme cold or threat by a predator, and the presence of a potential mate as signaled by releaser sounds, odors, or other stimuli. An animal cannot guide its actions or make second-by-second decisions through the employment of hormones. It must rely on quicker, more direct cues to provide a finer tuning of motivational states and to trigger specific actions. The second generalization is the intimate relationship, revealed by new techniques in microsurgery and histochemistry during the past 20 years, that exists between the behaviorally potent hormones and specific blocks of cells in the central nervous systems.

Both of these features of hormone-behavioral interaction are well illustrated by the role of estrogen in the sexual behavior of female cats. An estrous female responds to the approach of a male by crouching, elevating her rump, deflecting her tail sidewise to expose the vulva, and pawing the ground with treading movements of her hind legs. She readily submits to being mounted. If not in estrus, she instead reacts aggressively to the close approach of a male. It is well known that estrus is initiated by the rise of the estrogen titer of the blood. But in what way does estrogen prime the animal for sexual behavior? Not, it turns out, by the estrogen-mediated growth of the reproductive tract. When castrated females are injected repeatedly with small doses of estrogen over a long period of time, the reproductive tract develops completely, yet sexual behavior is still not induced (Michael and Scott, 1964). The female sexual response depends on a more direct action of the hormone. When needles tipped in slowly dissolving estrogen are inserted into certain parts of the hypothalamus, the castrated cats display typical estrous behavior, even though their reproductive tracts remain undeveloped (Harris and Michael, 1964). Michael (1966) also discovered that radioactively labeled estrogen injected into the bloodstream is preferentially absorbed by neurons in just those areas of the hypothalamus most sensitive to direct applications of the hormones by needle.

To about the same degree that hormones control some aspects of behavior, behavior controls the release of hormones. The signals exchanged by members of the same species frequently act not only to induce overt behavioral responses in others, but also to prime their physiology. Once modified in this way, the recipient animal responds to further signals with an altered behavioral repertory. The courtship of ring doves, for instance, depends on an exact marching order of hormones timed by the perception of external signals. When a pair is placed together in a cage, the male begins to court immediately. He is the initiator because his testes are active and probably secreting testosterone. He faces the female and repeatedly bows and coos. The sight of the displaying male activates mechanisms in the female's brain, which in turn instruct the pituitary gland to release gonadotropins. These hormones stimulate the growth of the female's ovaries, which begin to manufacture eggs and to release estrogen into the bloodstream. The essential steps are thus concluded for the successful initiation of nest building and mating (Lehrman, 1964, 1965).

The release of reproductive hormones into the bloodstream of female mice is also sensitive to signals from other members of the same species. In the manner of the medical sciences, the different kinds of physiological change are often called after their discoverers:

1. *Bruce Effect.* Exposure of a recently impregnated mouse female to a male with an odor sufficiently different from that of her stud results in failure of the implantation and rapid return to estrus. The adaptive advantage to the new male is obvious, but it is less easy to see why it is advantageous to the female and therefore how the response could have been evolved by direct natural selection.

2. *Lee-Boot Effect.* When about four or more female mice are grouped together in the absence of a male, estrus is suppressed and pseudopregnancies develop in as many as 61 percent of the individuals. The adaptive significance of the phenomenon is unclear, but it is evidently one of the devices responsible for the well-known reduction of population growth under conditions of high population density.

3. *Ropartz Effect.* The odor of other mice alone causes the adrenal glands of individual mice to grow heavier and to increase their production of corticosteroids; the result is a decrease in reproductive capacity of the animal. Here we have part but surely

not all of the explanation of the well-known stress syndrome. Some ecologists have invoked the syndrome as the explanation of population fluctuation, including the occasional "crashes" of overly dense populations of snowshoe hares and possibly some other small mammals.

4. *Whitten Effect.* An odorant found in the urine of male mice induces and accelerates the estrous cycle of the female. The effect is most readily observed in females whose cycles have been suppressed by grouping; the introduction of a male then initiates their cycles more or less simultaneously, and estrus follows in three or four days.

Until the pheromones are identified chemically, the number of signals involved in the various effects cannot be known with certainity. Bronson (1971) believes that as few as three substances can account for all the observed physiological changes: an estrus-inducer, an estrus-inhibitor, and an adrenocortical activator.

Learning

The Directedness of Learning

Viewed in a certain way, the phenomenon of learning creates a major paradox. It seems to be a negating force in evolution. How can learning evolve? Unless some Lamarckist process is at work, individual acts of learning cannot be transmitted to offspring. If learning is a generalized process whereby each brain is stamped afresh by experience, the role of natural selection must be solely to keep the tabula rasa of the brain clean and malleable. To the degree that learning is paramount in the repertory of a species, behavior cannot evolve. This paradox has been resolved in the writings of Niko Tinbergen, Peter Marler, Sherwood Washburn, Hans Kummer, and others. What evolves is the directedness of learning —the relative ease with which certain associations are made and acts are learned, and others bypassed even in the face of strong reinforcement. Pavlov was simply wrong when he postulated that "any natural phenomena chosen at will may be converted into conditioned stimuli." Only small parts of the brain resemble a tabula rasa; this is true even for human beings. The remainder is more like an exposed negative waiting to be dipped into developer fluid. This being the case, learning also serves as a pacemaker of evolution. When exploratory behavior leads one or a few animals to a breakthrough enhancing survival and reproduction, the capacity for that kind of exploratory behavior and the imitation of the successful act are favored by natural selection. The enabling portions of the anatomy, particularly the brain, will then be perfected by evolution. The process can lead to greater stereotypy—"instinct" formation—of the successful new behavior. A caterpillar accidentally captured by a fly-eating sphecid wasp might be the first step toward the evolution of a species whose searching behavior is directed preferentially at caterpillars. Or, more rarely, the learned act can produce higher intelligence. As Washburn has said, a human mind can easily guide a chimpanzee to a level of performance that lies well beyond the normal behavior of the species. In both species, the wasp and man, the structure of the brain has been biased in special ways to exploit opportunities in the environment.

The documentation of the directed quality of learning has been extensive. Consider the laboratory rat, often treated by experimental psychologists in the past as if it were a tabula rasa. Garcia et al. (1968) found that when rats are made ill from x-rays at the time they eat food pellets (and not given any other unpleasant stimulus), they subsequently remember the flavor but not the size of the pellets. If they are negatively reinforced by a painful electric shock while eating (and not treated with x-rays), they remember the size of the pellet associated with the unpleasant stimulus but not the flavor. These results are not so surprising when considered in the context of the adaptiveness of rat behavior. Since flavor is a result of the chemical composition of the food, it is advantageous for the rat to associate flavor with the after-effects of ingestion. Garcia and his coworkers point to the fact that the brain is evidently wired to this end: both the gustatory and the visceral receptors send fibers that converge in the nucleus of the fasciculus solitarius. Other sensory systems do not feed fibers directly into this nucleus. The tendency to associate size with immediate pain is equally plausible. The cues are visual, and they permit the rat to avoid such dangerous objects as a poisonous insect or the seed pod of a nettle before contact is made.

Very young animals display an especially sharp mosaic of learning abilities. The newborn kitten is blind, barely able to crawl on its stomach, and generally helpless. Nevertheless, in the several narrow categories in which it must perform in order to survive, it shows an advanced ability to learn and perform. Using olfactory cues, it learns in less than one day to crawl short distances to the spot where it can expect to find the nursing mother. With the aid of either olfactory or tactile stimuli it memorizes the route along the mother's belly to its own preferred nipple. In laboratory tests it quickly comes to tell one artificial nipple from another by only moderate differences in texture (Rosenblatt, 1972). Still other examples of constraints on learning are reviewed by Shettleworth (1972).

The process of learning is not a basic trait that gradually emerges with the evolution of larger brain size. Rather, it is a diverse array of peculiar behavioral adaptations, many of which have been evolved repeatedly and independently in different major animal taxa. In attempting to classify these phenomena, comparative psychologists have conceived categories that range from the most simple to the most complex. They have coincidentally provided a rank ordering of phenomena according to the qualities of flexibility in behavior, its precision, and its capacity for tracking increasingly more detailed changes in the environment. Excellent recent reviews of this rapidly growing branch of science have been provided by Hinde (1970), P. P. G. Bateson (1966), and Immelmann (1972).

The Ontogeny of Bird Song

The songs by which male birds advertise their territories and court females are particularly favorable for the analysis of learning and other processes of behavioral development. The songs are typically complex in structure and differ strongly at the level of the species. Considerable variation among individual birds also exists, some of it subject to easy modification by laboratory manipulation. Following the pioneering work of W. H. Thorpe, who began his studies in the early 1950's, biologists have investigated every aspect of the phenomenon from its neurological and endocrine basis to its role in speciation. This advance has been made possible by a single technical breakthrough—the sound spectrograph, by which songs can be recorded, dissected into their components, and analyzed quantitatively. Perhaps the single most important result has been the demonstration of the programmed nature of learning in the ontogeny of song, a lock-step relation that exists between particular stimuli, particular acts of learning, and the short sensitive periods in which they can be linked to produce normal communication. Complete reviews have been provided by Hinde and his coauthors (Hinde, ed., 1969; Hinde, 1970) and by Marler and Mundinger (1971).

One of the more discerning studies has been conducted on the white-crowned sparrow *Zonotrichia leucophrys* of North America. The male song consists of a plaintive whistle pitched at 3 to 4 kiloherz, followed by a series of trills or *chillip* notes. Many variations occur, particularly in the form of "dialects" that distinguish geographic populations. Under normal circumstances full song develops when the birds are 200 to 250 days old, but Marler and Tamura (1964) showed that this capacity is present much earlier. Young birds captured at an age of one to three months in the area where they were born and kept in acoustical isolation later sang the song in the dialect of their region. Others removed from the nest at 3 to 14 days of age and raised by hand in isolation also developed a song; it possessed some though not all of the basic simplified structure characteristic of the species as a whole and had none of the distinctive features of the regional dialect. Evidently, then, the dialect is learned from the adult birds during rearing and before the young birds themselves attempt any form of song. Hand-raised sparrows will sing the dialect of their region or another region if taped songs of wild birds are played to them from the age of about two weeks to two months. Thus the species-specific skeleton of the song seems fully innate in the looser usage of that term, while the population-specific overlay is acquired by tradition. It turns out, however, that even the skeleton requires some elements of learning, albeit highly directed in character and virtually unalterable under normal conditions. Konishi (1965) found that when birds are taken from the nest and deafened by removal of the cochlea, they can produce only a series of unconnected notes when they attempt to sing. This remains true when the birds have been trained by exposure to adult calls. In order to put together a normal call, even the skeletal arrangement of the species, it is essential for the white-crowned sparrows to hear themselves as they sing the elements previously learned.

The infiltration of learning into the evolution of bird song introduces a closer fit of the individual's repertory to its particular environment. Learning permits the immediate satisfaction of communicative needs without recourse to the tedious process of selection over several generations. An individual bird achieves its vocal niche quickly in a complex environment of sound. As a result it can distinguish a potential mate of the same species from among a confusing array of related species. Where regional dialects and their recognition are based to some extent on adult learning, the bird can utilize familiarity with old neighbors to eliminate unnecessary hostile behavior. In the case of convergent duet singing, the bird can perfect communication with its mate and reduce the chance of being distracted by other members of the species.

The Relative Importance of Learning

The slow phylogenetic ascent from highly programmed to flexible behavior is nowhere more clearly delineated than in the evolution of sexual behavior. The center of copulatory control in male insects is in the ganglia of the abdomen. The role of the brain is primarily inhibitory, with the input of sexual pheromones and other signals serving to disinhibit the male and guide him to the female. The total removal of the brain of a male insect—chopping off the head will sometimes do—triggers copulatory movements by the abdomen. Thus a male mantis continues to mate even after his cannibalistic mate has eaten away his head. Entomologists have used the principle to force matings of butterflies and ants in the laboratory. The female is lightly anesthetized to keep her calm, the male is beheaded, and the abdominal tips of the two are touched together until the rhythmically moving male genitalia achieve copulation. A similar control over oviposition is invested in the abdominal ganglia of female insects. The severed abdomens of gravid female dragonflies and moths can expel their eggs in a nearly normal fashion.

The sexual behavior of vertebrates differs from that of insects in being controlled almost wholly by the brain, particularly regions of the cerebral neocortex. Furthermore, there exists within the vertebrates as a whole a correlation between the relative size of the brain—a crude indicator of general intelligence (Rumbaugh, 1970)—and the dependence of male sexual behavior on the cerebral neocortex and social experience. As much as 20 percent of the cortex of male laboratory rats can be removed without visible impairment of their sexual performance. When 50 percent is removed, more than one-fifth of the animals still mate normally. In male cats, however, extensive bilateral injury to the frontal cortex alone causes gross abnormalities of sensorimotor adjustment. The animals display signs of intense sexual excitement in the presence of estrous females, but they are usually unable to make the body movements necessary for successful intromission. Higher primates, particularly chimpanzees and man, have prolonged, personalized sexual behavior which is even more vulnerable to cortical injury (Beach, 1940, 1964). The importance of social experience in sexual practice also increases with brain size, while the effectiveness of hormones in initiating or preventing it declines.

Socialization

Socialization is the sum total of social experiences that alter the development of an individual. It consists of processes that encompass most levels of organismic responses. The term and the set of diffuse ideas that enshroud it orginated in the social sciences (Clausen, ed., 1968; Williams, 1972) and have begun gradually to penetrate biology. In psychology socialization ordinarily means the acquisition of basic social traits, in anthropology the transmission of culture, and in sociology the training of infants and children for future social performance. Margaret Mead (1963), recognizing the different levels of organismic response implicit in the phenomenon, suggested that a distinction be made between true socialization, the development of those patterns of social behavior basic to every normal human being, and enculturation, the act of learning one culture in all its uniqueness and particularity. Vertebrate behaviorists who use the word *socialization* usually have in mind only the learning process (Poirier, 1972a), but if comparative studies are to be made in all groups of animals, its definition will have to embrace all the range of socially induced responses that occur within the lifetime of one individual. If that proposition is accepted, the following three categories can be recognized:

1. Morphogenetic socialization, for example caste determination
2. Learning of species-characteristic behavior
3. Enculturation

Socialization has resisted deep analysis because of two imposing difficulties encountered by both social scientists and zoologists. The first is the considerable technical problem of distinguishing behavioral elements and combinations that emerge by maturation, that is, unfold gradually by neuromuscular development independently of learning, and those that are shaped at least to some extent by learning. Where both processes contribute, their relative importance under natural conditions is extraordinarily difficult to estimate. The second major problem is, of course, the complexity and fragility of the social environment itself.

In spite of this, experimental research has now been pursued to the point that a few interesting generalizations are beginning to emerge. As expected, the form of socialization is roughly correlated among species with the size and complexity of the brain and the degree of involvement of learning. Members of colonies of lower invertebrates and social insects are socialized principally by the physiological and behavioral events that determine their caste during early development. The specialized zooids of colonial coelenterates and bryozoans may be established solely by morphogenetic change imposed by their physical location among other zooids. Although the development of "social behavior" has not been analyzed in these animals, the visible responses are so elementary and stereotyped that learning seems unlikely to play any important role. Caste determination in social insects is achieved mainly through the physiological influence of adult colony members on the developing individual. Often, as in some ant species, it is a matter of the amount (and perhaps quality) of food given to the larva. In the honeybees, the quality of food is paramount, depending on the presence or absence of certain unidentified elements in the royal jelly, which is fed to a few larvae sequestered in royal cells. In termites, inhibitory substances (pheromones) produced by the kings and queens force the development of the great majority of nymphs into one of the sterile worker castes.

Once a social insect has matured into a particular caste, it launches into the complex repertory of behaviors peculiar to that form. In the 10 days after a typical honeybee worker emerges as an adult winged insect from the pupa, she engages expertly in a wide variety of tasks that include at least some of the following: polishing and cleaning cells in the honeycomb and brood area, constructing new hexagonal cells out of wax to a precision of a tenth of a millimeter or better, attending the queen, flying outside the hive, ripening nectar into honey and storing it, feeding and grooming larvae, fanning on the comb to aid in thermoregulation, conducting and following waggle dances, and regurgitating with other workers. At 30 days the worker is old, her repertory largely behind her, and she has only a few more days of service as a forager left.

The role of learning in this brief but remarkable career has never been investigated, but is must be narrowly directed and stereotyped at best. We know that honeybees learn the odor of nestmates and the location of their hive and food sources. Tasks can be memorized and performed in a sequence, including the often complicated schedules of visits to flowers at specific times of the day. Isolated worker bees can be trained to walk through relatively complex mazes, taking as many as five turns in sequence in response to such clues as the distance between two spots, the color of a marker, and the angle of a turn in the maze. After associating a given color once with a reward of 2-molar sucrose solution, they are able to remember the color for at least two weeks. The location of a food site in the field can be remembered for a period of six to eight days; on one occasion bees were observed dancing out the location of a site following two months of winter confinement. Nevertheless, these feats become less impressive when it is realized how narrowly and immediately functional they really are. Like the minor song dialects of finches, honeybee learning represents lesser variation that overlays basic behavior patterns that either develop regardless of experience or else are learned along strict channels during brief sensitive periods (Lindauer, 1970).

Socialization has been much more intensively studied in primates than in other kinds of animals. The circumstance is fortunate for two reasons: the phylogenetic affinity of the Old World monkeys and apes to man plus the fact that socialization by learning appears to be deepest and most elaborate in these animals. Before describing the actual process as it is now understood, it is useful to outline the techniques of experimentation in a way that attempts to reflect the philosophy of the biologists who have conducted it. Their approach can be understood more readily if we draw an analogy between socialization and the biology of vitamins. In the evolution of a given species, a nutrient compound becomes a vitamin when it is so readily available in the normal diet that members of the species no longer need to synthesize it from simple components. In obedience to the principle of metabolic conservation, the species then tends to eliminate the biochemical steps required for the synthesis of the substance, thus permitting enzymatic protein and energy to be diverted for other, more urgent functions. At this point the molecule becomes "essential"—that is, a vitamin—in the sense that it must be included in the diet thereafter in order for the organism to thrive. Vitamin D, which regulates the absorption of calcium from the intestine by influencing membrane permeability or active transport, is a sterol produced from other sterols by irradiation with ultraviolet light. The human body obtains it easily, either in the diet or by transformation of dietary sterols. The existence of such vitamins can be discovered by systematically withholding suspected compounds from completely defined synthetic diets. Their role can be ascertained by a thorough study of physiology of vitamin-starved animals. Important additional effects, sometimes harmful and sometimes beneficial, can be induced by enriching the diet with abnormally large amounts.

Through an analogous form of evolutionary decay, behavioral elements involved in socialization become increasingly dependent on experience for normal development. These elements are most easily identified by noting their reduction or disappearance when various forms of normal social experience are withheld. This is the method of *environmental deprivation*. Sometimes the same or additional elements can be discovered and characterized in part by increasing the amount of stimuli above the normal laboratory level and observing modifications in the opposite direction. This is the method of *environmental enrichment*.

Studies of socialization in primates, particularly in the rhesus monkeys as the most favored species, have relied heavily on the technique of environmental deprivation. The rather involved results of these studies can be better understood if we order the experiments according to the amount of deprivation imposed, and hence the number and degree of perturbations they usually produce. The following list of experiences proceeds from the most to the least drastic.

Descending Degrees of Social Deprivation

1. The young monkey is raised by an artificial mother made of cloth and denied its real mother, peers, and all other social partners until maturity.
2. The young monkey is kept with its natural mother for part or all of its development but is not allowed contact with other monkeys until maturity.
3. The young animal is separated from its mother but allowed to associate with other monkeys of the same age.
4. The young animal is raised by its natural mother in the midst of a troop but is temporarily separated as an infant for short periods of time.
5. The young animal is raised with a normal social group but in a restricted laboratory environment as opposed to a natural or seminatural habitat.
6. The young animal is reared with its natural mother in as normal and complete a social setting as possible. The developmental schedule provides a control for the deprivation experiments. But differences among individuals inevitably arise owing to variation in heredity, social rank of the mother, illnesses and other events during development, and other uncontrolled circumstances. By careful clinical studies of individual case histories, considerable insight can be gained into the relative importance and interaction of the factors, although the system is too complex to permit quantitative assessments such as parameter estimates in multiple regression analyses.

The story of socialization in monkeys and apes revealed by these studies is one of a gradual release of the young animal from the bosom of its mother into the increasingly chancy social milieu of the surrounding troop. Day by day, week by week, the infant reduces the amount of time it spends asleep or attached to its mother's nipple while lengthening the duration of its tentative explorations away from her and increasing the number of contacts made with other members of the troop.

Poirier (1972b), while conceding the continuous nature of social development, has suggested that it can be conveniently divided into four arbitrarily defined periods. In the first, the *neonatal period,* the animal is a helpless infant, limited to the ingestion of milk and forms of locomotion that hold it close to the mother's body. Contact with the mother is continuous and close. In the *transition period* the infantile movements are supplemented by adult locomotor and feeding patterns. The animal is still closely associated with the mother but leaves her for increasingly long periods to play and to feed on its own. For most primate species the transition period lasts for several months, ending when the company of the mother is no longer frequently sought. The monkey now passes into the time of *peer socialization,* when much of the contact is with group members other than the mother. The most frequently sought peers are the mother's prior offspring, older females, and other youngsters of about the same age. The key events of this period are the completion of weaning and the gradual subsidence of the infantile behavior patterns. Finally, the animal enters the *juvenile-subadult period,* during which the infantile patterns disappear entirely and adult patterns, including sexual behavior, are first practiced. Females reach full adult status sooner than males, and both sexes mature more quickly in short-lived species than in long-lived species.

The spreading nexus of relationships that characterizes the later periods of socialization has been analyzed by van Lawick-Goodall (1968a), Burton (1972), Rowell (1972), and Hinde (1974). Only a few species have been studied in sufficient detail for long enough periods to draw firm conclusions. These include anubis baboons, macaques (Barbary, bonnet, Japanese, rhesus, and pigtail), vervets, and chimpanzees. The first contacts made by the youngster beyond its mother are normally with the maternal siblings. Even exceptionally restrictive, fearful mothers allow their infants to be approached by her older children, and sisters and half-sisters are the most favored of all. In the olive baboon, only the young females concern themselves with infants, while juvenile and subadult males restrict themselves to other young males in these age classes. Sibling relationships often endure into adulthood and form a principal basis for grooming partnerships and cliques. When rhesus males migrate to new troops, they sometimes penetrate these host societies by joining forces with brothers who have preceded them.

Young baboons and macaques begin to interact with peers when they are about six weeks old. They wander away from their mothers for long intervals and spend most of their waking time at play with other infants and juveniles. Both siblings and nonsiblings are now included in the widening circle of acquaintanceship. Virtually all of the components of aggressive and sexual behavior make their first appearance and are then strengthened and perfected by frequent practice. Initially, the behavior patterns are almost always expressed in the context of play, and as such they consist of improperly connected, nonfunctional fragments. Later, during the juvenile-subadult period, these fragments are linked together to form the full repertories of serious aggressive and sexual behavior. The choice of playmates and the relationships forged with them during play often persist into adulthood and hence are crucial to the future social status of the individual. As Rowell has pointed out, the tense, formal relationships between the adult males of macaque troops do not arise simply in vacuo by random interactions at puberty. They grow gradually from the existing relationships of the juvenile males that seem at first to be relaxed and playful.

The younger the animal, the more traumatic is the effect of a given type of social deprivation. Thus isolation for six months will irreparably damage the social capacity of an infant monkey or ape, but it will have only a minor, temporary effect on a mature male. Furthermore, the greater the deprivation, the deeper and more enduring the result. Total deprivation is almost wholly crippling to the infant's development; it can be partly erased by permitting the infant access to peers for short but frequent intervals. Provided it lives with a normally constituted social group, a monkey raised in a restricted laboratory environment differs from a feral animal in only quantitative ways, such as the time required to achieve normal sexual behavior. Both of these principles have been well documented by the results of the many deprivation experiments that have been performed on rhesus monkeys. The results are briefly summarized in Figure 7-2.

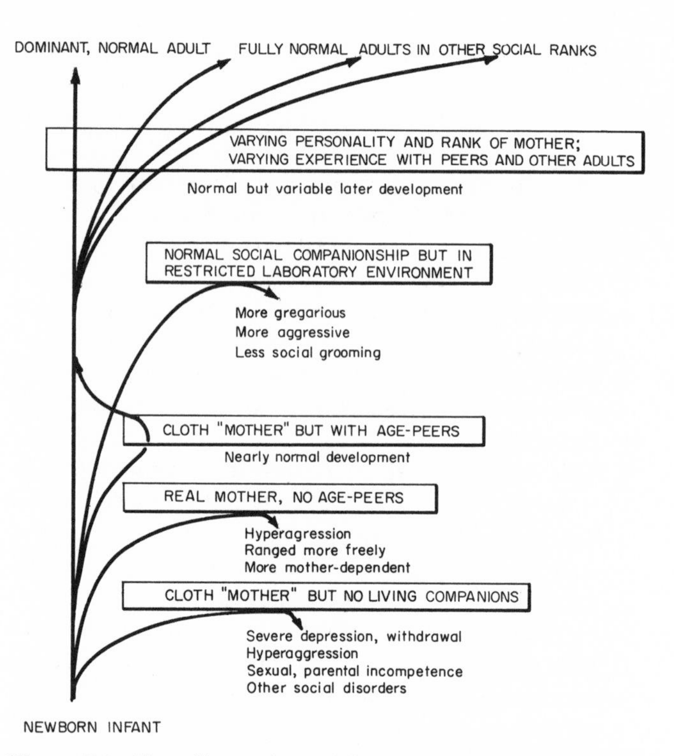

Figure 7-2 The effects of social deprivation on behavior in the rhesus monkey.

The trauma of extreme deprivation was first clearly revealed in Harry F. Harlow's famous experiments on "mother love" and other aspects of socialization in the rhesus monkey. Infants were removed from their mothers and given a choice of two crude substitute models, one constructed of wire and the other of terry cloth. The young animals strongly preferred the cloth model, which they hugged and clutched much of the time. The softness of the material proved crucial; the models were accepted even when they were supplied with huge round eyes made of bicycle lamps and bizarre faces that made them seem more like toys or gargoyles than real monkeys. The young rhesus monkeys physically thrived when they were supplied with milk in ordinary baby bottles attached to the front of the dummies. In fact, it seemed at first that a superior substitute had been found for real mothers. The cloth models never moved, never rebuked, and were an absolutely dependable supply of food. But as the monkeys grew up and were permitted to join other monkeys, their social behavior proved abnormal, to such an extent that comparable deviations in human societies would be ruled psychotic. They were sometimes hyperaggressive and sometimes autistic; in the latter state they sat withdrawn while rocking silently back and forth. They also cried a great deal and sucked their own fingers and toes. As Harlow and W. A. Mason showed later, the males grew up sexually incompetent. They tried to copulate with estrous females but were unable to assume the normal position, sometimes mounting the female from the side and sometimes thrusting against her back above the tail. Females proved equally abnormal. In estrus they refused to be mounted. When "raped" by experienced males, they bore infants but badly mistreated them, stepping on them and rejecting their attempts to nurse. Some of the infants managed to survive by persistence and ingenuity, but others had to be removed to save their lives. The degree of abnormality increased with the duration of the deprivation. Isolation for three months was followed by depression, but the monkeys recovered with no marked deviations. Isolation for six months resulted in extensive permanent damage and isolation for a year caused virtually total impairment.

Play

Play, virtually all zoologists agree, serves an important role in the socialization of mammals. Furthermore, the more intelligent and social the species, the more elaborate the play. These two key propositions having been stated, we must now face the question: What is play? No behavioral concept has proved more ill-defined, elusive, controversial, and even unfashionable. Largely from our personal experience, we know intuitively that play is a set of pleasurable activities, frequently but not always social in nature, that imitate the serious activities of life without consummating serious goals. Vince Lombardi, the great coach of the Green Bay Packers, was once dismissed by a critic of football as someone who taught men how to play a boy's game. That was unfair. Human beings are in fact so devoted to play that they professionalize it, permitting the lionized few who turn it into serious business to grow rich.

The question before us then is to what extent animal play is also serious business. In other words, how can we define it biologically? Robert M. Fagen (1974) has pointed out that most of the confusion about play stems from the existence of two wholly different orientations in general writings on the subject. On the one hand are the *structuralists*, who are concerned only with the form, appearance, and physiology of play. Structuralists, such as Fraser Darling (1937), Caroline Loizos (1966, 1967), and Corinne Hutt (1966), define play as any activity that is exaggerated or discrepant, divertive, oriented, marked by novel motor patterns or combinations of such patterns, and that appears to the observer to have no immediate function. *Functionalists*, on the other hand, define play as any behavior that involves probing, manipulation, experimentation, learning, and the control of one's own body as well as the behavior of others, and that also, essentially, serves the function of developing and perfecting future adaptive responses to the physical and social environment. For the functionalist, the wars of England were indeed won on the playing fields of Eton.

Whether the functionalist hypothesis is incorporated into the definition or not, play can be usefully distinguished from pure *exploration*. To explore is to learn about a new object or a strange part of the environment. To play is to move the body and to manipulate portions of a known object or environment in novel ways. As Hutt says, the goal of exploration is getting to know the properties of the new object, and the particular responses of investigation are determined by its properties. True play proceeds only within a known environment, and it is largely manipulative in quality. In passing from exploration to play, the animal or child changes its emphasis from "What does this object do?" to "What can I do with this object?" Play can also be separated from pure problem solving, especially when the latter has a simply functional goal and does not entail the pleasurable learning of rules and variations. Jerome Bruner (1968) has attempted to capture this distinction in the following epigram: play means altering the goal to suit the means at hand, whereas problem solving means altering the means to meet the requirements of a fixed goal.

Robert Fagen (1977) has pursued the functionalist interpretation with a mathematical model of the costs and benefits of play throughout the life cycle. In this formulation play commonly entails an immediate cost in fitness due to such functions as the useless expenditure of energy, the increased vulnerability to watching predators, the risking of dangerous episodes with adults, and so forth. But fitness at *later* life stages is enhanced by the experience and the improved status that play confers. In more exact form, the proposition says that the $l_y m_y$ values are lowered at age $y = x$, when the play takes place, but they are raised at some age beyond x as a result. The play schedule—the intensities of play programmed for each age y—will evolve in such a way that the summed gains in $l_y m_y$ will exceed the summed losses over all ages y, that is, throughout the potential life span. By experimentally inserting a priori values of losses and gains into a numerical model, Fagen proved that play can be eliminated entirely from the behavioral repertory by natural selection. The additional constraints on its evolution are as follows. The amount of play can decrease monotonically from birth, peak unimodally at some later age, or even peak bimodally at two later ages. But if play exists at all, it is present at age 0. In more realistic terms we can translate this last result to read that animals belonging to a playful species can be expected to start playing as soon as they have developed coordinated movements of their bodies and limbs. Also, under a wide range of conditions, play will be most prominent at a relatively early age.

Play appears to be strictly limited to the higher vertebrates. No case has been documented in the social insects (Wilson, 1971a), and the phenomenon must be very scarce or altogether absent throughout the remainder of the invertebrates. To my knowledge no example has been documented in the cold-blooded vertebrates, including the fishes, amphibians, and reptiles. The sole dubious exception is the Komodo dragon *Varanus komodoensis*, the world's largest lizard. Craven Hill (1946) reported that a large individual in the London Zoo "played" repeatedly with a shovel by pushing it noisily over the stony floor of its cage. This behavior might just as well be interpreted as a redirection of foraging movements, in which logs or other objects are pushed aside in the search for prey. On the basis of Hill's single anecdote play cannot be said to have been conclusively demonstrated in reptiles. Unequivocal play behavior has been reported in a few species of birds, and appears to be especially well developed in crows, ravens, jackdaws, and other members of the family Corvidae, which are noted for their relatively high intelligence and unspecialized behavior. Hand-reared ravens (*Corvus corax*) display several patterns that the structuralists would classify as play, including repeated episodes of hanging from horizontal ropes by one leg while performing acrobatic stances with the head and free leg (Gwinner, 1966). Finally, play, directed at both social companions and inanimate objects, occurs virtually throughout the mammals, from fruit bats to chimpanzees.

As this phylogenetic distribution alone suggests, play is associated with a large brain complex, generalized behavior, and, most especially, a large role for learning in the development of behavior. The play activities of a kitten, an animal not very high on this scale, are direct and to the point. Much of it consists of mock-aggressive rushes and rough-and-tumble play with the mother and other kittens, clearly portending the territorial and dominance aggression of adult life. The prolonged and elaborate patterns—the ones that make kittens so fascinating to watch—are the forerunners of the three basic hunting maneuvers of the adult cat. When the kitten spots the trailing end of a string, it slithers along the floor, tail twitching lightly, and suddenly pounces to press the string down to the ground with its claws. These are close to the exact motions by which an adult catches a mouse or some other small ground-dwelling animal. When a string is dangled in the air—sometimes even a mote of dust dancing in a beam of sunlight will do—the kitten chases it as an adult cat does a bird taking flight. Springing upward, it spreads its paws out and claps them together to seize the object in midair. Kittens also stand over objects and scoop them upward and to one side with a sweep of the paw. This last maneuver is quite possibly a rehearsal of the technique used later to capture small fish.

Not unexpectedly, the animal species indulging in the most elaborate and free-ranging forms of play is the chimpanzee, the most intelligent of the anthropoid apes and man's closest living relative (van Lawick-Goodall, 1968a). Sessions are initiated by one or the other of two special invitation signals: the play-walk, in which the chimpanzee hunches its back into a rounded form, pulls its head slightly down and back between the shoulders, and takes small, stilted steps; and the play-face, a distinctive expression assumed by opening the mouth in a form neither aggressive nor fearful while partially or wholly exposing the teeth. In addition to conventional forms of chasing, mounting, and rough-and-tumble play, young chimps inprovise a variety of games, including finger wrestling and "stealing" sprays of vegetation from one another.

A notable characteristic of animal play is the freedom with which behavioral elements are concatenated. The elements themselves can be well defined and are more or less consistent in form; they may even be closely similar to the serious adult behaviors they foreshadow. But the sequence in which they are put together is very variable and idiosyncratic—one might even say whimsical. It is possible that this trait of looseness is vital to the very process of environmental tracking itself. Play is the means by which the most appropriate combinations are identified, reinforced, and hence established as the future adult repertory.

Among the higher mammals, where it is most free-ranging, play loosens the behavioral repertory in each generation and provides the individual with opportunities to depart from the traditions of its family and society. Like sexual reproduction and learning generally, it is evidently one of the very broad adaptive devices sustained by second-order natural selection. At its most potent, in human beings and in a select group of other higher primates that includes the Japanese macaques and chimpanzees, playful behavior has led to invention and cultural transmission of novel methods of exploiting the environment. It is a fact worrisome to moralists that Americans and other culturally advanced peoples continue to devote large amounts of their time to coarse forms of entertainment. They delight in mounting giant inedible fish on their living room walls, idolize boxing champions, and sometimes attain ecstasy at football games. Such behavior is probably not decadent. It could be as psychologically needed and genetically adaptive as work and sexual reproduction, and may even stem from the same emotional processes that impel our highest impulses toward scientific, literary, and artistic creation.

Tradition, Culture, and Invention

The ultimate refinement in environmental tracking is tradition, the creation of specific forms of behavior that are passed from generation to generation by learning. Tradition possesses a unique combination of qualities that accelerates its effectiveness as it grows in richness. It can be initiated, or altered, by a single successful individual; it can spread quickly, sometimes in less than a generation, through an entire society or population; and it is cumulative. True tradition is precise in application and often pertains to specific places and even successions of individuals. Consequently, families, societies, and populations can quickly diverge from one another in their traditions, the phenomenon defined in Chapter 2 as "tradition drift." The highest form of tradition, by whatever criterion we choose to judge it, is of course human culture. But culture, aside from its involvement with language, which is truly unique, differs from animal tradition only in degree.

Some dialects in animal communication are learned, and to that extent they represent an elementary form of tradition. Local populations become differentiated through tradition drift, with such potentially adaptive consequences as the greater compatibility of mated pairs and enhanced efficiency of communication between the male and female, and the partial reproductive isolation of each deme distinguished by a dialect, which preserves the closeness of fit of its gene pool to the particular conditions of its local environment. All other factors being equal or negligible, the average geographic range of the dialects of a species diminishes as the behavioral plasticity of the species increases. The dialect ranges of the Indian hill mynah (*Gracula religiosa*), which is strongly imitative and has an unusually plastic call, do not extend beyond 17 kilometers (Bertram, 1970). Dialect formation based at least in part on learning between generations is widespread among species of songbirds (Thielcke, 1969). Geographic variation in vocalizations has been reported in several kinds of mammals, including pikas, pothead whales, and squirrel monkeys. Its significance is unknown, since the variation might be based on genetic differences and thus not constitute true tradition.

Most tradition in the best-investigated animals is concerned with *Ortstreue*—fidelity to place—a German phrase for which there is no precise English word equivalent. *Ortstreue* is the tendency of individuals to return to the places used by their ancestors in order to reproduce, to feed, or simply to rest. Its most striking manifestation is in the fixed migration routes of birds and mammals. Each year ducks, geese, and swans migrate hundreds or thousands of kilometers, following the same traditional flyways, stopping at the same resting places, and ending at the same breeding and overwintering sites. Since these birds fly in flocks of mixed ages, there is abundant opportunity for the young to learn the travel route from their elders. The greater the fidelity shown by the birds to the flyways, the less the gene flow between local breeding populations, and consequently the stronger the geographic variation within the species (Hochbaum, 1955).

Also traditional are the game trails used by mammals. Those of deer persist for generations, and where they follow paths smoothed into rocks, perhaps for centuries. Galápagos tortoises (*Geochelone elephantopus*) follow established trails during their annual migrations. At the beginning of the rainy season they descend from the moist habitats and waterholes of the highlands to lower elevations to feed and to lay eggs. Later they climb back to the highland refuges. In some places the tortoise trails run for kilometers and require days to follow, and they have almost certainly lasted for generations (Van Denburgh, 1914).

The breeding grounds of colonial birds are also traditional. Wynne-Edwards (1962) has called attention to the Viking names of certain of the British Isles based on the kinds of birds that nested there in the eighth to tenth centuries. The names are still appropriate: for example, Lundy, "isle of puffins," and Sulisgeir, "gannets' rock." Even nests and roosts can be passed from one generation to the next. Osprey nests and the mud workings of swallows sometimes persist for decades. The lodges of muskrats and beavers last for at least several generations, while a few earthen dens of the European badger are said to be centuries old (Neal, 1948).

Among the higher primates, traditions sometimes shift in a qualitative manner. Poirier (1969) observed changes in diet and foraging behavior in langurs (*Presbytis johnii*) of southern India after the monkeys' environment had been changed by human activity. One troop was forced into a new area when the habitat of its original home range was destroyed. Subsequently it altered its diet from *Acacia* to *Litasae* and *Loranthus*. Other troops have begun to shift to *Eucalyptus globulus*, an Australian tree that is being deliberately planted in place of the natural woodlands favored by the langurs. Although the adults are reluctant to eat anything but the leaf petioles of these aromatic trees, the infants sometimes consume the entire leaf. Poirier predicts that ultimately entire troops will incorporate eucalyptus as a principal food plant. Elsewhere, langurs are in the process of accommodating the encroachment of agriculture. In the Nilgiri area of India, potatoes and cauliflower were introduced not more than 100 years ago and are gradually replacing the natural woodland. The langurs come out of their refuges in the remaining forest patches to raid the crops. Not only do they feed on the vegetables in quantity, but they have learned to dig into the soil with their hands and to pull up the entire plants—a behavior pattern not yet seen in other langur troops.

The most carefully documented case histories of invention and tradition in primates have come from studies of the Japanese macaque *Macaca fuscata*. Since 1950 biologists of the Japanese Monkey Center have kept careful records of the histories of individuals in wild troops located at several places: Takasakiyama, near the northern end of Kyushu; Koshima, a small island off the east coast of Kyushu; and Minoo and Ohirayama on Honshu. More casual studies have been made at still other localities. At an early stage the Japanese scientists encountered differences between troops in the traditions of food-gathering behavior. The monkeys at Minoo Ravine had learned how to dig out roots of plants with their hands, while those at Takasakiyama, although living in a similar habitat, apparently never used the technique. The population at Syodosima regularly invaded rice paddies to feed on the plants, but the troops at Takagoyama were never observed to do so, despite the fact that they had lived for many years in hills surrounded by paddies and occasionally passed through them during their nomadic wanderings (Kawamura, 1963).

When the biologists offered new foods to the monkeys, they directly observed both dietary extensions and the means by which these changes are transmitted through imitation. At Takasakiyama caramels were accepted readily by monkeys under three years of age, and candy eating then spread rapidly through this age class. Mothers picked up the habit from the juveniles and passed it on to their own infants. A few adult males most closely associated with infants and juveniles eventually accepted caramels also. Propagation of the habit was most rapid among young animals, and slowest among the subadult males who were farthest removed socially from the young and their parents. After 18 months, 51.2 percent of the troop had been converted to candy eating (Itani, 1958).

The scientists who summarized the early findings, including Kinji Imanishi (1958, 1963) and Syunzo Kawamura (1963), spoke of the macaque society as a "subhuman culture" or "preculture" and the dietary shifts as enculturation. If these terms were at all justified, they became much more so by the remarkable series of events witnessed about this time in a single troop on Koshima Island. Starting in 1952, the biologists began to scatter sweet potatoes on the beach in an attempt to supplement the diet of the monkeys. The troop then ventured out of the forest to accept the gift, and in so doing it extended its activities to an entirely new habitat. The following year Kawamura (1954) observed the beginnings of a new behavior pattern associated with this habitat shift: some of the monkeys were washing sand off the potatoes by employing one hand to brush the sand away and the other to dip the potato into water. This and other subsequent behavioral changes were followed in detail during the ensuing ten years by Masao Kawai, who summarized the history of the population in 1965.

Potato washing was invented by a 2-year-old female named Imo. Within ten years the habit had been acquired by 90 percent of the troop members in all age classes, except for infants a year old or less and adults older then 12 years. During the same period, the washing was transferred from the fresh water of the brook to the salt water of the sea. The behavior was most readily learned by juveniles between 1 and 2½ years old, Imo's own age class. By 1958, five years after Imo invented it, potato washing was practiced by 80 percent of monkeys from 2 to 7 years in age. Older monkeys remained conservative; only 18 percent, all of them females, learned the behavior.

In 1955 Imo, the monkey genius, invented another food-gathering technique. The biologists had originally given wheat to the Koshima troop simply by scattering it onto the beach. The monkeys were then required to pick out the grains singly from among the particles of sand. Imo, now four years old, somehow learned to scoop handfuls of the mixed sand and wheat, carry them to the edge of the sea, and cast the mixture onto the water surface. When the sand sank, the lighter wheat grains were skimmed off the surface and eaten. The pattern by which this new tradition spread through the troop resembled that for sweet-potato washing. Juveniles passively taught their mothers and age-peers, and mothers their infants, but adult males largely resisted learning the technique. One important difference emerged, however: unlike potato washing, which spread most rapidly among monkeys one to two and a half years in age, wheat flotation was picked up most efficiently by members of the two-to four-year-old class, to which Imo herself belonged. The explanation of this difference in performance may lie in the relative complexity of the two tasks. Potato washing is only a slight modification of the procedure the macaques routinely follow when they pick up tubers and fruits from the ground with one hand and brush off dirt with the other. But the "placer mining" of wheat involves a qualitatively new element: throwing the food temporarily away and waiting a short period before retrieving it. It may well be that young animals are normally the inventors of new behavior patterns, but that only those with several years of experience can manage the most complex tasks. This notion has received support from experiments by Atsuo Tsumori and his associates on the troops at Koshima, Ohirayama, and Takasakiyama (Tsumori et al., 1965; Tsumori, 1967). Peanuts were buried in the sand to a depth of 6–7 centimeters in full sight of the troop. At each place, a minority of the individuals succeeded at the first try in the moderately difficult task of digging up the peanuts. Thereafter, the habit spread through most of the remainder of each troop. The most innovative animals were young, with the best performance coming from those four to six years of age.

The innovations of the Koshima troop have also provided a graphic illustration of the potential role of learned behavior as an evolutionary pacemaker. The food presented to the monkeys on the beach attracted them to a new habitat and presented them with opportunities for further change never envisioned by the Japanese biologists. Young monkeys began to enter the water to bathe and splash, especially during hot weather. The juveniles learned to swim, and a few even began to dive and to bring seaweed up from the bottom. One left Koshima and swam to a neighboring island. By a small extension in dietary opportunity, the Koshima troop had adopted a new way of life, or more accurately, grafted an additional way onto the ancestral mode. It is not too much to characterize such populations as poised on the edge of evolutionary breakthroughs, even though probably very few ever complete the process.

CHAPTER 8

Communication: Basic Principles

What is communication? Let me try to cut through the Gordian knot of philosophical discussion that surrounds this word in biology by defining it with a simple declarative sentence. Biological communication is the action on the part of one organism (or cell) that alters the probability pattern of behavior in another organism (or cell) in a fashion adaptive to either one or both of the participants. By adaptive I mean that the signaling, or the response, or both, have been genetically programmed to some extent by natural selection. Communication is neither the signal by itself nor the response; it is instead the relation between the two. Even if one animal signals and the other responds, there still has been no communication unless the probability of response was altered from what it would have been in the absence of the signal.

We know that in human beings communication can occur without an outward change of behavior on the part of the recipient. Trivial or otherwise useless information can be received, mentally noted, and never used. But in the study of animal behavior no operational criterion has yet been developed other than the change in patterns of overt behavior, and it would be a retreat into mysticism to try to add mental criteria. At the same time there exist certain probability-altering actions which common sense forbids us to lable communications. An attack by a predator certainly alters the behavior patterns of the intended victim, but there is no communicating in any sense in which we would care to use the word. Communication must also be consequential to some reasonable degree. If one animal simply pauses to watch as another moves by unknowingly at a distance, the passing animal has altered the behavior pattern of the first. But the passing animal was not really communicating in any way that could alter its own behavior or affect its relationship to the observing animal in the future. Perception occurred in this case, but not communication.

Human versus Animal Communication

The great dividing line in the evolution of communication lies between man and all of the remaining ten million or so species of organisms. The most instructive way to view the less advanced systems is to compare them with human language. With our own unique verbal system as a standard of reference we can define the limits of animal communication in terms of the properties it rarely—or never—displays. Consider the way I address you now. Each word I use has been assigned a specific meaning by a particular culture and transmitted to us down through generations by learning. What is truly unique is the very large number of such words and the potential for creating new ones to denote any number of additional objects and concepts. This potential is quite literally infinite. To take an example from mathematics, we can coin a nonsense word for any number we choose (as in the case of the "googol," which designates a 1 followed by 100 zeros). Human beings utter their words sequentially in phrases and sentences that generate, according to complex rules also determined at least partly by the culture, a vastly larger array of messages than is provided by the mere summed meanings of the words themselves. With these messages it is possible to talk about the language itself, an achievement I am utilizing here. It is also possible to project an endless number of unreal images: fiction or lies, speculation or fraud, idealism or demagoguery, the definition depending on whether or not the communicator informs the listener of his intention to speak falsely.

Now contrast this with one of the most sophisticated of all animal communicative systems, the celebrated waggle dance of the honeybee (*Apis mellifera*), first decoded in 1945 by the German biologist Karl von Frisch. When a foraging worker bee returns from the field after discovering a food source (or, in the course of swarming, a desirable new nest site) at some distance from the hive, she indicates the location of this target to her fellow workers by performing the waggle dance. The pattern of her movement is a figure eight repeated over and over again in the midst of crowds of sister workers. The most distinctive and informative element of the dance is the straight run (the middle of the figure eight), which is given a particular emphasis by a rapid lateral vibration of the body (the waggle) that is greatest at the tip of the abdomen and least marked at the head.

The complete back-and-forth shake of the body is performed 13 to 15 times per second. At the same time the bee emits an audible buzzing sound by vibrating her wings. The straight run represents, quite simply, a miniaturized version of the flight from the hive to the target. It points directly at the target if the bee is dancing outside the hive on a horizontal surface. (The position of the sun with respect to the straight run provides the required orientation.) If the bee is on a vertical surface inside the darkened hive, the straight run points at the appropriate angle away from the vertical, so that gravity temporarily replaces the sun as the orientation cue. (See Figure 8-1.)

The straight run also provides information on the distance of the target from the hive, by means of the following additional parameter: the farther away the goal lies, the longer the straight run lasts. In the carniolan race of the honeybee a straight run lasting a second indicates a target about 500 meters away, and a run lasting two seconds indicates a target 2 kilometers away. During the dance the follower bees extend their antennae and touch the dancer repeatedly. Within minutes some begin to leave the nest and fly to the target. Their searching is respectably accurate: the great majority come down to search close to the ground within 20 percent of the correct distance.

Superficially, the waggle dance of the honeybee may seem to possess some of the more advanced properties of human language. Symbolism occurs in the form of the ritualized straight run, and the communicator can generate new messages at will by means of the symbolism. Furthermore, the target is "spoken of" abstractly: it is an object removed in time and space. Nevertheless, the waggle dance, like all other forms of nonhuman communication studied so far, is severely limited in comparison with the verbal language of human beings. The straight run is after all just a reenactment of the flight bees will take, complete with wing buzzing to represent the actual motor activity required. The separate messages are not devised arbitrarily. The rules they follow are genetically fixed, and always designate, with a one-to-one correspondence, a certain direction and distance.

In other words, the messages cannot be manipulated to provide new classes of information. Moreover, within this rigid context the messages are far from being infinitely divisible. Because of errors both in the dance and in the subsequent searches by the followers, only about three bits of information are transmitted with respect to distance and four bits with respect to direction. This is the equivalent of a human communication system in which distance would be gauged on a scale with 8 divisions and direction would be defined in terms of a compass with 16 points. In reading single messages, northeast could be distinguished from east by northeast, or

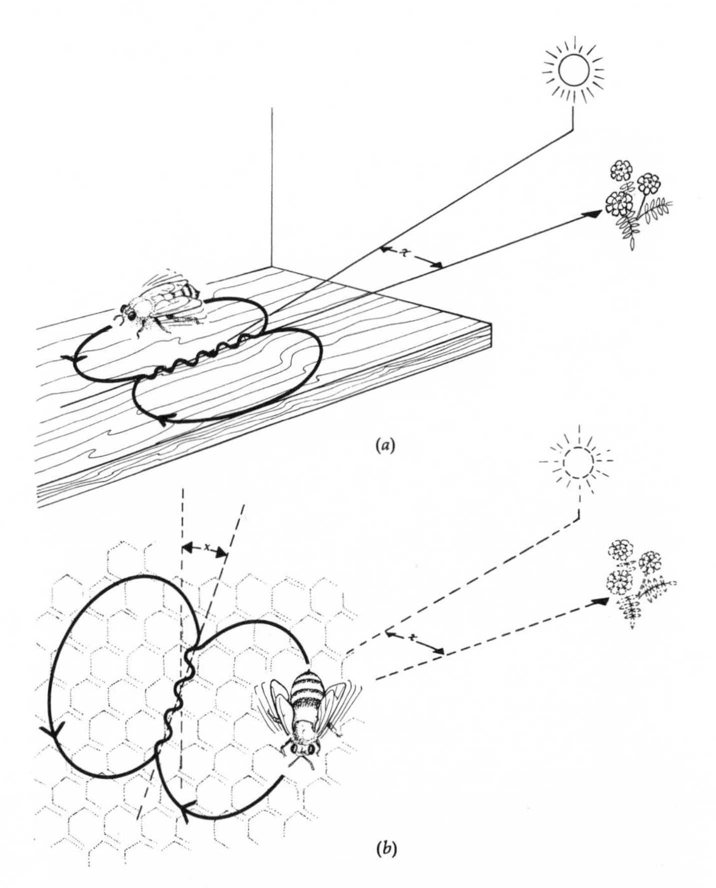

Figure 8-1 The waggle dance of the honeybee. As the bee passes through the straight run she vibrates ("waggles") her body laterally, with the greatest movement occurring in the tip of the abdomen and the least in the head. At the conclusion of the straight run, she circles back to about the starting position, as a rule alternately to the left and right. The follower bees acquire the information about the food find during the straight run. In the examples shown here the run indicates a food find 20% to the right of the sun as the bee leaves the nest. If the bee performs the dance outside the hive (*a*), the straight run of the dance points directly toward the food source. If she performs a dance inside the hive (*b*), she orients herself by gravity, and the point directly overhead takes the place of the sun. The angle *x* (= 20%) is the same for both dances. (From Curtis, 1968; based on von Frisch.)

west from west by southwest, but no more refined indication of direction would be possible.

Discrete versus Graded Signals

Animal signals can be partitioned roughly into two structural categories: discrete and graded, or, as Sebeok (1962) designated them, digital and analog. Discrete signals are those that can be presented in a simple off-or-on manner, signifying yes or no, present or absent, here or there, and similar dichotomies. They are most perfectly represented in the act of simple recognition, particularly during courtship. The steel-blue back and red belly of the male three-spined stickleback (*Gasterosteus aculeatus*) is an example of a discrete signal. Another is the ritualized preening of the male Mandarin duck (*Aix galericulata*), who whips his head back in a striking movement to point at the bright orange speculum of his wing. Still other examples are provided by the bioluminescent flashing sequences of fireflies. Discreteness of form also characterizes the communion signals by which members of a group identify one another and stay in contact, such as the duetting of birds and certain grunting calls of ungulates. Discrete signals become discrete through the evolution of "typical intensity" (Morris, 1957). That is, the intensity and duration of a behavior becomes less variable, so that no matter how weak or strong the stimulus evoking it, the behavior always stays about the same.

In contrast, graded (analog) signals have evolved in a way that increases variability. As a rule, the greater the motivation of the animal or the action about to be performed, the more intense and prolonged the signal given. The straight run of the honeybee waggle dance denotes rather precisely the distance from the hive to the target. The "liveliness" or

Figure 8-2 Graded signals in the aggressive displays of a rhesus monkey (top) and a green heron (bottom). In the rhesus what begins as a display of low intensity, a hard stare (left), is gradually escalated as the monkey rises to a standing position (middle) and then, with an open mouth, bobs its head up and down (right) and slaps the ground with its hands. If the opponent has not retreated by now, the monkey may actually attack. A similarly graduated aggressive display is characteristic of the green heron. At first (middle) the heron raises the feathers that form its crest and twitches the feathers of its tail. If the opponent does not retreat, the bird opens its beak, erects its crest fully, ruffles all its plumage to give the illusion of increased size, and violently twitches its tail (right). Thus in both animals the more probable the attack, the more intense the aggressive display. (Based on Altmann, 1962a, and Meyerriecks, 1960; from Wilson, 1972b. From "Animal Communication" by E. O. Wilson.)

"vivacity" of the dance and its overall duration increase with the quality of the food find and the favorableness of the weather outside the hive. Graded communication is also strikingly developed in aggressive displays among animals. In rhesus monkeys, for example, a low-intensity aggressive display is a simple stare. The hard look a human being receives when he approaches a caged rhesus is not so much a sign of curiosity as it is a cautious display of hostility. Rhesus monkeys in the wild frequently threaten one another not only with stares but also with additional displays on an ascending scale of intensity. To the human observer these displays are increasingly obvious in their meaning. The new components are added one by one or in combination: the mouth opens, the head bobs up and down, characteristic sounds are uttered, and the hands slap the ground. By the time the monkey combines all these components, and perhaps begins to make little forward lunges as well, it is likely to carry through with an actual attack (Figure 8-2). Its opponent responds by retreating or by escalating its own displays. These hostile exchanges play a key role in maintaining dominance relationships in the rhesus society.

Gradation in one form or another characterizes most of the major categories of communication in animal societies. Birds and mammals transmit a rich array of messages, some of which are qualitatively different in meaning, by gradually varying postures and sounds (Andrew, 1972). Ants, to cite a very different kind of organism, release quantities of alarm substances in approximate relation to the degree to which they have been stimulated. The amplification of a signal can be accomplished simply by the gradual increase of the power output, movement, melanophore contraction, or whatever component contains the information. Or it can be achieved by adding wholly new components. A striking example of the second method is found in the mobbing calls of certain birds, such as the European blackbird, which add components of higher frequencies as the mobbing attacks on other birds intensify.

Signal Specificity

The communication systems of insects, of other invertebrates, and of the lower vertebrates (such as fishes and amphibians) are characteristically stereotyped. This means that for each signal there is only one response or very few responses, that each response can be evoked by only a very limited number of signals, and that the signaling behavior and the responses are nearly constant throughout entire populations of the same species. An extreme example of this rule is seen in the phenomenon of chemical sex attraction in moths. The female silkworm moth draws males to her by emitting minute quantities of a complex alcohol from glands at the tip of her abdomen. The secretion is called bombykol (from the name of the moth, *Bombyx mori*), and its chemical structure is *trans*-10-*cis*-12-hexadecadienol.

Bombykol is a remarkably powerful biological agent. According to estimates made by Dietrich Schneider and his coworkers at the Max Planck Institute for Comparative Physiology at Seewiesen in Germany, the male silkworm moths start searching for the females when they are immersed in as few as 14,000 molecules of bombykol per cubic centimeter of air. The male catches the molecules on some 10,000 distinctive sensory hairs on each of his two feathery antennae. Each hair is innervated by one or two receptor cells that lead inward to the main antennal nerve and ultimately through connecting nerve cells to centers in the brain. The extraordinary fact that emerged from the study by the Seewiesen group is that only a single molecule of bombykol is required to activate a receptor cell. Furthermore, the cell will respond to virtually no stimulus other than molecules of bombykol. When about 200 cells in each antenna are activated per second, the male moth starts its motor response (Schneider, 1969). Tightly bound by this extreme signal specificity, the male performs as little more than a sexual guided missile, programmed to home on an increasing gradient of bombykol centered on the tip of the female's abdomen—the principal goal of the male's adult life.

Such highly stereotyped communication systems are particularly important in evolutionary theory because of the possible role they play in the origin of new species. Conceivably, one small change in the sex-attractant molecule induced by a genetic mutation, together with a corresponding change in the antennal receptor cell, could result in the creation of a population of individuals that would be reproductively isolated from the parental stock. Persuasive evidence for the feasibility of such a mutational change has been adduced by Roelofs and Comeau (1969). They found two closely related species of moths (members of the genus *Bryotopha* in the family Gelechiidae) whose females' sex attractants differ by only the configuration of a single carbon atom adjacent to a double bond. In other words, the attractants are simply different geometric isomers. Field tests showed not only that a *Bryotopha* male responds solely to the isomer of his own species but also that his response is inhibited if some of the other species' isomer is also present.

Other kinds of signals are clearly not designed to impart specificity. The alarm substances of ants, termites, and social bees consist of an astonishing diversity of terpenes, hydrocarbons, and esters, most of which have low molecular weights. In spite of the fact that they differ in composition and proportionality from one species to another, they are generally active across broad taxonomic groups. When an agitated honeybee worker discharges isoamyl acetate or 2-heptanone, it alarms not only her nestmates but also any ant or termite that happens to be in the near vicinity. This phenomenon is precisely what the evolutionist would expect. Privacy is not a requirement of alarm communication, and when the communication is coupled with interspecific aggressive behavior, signals should be expected to affect enemies as well as nestmates. The same differences in breadth of activity are found among the communication systems of birds and are subject to the same explanation. Territorial and courtship displays, including advertising songs, are characteristically elaborate and species-distinct. Their exceptional complexity and repetitive patterning are in fact the reasons why human beings consider them beautiful. But esthetics are not the primary consideration for birds. The displays are sufficient, in the great majority of cases, to isolate the members of each species of bird from all other species breeding in the same area. Where "mistakes" occur, resulting in interspecific territorial combat or hybrids, they are usually limited to closely related species and most often to those that have come into close contact only in the most recent geologic time. In contrast, the mobbing calls of small birds, which assemble other birds for cooperation in driving a predator from the neighborhood, are very similar from one species to another and are understood by all.

Signal Economy

When evaluated by human standards, the number of signals employed by each species of animal seems severely limited. In order to establish some quantitative measure pertaining to this intuitive generalization, let us define a signal as any behavior that conveys information from one individual to another, regardless of whether it serves other functions as well. Most communication in animals is mediated by displays, which are behavior patterns that have been specialized in the course of evolution to convey information. In other words, a display is a signal that has been changed in ways that uniquely enhance its performance as a signal. The hawk warning call of a songbird, the hostile eyelid flashing of a male baboon, the zigzag dance of a courting male stickleback, and the release of sex attractant by a female moth are all examples of displays. By confining our attention for the moment to displays, we can delimit the set of the most important, easily diagnosed signals.

Recent field studies have established the curious fact that even the most highly social vertebrates have no more than 30 or 40 separate displays in their entire repertory. Data compiled by Martin H. Moynihan (1970a) indicate that throughout the vertebrates the number of displays varies from species to species by a factor of only three or four. More precisely the number ranges from a minimum of 10 in certain fishes to a maximum of 37 in the rhesus monkey, one of the primates closest to man in complexity of social organization. The full significance of this rule of relative inflexibility is not yet clear. It may be that the maximum number of displays any animal needs in order to be fully adaptive in any ordinary environment, even a social one, is 10 to 40. Or it may be, as Moynihan has suggested, that each number represents the largest amount of signal diversity the particular animal's brain can handle efficiently in quickly changing social interactions.

The general paucity of signal diversity in animal communication contrasts sharply with the seemingly endless productivity of human language. Yet certain intriguing parallels exist between man and other organisms. The paralinguistic signals of each human culture, including hand gestures and eyebrow raises, are roughly comparable in number to the displays of animals; the average person uses about 150–200 of these "typical" nonverbal gestures while communicating. The sound structure of language is based upon 20 to 60 phonemes, the precise number varying according to culture. Perhaps 60 phonemes represent the maximum number of simple discrete vocalizations that the ear can distinguish, just as 30 or 40 displays are perhaps the most that an animal can distinguish efficiently. Human language is created by the sequencing of these sounds into an ascending hierarchy of morphemes, words, and sentences that contain sufficient redundancy to make them easily distinguishable.

The Increase of Information

Although the number of displays catalogued by ethologists is 50 per species or less, the actual number of messages may be far greater. In the simplest systems a display may have only a single meaning, with no nuances permitted. Sexual communication in insects and other invertebrates is often of this kind. The whine of a female mosquito in flight, the ultraviolet flash of a male sulfur butterfly's wings, the release of *cis*-11-tetradecenyl acetate by a female leaf roller moth—each occurs in only one context and conveys the single unalterable message of sexual advertisement. However, in some invertebrate and the great majority of vertebrate systems, the number of messages that can be conveyed by one signal is increased by enrichment devices. The signal can be graded; it can be combined with other signals either simultaneously or in various sequences to provide new meaning; or it can be varied in meaning according to the environmental context. The extremes of enrichment are found, as might be expected, in the higher primates. Furthermore, the elementary concept of the social releaser, first developed in studies of sexual and aggressive behavior of birds and insects, has tended to break down most dramatically in mammals and particularly the higher primates. A full understanding of animal communication therefore depends on a systematic account of the enrichment devices, a brief version of which follows.

Adjùstment of Fading Time

Any signal that is limited in space and time potentially provides information about both of these parameters. A predator-warning call designed to conceal the position of the signaler transmits only the information that a disturbing object has been sighted. In contrast, a territorial advertisement call, which can be easily localized, conveys both the challenge and the location of a part of the territory. This general mode of enrichment is especially clear in cases of chemical communication. The interval between discharge of the pheromone and the total fade-out of its active space can be adjusted in the course of evolution by altering the Q/K ratio, that is, the ratio of the amount of pheromone emitted (Q) to the threshold concentration at which the receiving animal responds (K). Q is measured in number of molecules released in a burst, or in number of molecules emitted per unit of time, while K is measured in molecules per unit of volume (Bossert and Wilson, 1963). Where location of the signaler is relevant, the rate of information transfer can be increased by lowering the emission rate (Q) or raising the threshold concentration (K), or both. This adjustment achieves a shorter fade-out time and permits signals to be more sharply pinpointed in time and space. A lower Q/K ratio characterizes both alarm and trail systems.

In the case of ingested pheromones, the duration of the signal can be shortened by enzymatic deactivation of the molecules. When Johnston et al. (1965) traced the metabolism of radioactive *trans*-9-keto-2-decenoic acid fed to worker honeybees, they found that within 72 hours more than 95 percent of the pheromone had been converted into inactive substances consisting principally of 9-ketodecanoic acid, 9-hydroxydecanoic acid, and 9-hydroxy-2-decenoic acid.

Increase in Signal Distance

If part of the message is the location of the signaler, the information in each signal increases as the logarithm of the square of the distance over which the signal travels. In chemical systems it is the active space, or the space within which the concentration of the pheromone is at or above threshold concentration, that must be expanded. An increase in active space can be achieved either by increasing Q or by decreasing K. The latter is more efficient, since K can be altered over many orders of magnitude by changes in the sensitivity of the chemoreceptors, while a comparable change in Q requires enormous increases or decreases in pheromone production and capacity of the glandular reservoirs. The reduction of K has been especially prevalent in the evolution of airborne insect sex pheromones, where threshold concentrations are sometimes on the order of only hundreds of molecules per cubic centimeter. When the pheromone is expelled downwind, a relatively small amount is required to create very long active spaces, because orientation can be achieved by anemotaxis, or movement against or with the wind, rather than by a more laborious movement up or down the odor gradient. As a consequence, Q can be kept small. The rate of information transfer is kept down, in the sense that signals cannot be turned on or off as rapidly. But the total amount of information eventually transmitted is increased, since a very small target can be pinpointed within a very large space.

The signaler may also identify the location of an object in space. When a sentinal male baboon barks to alert his troop, the troop members first look at him, then follow the direction of his gaze in an attempt to locate the disturbing stimulus (Hall and DeVore, 1965). The straight run of the waggle dance of the honeybee conveys detailed information on the location of targets and hence is more efficient than the round dance, which merely alerts bees to the existence of food somewhere near the hive (von Frisch, 1967).

Increase in Signal Duration

When a signal is broadcast continuously, the potential amount of information transferred increases evenly with time. Anatomical structures used in courtship are examples of more or less effortless sustained signals. The antlers of a male deer, the swollen buttocks of an estrous chimpanzee female, and the brighly colored legs of a sexually active male heron all continuously broadcast the reproductive status of their owners. Scent posts left by territorial mammals also signal continuously, and in addition, as the scent weakens by diffusion or changes chemically, its concentration provides further information on the age of the signal and the probability that the signaler is still in the vicinity.

Structures built by animals can provide the most durable signal source of all. When the larvae of stem-dwelling eumenine wasps pupate, their bodies must be aligned so that the insects face the outer, open end of the hollow stem. If the pupae are accidentally pointed in the opposite direction, the newly eclosing adult wasps try to dig down through the pith of the stem toward the trunk of the bush or tree—and they die in the process. How does a larva know the correct direction to face when it is about to turn into a pupa? Kenneth Cooper (1957) found that the mother wasp provides the necessary information for her offspring when she first constructs the cell in the hollow stem. She makes the texture and concavity (versus convexity) of the terminal wall leading to the outside different from the texture and concavity of the wall leading inward toward the tree trunk. The larva instinctively uses this information to orient its body at pupation, even though it has never had direct contact with the mother or access of any kind to the outside world. Cooper was able to change the orientation of larvae at will by placing them in artificial stems containing cells of deliberately varied construction.

Gradation

All other circumstances being equal, graded messages convey more information than equivalent discrete messages. Consider the simplest possible case, in which one discrete signal is compared with a single signal selection from along a point in a gradient. The discrete signal can only exist or not exist. In the absence of alternate signals in the same message category, it conveys at most one bit of information. The graded signal, in contrast, exists or does not exist, and when it exists further designates a point on the gradient. The additional number of bits yielded is a function of the logarithm of the total number of points on the gradient that can be discriminated. Now suppose that the two systems being compared are (1) a set of discrete signals arrayed along a gradient, labeled, say, 1 to 10 along a scale of rising intensity, and (2) continuously varying signal arrayed along the same scale. Let the precision of emission and reception be the same in both cases. It can be shown that the continuously varying system will always carry more information than the one divided into discrete steps. This principle can be seen more clearly by comparing the honeybee waggle dance, which is a continuously varying system, with an imaginary equivalent system divided into discrete messages. We know that because of errors in both the dance and the execution of the outward flights directed by the dance, the amount of information conveyed about direction is rather limited, consisting of about four bits. This is the equivalent of pinpointing any target within one or other of 2^4, or 16, equiprobable compass sectors. Suppose that the imaginary competing system has 16 discrete signals representing the same compass sectors. If the precision of transmission is the same as in the real waggle dance, the information transmitted will be less than four bits per dance. This is because some of the bees will inevitably end up in sectors other than the one designated by the signal, and the probability and degree of their errors would then have to be translated into bits and deducted from the four bits that represent the maximum in a perfect discrete system.

Composite Signals

By combining signals it is possible to give them new meanings. The theoretical upper limit of combinatorial messaage is the "power set" of all of its components, or the set of all possible combinations of subsets. Thus, if *A*, *B*, and *C* are three discrete signals, each with a different meaning, and each combination produces still one more message, the total ensemble of messages possible is the power set consisting of seven elements: *A*, *B*, *C*, *AB*, *AC*, *BC*, and *ABC*. No animal species communicates in just this way, but many impressive examples have been found in which conspicuous signals are used effectively in different combinations to provide different meanings. A case from the horse family (Equidae) embracing both discrete and graded signals is shown in Figure 8-3. A zebra or other equid shows hostility by flattening its ears back and friendliness by pointing them upward (discrete signals). In both postures the intensity is indicated by the degree to which the mouth is opened simultaneously (a graded signal). The mare is able to produce a third message by adding two more components: when ready to mate, she presents the stallion with the threat face but at the same time raises her hindquarters and moves her tail aside.

Chemical communication, like visual communication, lends itself easily to the production of composite messages. Many species of insects and mammals possess multiple exocrine glands, each of which produces pheromones with a different meaning. Kullenberg (1956), for example, found that females of certain aculeate wasps release simple attractants from the head that act in concert with sexual excitants released from the abdomen. Different substances with different meanings can also be generated by the same gland. A minimum of 32 compounds have been

Figure 8-3 Composite facial communication in zebras (*Equus burchelli*). Threat is indicated by laying the ears back (a discrete signal) and opening the mouth ever wider to indicate increasing amounts of hostility (a graded signal). When making a friendly greeting, the zebra opens its mouth variably in the same way, but points the ears upward. (Modified from Trumler, 1959.)

detected in the heads of honeybee queens. Most or all are present in the mandibular gland secretion. The biological significance of most of these substances is still unknown. Some are undoubtedly precursors to pheromones, but at least two are known pheromones with contrasting effects. The first, 9-keto-2-decenoic acid, is basically an inhibitor. Operating in conjunction with additional scents produced elsewhere in the body, it reduces the tendency of the worker bees to construct royal cells and to rear new queens, who would then be rivals of the mother queen. It also inhibits ovarian development in the workers, in effect preventing them from entering into rivalry with the queen. The second mandibular gland pheromone, 9-hydroxy-2-decenoic acid, causes clustering and stabilization of worker swarms and helps to guide the swarms from one nest site to another.

Syntax

True syntax in the sense of human linguistics, wherein the meaning of combinations of signals depends on the order of appearance, has not yet been demonstrated in animals. The one possible exception is play invitation, to be described later in a discussion of metacommunication, and even this is at best a marginal case. True syntax occurs when separate signals, say, *A*, *B*, and *C*, that have distinct meanings when alone create new messages when presented in various orders: *ABC*, *CBA*, *CAB*, and so forth. In human speech, each of the three permutations "George hunts," "George hunts the bear," and "The bear hunts George" has a very different meaning. No comparable process of message formation is known to exist in animal communication.

Even so, the distinctiveness of a *single* message often depends on the ordering of the elements that compose it. The head-bobbing movements of spiny lizards (*Sceloporus*) are sequenced through time in ways that permit females to recognize males of their own species. Experimental models made to bob according to various of the species patterns attract watching females of the species being imitated (Hunsaker, 1962). But the sequence of bobs is not a sentence that individual lizards break down and rearrange; it is a solitary, unalterable signal. Similarly, Brémond (1968) found that the sequence of notes sung by males of the European robin (*Erithacus rubecula*) is important for recognition by other members of the same species. In contrast, sequence of notes is not important in the songs of the wood lark (*Lullula arborea*) and indigo bunting (*Passerina cyanea*) (Tretzel, 1966; S. T. Emlen, 1972). Like lizard nodding, robin song falls short of syntactic organization. It is merely a case of temporal cues being added to pitch, duration, and other components that impart distinctiveness to unit signals.

Metacommunication

A peculiar form of composite signaling is metacommunication, or communication about the meaning of other acts of communication (Bateson, 1955). An animal engaged in metacommunication alters the meaning of signals belonging to categories other than the original signals that are being transmitted either simultaneously or immediately afterward. Altmann (1962a,b), who first applied this concept extensively to the behavior of nonhuman primates, recognized two circumstances in which metacommunication occurs. The first is status signaling. A dominant male rhesus monkey can be recognized by his brisk, striding gait; his lowered, conspicuous testicles; the posture of his tail, which is held erect and curled back at the tip; and his calm "major-domo" posture, during which he gazes in a confident, unhurried manner at any other monkey catching his attention. A subordinate male displays the opposite set of signals (Figure 8-4). Similar signaling has been recorded in other species of macaques and baboons. Altmann's hypothesis is that the displaying animal communicates its own knowledge of its status and therefore the likelihood that it will attack or retreat if confronted. Since the individual troop members know one another personally, they can judge for themselves whether particular rivals are prepared to alter the dominance order. They evaluate the general "attitude" of the other members of the society. This ex-

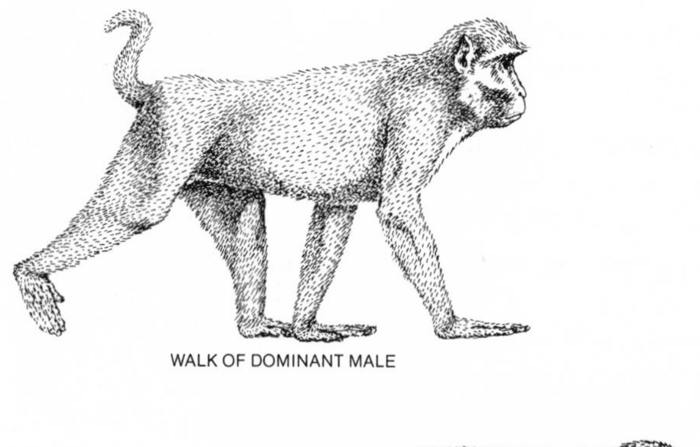

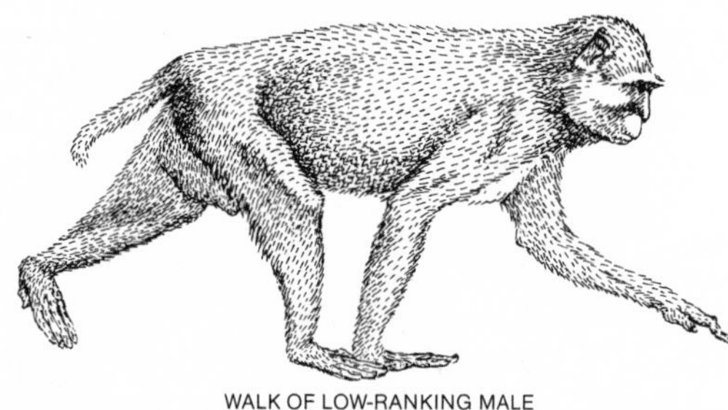

Figure 8-4 Metacommunication in rhesus monkeys includes status signals. The postures and movements of individuals indicate the rank they occupy in the dominance order. (From Wilson et al., 1973; based on S. A. Altmann.)

planation is eminently plausible but has not yet been subjected to any convincing test.

The second form of primate metacommunication is play invitation. The play of rhesus monkeys, like that of most other mammals, is devoted largely to mutual chasing and mock fighting. The invitation signals consist of gamboling and gazing at playmates from between or beside their own legs with their heads upside down. In the play that ensues, the monkeys wrestle and mouth one another vigorously. Although easily capable of hurting one another, the monkeys seldom do. Real damage will result later from escalated versions of the same behavior during bouts of intense aggression. Play signaling says approximately the same thing as this simple human message: "What I am doing, or about to do, is for fun; don't take it seriously. In fact—join me!"

Chimpanzees, baboons, and Old World monkeys invite play by presenting the "play-face" or "relaxed open-mouth display" (Andrew, 1963b; van Hooff, 1972): the mouth is opened widely with the lips still covering most or all of the teeth, and the mouth corners are not pulled forward as in the overtly aggressive bared-teeth display (see Figure 8-5.) The body and eyes continue to move in a relaxed manner, while breathing becomes quick and shallow. In chimpanzees the accelerated breathing is vocalized as a series of hoots that sound like "ahh ahh ahh." The play-face, in fact, may be homologous to the relaxed grin of human beings. A man who taps a friend on the arm or punches him lightly on the chest is unlikely to receive a hostile response if he remembers to grin broadly at the same time. In Western cultures the combined gestures are routinely an invitation to friendly banter.

Context

Even though an animal is limited to a small signal repertory, it can greatly increase the information transmitted by presenting each signal in different contexts. The meaning of the signal then depends on the other stimuli impinging simultaneously on the receiver. Consider an imaginary extreme case in which an animal is limited to one signal that generally alerts other members of the same species. Particularity is added as follows: when presented in the face of danger, the signal serves to alarm; given while the animal is in its own territory, it is a threat to sexual rivals or an invitation to potential mates;

Figure 8-5 The relaxed open-mouth display of the crab-eating monkey (*Macaca fascicularis*) on the right is a play invitation signal of a kind widely used by Old World monkeys and apes, and may be homologous to the grin of human beings. The two monkeys shown here are engaged in play-fighting. (Redrawn from van Hooff, 1972.)

when presented to offspring, it means that food is about to be offered—and so on.

W. J. Smith (1969a,b) has stressed the importance of contextual change for the enrichment of communication in birds. The male of the eastern kingbird *Tyrannus tyrannus,* for example, emits a general purpose call that sounds roughly like "Kitter!" The Kitter is evoked in a variety of contexts when the bird experiences indecisiveness or interference in its attempts to approach some object—a perch, a mate, or another bird. When a lone male flies from perch to perch in his newly delimited territory, the Kitter is employed to attract a female and to warn off potential rivals. Later, the same signal is evidently used as an appeasement signal by the male when approaching his mate.

The social insects have developed some extreme forms of contextual enrichment. The honeybee queen substance, 9-keto-2-decenoic acid, functions as a caste-inhibitory pheromone inside the hive, as the primary female sex attractant during the nuptial flight, and as an assembly scent for the colony during swarming. The waggle dance of the honeybee guides workers to new food finds; it also directs the swarm to new nest sites. The Dufour's gland secretion of the fire ant *Solenopsis invicta* is an attractant that is effective on members of all castes during most of their adult lives. Under different circumstances it serves variously to recruit workers to new food sources, to organize colony emigration, and—in conjunction with a volatile cephalic secretion—to evoke oriented alarm behavior.

Mass Communication

Many of the most highly organized communication systems of social insects contain components of information that cannot be passed from one individual to another, but only from one group to another. This is the phenomenon which I have called mass communication (Wilson, 1962, 1971a). As an example, consider how the number of fire ant workers leaving the nest is controlled by the amount of trail substance emitted by nestmates already in the field. Tests involving the use of enriched trail pheromone have shown that the number of individuals attracted outside the nest is a linear function of the amount of substance presented to the colony as a whole. Under natural conditions this quantitative relation results in the adjustment of the outflow of workers to the level needed at the food source. Equilibration is then achieved in the following manner. The initial buildup of workers at a newly discovered food source is exponential, and it decelerates toward a limit as workers become crowded on the food mass because workers unable to reach the mass turn back without laying trails and because trail deposits made by single workers decline to below threshold concentrations within a few minutes. As a result, the number of workers at food masses tends to stabilize at a level that is a linear function of the area of the food mass. Sometimes, for example when the food find is of poor quality or far away, or when the colony is already well fed, the workers do not cover the find entirely, but equilibrate at a lower density. This additional mass communication of quality is achieved by means of an "electorate" response, in which individuals choose whether to lay trails after inspecting the food find. If they do lay trails, they adjust the quantity of pheromone according to circumstances (Hangartner, 1969a). The more desirable the food find, the higher the percentage of positive responses, the greater the trail-laying effort by individuals, the greater the amount of trail pheromone presented to the colony, and hence the greater the number of newcomer ants that emerge from the nest. Consequently, the trail pheromone, through the mass effect, provides a control that is more complex than could have been assumed from knowledge of the relatively elementary forms of individual behavior alone.

CHAPTER 9

Communication: Functions and Complex Systems

The total analysis of a communication system is a relatively simple concept to create—and a forbiddingly difficult task to accomplish. The analysis falls into three parts: the identification of the *function* of the message, that is, what it means to the communicants and therefore ultimately what role it plays in altering genetic fitness; the inference of the *evolutionary* or *cultural derivation* of the message; and the full specification of the *channel,* from the neurophysiological events that initiate the signaling behavior to the processes by which the signal is emitted, conducted, received, and interpreted. Philosophers have not overlooked the fact that human thought is just a special case of communication. Some have viewed the study of communication as coextensive with logic, mathematics, and linguistics. C. S. Pierce, Charles Morris, Rudolf Carnap, and Margaret Mead, for example, have used the word *semiotic* (or *semiotics*) to designate the analysis of communication in the broadest sense. One of the few useful insights originating from these attempts at synthesis is the recognition that even in human language a word or phrase conveys only a minute fraction of the stimuli associated with the referent. "A tree," for example, alludes to a short list of properties, including certain general attributes of plants, woodiness, canopy atop trunk, and relatively large size. It does not specify details of molecular structure, principles of forest ecology, or any other of the expanding range of qualities of "treeness" that dendrologists have only begun to delineate. In short, even human language is concerned with what the ethologists designate as sign stimuli. T. A. Sebeok (1963, 1965), reflecting on zoology from the viewpoint of a linguist, recognized that animal communication deals far more explicitly than human language with signs and on that basis alone can provide useful guides for the deeper analysis of linguistics. He suggested that the study of animal communication be called "zoosemiotics," in recognition of the fact that it is compounded of two elements: first, the strongly evolutionary emphasis of ethology, which describes whole patterns of behavior under natural conditions and deduces their adaptive significance in the genetic sense; and second, the logical and analytic techniques associated with human-oriented semiotics. There have been other efforts to adapt the principles of human semiotics to the description of animal communication. Hockett and Altmann have systematically listed the design features of human speech and used them to reclassify certain phenomena in animal behavior. Other investigators have carried the mathematical techniques of information theory, developed originally to study human communication, into the study of animal systems. Marler (1961, 1967) has adapted the objective linguistic classification systems of Morris (1946) and Cherry (1957) in an attempt to further deanthropomorphize descriptions of animal behavior.

But there is danger in forcing too early a marriage of animal behavior studies and human linguistics. Human language has unique properties facilitated by an extraordinary and still largely unexplained growth of the forebrain. The deep grammars hypothesized by Chomsky and Postal, if they really exist, are likely to be as diagnostic a trait of *Homo sapiens* as man's bipedal stride and peculiar glottolaryngeal anatomy—and consequently constitute a *de novo* adaptation that cannot be homologized. The introduction of linguistic terminology into zoology, and the reverse, should be attempted only in an exploratory and heuristic manner, with no congruence between zoological and linguistic classifications being forced all the way. It is in this spirit that I wish to take a strongly phenomenological approach to animal communication, beginning with observed facts and classifying them tentatively by induction.

The Functions of Communication

Social behavior comprises the set of phenotypes farthest removed from DNA. As such it is very labile in evolution, the phenomenon most subject to amplification in the transcription of information from the genes to the phenotype of individuals. It is also the class of phenotypes most easily altered by addition or subtraction of otherwise unrelated components. Hence when communication is viewed over all groups of organisms, the included behaviors become so eclectic in nature as to be beyond all hope of homology, and so divergent in function as to defy simple classification.

By focusing on this matter of lability and heterogeneity in social behavior we come quickly to the crux of the problem. The key idea, seldom recognized explicitly in the past, is that the art of classification is central to the study of function in social behavior. In fact, zoosemiotics presents just one more case of the two classical problems of taxonomic theory: how to define the ultimate unit to be classified (in this case, the function), and how to cluster these units into a hierarchy of categories that serve as a useful shorthand while at the same time staying reasonably close to phylogeny. In the taxonomy of organisms the basic unit is the species. Groups of species judged by largely subjective criteria to most closely resemble one another owing to common descent are grouped into genera. Similar and related genera are clustered into families, families into orders, and so on upward to the phyla and kingdoms. In creating classifications of functions, animal behaviorists employ the message as the basic unit. Although not taxonomists working with species, they more or less consciously perform the same mental operations. A set of messages labeled a "message category" is the semiotic equivalent of a genus or family, and it is no more intuitively sound than the definition of the individual messages clustered within it. No a priori formula can give the message category crisper definition or deeper meaning. For this reason the best way to consider the significance of animal communications is to start with a simple, relatively finely defined catalog of functional categories, that is, the "species" of our semiotic classification, and then to proceed with clustering. The categories discussed below provide a balanced sample of existing knowledge, but they are not as finely divided as possible. Thus sexual signals could be broken down into at least six subcategories, many of which overlap broadly, while caste inhibition signals in social insects could be more precisely matched to many of the distinguishable castes, and so forth.

Facilitation and Imitation

Both the induction of behavior by the mere presence of another member of the species and the close imitation of another's behavior patterns can be construed as acts of communication in the broadest sense. The actions of members of family groups and closely knit societies are often highly coordinated, and it is advantageous to the leaders as well as to the followers that the group act as a unit. Components of locomotion have often been modified to serve as signals for inducing locomotion in groupmates: the swing step of hamadryas baboons, for example, or ritualized wing flicking by birds in flocks. Social insects have carried facilitation to an extreme in the coordination of group activity. Wasps departing on foraging trips tend to activate nearby wasps into flight also. Ants and termites initiating soil excavation and other nest construction activities attract nestmates, who join in the labor. The mere sight of another individual in rapid motion excites and attracts workers of large-eyed ant species. This form of communication, called kinopsis, aids in the capturing and subduing of prey. The result of such facilitation is the concentration of group effort in space and time, a form of coordination which is manifestly to the advantage of both the signaler and the responder.

Monitoring

A complementary function of facilitation and imitation is the persistent observation of the activities of other animals. The presence of food or water, the intrusion of a territorial rival, and the appearance of a predator can all be "read" from the actions of neighbors. Whether monitoring is true communication even in the most liberal sense can be debated.

Contact

Social animals use signals that serve in some circumstances just to keep members of a group in touch. The habit is particularly well developed in species that move about under conditions of poor visibility. South American tapirs (*Tapirus terrestris*) use a short "sliding squeal" to stay in touch in the dense vegetation of their rain forest habitat (Hunsaker and Hahn, 1965). The lemurlike sifaka (*Propithecus verreauxi*) uses a cooing sound for the same purpose (Alison Jolly, 1966). Duetting, during which pairs of animals exchange notes in rapid succession, functions as a contact-maintaining signal. The phenomenon occurs widely in frogs, in birds, and in at least two species of primates, the tree shrew *Tupaia palawanensis* and the siamang *Symphalangus syndactylus*.

Individual and Class Recognition

The capacity to recognize different castes is widespread in the social insects. Nest queens are treated in a preferential way, and workers easily distinguish them from virgin members of the same caste. Within the largest colonies these fecund individuals are as a rule heavily attended by nurse workers, who constantly lick their bodies and offer them food. Among honeybees at least, three unique pheromones have been implicated in this special treatment: *trans*-9-keto-2-decenoic acid and *trans*-9-hydroxy-2-decenoic acid from the mandibular glands and an unidentified volatile attractant from Koschevnikov's gland, located at the base of the sting.

In addition to identifying castes, workers of some species can detect differences among life stages. In the relatively primitive myrmicine ant genus *Myrmica*, workers are evidently not capable of distinguishing the tiny first instar larvae from eggs, so that when eggs hatch the larvae are left for a time in the midst of the egg pile. As soon as they molt and enter the second instar, however, the larvae are removed by the workers and placed in a separate pile (Weir, 1959). The tendency to segregate eggs, larvae, and pupae into separate piles is a nearly universal trait in ants. An identification substance can be extracted from larvae of the fire ant *Solenopsis invicta* and transferred to previously inert dummies, causing workers to carry the dummies to the larval piles (Glancey et al., 1970).

The ability to distinguish infants, juveniles, and adults is a universal trait of the vertebrates. Several sensory modalities are routinely employed, including particularly sound, vision, and smell. Often the response is quite specialized and insectlike in its stereotyped quality. The cichlid fish *Haplochromis bimaculatus* distinguishes larvae from fry by odor alone. The characteristic responses of the adults can be obtained by placing them either in "fry water" or "larva water" from which the immature stages have previously been removed. Parents of altricial birds recognize the nestlings at least in part by the distinctive appearance of their gaping maws. In a few species, such as the estrildid finches, the effect is enhanced by a strikingly colored mouth lining, which may be further embellished by special paired markings (Eibl-Eibesfeldt, 1970).

Among species of the higher vertebrates it is commonplace for individuals to be able to distinguish one another by the particular way they deliver signals. Indigo buntings, American robins, and certain other songbirds learn to discriminate the territorial calls of their neighbors from those of strangers that occupy territories farther away. When a recording of a song of a neighbor is played near them, they show no unusual reaction, but a recording of a stranger's song elicits an agitated aggressive response. Analyses by Falls (1969), Thielcke and Thielcke (1970), and Emlen (1972) have revealed the particular components of songs, such as absolute frequency (in the white-throated sparrow) and detailed phrase morphology (in the indigo bunting), that vary from individual to individual and are used by the birds to make identifications. A similar personalization of signals allows families of seabirds to keep together as a unit in the dense, clamorous breeding colonies.

Mammals are at least equally adept at discriminating among individuals of their own kind. A wide variety of cues are employed by different species to distinguish mates and offspring from outsiders. The faces of gorillas, chimpanzees, and red-tailed monkeys (*Cercopithecus nictitans*) are so variable that human observers can tell individuals apart at a glance. It is plausible that the equally visual nonhuman primates can do as well (Marler, 1965; van Lawick-Goodall, 1971). Some mammal species use secretions to impart a personal odor signature to their environment or to other members in the social group. As all dog owners know, a dog urinates at regular locations within its territory at a rate that seems to exceed physiological needs. What is less well appreciated is the communicative function this compulsive behavior serves: a scent included in the urine identifies the animal and announces its presence to potential intruders of the same species. Scent marking probably serves as a repulsion device in the ancestral wolf to keep the pack territory free of intruders. This behavior is widespread, if not universal, among other species of Canidae, and tigers and domestic cats establish scent posts and partial territories in approximately the same fashion.

Purists may argue that identification of an individual does not constitute true communication. Nevertheless, all of the rest of the social repertory is dependent on the constant input of such information. Slight alterations of this input cause a prompt change in the interactions of group members. If experimenters remove an ant larva from the brood pile and place it in an adjacent, less suitable chamber, the workers promptly pick it up and carry it back. If experimenters wash its cuticle lightly with a solvent to disturb the larva's odor, it is killed and eaten instead. When experimenters give a goat kid to a mother not imprinted on its personal odor, it is driven away and allowed to starve. In each of these cases other behavioral patterns are activated, but they are dependent in their timing and orientation on the constant reception of identification signals.

Status Signaling

Various peculiarities in appearance and signaling, often metacommunicative in nature, serve to identify the rank of individuals within dominance hierarchies. This subject was discussed in another context in Chapter 8.

Begging and Offering of Food

Elaborate systems of begging and feeding have evolved repeatedly in the birds and mammals. Nestling birds recognize returning parents by landing calls, the sight of the movement of the parents' bodies above the nest rim, the jarring of the nest as the adults alight, or combinations of such signals. They then respond by gaping. The visual releasers in the maw of the young bird induce the parent to drop pieces of food into it or to regurgitate to it. Other, more specific signals may accompany the exchanges. The conspicuous red dot on the lower beak of the adult herring gull guides the young to the exact spot of the parent's anatomy where they are most likely to receive regurgitated food (Tinbergen, 1951). As they grow older and more agile, the offspring commonly use conspicuous wing movements while begging. The bald ibis (*Geronticus eremita*) and Australian wood swallows (*Artamus*) spread their wings and wave them slowly, while songbirds quiver the wings (Immelmann, 1966; Wickler, 1972). Among precocial birds, begging and feeding are absent or else replaced by forms of feeding enticement. When the hen of a domestic fowl discovers food she lures her chicks to her side by clucking. She may also peck conspicuously at the ground, pick up bits of food and let them fall to the ground again (Wickler, 1972).

Mammals that feed their young primarily through lactation use relatively simple begging and feeding signals. Among deer, antelopes, and related ungulates, mothers that bear single young or twins stand in an open stiff-legged pose and let their young approach them from beneath for suckling. Those giving birth to multiple young, such as the pigs and their relatives (Suidae), lie on their side (Fraser, 1968). In both kinds of ungulates the young are strongly precocial; in the extreme case of the wildebeest and pigs, they are able to walk and follow the mother within an hour after birth. Jackals, African wild dogs, and wolves regurgitate to their young like birds, and the young have evolved an appropriate form of communication to initiate the behavior: they vigorously nuzzle the lips of the adults in an attempt to induce the regurgitation, sometimes forcing their heads inside the open jaws to take food directly from the parent's mouth (Mech, 1970; H. and J. van Lawick-Goodall, 1971).

The exchange of food reaches its extreme development in social insects, and in fact it is fundamental to the organization of their colonies. When the food is in liquid form, delivered by regurgitation from the crop or as secretions from special glands associated with the alimentary tract, the exchange is referred to as *trophallaxis*. Trophallaxis is very widespread but not universal among the higher social insects. It occurs generally through the eusocial wasps, including the socially rather primitive *Polistes*. It appears in a highly irregular pattern among the bees, reflecting both the phylogenetic position of the species concerned and the constraints placed on it by the food habits and nest forms of these insects. In the bumblebees, a primitively social group, the workers simply place pollen on the eggs or larvae, and very little direct contact occurs between adults and larvae; furthermore, the exchange of liquid food is extremely rare (Free, 1955b). Sealed brood cells prevent the adults of stingless bees (Meliponini) from feeding the larvae, but regurgitation among adults is a common event (Sakagami and Oniki, 1963). Although the brood cells of honeybee colonies are kept open and workers provision them continuously, they do not feed the larvae by direct regurgitation onto the mouthparts. Adult honeybees, by contrast, regurgitate to one another at a very high rate. Workers regurgitate water, nectar, and honey to one another out of their crops, but larvae and queens receive most of their protein from royal jelly or brood food secreted by the hypopharyngeal glands (Free, 1961b).

Trophallaxis is especially well developed in ants and termites. In ants it reflects phylogeny. In the primitive genera *Amblyopone* and *Myrmecia*, the habit is either rare or frequent but poorly executed. In the higher subfamilies (Aneuretinae, Myrmicinae, Dolichoderinae, Formicinae), the exchange is frequent, and in the last two groups it is prevalent enough to result in a fairly even distribution of liquid food throughout the worker force of the colony. Trophallaxis is practiced throughout the termites. In all the lower termite species examined to date, belonging to the families Kalotermitidae and Rhinotermitidae, the members of the colony feed one another with both "stomodeal food," which originates in the salivary glands and crop, and "proctodeal food," which originates in the hindgut. The stomodeal material is a principal source of nutriment for the royal pair and the larvae. It is a clear liquid, apparently mostly secretory in origin but with an occasional admixture of woody fragments. The proctodeal material is emitted from the anus. It is quite different from ordinary feces since it contains symbiotic flagellates that feces completely lack, and it has a more watery consistency.

Grooming

Grooming is an eclectic set of behaviors evolved in various combinations by many different phylogenetic lines of animals. Although the behaviors superficially resemble one another, they differ in many mechanical details and serve a diversity of functions. Therefore, the clumping of all kinds of grooming into a single functional category is frankly an artifice taken for convenience and partly as a concession to our imperfect knowledge of the adaptive significance of most of its individual variants.

One generalization can nevertheless be drawn about the meaning of grooming in both the vertebrates and social insects. Vertebrates use allogrooming (the grooming of other individuals) to some extent as a cooperative hygienic device, and this is likely to be its primitive function. However, allogrooming is one of the most easily ritualized of all social behaviors, and it has been repeatedly and consistently transformed into conciliatory and bonding signals. Often these social functions completely overshadow the hygienic function, which in extreme cases may be entirely absent. In social insects allogrooming is still largely a mysterious process. It could be basically hygienic, although direct evidence on this point is lacking. In some cases it distributes pheromones and may also serve to spread and imprint the colony odor. Therefore, in social insects as in vertebrates, allogrooming appears to have evolved at least to some extent into a group bonding device.

Allogrooming in birds, more precisely referred to as allopreening, is preeminently if not exclusively devoted to communication (Sparks, 1965; 1969; Harrison, 1965). The behavior has a scattered phylogenetic distribution within this group and occurs in only a minority of the species. It is limited almost wholly to species in which there is a great deal of bodily contact, such as waxbills (Estrildidae), babblers (Timaliidae), white-eyes (Zosteropidae), and parrots (Psittacidae). Advanced social behavior may be as important a correlate as bodily contact, because some crows and related birds (Corvidae) allopreen while maintaining individual distances. Allopreening usually functions as an appeasement display: when birds respond to threats or attacks as if they were about to be preened, the attack is typically inhibited.

Allogrooming is widespread in rodents, where it consists of a gentle nibbling of the fur. The behavior is most frequently observed in conflict situations, although it also occurs under other circumstances where neither participating animal appears to be aggressive or tense. Allogrooming is found sporadically in other mammals. Perhaps the most ritualized form is displayed by the mouflon (*Ovis ammon*), a mountain sheep of the southern Mediterranean region. Soon after two males have fought for dominance, the loser performs an appeasement ceremony in which he licks the winner on the neck and shoulders. Often the dominant animal kneels forward on his carpal joints, that is, on his "wrists," in order to assist the process (Pfeffer, 1967).

Among the primates, allogrooming is a way of life. By passing from phylogenetically lower to higher groups, one can detect a marked shift from a dependence on the mouth and teeth to a nearly exclusive use of the hands. Tree shrews groom with their teeth and tongues, employing the procumbent lower incisors as a "tooth comb." Lemurs, the most primitive living animals that are indisputably primates, employ the teeth, tongue, and hands in close coordination (Buettner-Janusch and Andrew, 1962). In higher primates the hands are the principal grooming instruments. The basic movements consist of drawing hair through the thumb and forefinger, rubbing the thumb in a variable rotary pattern against the direction of the hair tracts, and lightly scratching and raking the hair and skin with the nails. Objects loosened by these actions are conveyed to the mouth to be tasted and sometimes eaten (Sparks, 1969).

Allogrooming in higher primates serves at least in part to clean fellow troop members. Parasites are systematically removed by hand movements, while wounds are cleaned by hand and sometimes even licked. At the same time, most primate species employ allogrooming in a strongly social role. During moment-to-moment encounters between troop members, grooming is reciprocally related to aggression: as one interaction goes up in frequency and intensity, the other goes down. In tense situations animals either offer to groom or present their bodies for grooming. These gestures are seldom followed by serious further threat or fighting, and in fact they seem to avert aggression. In most higher primate species, the dominant animals seem to be groomed disproportionately by their subordinates. The relationship, if borne out by future studies, is wholly in accord with our idea of the primitive role of the behavior, since it is the recipient who receives the greatest benefit.

Workers of a majority of social insects groom nestmates with their glossae (tongues) and, much less frequently, their mandibles, which are the functional equivalents of the primate hands. At least some of the social wasps, for example *Polistes* (Eberhard, 1969), engage in grooming. But the phenomenon is only occasional in the meliponines (Sakagami and Oniki, 1963) and evidently very rare or absent in the bumblebees (Free and Butler, 1959) and the primitively social halictine bee *Dialictus zephyrus* (Batra, 1966). The significance of allogrooming in social insects is not really understood. We can only guess that its cleansing action is in some way beneficial. Probably allogrooming plays some role in the transfer of colony odors and pheromones. It is known, for example, that the queen substance of the honeybee, 9-ketodecenoic acid, is initially transmitted in this fashion from the queen to the worker (Butler, 1954a,b).

Alarm

To alarm a groupmate is to alert it to any form of danger. As a rule, the danger is an approaching predator or territorial invader. But it can be anything else: in termites, for example, alerting trail substances are released in the presence of a breach in the nest wall. Although most alarm signals are general in the scope of their designation, a few are narrowly specific. According to Eberhard (1969), paper wasps (*Polistes*) respond to certain parasites of their own brood in a unique way. In particular, when an ichneumonid wasp of the genus *Pachysomoides* is detected on or near the nest, the paper wasps launch into an intense bout of short runs and wing flipping, which quickly spread through the entire colony. Mammalian alarm calls are mostly nonspecific, but the vervet (*Cercopithecus aethiops*), an arboreal African monkey, uses a lexicon of at least four or five sounds to identify enemies. A snake evokes a special chutter call and a minor bird or mammalian predator an abrupt *uh*! or *nyow!* As soon as a major bird predator is seen, the vervets emit a call that sounds like *rraup;* when either the bird or a major mammal predator is close, the monkeys chirp and produce the threat-alarm bark. The responses of the vervets vary according to the different signals. The snake chutter and minor-predator calls direct the attention of the monkeys to the danger. The *rraup* call, indicating the presence somewhere of a large bird, causes the monkeys to scatter out of the open areas and treetops and into closed vegetation. The other alarm calls cause the monkeys to look at the predator while retreating to cover (Struhsaker, 1967c).

Responses to alarm signals differ markedly among species and according to circumstances in which individuals find themselves. In their studies of formicine ants, Wilson and Regnier (1971) classified species roughly into those that display predominantly aggressive alarm, in which workers orient aggressively toward the center of disturbance, and those that react with panic alarm, in which the workers scatter in all directions while attempting to rescue larvae and other immature stages. There is good evidence that aggressive alarm is the more general form and has evolved as part of an "alarm-defense system," in which the spraying of defensive chemicals and other forms of attack on enemies come to serve as increasingly effective alerting signals for nestmates as well.

Distress

The young of a diversity of bird and mammal species utilize special distress calls to attract adults to their sides. The chicks of precocial birds, such as domestic fowl, ducks, and geese, pipe in a way that is indistinguishable from the call emitted when they are cold or hungry (Lorenz, 1970). The pups of African wild dogs give a special "lamenting call" (*Klage*) when deserted (Kühme, 1965). Stridulation in leaf-cutting ants, a squeaking sound created by scraping a ridge on the third abdominal segment against a row of finer ridges on the fourth segment, appears to serve primarily or exclusively as a distress signal. The ants begin squeaking when trapped in close quarters, particularly when they are pinned down by a predator or caught in a cave-in. The sound alone brings nestmates to their aid (Markl, 1968).

Assembly and Recruitment

No firm line exists between assembly and recruitment. Assembly can be defined crudely as the calling together of members of a society for any general communal activity. Recruitment is merely a special form of assembly, by which groupmates are directed to some point in space where work is required.

Assembling signals serve above all to draw societies into tighter physical configurations. The bright spotting and banding of coral fishes, called "poster coloration" by zoologists, is a case in point. Experiments by Franzisket (1960) showed that *Dascyllus aruanus* are attracted by the black-and-white banding pattern characteristic of their species, and that the response helps to hold individuals together in schools. The striking, individualistic colors of many species of coral fish may be required for quick, precise assembly and coordination of schools among the large numbers of other kinds of fish that crowd their habitat (W. J. Hamilton, III, personal communication). Among mammals, the howling of wolves collects members of the pack that have scattered over the large territory routinely patrolled by these animals (Mech, 1970); and chimpanzees have a functionally similar, booming call that alerts distant troop members when a food tree is discovered (Sugiyama, 1972).

The known techniques of assembly in social insects are almost entirely chemoreceptive in nature. Termites are attracted to one another over distances of at least several centimeters by the odor of 3-hexen-1-ol emitted from the hindgut, while fire ants find one another by moving up carbon dioxide gradients (Wilson, 1962; Verron, 1963; Hangartner, 1969b). Somewhat more complex forms of pheromone-meditated attraction and assembly have been found in the honeybee. When workers have discovered a new food source or have been separated from their companions for a long period of time, they elevate their abdomens, expose their Nasanov glands, and release a strong scent consisting of a mixture of geraniol, nerolic acid, geranic acid, and citral (von Frisch and Rösch, 1926; Butler and Calam, 1969); these pheromones draw other workers over considerable distances. It has also been demonstrated (Velthuis and van Es, 1964; Mautz et al., 1972) that swarming bees expel the Nasanov gland scent when they first encounter the queen, thus attracting other workers to the vicinity. The substances therefore function as true assembly pheromones. Evidently the discovery of food lowers the threshold of the response and turns the pheromones secondarily into recruitment signals.

True recruitment is a form of communication that is apparently limited to the social insects. Ants, bees, wasps, and termites have evolved a multitude of ingenious signaling devices to assemble workers for joint efforts in food retrieval, nest construction, nest defense, and migration (see Wilson 1971a).

Leadership

A few vertebrate and insect species use signals that seem explicitly designed to initiate and to direct the movement of groups. Parents and young of precocial birds use an elaborate system of signals to coordinate their travels. The mother mallard (*Anas platyrhynchos*), for example, walks ahead at a pace just slow enough for her ducklings to stay close behind, all the while emitting a special guiding call. When a duckling falls too far to the rear, it begins piping in distress. The mother immediately stops, extends her body, flattens her feathers, and calls more loudly. If the stragglers do not find their way to her in a short time, she runs back to them, momentarily forgetting the ducklings close to her. When she reaches the stragglers, they all exchange greeting and "conversation" calls (Lorenz, 1970).

A comparable signal is contained in the "swing step" of dominant male hamadryas baboons. These animals control the movements of their group to an unusual degree for primates. When they wish to depart, they take large, rapid steps while lifting their tails and swaying their buttocks rhythmically from side to side. These movements appear to induce subordinates to fall in behind (Kummer, 1968).

The honeybees have evolved two spectacular forms of leadership signal that exceed anything known in the nonhuman vertebrates. The first, of course, is the waggle dance. The second is the buzzing run, also called the breaking dance or *Schwirrlauf,* which is used by honeybee colonies to initiate swarming. Just before the swarm occurs, most of the bees are still sitting idly in the hive or outside the front of the entrance. As midday approaches and the air temperature rises, one or several bees begin to force their way through the throngs with great excitement, running in a zigzag pattern, butting into other workers, and vibrating their abdomens and wings in a fashion similar to that observed during the straight run of the waggle dance. The *Schwirrlauf* is swiftly contagious and leads to an explosive exodus of a large fraction of the colony.

Synchronization of Hatching (Embryonic Communication)

The young of precocial birds belonging to the same clutch have a strong incentive to hatch as close together in time as possible. The brooding mother and the first-hatched young will be on the move within hours; chicks left behind in the egg will perish. Synchronized hatching of entire broods, requiring at most one or two hours, is a general trait of precocial birds, including particularly pheasants, partridges, grouse, ducks, and rheas. When the eggs of these species are incubated separately, the hatching times are spread over a period of days; but when they are kept together, hatching is synchronous. Margaret Vince (1969) has obtained strong experimental evidence that the coordination is achieved by sound signals exchanged by the chicks while they are still in the eggs. The vocalizations become loudest and most persistent just prior to hatching. The most characteristic sound is a regular loud click, audible when the egg is held to the ear. It is not caused by a tapping against the shell, as biologists once widely believed, but is a true vocalization associated with breathing movements.

Play Invitation

The specialized signals used by mammals to initiate play with groupmates have been reviewed in Chapter 8.

Threat, Submission, and Appeasement

The complex, often graded systems of signals that mediate agonistic behavior have already been introduced in Chapter 8. They underlie the phenomena of territoriality and dominance (see Chapters 12 and 13).

Nest-Relief Ceremony

In bird species where both parents care for helpless young, one typically remains at the nest while the other forages. When the forager returns it then relieves the mate from nest duty. The changing of the guard is a delicate operation in which recognition of the mates is first established by personalized sounds and other signals, then mutual agreement to change is reached by ceremonies special to the occasion (Armstrong, 1947; Lorenz, 1971). In some species the ceremony is obviously related phylogenetically to appeasement behavior used in agonistic encounters, including the tense give and take of the original formation of the pair bond. The male grey heron (*Ardea cinerea*) relieves his mate with a series of typically conspicuous reciprocal communications. He first alights on the rim of the nest with a vigorous flapping of wings, to which the female responds by stretching her neck upward and crying out several times. Now the pair stand back to back while calling loudly. Finally, the male bends his head down with the crest raised, snaps his beak several times, and settles on the nest, after which the female departs. Sometimes the routine is varied as follows: the male stretches his neck and head upward, raises his crest, and flaps his wings, while the female performs a muted version of the same display. The male nightjar (*Caprimulgus europaeus*) flies in to the nest uttering a characteristic churring sound, and the female responds with the same note. He next settles close to her and, as they sway gently from side to side, eases her off the nest and takes her place. The female then flies off.

Sexual Behavior

The full course of sexual activity is a tightly orchestrated sequence of behaviors that differ radically in form and function while remaining channeled toward the single act of fertilization. At least five such classes of acts can be distinguished: sexual advertisement, courtship, sexual bonding, copulatory behavior, and postcopulatory displays. In addition a few signals are known with the explicit function of inhibiting reproduction. These categories will be treated later in the chapter on sexual behavior (Chapter 15).

Caste Inhibition

The queens of the most advanced social insects secrete pheromones that inhibit the development of immature stages into new queens. The result is the production of a high proportion of infertile workers that protect and feed the mother queen. The consort males of termites also produce a substance that inhibits male nymphs from developing into their own caste. In honeybees the female substance has been identified as the ubiquitous *trans*-9-keto-2-decenoic acid, which is secreted by the hypertrophied mandibular glands of the hive queen. The odor of the pheromone prevents workers from constructing the enlarged royal cells in which new queens can be reared from early larval stages. Each spring the ketodecenoic acid production of the hive queen is lowered, permitting the production of a few new queens and the subsequent multiplication of the colony by fission.

Complex Systems

It is a common misconception, held even by zoologists, that most animal communication consists of simple signals that reciprocate as stimuli and responses. Such a digital simplicity does indeed occur among microorganisms and many of the lower metazoan invertebrates. But where animals possess brains containing, say, on the order of ten thousand neurons or better, their social behavior tends to be much more devious and subtle. This generalization can best be supported with examples. I will start with a well-analyzed "ordinary" communication system, courtship in doves, to show how intricate such behavioral exchanges can be, and then examine several of the most advanced animal systems so far discovered, in order to give a sense of the upper limit reached by animal communication systems as a whole.

Reproduction in the Ring Dove

The reproductive behavior of ring doves (*Streptopelia risoria*) appears on casual observation to be mediated by a relatively few simple signals exchanged between the mated pair over a period of several weeks. In fact, as the careful researches of D. S. Lehrman and his associates have shown, it is a physiological drama that unfolds through the precise orchestration of communication, external stimuli, and hormone action (Lehrman 1964, 1965). The full cycle runs six to seven weeks (Figure 9-1). As soon as an adult male and female are placed together in a cage containing nesting materials, the male begins to court by bowing and cooing. After a few hours the birds select a concave nesting site (a bowl works well in the laboratory) and crouch in it, uttering a characteristic cooing sound. Soon afterward the two birds carry material to the site and build a loose nest with it. After several days of building activity, the female becomes closely attached to the nest, and soon afterward lays two eggs. Thereafter the two birds take turns incubating. Experiments by Lehrman and his coworkers indicate that the sight and sound of the mate alone stimulates the pituitary gland to secrete gonadotropins. These substances induce an increase in estrogen, which triggers nest-building behavior, and progesterone, which initiates incubation behavior. Another pituitary hormone, prolactin, causes growth in the epithelium of the crop. The sloughed-off epithelium functions as a kind of "milk" which is

Figure 9-1 Programmed reproductive communication in the ring dove. The reproductive cycle takes six to seven weeks and is mediated by interacting stimuli from the mate, the nest materials, and several hormones secreted in sequence. (From Wilson et al., 1973; based on D. S. Lehrman.)

regurgitated to the squabs. Prolactin also sustains incubation behavior. When the squabs reach two to three weeks of age, the parents begin to neglect them, and soon the parents initiate a new endocrine-behavioral cycle. In the laboratory the process recycles continuously around the year.

Extreme Courtship Displays in Insects and Vertebrates

Although the brains of insects are orders of magnitude smaller than those of vertebrates, their most elaborate displays are at least equally complex. This generalization is illustrated by the waggle dance of the honeybee and the combined odor trails and tactile displays of certain ants. It is further exemplified by the courtship displays of many kinds of insects. Probably the most complex pattern known is that of the acridid grasshoppers belonging to the genus *Syrbula* (see Figure 9-2). As described by Otte (1972), the displays used in the sequences are mostly composed of one or the other of several kinds of sounds made by stridulation, combined with special caresses with the antennae and wings. Perhaps the most elaborate courtship process known in vertebrates is that of the ruff *Philomachus pugnax*. Males perform on leks in which they are positioned according to their status in a dominance hierarchy. A total of at least 22 visual displays are employed, with males of different ranks distinguished by the subsets of signals they employ (Hogan-Warburg, 1966; Rhijn, 1973). My subjective impression is that the courtship repertories of the insect *Syrbula* and the bird *Philomachus* are roughly comparable in complexity.

Whale Songs

The most elaborate *single* display known in any animal species may be the song of the humpback whale *Megaptera novaeangliae*. First recognized by W. E. Schevill and later analyzed in some detail by Payne and McVay (1971), the song lasts for intervals of 7 to more than 30 minutes' duration. The really extraordinary fact established by Payne and McVay is that

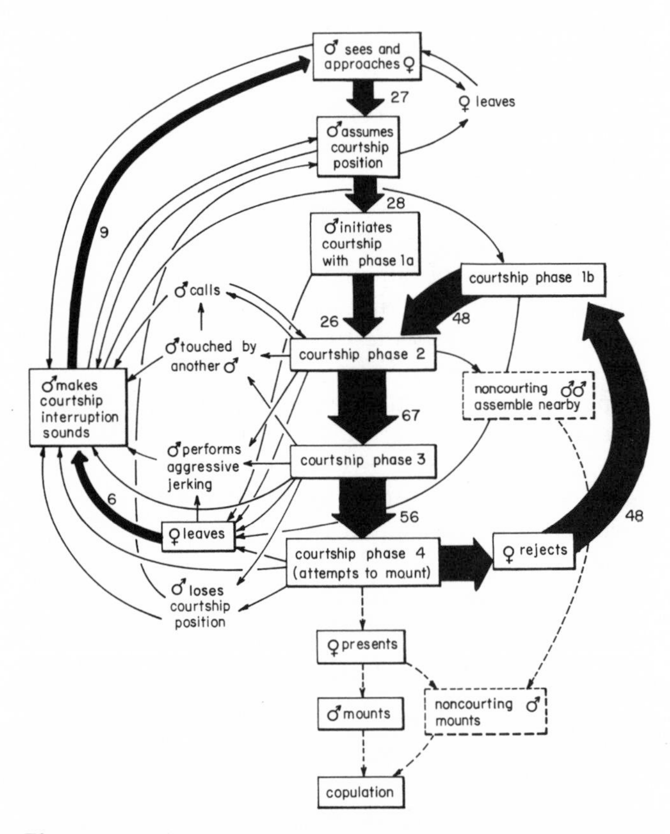

Figure 9-2 The most complex courtship procedure known in insects is that used by grasshoppers of the genus *Syrbula*, in this case *S. admirabilis*. The number of observations of each transition between steps is given next to the arrows and is further indicated by the thickness of the arrows. The separate signals, including those labeled phases, consist of combinations of vocalizations and movements of body parts. (From Otte, 1972.)

each whale sings its own particular variation of the song, consisting of a very long series of notes, and it is able to repeat the performance indefinitely. Few human singers can sustain a solo of this length and intricacy. The songs are very loud, generating enough volume to be heard clearly through the bottoms of small boats at close range and by hydrophones over distances of kilometers. The notes are eerie yet beautiful to the human ear. Deep basso groans and almost inaudibly high soprano squeaks alternate with repetitive squeals that suddenly rise or fall in pitch. The functions of the humpback whale song are still unknown. There is no evidence that special information is encoded in the particular sequence of notes. In other words, the song evidently does not contain sentences or paragraphs but consists of just one very lengthy display. The most plausible hypothesis is that it serves to identify individuals and to hold small groups together during the long annual transoceanic migrations. But the truth is not known, and the phenomenon may yet hold some real surprises. Other whale species use vocalizations, some of them similar to a few of the components of the humpback whale song (Schevill and Watkins, 1962; Schevill, 1964), but none is known to approach the complexity of this one species.

Gorilla and Chimpanzee Displays

Among land animals the most complicated single displays are probably those performed by the two evolutionarily most advanced and intelligent species of anthropoid apes, the gorilla and the chimpanzee. The famous chest-beating display of the gorilla occurs infrequently and is given only by the dominant silver-backed males. According to Schaller (1965a), the entire display consists of nine acts, which may be presented singly or in any combination of two or more. When in combination there is a tendency for the behaviors to appear in the following predictable sequence:

1. To start, the gorilla sits or stands while emitting from 2 to 40 clear hoots, at first distinct but then becoming slurred as their tempo increases.
2. The hoots are sometimes interrupted as the gorilla plucks a leaf or branch from the surrounding vegetation and places it between his lips in what appears to be a ritualized form of feeding.
3. Just before reaching the climax of the display, the animal rises on his hind legs and remains bipedal for several seconds.
4. While rising, he often grabs a handful of vegetation and throws it upward, sideways, or downward.
5. The climax of the display is chest beating, in which the standing gorilla raises his bent arms laterally and slaps his chest alternately with open, slightly cupped hands from 2 to 20 times. The beats are very fast, about 10 a second. Gorillas sometimes beat their abdomens and thighs, as well as branches and tree trunks.
6. A leg is sometimes kicked into the air while the chest is being drummed.
7. During or immediately following chest beating, the gorilla runs sideways, first a few steps bipedally and then quadrupedally, for 3 to 20 or more meters.
8. While running, the gorilla sweeps one arm through the vegetation, swats the undergrowth, shakes branches, and breaks off trees in its path.
9. The final act of a full display consists of thumping the ground with one or both palms.

The display appears to function very generally in advertisement and threat. It is seen most frequently when the male encounters a man or another gorilla troop, or when some other member of the troop begins to display. But is also occurs during play and sometimes even without any outside stimulus evident to the observer.

Even stranger are the "carnivals" of chimpanzee troops. From time to time, at unpredictable periods of the day or night, groups of the apes unleash a deafening outburst of noise—shouting at maximum volume, drumming trunks and buttresses of trees with their hands, and shaking branches, all the while running rapidly over the ground or brachiating from branch to branch (Sugiyama, 1972). The awed human observer feels he is in the presence of pandemonium. Unlike gorilla chest beating, the chimpanzee choruses are communal in nature. Far from serving to intimidate and disperse animals, they appear to keep scattered troops in touch and even to bring them closer together. The frenzies occur most often when the apes are on the move or have gathered for the first time in a feeding area. Sugiyama and the Reynolds believe that they serve in part to recruit other chimpanzees to newly discovered fruit trees, but the evidence is thin. The possibility remains open that the display serves other, perhaps wholly unexpected functions.

CHAPTER 10

Communication: Origins and Evolution

Where do the communication codes of animals come from in the first place? By comparing the signaling behavior of closely related species, zoologists can sometimes link together the evolutionary steps that lead to even the most bizarre communication systems. Any evolutionary change that adds to the communicative function has been called "semanticization" by Wickler (1967a). At one conceivable extreme of the semanticizing process, only the response evolves. Thus the sensory apparatus and behavior of the species is altered in such a way as to provide a more adaptive response to some odor, movement, or anatomical feature that already exists and that itself does not change. Male lobsters and decapod crabs, for example, respond to the molting hormone (crustecdysone) of the female as if it were a sex attractant. It is possible, although not yet proven, that crustecdysone has assumed a signaling function entirely through an evolved change in male behavior. The vast majority of known cases of semantic alteration, however, involve *ritualization,* the evolutionary process by which a behavior pattern changes to become increasingly effective as a signal. Commonly and perhaps invariably, the process begins when some movement, anatomical feature, or physiological trait that is functional in quite another context acquires a secondary value as a signal. For example, members of a species can begin by recognizing an open mouth as a threat or by interpreting the turning away of an opponent's body in the midst of conflict as an intention to flee. During ritualization such movements are altered in a way that makes their communicative function still more effective. Typically, they acquire morphological support in the form of additional anatomical structures that enhance the conspicuousness of the movement. They also tend to become simplified, stereotyped, and exaggerated in form. In extreme cases the behavior pattern is so modified from its ancestral state that its evolutionary history is all but impossible to decipher. As with the epaulets, shako plumes, and piping that garnish military dress uniforms, the practical functions that originally existed have long since been obliterated in order to maximize efficiency in communication.

Ritualized biological traits are referred to as displays. A special form of display recognized by zoologists is the *ceremony,* a highly evolved set of behaviors used to conciliate and to establish and maintain social bonds. We are all familiar with ceremonies in our own social life. Although the American culture is still too young to have many rituals that are truly indigenous, an interesting set can be seen in the yearly commencement at Harvard University. During this seventeenth-century affair the governor of Massachusetts is escorted by mounted lancers, the sheriffs of Middlesex and Essex counties appear in formal dress to represent civil authority, and a student gives a Latin oration. Each of the performances has lost its original function and is perpetuated only as ceremony in the truest sense. In a closely parallel fashion, animals use ceremonies to reestablish sexual bonds, to change position at the nest, and to avoid or reduce aggression during close interactions. Ceremony, to use Edward Armstrong's phrase, is the evolved antidote to clumsiness, disorder, and misunderstanding.

The ritualization of vertebrate behavior often begins in circumstances of conflict, particularly when an animal is undecided whether to complete an act. Hesitation in behavior communicates to onlooking members of the same species the animal's state of mind or, to be more precise, its probable future course of action. The advertisement may begin its evolutionary transformation as a simple intention movement. Birds intending to fly typically crouch, raise their tails, and spread their wings slightly just before taking off. Many species have independently ritualized one or more of these components into effective signals (Daanje, 1950; Andrew, 1956). In some species white rump feathers produce a conspicuous flash when the tail is raised. In others the wing tips are flicked repeatedly downward to uncover conspicuous areas on the primary feathers of the wings. In their more elementary forms the signals serve to coordinate the movement of flock members and perhaps also to warn of approaching predators. When hostile components are added, such as thrusting the head forward or spreading the wings as the bird faces its opponent, flight intention movements become ritualized into threat signals. But the more elaborate and extreme manifestations of this form of ritualization occur where the basic movements are incorporated into courtship displays (see Figure 10-1).

Signals also evolve from the ambivalence created by the conflict between two or more behavioral tendencies (Tinbergen, 1952). When a male faces an opponent, undecided whether to attack or to flee, or approaches a potential mate with strong tendencies both to intimidate and to court, he may at first choose neither course of action. Instead he performs a third, seemingly irrelevant act. He redirects his aggression at some object nearby, such as a pebble, a blade of grass, or a bystander, who then serves as a scapegoat. Or the animal may abruptly switch to a *displacement activity:* a behavior pattern with no relevance whatever to the circumstance in which the animal finds itself. The animal preens itself, for example, or launches into ineffectual nest-building movements, or pantomimes feeding and drinking. Such redirected and displacement activities have often been ritualized into strikingly clear signals used in courtship. As Tinbergen expressed the matter, these new signals are derived from preexisting

Figure 10-1 Flight intention movements have been ritualized to serve as a courtship display in the male European cormorant *Phalacrocorax carbo.* In the presence of the female, the male performs a conspicuous but nonfunctional modification of the take-off leap. (From Hinde, 1970, after Kortlandt, 1940. From *Animal Behavior,* by R. A. Hinde. Copyright © 1970 by McGraw-Hill Book Company. Used with permission.)

motor patterns—they have been "emancipated" in evolution from the old functional context.

Many signals do evolve from ritualized intention and displacement activities. But ritualization is a pervasive, highly opportunistic evolutionary process that can be launched from almost any convenient behavior pattern, anatomical structure, or physiological change—not just from displacement and intention activities. To discover the sources of signals, we must closely analyze them with respect to the immediate biological context in which they occur. When we do that, it becomes clear that all manner of biological processes, from blushing and sweating to mucus secretion and defecation, some of them under the control of the autonomic nervous system, have been appropriated by one species or another. The following examples illustrate the almost protean nature of the process.

Ritualized predation. As part of his courtship ceremonies, the male grey heron (*Ardea cinerea*) routinely performs what is clearly a modified fishing movement. With his crest and certain body feathers erected, he points his head down as though striking at an object in front of him and snaps the mandibles with a loud clash (Verwey, 1930).

Ritualized food exchange. Billing in birds serves multiple functions centered on the establishment and maintenance of bonds (Wickler, 1972). In some species, such as the masked lovebird *Agapornis personala,* it is used by mated pairs as a greeting ceremony or to end quarrels. In others, for example the Canada jay *Perisoreus canadensis,* billing is an appeasement signal employed by subordinate birds in flocks. The display has evidently originated as a ritualized variant of food exchange between young and adults. When a subordinate bird employs it in appeasement, it is usually similar or identical to the begging motions of a young bird, which include a squatting body posture and wing quivering. The billing of mated pairs is often accompanied by actual feeding of one bird by the other. The male masked lovebird regularly feeds the female, who remains at the nest to care for the brood. Male terns of the genus *Sterna* feed their partners just before or during copulation through motions that appear identical to the feeding of young (Nisbet, 1973).

Smiling and laughing. Van Hooff (1972) believes that smiling and laughter in human beings can be homologized in a straightforward way with similar and equally complex displays used by the other higher primates. Smiling, according to van Hooff's hypothesis, was derived in evolution from the "bared-teeth display," one of the phylogenetically most primitive social signals. The members of most primate species assume this expression when they are confronted with an aversive stimulus and have a moderate to strong tendency to flee. The display intensifies when escape is thwarted. In higher primates the bared-teeth display is commonly silent in expression. Among chimpanzees it is furthermore graded in intensity and is used flexibly to establish friendly contacts within the troop. The "relaxed open-mouth display," often accompanied by a short expirated vocalization, is a signal ordinarily associated with play. In man these two signals, the silent bared-teeth display and the relaxed open-mouth display, appear to have converged to form two poles in a new, graded series ranging from a general friendly response (smile) to play (laughter). A third kind of signal that developed from the archaic facial expressions is the bared-teeth scream display. This behavior, which is widespread in primates but missing in man, indicates extreme fear and submission, as well as readiness to attack if the animal is pressed further. (See Figure 10-2).

Ritualized flight. The male courtship of some bird species features a labored, conspicuous form of flight during which special plumage patterns are revealed to maximum advantage. An example is given in Figure 10-3. The males of many species or oedipodine grasshoppers perform display flights that appear to attract females watching from the ground. During the performance they fly upward while flashing their

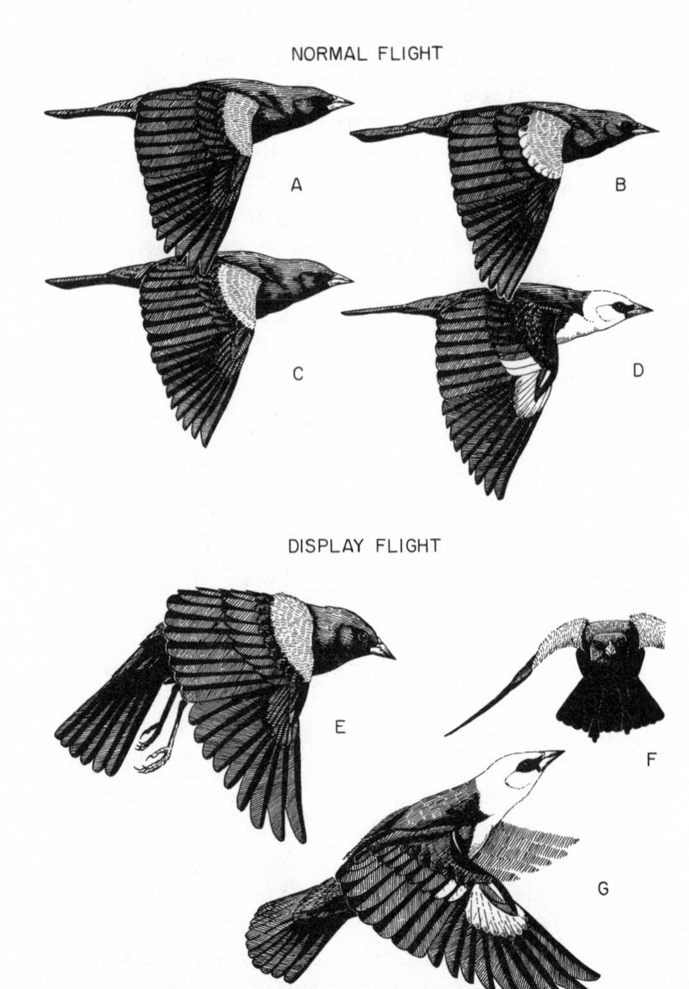

Figure 10-3 The ritualization of flight in male blackbirds. Normal flight is shown in three races of the red-winged blackbird *Agelaius phoeniceus* (*A–C*) and yellow-headed blackbird *Xanthocephalus xanthocephalus* (*D*). *E* and *F* present two views of a male red-winged blackbird in ritual flight, and *G* a side view of a male yellow-headed blackbird in the same form of display. (From Orians and Christman, 1968. Originally published by the University of California Press; reprinted by permission of the Regents of the University of California.)

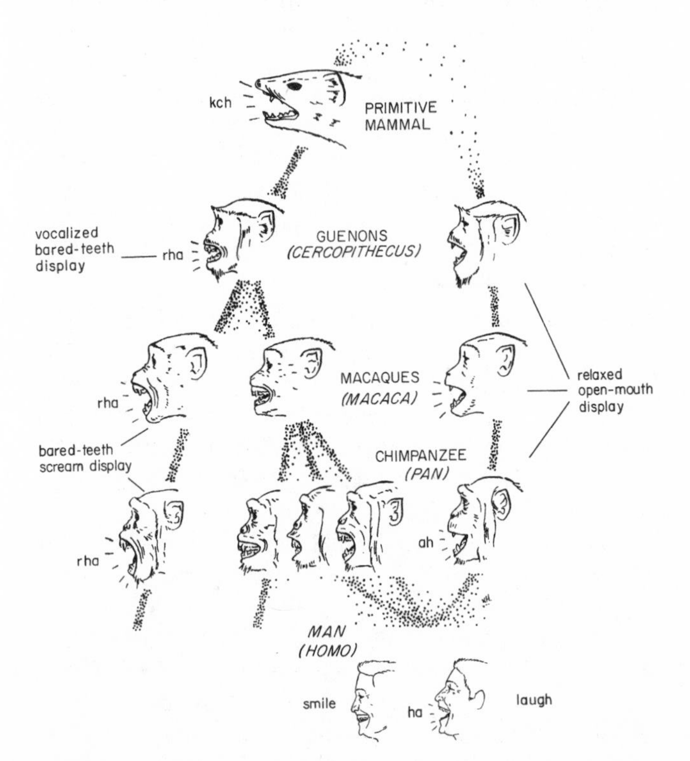

Figure 10-2 Phylogeny of facial signals in primates. (Modified from van Hooff, 1972.)

brightly colored hindwings or rapidly snapping their hindwings to create a peculiar vocalization called "crepitation" by entomologists (Otte, 1970).

Ritualized excretion and secretion. In order to mark their scent posts, various mammals use the metabolic breakdown products released in urination and defecation, as well as special glandular products released by glands associated with the urethra and anus. Some of the species, such as the giant rat of Africa (*Cricetomys gambianus*), the mongoose, and other viverrids, employ hand stands and other movements distinct from basic urination and defecation in order to deposit the scent on tree trunks and other objects above ground level. Odorous components of urine in mice have taken on a regulatory function in reproduction, serving to block or to coordinate estrus and pregnancy according to circumstances. Domestic swine boars release a substance with the urine that induces lordosis in the sow. In a parallel fashion, formicine ants lay odor trails from material in the hindgut. Although trail marking is a wholly distinctive behavior, it is reasonably hypothesized to have originated as a ritualized form of defecation.

Ritualization of waste products need not be confined to feces and urine. The sex attractant of female rhesus monkeys emanates from the vagina. It has recently been found to consist of a mixture of at least five short-chain fatty acids (Curtis et al., 1971). These substances are ordinary products of lipid metabolism and might well have been appropriated in evolution by a partial ritualization of such materials excreted at low concentrations through the epidermis.

The Sensory Channels

The concept of ritualization and its aftermath have left us with a picture of extreme opportunism in the evolution of communication systems, in which signals are molded from almost any biological process convenient to the species. It is therefore legitimate to analyze the advantages and disadvantages of the several sensory modalities as though they were competing in an open marketplace for the privilege of carrying messages. Put another, more familiar way, we can reasonably hypothesize that species evolve toward the mix of sensory cues that maximizes either energetic or informational efficiency, or both. Let us now examine the principal sensory modalities with special reference to their competitive ability, meaning the relative advantages and disadvantages of their physical properties.

Chemical Communication

Pheromones, or substances used in communication between members of the same species, were probably the first signals put to service in the evolution of life. Whatever communication occurred between the ancestral cells of blue-green algae, bacteria, and other procaryotes was certainly chemical, and this mode must have been continued among the eucaryotic protozoans descended from them. At this state of our knowledge it is still reasonable to speculate with J. B. S. Haldane that pheromones are the lineal ancestors of hormones. When the metazoan soma was organized in evolution, hormones appeared simply as the intercellular equivalent of the pheromones that mediate behavior among the single-celled organisms. With the emergence of well-formed organ systems among the platyhelminths, coelenterates, and other metazoan phyla, it was possible to create more sophisticated auditory and visual receptor systems equipped to handle as much information as the chemoreceptors of the single-celled organisms. Occasionally these new forms of communication have overridden the original chemical systems, but pheromones remain the fundamental signals for most kinds of organisms. This important fact was not fully appreciated in the early days of ethology, when attention was naturally drawn to the visual and auditory systems of birds and other large vertebrates whose sensory physiology most closely resembles our own. But now chemical systems have been discovered in many microorganisms and lower plants and in most of the principal phyla. They continue to turn up with great dependability in species whenever a deliberate search is made for them, to an extent that makes it reasonable to conjecture that chemical communication is virtually universal among living organisms. Moreover, chemical systems are at least as diverse in function as visual and auditory systems.

Chemical communication of a high degree of sophistication also occurs in exchanges between species that are closely adapted to each other, particularly between symbionts and predators and their prey. The term *allomone* has been coined by W. L. Brown and Thomas Eisner (in Brown, 1968) for interspecific chemical signals. Later Brown et al. (1970) muddied the nomenclatural water a bit by recommending a distinction between allomones, which are adaptive to the sender, and "kairomones," which are adaptive to the receiver. This is a difficult and occasionally impossible choice to make in practice, and the prudent course would seem to be to drop the latter term and continue to use "allomone" in the broader sense.

Chemical signals possess several outstanding advantages. They transmit through darkness and around obstacles. They have potentially great energetic efficiency. Less than a microgram of a moderately simple compound can produce a signal that lasts for hours or even days. Pheromones are energetically cheap to biosynthesize, and they can be broadcast by an operation as simple as opening a gland reservoir or everting a glandular skin surface. They have the greatest potential range in transmission of any kind of signal used by animals. At one extreme, pheromones are conveyed by contact chemoreception or over distances of millimeters or less, which makes them ideal for communication among microorganisms. At the other extreme, and without radical alteration of design in biosynthesis and reception, they can generate active spaces as much as several kilometers in length. The potential life of chemical signals is very great, rivaled in animal systems only by the structural cues present in nest architecture. When put down as scent posts or odor trails, pheromones also have a strange capacity for transmitting into the future. Even the animal that created the signal has the opportunity to come back and make use of it at a later time.

The outstanding disadvantages of chemical communication are slowness of transmission and fade-out. Because pheromones must be diffused or carried in a current, the animal cannot convey a message quickly over long distances, nor can it abruptly switch from one message to another. Although rats are able to distinguish the odors of dominant animals from the odors of submissive ones (Krames et al., 1969), no evidence exists of pheromones that transmit rapid changes in aggressiveness and status in the manner that is routine in auditory and visual communication. Furthermore, no case of information transfer by frequency and amplitude modulation of chemical emissions has been reported in any kind of animal, and there is abundant evidence that animals have not in general relied on the modulation of single chemical signals.

Yet animals have countered this disadvantage by resorting to another strategy—the multiplication of glands or other principal biosynthetic sites to permit the independent discharge of pheromones with different meanings. The most olfactory mammals are covered with such signal sources. The black-tailed deer *Odocoileus hemionus,* for example, produces pheromones in at least seven sites: feces, urine, tarsal glands, metatarsal glands, preorbital glands, forehead "gland," and interdigital glands. Insofar as they have been analyzed experimentally, the substances from each source have a different function (Müller-Schwarze, 1971). Additional pheromone-producing glands occur elsewhere in other kinds of mammals: the body flanks, the chin, the perineum, the pouches of female marsupials, and so on. The social insects have carried this method of informational enrichment to still greater lengths. The workers and queens of the most advanced social Hymenoptera are walking batteries of exocrine glands (see Figure 10-4).

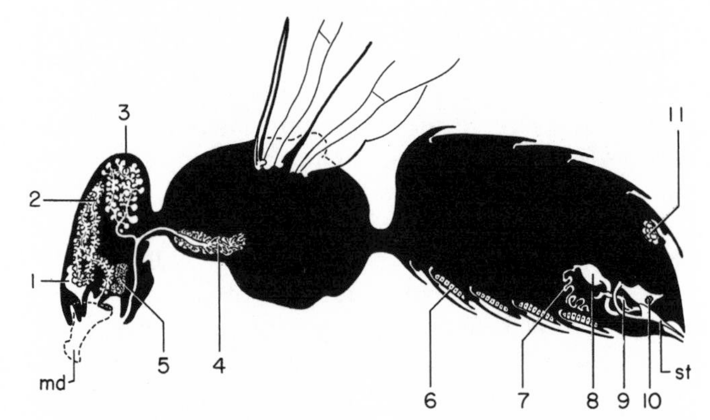

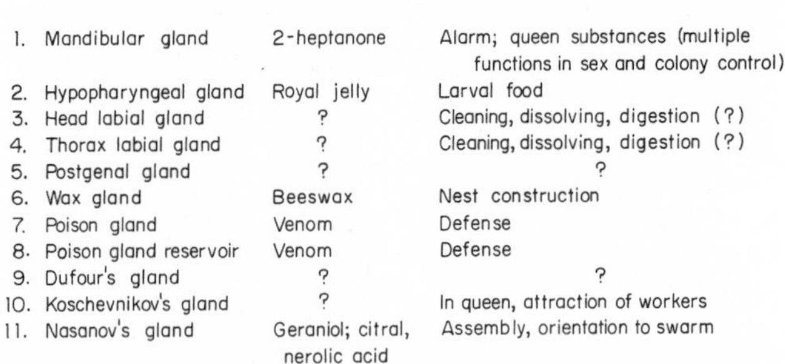

1. Mandibular gland	2-heptanone	Alarm; queen substances (multiple functions in sex and colony control)
2. Hypopharyngeal gland	Royal jelly	Larval food
3. Head labial gland	?	Cleaning, dissolving, digestion (?)
4. Thorax labial gland	?	Cleaning, dissolving, digestion (?)
5. Postgenal gland	?	?
6. Wax gland	Beeswax	Nest construction
7. Poison gland	Venom	Defense
8. Poison gland reservoir	Venom	Defense
9. Dufour's gland	?	?
10. Koschevnikov's gland	?	In queen, attraction of workers
11. Nasanov's gland	Geraniol; citral, nerolic acid	Assembly, orientation to swarm

Figure 10-4 The numerous exocrine glands of the honeybee that are devoted to social organization. The honeybee is an example of a species that has enlarged its chemical "vocabulary" by the involvement of additional glands in the production of pheromones. Not indicated in this diagram is a site at the base of the worker sting (*st*) that produces isoamyl acetate, an alarm substance. The outline of the mandible (*md*) is indicated by a dashed line.

The size of pheromone molecules that are transmitted through air can be expected to conform to certain physical rules (Wilson and Bossert, 1963). In general, they should possess a carbon number between 5 and 20 and a molecular weight between 80 and 300. The a priori arguments that led to this prediction are essentially as follows. Below the lower limit, only a relatively small number of kinds of molecules can be readily manufactured and stored by glandular tissue. Above it, molecular diversity increases very rapidly. In at least some insects, and for some homologous series of compounds, olfactory efficiency also increases steeply. As the upper limit is approached, molecular diversity becomes astronomical, so that further increase in molecular size confers no further advantage in this regard. The same consideration holds for intrinsic increases in stimulative efficiency, insofar as they are known to exist. On the debit side, large molecules are energetically more expensive to make and to transport, and they tend to be far less volatile. However, differences in the diffusion coefficient due to reasonable variation in molecular weight do not cause much change in the properties of the active space, contrary to what one might intuitively expect. Wilson and Bossert further predicted that the molecular size of sex pheromones, which generally require a high degree of specificity as well as stimulative efficiency, would prove higher than that of most other classes of pheromones, including, for example, the alarm substances. The empirical rule displayed by insects is that most sex attractants have molecular weights that are between

200 and 300, while most alarm substances fall between 100 and 200. Some of the evidence for this last statement, together with a discussion of the exceptions, has been reviewed by Wilson (1968b).

Auditory Communication

Like pheromones, sonic signals flow around obstacles and can be broadcast day and night in all weather conditions. They are intermediate in energetic efficiency between pheromones, which require little effort to transmit, and visual displays, many of which require extensive movement of the entire body. Sounds have considerable potential reach, exceeding the capacity of pheromones and light under a wide range of real conditions. Fraser Darling (1938) noted that the calls of gulls and other colonial seabirds are heard by other birds in the breeding colonies for distances of up to 200 meters. This is also approximately the reach of a great many species of songbirds and calling insects living in various habitats. Under the best of conditions, the most vocal of vertebrates can be heard for much greater distances. The roaring of male colobus and howler monkeys can be heard by human observers for over 1 kilometer. The champions among birds, and possibly terrestrial animals as a whole, are the colonially breeding grouse (Tetraonidae). The booming calls of the males reach for over a kilometer in the open country around the display grounds; and in a few species, such as the black grouse *Lyrurus tetrix* of Europe and the greater prairie chicken *Tympanuchus cupido* of North America, they can be heard for 3 to 5 kilometers. The visual displays of the same species, in contrast, can be seen for distances of no more than 1 kilometer and usually for much less (Hjorth, 1970).

This is not to say that animals evolve so as to shout over the greatest possible distances. On the contrary, the volume and frequency of animal calls seems designated to reach just those individuals of concern to the signaler and no others. To broadcast beyond them is to provide an unnecessary and dangerous homing beacon for predators. In some cases, of course, it is to the animal's advantage to project signals as far as possible. Males displaying on their leks, infants lost and in distress, and social animals calling in alarm while fleeing from a predator all seek maximum volume and transmission distance. The mobbing calls of birds provide an excellent example. They appear to be designed to carry long distances and to permit easy localization of the predatory birds that the mobs are attacking. In contrast, mothers calling young to their sides, group members maintaining contact in thick vegetation, and mated pairs performing nest-changing ceremonies use more modest, private signals that seldom reach past the ears of the intended receivers.

Moynihan (1969) has utilized this principle to explain the difference in pitch found in the calls of various species of New World monkeys. The distance champions among these animals, the howler monkeys (*Alouatta*), use low-pitched roars to signal to rival groups far out of sight in the dense canopy of the rain forest. Other species, including the tamarins and night monkeys, emit high-pitched calls, which dissipate energy in the air more rapidly than sounds of lower frequency and hence die out in shorter time. These high-frequency calls are also restricted by their greater tendency to scatter when they strike the numerous leaves and branches surrounding the signaler animal. The calls are used in a variety of circumstances, including short-range contact between neighboring troops. Moynihan has argued from several lines of evidence that the higher pitch of the signals is not an automatic outcome of the smaller size of the monkeys but is a specially evolved trait that enhances privacy and hence relief from the more intense predation generally suffered by small animals.

By far the most favorable design feature of vocal communication, the one that surely led to its adoption in the evolution of human linguistics, is its flexibility. Where pheromones must be deployed among multiple gland reservoirs to increase the rate of information transfer to any appreciable degree, all of the requisite sound signals can be generated from a single organ. Simple mechanical adjustments of the organ permit it to vary the volume, pitch, harmonic structure, and sequencing of notes that in combination create a vast array of distinguishable signals. The rapidity of the transmission of sounds and the equal quickness of their fade-out provide the basis for a very high rate of information transfer.

Tactile Communication

Communication by touch is maximally developed just where we would expect to find it, in those intimate sequences of aggregation, conciliation, courtship, and parent-offspring relations that bring animals into the closest bodily contact. For tightly aggregating species, such as hibernating beetles and pods of moving fish, body contact is both the evident goal and the signal that terminates the searching behavior. But in some instances it triggers other physiological and behavioral changes that lead animals into new modes of existence. Tactile stimulation in aphids is the dominant cue that transforms these insects from wingless into winged forms. The alates reproduce sexually and disperse much more easily, thus alleviating population pressure in the mother colony while founding new colonies on additional host plants (Lees, 1966). Locust nymphs reared in groups learn to distinguish their fellow hoppers from other dark objects of equal size, and they greet them with the typical social responses of the species—kicking their own hind legs, twirling their antennae, and inspecting the bodies of the other locusts with their palps and antennae. Peggy Ellis (1959) simulated the socialization process by rearing nymphs in isolation but in constant contact with fine, constantly moving wires. The locusts achieved a normal level of response by this form of tactile stimulation alone.

Visual Communication

Directionality is the paramount feature of systems of visual communication. Visual images are instantly pinpointed in space: the honeybee, a typical large-eyed insect, can distinguish two points that subtend an angle of approximately 1°, while the human eye, which is typically mammalian in construction, has an angle of resolution of 0.01°. Light signals lend themselves to either one or the other of two opposite strategies of signal duration. At one extreme, patterns of shading and coloration can be grown more or less permanently into the surface, or else added temporarily by special pigment deposition, chromatophore expansion and contraction, and so forth, providing signals of long duration at minimal energy cost. Hence whenever vision is possible, optic signals are found to be paramount in the identification of species as individuals, as well as of the status of individuals within dominance systems. At the opposite extreme, visual signals can be designed in such a way as to provide rapid fade-out and turnover. Consequently, they are routinely coupled in evolution with acoustic signals to transmit the most rapidly fluctuating moods of courtship and aggressive encounters.

But the distinctive features of light signals are advantageous only under limited conditions. In the absence of light, visual communication fails unless the animals can generate their own signals by bioluminescence. Visual communication further works only when the signals are directed at the photic receptors. In order to communicate with any precision, two animals must not only perform the appropriate actions but orient themselves correctly for each transmission. This probably explains the fact that although many animal species are known whose systems are wholly chemical, and many others whose systems are almost exclusively auditory, there are few if any that depend to a comparable degree on vision.

Electrical Communication

Sharks and rays, catfish, common eels (Anguillidae), and electric fish (Gymnotidae, Mormyridae, Gymnarchidae) are capable of sensing and orienting to low-frequency, feeble voltage gradients (Kalmijn, 1971; Bullock, 1973). Electroreception is widely used as a prey-seeking device. By means of feeble, steady electric fields that leak out of flatfish, sharks are able to locate these prey even when they are buried in sand. Furthermore, electric fish generate their own fields by means of electric organs consisting of highly modified muscle tissue. When prey or other objects in the water disturb the field, their presence is betrayed to the fish even when all other sensory cues are lacking (Lissmann, 1958). In view of this degree of sophistication, it is perhaps not surprising to find that at least some of the electric fish also use their fields to communicate with one another. For example, Black-Cleworth (1970) showed that individuals of *Gymnotus carapo* recognize and tend to avoid the normal electrolocating pulses of members of their own species. Attacks are also preceded and accompanied by sudden increases in the discharge frequencies, a pattern similar to the acceleration of pulses triggered when prey are located.

We do not know whether electrocommunication occurs in animals other than the electric fish because the phenomenon can only be revealed by special techniques. The advantages of this sensory channel are considerable. Like sound, electric fields can be detected in the dark, and they flow around ordinary obstacles. They are also strongly directional, and, insofar as they prove to be used by a relatively few species, they provide a high degree of privacy. At the same time, they can be used only in relatively quiet water and can be employed only over a short range.

Evolutionary Competition among Sensory Channels

If the theory of natural selection is really correct, an evolving species can be metaphorized as a communications engineer who tries to assemble as perfect a transmission device as the materials at hand permit. Microorganisms, sponges, fungi, and the lowest metazoan invertebrates are all stuck with chemoreception and tactile responses. Visual and auditory systems require multicellular receptor organs, and auditory signals require special sound-producing organs as well. Electric systems also depend on multicellular signaling and receiving devices. In general, the more primitive the organism and the simpler its body plan, the more it depends on chemical communication.

The effects of phylogenetic constraint on the selection of sensory channels are apparent to a lesser degree throughout the higher invertebrates and vertebrates. For example, consider why butterflies are colorful and silent. They seem bright and cheerful to

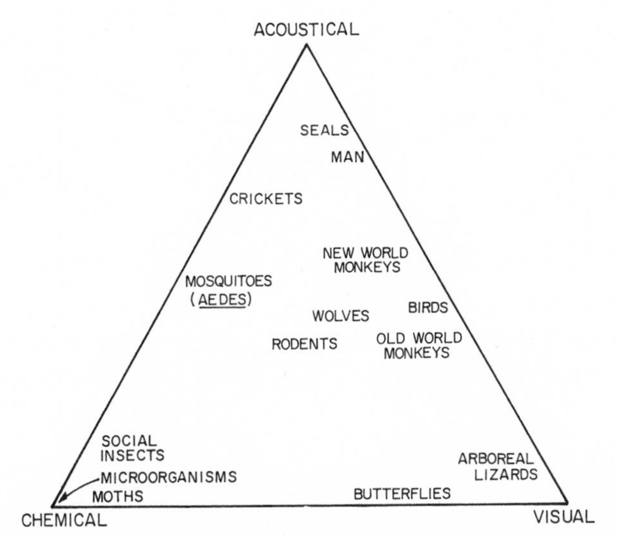

Figure 10-5 The relative importance of sensory channels in selected groups of organisms. The nearness of the group to each apex indicates, by wholly subjective and intuitive criteria, the proportionate usage of the channel in the species signal repertory. Tactile and electrical channels are not included.

us in large part because we are vertebrates heavily dependent on vision, and butterflies have tended to develop poisonous and distasteful substances to repel vertebrate predators while simultaneously evolving audacious color patterns to provide warnings about their unpalatable condition (Brower, 1969). They have also evolved distinct ultraviolet wing and body patterns, visible to one another but not to vertebrates, which serve as the medium of much of their private communication (Silberglied and Taylor, 1973). Why haven't they also evolved elaborate acoustical signals like birds? Butterflies and birds live in the same environments, fly at approximately the same heights, and communicate over comparable distances. The answer appears to be that the bodies of adult butterflies, unlike those of birds, are too small and constructed on too delicate a plan to allow the development of the noisy sound-producing machinery required to transmit effectively over long distances.

Within their personal phylogenetic constraints, species have chosen and molded sensory channels in astonishingly diverse combinations (see Figure 10-5). They have also attained an efficiency of design that should impress any human engineer. Consider butterflies again: like moths, they use sex pheromones extensively, but unlike moths, they transmit the pheromones principally by contact or through air over distances of no more than a few centimeters. The reason for this curtailment may well be that the thermal updrafts and turbulence of the daytime atmosphere preclude the formation of long active spaces. Ethologists have had no difficulty making such correlations between environment and sensory modes across widely separated phylogenetic groups. Some of the best evolutionary reconstructions have traced shifts from one modality to another at the species level and are based on enough detail to be fully persuasive.

CHAPTER 11

Aggression

What is aggression? In ordinary English usage it means an abridgment of the rights of another, forcing him to surrender something he owns or might otherwise have attained, either by a physical act or by the threat of action. Biologists cannot improve on this definition, even in the narrow context of animal behavior, except to specify that in the long term a loss to a victim is a real loss only to the extent that it lowers genetic fitness. In an attempt to be more precise, many writers have turned to the word *agonistic*, coined by Scott and Fredericson (1951) to refer to any activity related to fighting, whether aggression or conciliation and retreat. However, agonistic behavior cannot be defined any more precisely than aggressive behavior or fighting behavior in particular cases, and the term is ordinarily useful only in pointing to the close physiological interrelatedness of aggressive and submissive responses. But we should not worry too much about terminology. The essential fact to bear in mind about aggression is that it is a mixture of very different behavior patterns, serving very different functions. Here are its principal recognized forms.

1. *Territorial aggression.* The territorial defender utilizes the most dramatic signaling behavior at its disposal to repulse intruders. Escalated fighting is usually employed as a last resort in case of a stand-off during mutual displays. The losing contender has submission signals that help it to leave the field without further physical damage, but they are ordinarily not so complex as those employed by subordinate members of dominance orders. By contrast, females of bird species entering the territories of males often use elaborate appeasement signals to transmute the aggressive displays of the males into conciliation and courtship.

2. *Dominance aggression.* The aggressive displays and attacks mounted by dominant animals against fellow group members are similar in many respects to those of the territorial defenders. However, the object is less to remove the subordinates from the area than to exclude them from desired objects and to prevent them from performing actions for which the dominant animal claims priority. In some mammalian species, dominance aggression is further characterized by special signals that designate high rank, such as the strutting walk of lemmings, the leisurely "major-domo" stroll with head and tail up of rhesus macaques, and the particular facial expressions and tail postures of wolves. Subordinates respond with an equally distinctive repertory of appeasement signals.

3. *Sexual aggression.* Males may threaten or attack females for the sole purpose of mating with them or forcing them into a more prolonged sexual alliance. Perhaps the ultimate development in higher vertebrates is the behavior of male hamadryas baboons, who recruit young females to build a harem and continue to threaten and harass these consorts throughout their lives to prevent them from straying.

4. *Parental disciplinary aggression.* Parents of many kinds of mammals direct mild forms of parental aggression at their offspring to keep them close at hand, to urge them into motion, to break up fighting, to terminate unwelcome suckling, and so forth. In most but not all cases the action serves to enhance the personal genetic fitness of the offspring.

5. *Weaning aggression.* The parents of some mammal species threaten and even gently attack their own offspring at the weaning time, when the young continue to beg for food beyond the age when it is necessary for them to do so. Recent theory (see Chapter 16) suggests that under a wide range of conditions there exists an interval in the life of a young animal during which its genetic fitness is raised by continued dependence on the mother, while the mother's fitness is simultaneously lowered. This conflict of interests is likely to bring about the evolution of a programmed episode of weaning aggression.

6. *Moralistic aggression.* The evolution of advanced forms of reciprocal altruism carries with it a high

probability of the simultaneous emergence of a system of moral sanctions to enforce reciprocation (see Chapter 5). Human moralistic aggression is manifested in countless forms of religious and ideological evangelism, enforced conformity to group standards, and codes of punishment for transgressors.

7. *Predatory aggression*. There has been some question about whether predation can be properly classified as a form of aggression (for example, Davis, 1964). Yet if one considers that cannibalism is practiced by many animal species, sometimes accompanied by territoriality and other forms of aggression, and sometimes not, it is hard to regard predation as an entirely different process.

8. *Antipredatory aggression*. A purely defensive maneuver can be escalated into a full-fledged attack on the predator. In the case of mobbing the potential prey launches the attack before the predator can make a move. The intent of mobbing is often deadly and in rare instances brings injury or death to the predator.

Previous authors have stressed the eclectic nature of "aggression." Aggressive behavior serves very diverse functions in different species, and different functional categories evolve independently in more than one control center of the brain. Barlow (1968) cited an illuminating example of multiple forms of aggression coexisting in the behavior of rattlesnakes. When two males compete, they intertwine their necks and wrestle as though testing each other's strength, but they do not bite. In contrast, the snake stalks or ambushes prey—it strikes from any of a number of positions. Also, it does not give warning with its rattle. When confronted by an animal large enough to threaten its safety, the rattlesnake coils, pulls its head forward to the center of the coil in striking position, and raises and shakes its rattle. It may also rear the head and neck into a high S-shaped posture. However, if the intruder is a king snake, a species specialized for feeding on other snakes, the rattlesnake switches to a wholly different maneuver: it coils, hides its head under its body, and slaps at the king snake with one of the raised coils.

Aggression and Competition

The largest part of aggression among members of the same species can be viewed as a set of behaviors that serve as competitive techniques. Competition, as most ecologists employ the word, means the active demand by two or more individuals of the same species (intraspecific competition) or members of two or more species at the same trophic level (interspecific competition) for a common resource or requirement that is actually or potentially limiting. This definition is consistent with the assumptions of the Lotka-Volterra equations, which still form the basis of the mathematical theory of competition. The theory of population biology suggests that competitive phenomena are meaningfully divided into two large classes: *sexual competition* and *resource competition*. The former is exemplified by the violent *machismo* of males in the breeding season and especially upon the communal display grounds: the horn fighting of male sheep, deer, and antelopes, the spectacular displays and fighting among males of grouse and other lek birds, the heavyweight battles of elephant seals for the possession of harems, and others. The struggle for possession of multiple females is competition for a very special kind of resource. It becomes a significant part of the repertory when *r* selection is paramount or when other environmental pressures are relaxed to the extent that males can afford to invest the large amounts of time and energy required to be a polygamist.

Nonsexual aggression practiced within species serves primarily as a form of competition for environmental resources, including especially food and shelter. It can evolve when shortages of such resources become density-dependent factors (see Table 11-1 and the introductory discussion of density dependence in Chapter 4). However, even in this circumstance aggression is only one competitive technique among many that can emerge. For reasons that we are only beginning to understand, species may elect to compete by means of scrambling methods that do not include aggressive encounters. The following generalizations about competition in animals also pertain to the evolution of aggressive behavior (Wilson, 1971b).

1. The mechanisms of competition between individuals of the same species are qualitatively similar to those between individuals of different species.

Table 11-1 A simplified classification of the density-dependent factors that reduce population growth rates. The factors grouped under contest competition are asterisked to stress that aggressive behavior constitutes only one alternate outcome in the evolution of density-dependent controls.

A. Competition
 *1. Contest competition
  *a. Fighting and cannibalism
  *b. Territoriality
  *c. Dominance orders
 2. Scrambling competition
B. Predation and disease
C. Emigration
D. Noncompetitive modification of the environment

2. There is nevertheless a difference in intensity. Where competition occurs at all, it is generally more intense within species than between species.

3. Several theoretical circumstances can be conceived under which competition is perpetually sidestepped (Hutchinson, 1948, 1961). Most involve the intervention of other density-dependent factors of the kind just outlined or fluctuations in the environment that regularly halt population growth just prior to saturation.

4. Field studies, although still very fragmentary in nature, have tended to verify the theoretical predictions just mentioned. Competition has been found to be widespread but not universal in animal species. It is more common in vertebrates than in invertebrates, in predators than in herbivores and omnivores, and in species belonging to stable ecosystems than in those belonging to unstable ecosystems. It is often forestalled by the prior operation of other density-dependent controls, the most common of which are emigration, predation, and disease.

5. Even where competition occurs, it is frequently suspended for long periods of time by the intervention of density-independent factors, especially unfavorable weather and the frequent availability of newly created empty habitats.

6. Whatever the competitive technique used—whether direct aggression, territoriality, nonaggressive "scrambling," or something else—the ultimate limiting resource is usually food. Although the documentation for this statement (Lack, 1966; Schoener, 1968a) is still thin enough to be authoritatively disputed (Chitty, 1967), there still seem to be enough well-established cases to justify its provisional acceptance as a statistical inference. It is also true, however, that a minority of examples involve other limiting resources: growing space in barnacles and other sessile marine invertebrates; nesting sites in the pied flycatcher and Scottish ants; resting places of high moisture in salamanders and of shade in the mourning chat in African deserts; nest materials in rooks and herons.

The Mechanisms of Competition

If aggressive behavior is only one form of competitive technique, consider now a series of cases that illustrate the wide variation in this technique actually recorded among animal species. We will start with aggression in its direct and most explicit form and then look at some more subtle and indirect forms.

Direct Aggression

Fighting is common in some insect species. Ant colonies, for instance, are notoriously aggressive toward one another, and colony "warfare" both within and between species has been witnessed by many entomologists. Pontin (1961, 1963) found that the majority of the queens of *Lasius flavus* and *L. niger* attempting to start new colonies in solitude are destroyed by workers of their own species. Colonies of the common pavement ant *Tetramorium caespitum* defend their territories with pitched battles conducted by large masses of workers. The adaptive significance of the fighting has been made clear by the recent discovery that the average size of the worker and the production of winged sexual forms at the end of the season, both of which are good indicators of the nutritional status of the colony, increase with an increase in territory size (Brian, Elmes, and Kelly, 1967).

Even cannibalism is normal among the members of some insect species. In the life cycle of certain species of parasitic wasps belonging to the families Ichneumonidae, Trigonalidae, Platygasteridae, Diapriidae, and Serphidae the larvae undergo a temporary transformation into a bizarre fighting form that kills and eats other conspecific larvae occupying the same host insect. This reduces the number of parasites so that the less numerous survivors can more easily grow to the adult stage on the limited host tissue available.

Murder and cannibalism are also commonplace in the vertebrates. Lions, for example, sometimes kill other lions. In his study of the Serengeti prides, Schaller (1972) observed several fights between males that ended fatally. He also recorded a case of the killing and cannibalism of cubs after one of the protector males died and the territory was invaded by several other prides. Less severe fighting is more frequent, and it results in injuries and infections that ultimately shorten the lives of many individuals. Hyenas are truly murderous by human standards, and are also habitual cannibals. Mothers must stand guard while their cubs are feeding on a carcass in order to prevent them from being eaten by other members of the clan; and neighboring clans sometimes engage in pitched battles over carcasses of prey that one or the other of the groups has killed. Male Japanese and pig-tailed macaques have been seen to kill one another under seminatural and captive conditions when fighting for supremacy. When a new group of Barbary macaques was introduced into the Gibraltar population, severe fighting broke out that resulted in some deaths (Keith, 1949). In central India, roaming langur males sometimes invade established troops, oust the dominant male, and kill all of the infants (Sugiyama, 1967). Young black-headed gulls that wander from the parental nest territory are attacked and sometimes killed by other gulls (Armstrong, 1947), while in the brown booby (*Sula leucogaster*) of Ascension Island, the first-hatched young routinely thrusts its second-hatched sibling from the nest (Simmons, 1970). A case of cannibalism of an infant by an adult has even been reported in the chimpanzee, but the event does appear to be truly rare (Suzuki, 1971).

The evidence of murder and cannibalism in mammals and other vertebrates has now accumulated to the point that we must completely reverse the conclusion advanced by Konrad Lorenz in his book *On Aggression,* which subsequent popular writers have proceeded to consolidate as part of the conventional wisdom. Lorenz wrote, "Though occasionally, in the territorial or rival fights, by some mishap a horn may penetrate an eye or a tooth an artery, we have never found that the aim of aggression was the extermination of fellow members of the species concerned." On the contrary, murder is far more common and hence "normal" in many vertebrate species than in man. I have been impressed by how often such behavior becomes apparent only when the observation time devoted to a species passes the thousand-hour mark. But only one murder per thousand hours per observer is still a great deal of violence by human standards. In fact, if some imaginary Martian zoologist visiting Earth were to observe man as simply one more species over a very long period of time, he might conclude that we are among the more pacific mammals as measured by serious assults or murders per individual per unit time, even when our episodic wars are averaged in. If the visitor were to be confined to George Schaller's 2,900 hours and one randomly picked human population comparable in size to the Serengeti lion population, to take one of the more exhaustive field studies published to date, he would probably see nothing more than some play-fighting—almost completely limited to juveniles—and an angry verbal exchange or two between adults. Incidentally, another cherished notion of our wickedness starting to crumble is that man alone kills more prey than he needs to eat. The Serengeti lions, like hyenas, sometimes kill wantonly if it is convenient for them to do so. As Schaller concludes, "the lion's hunting and killing patterns may function independently of hunger."

There is no universal "rule of conduct" in competitive and predatory behavior, any more than there is a universal aggressive instinct—and for the same reason. Species are entirely opportunistic. Their behavior patterns do not conform to any general innate restrictions but are guided, like all other biological traits, solely by what happens to be advantageous over a period of time sufficient for evolution to occur. Thus if it is of even temporary selective advantage for individuals of a given species to be cannibals, at least a moderate probability exists that the entire species will evolve toward cannibalism.

Mutual Repulsion

When workers of the ants *Pheidole megacephala* and *Solenopsis globularia* meet at a feeding site, some fighting occurs, but the issue is not settled in this way. Instead, dominance is based on organizational ability. Workers of both species are excitable and run off the odor trails and away from the food site when they encounter an alien. The *Pheidole* calm down, relocate the odor trails, and assemble again at the feeding site more quickly than the *Solenopsis.* Consequently, they build up their forces more quickly during the clashes and are usually able to control the feeding sites. *S. globularia* colonies are nevertheless able to survive by occupying nest sites and foraging areas in more open, sandy habitats not penetrated by *P. megacephala* (Wilson, 1971b).

Other examples are known in which competition for resources is conducted by indirect forms of repulsion. Females of the tiny wasp *Trichogramma evanescens* parasitize the eggs of a wide variety of insect host species by penetrating the chorion with their ovipositors and inserting their own eggs inside. Other females of the same species are able to distinguish eggs that have already been parasitized, evidently through the detection of some scent left behind in trace amounts by the first female; thus alerted, they invariably move on to search for other eggs (Salt, 1936).

The Limits of Aggression

Why do animals prefer pacifism and bluff to escalated fighting? Even if we discount the very large number of species in which density-dependent controls are sufficiently intense to prevent the populations from reaching competitive levels, it still remains to be explained why overt aggression is lacking among most of the rest of the species that do compete. The answer is probably that for each species, depending on the details of its life cycle, its food preferences, and its courtship rituals, there exists some optimal level of aggressiveness above which individual fitness is lowered. For some species this level must be zero, in other words the animals should be wholly nonaggressive. For all others an intermediate level is optimal.

There are at least three kinds of constraints on the evolutionary increase of aggressiveness. First, a danger exists that the aggressor's hostility will be directed against unrecognized relatives. If the rates of survival and reproduction among relatives are thereby lowered, then the replacement rate of genes held in common between the aggressor and its relatives will also be lowered. Since these genes will include the ones responsible for aggressive behavior, such a reduction in inclusive fitness will work against aggressive behavior as well. This process will continue until the difference between the advantage and disadvantage, measured in units of inclusive fitness, is maximized. Second, an aggressor that attacks an opponent with intent to destroy can expect to receive an all-out defense in return, with the added chance that both will be injured or killed. The optimum level of hostility is the minimum that will guarantee victory (Maynard Smith and Price, 1973). Third, an aggressor spends time in aggression that could be invested in courtship, nest building, and the feeding and rearing of young. Dominant white leghorn hens, for example, have greater access to food and roosting space than subordinates, but they present less to cocks and hence are mated fewer times.

In addition, the particularities of the environments of different species can sometimes be related directly to the forms and intensities of aggressive behavior that characterize them. Species of chipmunks (*Tamias* and *Eutamias*), for example, vary notably in the amount of territorial defense they display. According to Heller (1971), territorial intensity is determined in evolution by the interaction of the magnitude of the need to gain absolute control over the territory and the cost of defending the territory in terms of energy loss and risk from predators. These factors differ greatly from one habitat to another, sufficiently, according to Heller, to account for the fact that some *Eutamias* species are strongly territorial while others are apparently nonterritorial.

The Proximate Causes of Aggression

Aggression evolves not as a continuous biological process as the beat of the heart, but as a contingency plan. It is a set of complex responses of the animal's endocrine and nervous system, programmed to be summoned up in times of stress. Aggression is genetic in the sense that its components have proved to have a high degree of heritability and are therefore subject to continuing evolution. The documentation for this statement is substantial and has been reviewed by Scott and Fuller (1965) and McClearn (1970). Aggression is also genetic in a second, looser sense, meaning that aggressive and submissive responses of some species are specialized, stereotyped, and highly predictable in the presence of certain very general stimuli. The adaptive significance of aggression, its ultimate causation and the environmental pressures that guide the natural selection of its genotypic variation, should be an object of analysis whenever aggressive or submissive components are discerned in any form of social behavior.

The proximate causes of the variation will now be examined. They are most easily understood when classified into two sets of factors. The first is the array of external environmental contingencies to which the animal must be prepared to respond, including encounters with strangers from outside the social group, competition for resources with other members of its own group, and daily and seasonal changes in the physical environment. All of these exigencies provide stimuli to which the animal's aggressive scale must be correctly adjusted. The second set of stimuli is the internal adjustments through learning and endocrine change by which the animal's aggressive responses to the external environment are made more precise.

External Environmental Contingencies

Encounters outside the group. The strongest evoker of aggressive response in animals is the sight of a stranger, especially a territorial intruder. This xenophobic principle has been documented in virtually every group of animals displaying higher forms of social organization. Male lions, normally the more lethargic adults of the prides, are jerked to attention and commence savage rounds of roaring when strange males come into view. Nothing in the day-to-day social life of an ant colony, no matter how stressful, activates the group like the introduction of a few alien workers. The principle extends to the primates. Southwick (1967, 1969) conducted a series of controlled experiments on confined rhesus monkeys in order to weigh the relative importance of several major factors in the evocation of aggression. Food shortages actually caused a decrease in aggressive-submissive interactions, since the animals reduced all social exchanges and began to devote more time to slow, tedious explorations of the enclosure. Crowding of the monkeys induced a somewhat less than twofold increase in aggressive interactions. The introduction of strange rhesus monkeys, however, caused a fourfold to tenfold increase in such interactions. The experiment put a more precise measure on what is observed commonly in the wild. The rate of aggression displayed when two rhesus groups meet, or a stranger attempts to enter the groups, far exceeds that seen within the troops as they pass through the stressful episodes of their everyday life.

Food. The relation of aggressive behavior to the supply and distribution of food is generally complex in animals and difficult to predict for any particular species. In general, aggressive-submissive exchanges increase sharply when food is clumped instead of scattered and domination of one piece of the food or of a small area of ground on which food is concentrated becomes profitable. Baboons ordinarily forage like flocks of birds, fanning out in a search for small vegetable items that are picked off the ground and eaten quickly. The troop members seldom challenge one another under these circumstances. But when a clump of grass shoots is discovered in elephant dung, or a small animal is killed, the baboons threaten one another and may even fight over the food.

Crowding. As animals move into ever closer proximity, the rate at which they encounter one another goes up exponentially. All other things being equal, the frequency of aggressive interactions goes up at the same exponential rate. The space-aggression curves of some species, however, are more complex. At intermediate densities crayfish (*Orconectes virilis*) form territories, but at extremely high densities they collapse into peaceful aggregations (Bovbjerg and Stephen, 1971). When individuals of *Dascyllus aruanus,* an Australian reef fish, are crowded around pieces of synthetic coral in increasing densities, the rate of their aggressive encounters first rises, then drops. Aggression in these animals also rises as a function of group size quite independently of density (Sale, 1972).

Seasonal change. The aggressive interactions of most animal species peak in the breeding season. Fighting among tigers, for example, is limited to the contest between males for estrous females. Baikov (1925) described his experience with the Manchurian tiger as follows: "I have spent many nights in the taiga alone with my fellow hunters, sitting by the fire and listening to tigers challenging their rivals—resounding through the frost-bound forests; but though the battle ground is invariably drenched with blood, such encounters never end in death." The sifaka (*Propithecus verreauxi*), a Madagascan lemur, is placid through most of the year but erupts in savage fighting during the breeding season (Alison Jolly, 1966). The female reindeer is a passive animal during most of her life. But just before and after giving birth she becomes aggressive toward other herd members, especially toward the yearlings. Other seasonal patterns can be cited at length from the literature on the life histories of both the vertebrates and invertebrates.

Learning and Endocrine Change

Previous experience. A variety of experiences in the life of an animal can influence the form and intensity of its aggressive behavior. Aggression in laboratory rats can be increased by straightforward instrumental training. The behavior amplified in these studies is the "pain-aggression" response: when two rats are presented with certain painful stimuli, such as an electric shock, they attack each other by standing face-to-face on their hind legs, thrusting their heads forward with mouths open, and vigorously lunging and biting at each other.

Instrumental amplification of aggressive behavior is a laboratory manifestation of the socialization by which animals learn their place in territorial and dominance relationships under natural conditions. As animals move up in rank, their readiness to attack increases, particularly when they encounter rivals who have been defeated previously. Animals that are defeated consistently in encounters with one set of opponents become psychologically "down," display timidity when they encounter new sets of opponents, and thus are more likely to retain their low rank than others who have known early triumphs (Ginsburg and Allee, 1942; McDonald et al., 1968). Although the effect is generally thought of as a vertebrate trait, it also occurs in such insects as bumblebees and crickets (Alexander, 1961; Free, 1961a).

The normal aggressive responses of mammals are also influenced by the socialization process. Male house mice (*Mus musculus*) reared in isolation after weaning are less aggressive than those reared in social groups. The longer the mice are exposed to others, the greater their aggressiveness toward strangers at later times. The critical period is relatively long; in experiments by King (1957) isolation as late as 20 days still had a depressing effect on subsequent responses.

Hormones and aggression. The endocrine system of vertebrates acts as a relatively coarse tuning device for the adjustment of aggressive behavior. The interactions of the several hormones involved in this control are complicated (see Figure 11-1). However, they can be understood readily if the entire system is viewed as comprising three levels of controls: the first determines the state of preparedness (androgen, estrogen, and luteinizing hormone), the second the capacity for a quick response to stress (epinephrine), and the third the capacity for a slower, more sustained response to stress (adrenal corticoids).

The level of preparedness to fight is what we usually refer to as aggressiveness, in order to contrast it with the act of aggression. Aggressiveness, as Rothballer (1967) has said, is a threshold. It can be measured either by the amount of the provoking stimulus required to elicit the act or by the intensity and prolongation of the act in the face of a given stimulus. The class of hormones most consistently associated in the vertebrates with heightened aggressiveness is the androgens, which are 19-carbon

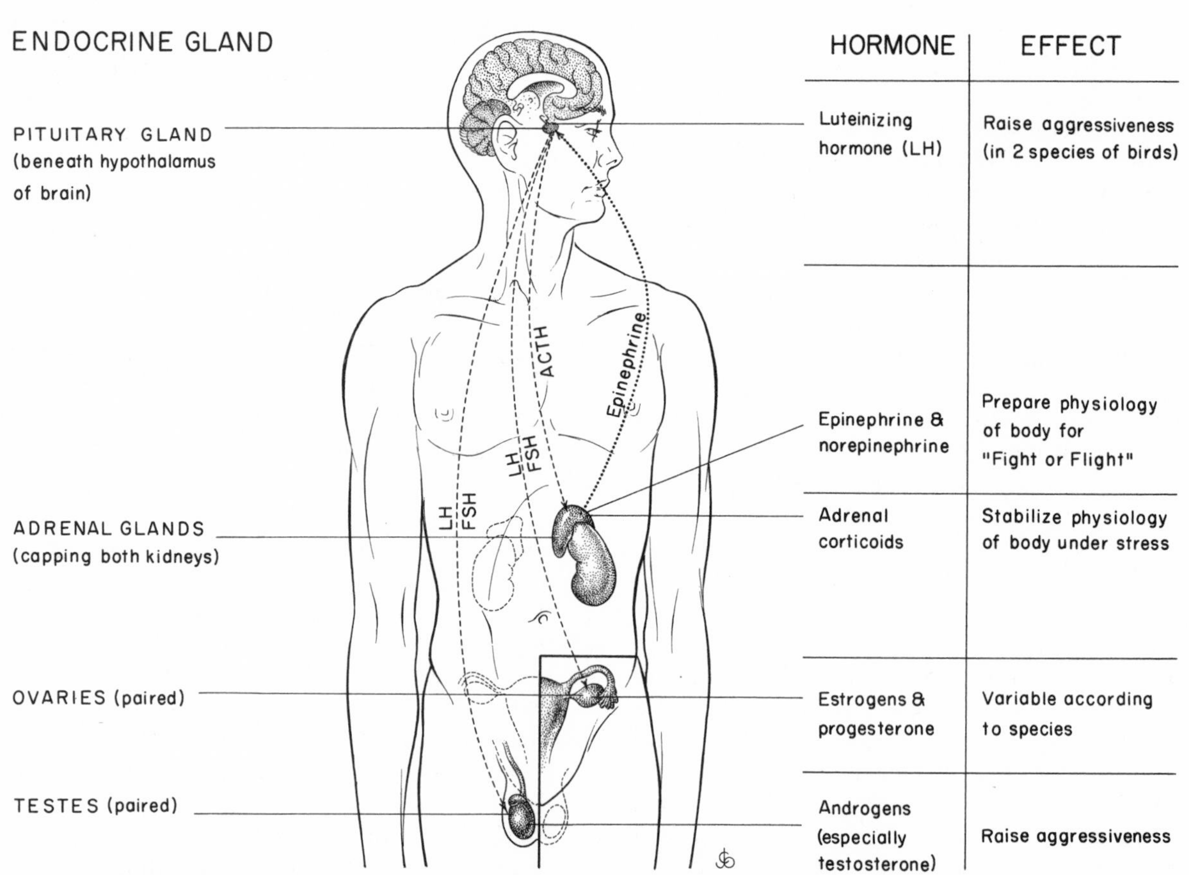

Figure 11-1 The principal hormones that affect aggressive behavior in mammals. The pituitary gland, stimulated by impulses from the hypothalamus and to a lesser degree by epinephrine ("adrenalin"), releases corticotrophin (ACTH), which enlarges the cortex of the adrenal gland and raises the output of the adrenal corticoids. The pituitary gland also releases the luteinizing hormone (LH), which stimulates the production of androgens in the testes of the male. In the female, LH acts synergistically with the follicle-stimulating hormone (FSH) to promote the secretion of estrogen by the follicles of the ovary. Stimulation of certain neurons of the autonomic nervous system, controlled mainly by the hypothalamus, causes the medulla of the adrenal gland to release epinephrine. This scheme is based on results obtained piecemeal from a variety of vertebrate species, and the use here of the human system is for convenience only.

steroids, with methyl groups at C-10 and C-13, secreted by the Leydig cells of the testes. The behaviorally most potent androgen is evidently testosterone. It has been known since the experiments of Arnold Berthold in 1849 that roosters stop crowing and fighting when they are castrated, but retain these behaviors if testes from other roosters are implanted in their abdominal cavities. In recent years it has been demonstrated that the behaviors can be restored by injection of appropriate amounts of testosterone proprionate. A similar effect has been demonstrated in a wide range of species, including swordtail fish, gobies (*Bathygobius*), anolis lizards, fence lizards (*Sceloporus*), painted turtles (*Chrysemys*), night herons, doves, songbirds, quail, grouse, deer, mice, rats, and chimpanzees. Immature males, including boys, can be brought into maturity more quickly by injections of testosterone proprionate, and in some species even the behavior of females can be strongly masculinized.

The effects of the androgens extend deeply into physiological and social traits that are coupled to aggressiveness. When Mongolian gerbils (*Meriones unguiculatus*) of either sex are castrated, they resorb the ventral sebaceous glands with which territories are marked. The glands are regenerated and territorial behavior resumed when testosterone proprionate is injected (Thiessen, Owen, and Lindzey, 1971). As Allee, Collias, and Lutherman first showed in 1939, hens given small doses of testosterone become more aggressive and move up in rank within the dominance hierarchies of the flocks. Watson and Moss (1971) reported that red grouse cocks (*Lagopus lagopus*) implanted with androgen became more aggressive, nearly doubled their territory size, devoted more time to courtship, and mated with two hens instead of the usual one. Two nonterritorial cocks that had been in poor physical condition regained good condition and drove back territorial cocks to establish new territories on their own. Although they remained unmated that year, they survived the winter and set up territories the following year. One of the cocks was able to acquire a hen the following summer. Without the implant both cocks would almost certainly have perished during the winter.

Among the vertebrates, a seasonal rise in the androgen titer of males is generally associated with an increase in aggressiveness, an establishment or enlargement of territory in those species that are territorial, and the onset of sexual behavior. In short, the androgens initiate the breeding season. Also, dominance among males is correlated with their androgen level. The relation between the hormone and behavior is nevertheless much more complicated than a simple chemical reaction. In higher vertebrates dominance depends to a large extent on experience and on the deference shown by other members of the group on the basis of past performance. Birds are notably inconsistent in their reactions. Davis (1957), for example, showed that testosterone does not affect the rank of starlings in the roost hierarchies. Males of blackbirds (*Turdus merula*) and many other bird species continue to defend territories in the fall and winter, when the gonads are very small (Snow, 1958). The explanation of these inconsistencies could lie in the continued role of even low levels of androgen or in the overrriding effects of other hormones. A new range of possibilities was opened by Mathewson's discovery (1961) that injections of the luteinizing hormone (LH) increase aggressiveness and dominance rank in starlings where testosterone fails. One function of LH is to stimulate the production of testosterone. Davis (1964) has suggested that LH has the more fundamental role in controlling aggression and that the function has been shared with testosterone only as a later evolutionary development. However, the data are not yet adequate to test this hypothesis.

Among higher primates the relationship of androgens to aggressiveness may be more complex still. Rose et al. (1971) found that plasma testosterone levels are correlated with aggressiveness in male rhesus monkeys. However, the correlation with rank in the dominance order was less precise, since high-ranking males had lower levels. Equally surprising, the testosterone titers of the lowest-ranking males were higher than those of solitary-caged animals. The possibility exists that aggressiveness induces testosterone secretion, via the brain-pituitary-testis route, rather than the other way around. Or, equally likely, hormone production and the behavior pattern are both enhanced by other, as yet unidentified stimuli, such as experience and input from other hormones, that influence the pituitary-testis axis.

The estrogens induce a confusing variety of effects on aggressive behavior. The great bulk of these substances is produced by the ovaries, but small quantities are also found in the adrenals, placentae, testes, and even spermatozoans. Thus although the estrogens are primarily female hormones, they can also play some role in male physiology. In general, high estrogen levels promote feminization, female sexual responses, and hence less aggressiveness except when the individual is defending young or, less frequently, competing with other adults. Male vertebrates injected with estrogen typically become less aggressive. A red grouse cock implanted with estrogen by Adam Watson (Watson and Moss, 1971) lost his hen and eventually his territory. In contrast, castrated males of the golden hamster (*Mesocricetus auratus*) regain aggressive traits when injected with estrogen, while females remain unaffected (Vandenbergh, 1971). When female hamsters are given both estrogen and progesterone, which duplicate the conditions for a normal estrus, aggressiveness is stongly suppressed (Kislak and Beach, 1955). Female chimpanzees become more aggressive with estrogen treatment (Birch and Clark, 1946). In humans, the effects are either negligible or so subtle that they can be defined only by psychoanalytic study (Gottschalk et al., 1961). There does seem to be some diminution of anxiety and aggressiveness in women in the middle of the menstrual cycle, when ovulation is scheduled to occur (Ivey and Bardwick, 1968). At this time receptivity should be greatest, and both estrogen and progesterone levels are at their peak. In sum, we can infer from the fragmentary evidence that estrogen influences vertebrate aggressivity in ways that are highly conditional. When it is adaptive to be sexually submissive, particularly at estrus, when progesterone levels are high, aggressiveness is suppressed. At other times estrogen may actually raise aggressiveness in a way that helps a female to maintain status and to defend offspring. The inhibiting effect on male aggressiveness could well be a meaningless artifact.

Given that LH and the gonadal hormones maintain a vertebrate in a state of readiness appropriate to its rank and reproductive status, epinephrine is the hormone by which it makes a fine adjustment to emergencies that arise moment by moment. Epinephrine is a catecholamine, a derivative of tyrosine secreted primarily by the medullas of the adrenal glands. Release of epinephrine into the bloodstream is stimulated by sympathetic nerves and thus ultimately falls under the command of the hypothalamus. The substance is complementary to the other principal catecholamine, norepinephrine, which is coupled with the parasympathetic nervous system and has generally different, sometimes opposite physiological effects. Epinephrine acts quickly in conjunction with the sympathetic system to prepare the entire body for "fight or flight." The heart rate and systolic blood pressure go up. Vasodilation occurs over the body, and the eosinophil count rises. The blood flow through the skeletal muscle, brain, and liver increases as much as 100 percent. Blood sugar rises. Digestion and reproductive functions are inhibited. In man at least there is also an onset of a feeling of anxiety. Epinephrine is released whenever the vertebrate is placed in a stressful situation, whether low temperatures, a "narrow escape," or hostility from some other member of the species. The hormone does not itself cause the animal to be aggressive but instead prepares it to be more efficient during aggressive encounters. Under certain conditions epinephrine also acts to promote the release of corticotrophin (ACTH) from the anterior lobe of the pituitary, with a consequent release of adrenal corticoids and a gearing of the body for more prolonged adjustment to stress.

Norepinephrine is also released in response to general stress, but independently of epinephrine. Where epinephrine triggers a massive general response of the body, mobilizing glycogen as blood glucose and redistributing blood to the action centers, norepinephrine acts mainly to sustain blood pressure. It promotes heart action and vasodilation while having relatively little effect on the rate of blood flow or metabolism. Thus epinephrine conforms closely to the classical emergency theory of medullary adrenal action, with norepinephrine playing a secondary, principally regulatory role. A curious effect discovered in human beings is that violent participation in aggressive encounters induces the release of relatively large quantities of norepinephrine together with only moderate amounts of epinephrine, while the *anticipation* of aggressive interaction, in the form of anger or fear, favors only the release of epinephrine. Professional hockey players on the bench, for example, secrete only epinephrine at the same time that their teammates playing on the floor are secreting mostly norepinephrine.

The subject of behavioral endocrinology has been dominated almost exclusively by vertebrate studies. It will perhaps surprise the reader to learn that this circumstance may be wholly justified. No evidence exists, at least to my knowledge, of any hormonal system that regulates aggressive behavior in invertebrates, including insects (see also Barth, 1970, and Truman and Riddiford, 1974). Ewing (1967) has reported death in *Naupheta* cockroaches associated with aggression-induced stress, but this in no way implicates the endocrine system. Furthermore, although the development of sex differences in insects is mediated by hormones, the direct physiological effects are not known to include the alteration of aggressiveness.

Human Aggression

Is aggression in man adaptive? From the biologist's point of view it certainly seems to be. It is hard to believe that any characteristic so widespread and easily invoked in a species as aggressive behavior is in man could be neutral or negative in its effects on individual survival and reproduction. To be sure, overt aggressiveness is not a trait in all or even a majority of human cultures. But in order to be adaptive it is enough that aggressive patterns be evoked only under certain conditions of stress, such as those that might arise during food shortages and periodic high population densities. It also does not matter whether the aggression is wholly innate or is acquired partly or wholly by learning. We are now sophisticated enough to know that the capacity to learn certain behaviors is itself a genetically controlled and therefore evolved trait.

Such an interpretation, which follows from our information on patterned aggression in other animal species, is at the time very far removed from the sanguinary view of innate aggressiveness which was expressed by Raymond Dart (1953) and had so much influence on subsequent authors:

> The blood-bespattered, slaughter-gutted archives of human history from the earliest Egyptian and Sumerian records to the most recent atrocities of the Second World War accord with early universal cannibalism, with animal and human sacrificial practices or their substitutes in formalized religions and with the worldwide scalping, headhunting, body-mutilating and necrophiliac practices of mankind in proclaiming this common blood lust differentiator, this mark of Cain that separates man dietetically from his anthropoidal relatives and allies him rather with the deadliest of Carnivora.

This is very dubious anthropology, ethology, and genetics. It is equally wrong, however, to accept cheerfully the extreme opposite view, espoused by many anthropologists and psychologists (for example, Montagu, 1968a,b) that aggressiveness is only a neurosis brought out by abnormal circumstances and hence, by implication, nonadaptive for the individual. When T. W. Adorno, for example, demonstrated (in *Authoritarian Personality*) that bullies tend to come from families in which the father was a tyrant and the mother a submerged personality, he identified only one of the environmental factors affecting expression of certain human genes. Adorno's finding says nothing about the adaptiveness of the trait. Bullying behavior, together with other forms of aggressive response to stress and unusual social environments, may well be adaptive—that is, programmed to increase the survival and reproductive performance of individuals thrown into stressful situations. A revealing parallel can be seen in the behavior of rhesus monkeys. Individuals reared in isolation display uncontrolled aggressiveness leading frequently to injury. Surely this manifestation is neurotic and nonadaptive for the individuals whose behavioral development has been thus misdirected. But it does not lessen the importance of the well-known fact that aggression is a way of life and an important stabilizing device in the free-ranging rhesus societies.

This brings us to the subject of the crowding syndrome and social pathology. Leyhausen (1965) has graphically described what happens to the behavior of cats when they are subjected to unnatural crowding: "The more crowded the cage is, the less relative hierarchy there is. Eventually a despot emerges, 'pariahs' appear, driven to frenzy and all kinds of neurotic behaviour by continuous and pitiless attack by all others; the community turns into a spiteful mob. They all seldom relax, they never look at ease, and there is a continuous hissing, growling, and even fighting. Play stops altogether and locomotion and exercises are reduced to a minimum." Still more bizarre effects were observed by Calhoun (1962a,b) in his experimentally overcrowded laboratory populations of Norway rats. In addition to the hypertensive behavior seen in Leyhausen's cats, some of the rats displayed hypersexuality and homosexuality and engaged in cannibalism. Nest construction was commonly atypical and nonfunctional, and infant mortality among the more disturbed mothers ran as high as 96 percent.

Such behavior is obviously abnormal. It has its close parallels in certain of the more dreadful aspects of human behavior. There are some clear similarities, for example, between the social life of Calhoun's rats and that of people in concentration and prisoner-of-war camps. We must not be misled, however, into thinking that because aggression is twisted into bizarre forms under conditions of abnormally high density, it is therefore nonadaptive. A much more likely circumstance for any given aggressive species, and one that I suspect is true for man, is that the aggressive responses vary according to the situation in a genetically programmed manner. It is the total *pattern* of responses that is adaptive and has been selected for in the course of evolution.

The lesson for man is that personal happiness has very little to do with all this. It is possible to be unhappy and very adaptive. It we wish to reduce our own aggressive behavior, and lower our catecholamine and corticosteroid titers to levels that make us all happier, we should design our population densities and social systems in such a way as to make aggression inappropriate in most conceivable daily circumstances and, hence, less adaptive.

CHAPTER 12

Social Spacing, Including Territory

Some animals, planktonic invertebrates for example, drift through life without fixed reference points in space. They contact other members of the species only fleetingly as sexual partners and serve briefly, if at all, as parents. Other animals, including nearly all vertebrates and a large number of the behaviorally most advanced invertebrates, conduct their lives according to precise rules of land tenure, spacing, and dominance. These rules mediate the struggle for competitive superiority. They are enabling devices that raise personal or inclusive genetic fitness. In order to understand the rules it is necessary to begin with an elementary classification of special social relationships in which they are involved:

Total range: the entire area covered by an individual animal in its lifetime (Goin and Goin, 1962).

Home range: the area that an animal learns thoroughly and habitually patrols (Seton, 1909; Burt, 1943). In some cases the home range may be identical with the total range; that is, the animal familiarizes itself with one area and never leaves it. Many times the home range and the territory are identical, meaning that the animal excludes other members of the same species from all of its home range. In the great majority of species, however, the home range is larger than the territory, and the total range is much larger than both. Ordinarily, the home range is patrolled for food, but in addition it may contain familiar look-out positions, scent posts, and emergency retreats. It can also be shared jointly by the members of an integrated social group.

Core area: the area of heaviest regular use within the home range (Kaufmann, 1962). Core areas can be confidently delimited in such species as coatis and baboons, in which they are associated with sleeping sites located in a more or less central position in the home range. Fundamentally, however, the precise limits of both the home range and the core area are arbitrary, depending on the time the observer spends in the field and the minimum number of times he requires an animal to visit a given locality in order for the locality to be included. The solution to the problem, as Jennrich and Turner (1969) have pointed out, is simply to define a home range as the area of the smallest subregion of the total range that accounts for a specified proportion of the summed utilization. A smaller proportion can be selected to circumscribe the core area as any subregion in which the visitation is strongly disproportionate. The two specifications should prove most useful when comparing societies or the social systems of different species.

Territory: an area occupied more or less exclusively by an animal or group of animals by means of repulsion through overt defense or advertisement (Noble, 1939; J. L. Brown, 1964; Wilson, 1971b). As I will show in a later discussion, this definition is but one nuance of several that have been advanced during the past twenty-five years. It has been selected here because it fits the intuitive concept of most investigators and, more important, is the most practical in application. The territory need not be a fixed piece of geography. It can be "floating" or *spatiotemporal* in nature, meaning that the animal defends only the area it happens to be in at the moment, or during a certain time of day or season, or both.

Individual distance (social distance): the minimum distance that an animal routinely keeps between itself and other members of the same species (Hediger, 1941, 1955; Conder, 1949). Each species has a characteristic minimum distance that can be measured when animals are not on their territories. The measurement becomes meaningless on territories, since the minimum distance to the nearest neighbor then changes continuously as the animal moves about inside its territory. Outside the territory, individual distance varies from zero in aggregating species to a meter or more in some large birds and mammals. When this distance is greater than zero, the animal enforces the spacing by either retreating from the encroaching neighbor or threatening it away. Individual distance is not to be confused with *flight distance*, the minimum distance an animal will allow a predator to approach before moving away (Hediger, 1950).

Dominance: the assertion of one member of a group over another in acquiring access to a piece of food, a mate, a place to display, a sleeping site or any other requisite that adds to the genetic fitness of the dominant individual (see Chapter 13).

When the behaviors of many animal species are compared, their separate manifestations of home range, territory, individual distance, and dominance are seen to form a continuously graded series. Each species occupies its own position along the gradients. Some encompass a large segment of one or more of the gradients, utilizing them as behavioral scales to provide variable responses to a changing environment. Others are fixed at one point.

In this chapter I will present the most general qualities of territorial behavior adduced from field and laboratory studies, and attempt to account for at least some of them by newly developed concepts in population ecology.

A "Typical" Territorial Species

Figure 12-1 shows a mutual threat display between two male pike blennies, illustrating what zoologists consider to be some of the most general characteristics of aggressive behavior, and particularly of territorial behavior. For the reader unfamiliar with the natural history of territoriality, these fish will provide an interesting introduction. The behavior displayed is "typical" in most respects: (1) it is most fully developed in adult males; (2) there is a clearly delimited area within which each male begins to display to intruders of the same species, especially other adult males; (3) the resident male—or, as in these blennies, the larger male—usually wins the contest; (4) some of the most conspicuous and elaborate behaviors of the entire species' repertory occur during these particular exchanges; (5) the posturings make the animal appear larger and more dangerous; (6) the exchanges are mostly limited to bluffing, and even if fighting occurs it does not ordinarily result in injury or death. Almost all of these generalizations are violated by one species or another. For the moment, however, let us examine *Chaenopsis* a little more closely as a baseline example.

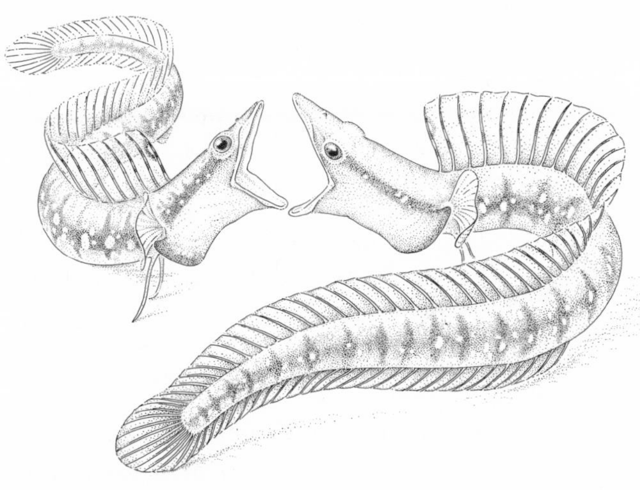

Figure 12-1 Territorial display and fighting between two male pike blennies (*Chaenopsis ocellata*). (From Wilson et al., 1973. Based on Robins, Phillips, and Phillips, 1959.)

Chaenopsis ocellata is a small fish, 7 centimeters in length, that dwells on the bottom of shallow water inshore from southeastern Florida to Cuba. The males occupy the abandoned burrows of annelid worms. The approach of any animal to within 25 centimeters excites the interest of the male, who lifts his head and erects his dorsal fin. If the intruder is another male pike blenny, the alert posture is escalated into a full-scale threat display, marked by a rapid increase in the respiratory rate, an intense darkening of the spinous portions of the dorsal fin and of the head, spreading of the pectoral fins, and finally a wide gaping of the mouth and spreading of the azure branchiostegal membranes. In most cases this dramatic transformation is enough to turn the intruder back. If the stranger persists in its onward course, however, the resident male carries through an attack. The two blennies meet snout to snout and then raise the anterior two-thirds of their bodies off the substrate, curling their tails on the bottom for support. The mouths are gaped widely and placed in contact with each other, the branchiostegal membranes are kept fully extended, and the pectoral fins are fanned rapidly to maintain position. If the two fish are nearly equal in size they may rise and fall in this ritualized combat several times without losing oral contact. If one male is smaller, he usually concedes after the completion of the first rising contact. The winning male is the one that suddenly shifts its mouth sideways and clamps down hard. The loser then abruptly folds in its dorsal fin and branchiostegal membranes, and contact is broken as both males drop to the bottom. Uninjured, the beaten male leaves the scene. Female pike blennies are not challenged by resident males. Very probably the tolerance toward them is the prelude to courtship during the breeding season (Robins et al., 1959).

The Multiple Forms of Territory

Territoriality, like other forms of aggression, has taken protean shapes in different evolutionary lines to serve a variety of functions. And like general aggression, it has proved difficult to define in a way that comfortably embraces all of its manifestations. The problem becomes simpler, however, when we notice that previous authors have tended to speak at cross purposes. A few have defined territory in terms of economic function: the territory is said to be the area which the animal uses exclusively, regardless of the means by which it manages its privacy. Pitelka (1959), for example, argued that "the fundamental importance of territory lies not in the mechanism (overt defense or any other action) by which the territory becomes identified with its occupant, but the degree to which it in fact is used exclusively by its occupant." A majority of biologists, in contrast, define territory by the mechanism through which the exclusiveness is maintained, without reference to its function. They follow G. K. Noble's (1939) simplification of Eliot Howard's concept by defining territory as any defended area. Or, to use D. E. Davis' alternate phrase, territorial behavior is simply social rank without subordinates.

I am convinced that this time the majority is right for practical reasons, that defense must be the diagnostic feature of territoriality. More precisely, territory should be defined as an area occupied more or less exclusively by animals or groups of animals by means of repulsion through overt aggression or advertisement. We know that the defense varies gradually among species from immediate aggressive exclusion of intruders to the subtler use of chemical signposts unaccompanied by threats or attacks.

Maintenance of territories by aggressive behavior has been well documented in a great many kinds of animals. Dragonflies of the species *Anax imperator*, for example, patrol the ponds in which their eggs are laid and drive out other dragonflies of their species as well as those of the similar-appearing *Aeschna juncea* by darting attacks on the wing (Moore, 1964). Orians (1961) found that the tri-colored blackbird *Agelaius tricolor* is excluded by the red-winged blackbird *A. phoeniceus* in the western United States by a different kind of interaction. Colonies of the former species do not defend territories and are consequently interspersed in the seemingly less favorable nesting sites not preempted by the aggressive *A. phoeniceus* males.

A somewhat less direct device of territorial maintenance consists of repetitious vocal signaling. Familiar examples include some of the more monotonous songs of crickets and other orthopteran insects, frogs, and birds. Such vocalizing is not directed at individual intruders but is broadcast as a territorial advertisement. An even more circumspect form of advertisement is seen in the odor "signposts" laid down at strategic spots within the home range of mammals. Leyhausen (1965) has pointed out that the hunting ranges of individual house cats overlap considerably, and that more than one individual often contributes to the same signpost at different times. By smelling the deposits of previous passersby and judging the duration of the fading odor signals, the foraging cat is able to make a rough estimate of the whereabouts of its rivals. From this information it judges whether to leave the vicinity, to proceed cautiously, or to pass on freely.

We do not have the information needed to decide whether occupied land is generally denied at certain times to other members of the species by means of chemical advertisement. Animal behaviorists have naturally focused their attention on the more spectacular forms of aggressive behavior that arise during confrontations. When such behavior is lacking, one is tempted to postulate that exclusion is achieved by advertisement of one form or other. It remains to be pointed out that the exclusive use of terrain must be due to one or the other of the following five phenomena: (1) overt defense, (2) repulsion by advertisement, (3) the selection of different kinds of living quarters by different life forms or genetic morphs, (4) the sufficiently diffuse scattering of individuals through random effects of dispersal, or (5) some combination of these effects. Where interaction among animals occurs, specifically in the first two listed conditions, we can say that the occupied area is a territory.

Territorial behavior is widespread in animals, and many studies have established the primary function of the territory in particular species with a reasonable degree of confidence. The principal resources defended by various species include food, shelter, space for sexual display, and nest sites. In the case of bird and ungulate leks, the territories are set up by males and used almost exclusively for breeding. They do not supply protection from predators. Indeed, males of ungulates such as the wildebeest are subject to the heaviest danger from lions and other predators while on their leks (Estes, 1969). Furthermore, the dense concentrations of animals around the display grounds make the land less favorable for feeding. Lek birds such as prairie chickens and turkeys even go elsewhere to feed. So the resource guarded on the lek territories is simply the space for sexual display, plus the females that respond to the male within that space.

Analyses of other forms of territories require different modes of inference but can be made just as positive. Female marine iguanas (*Amblyrhynchus cristatus*) in the Galápagos Islands are normally casual about egg laying, merely placing the eggs in loose soil and departing. On Hood Island, however, nest sites are scarce. The females compete for the limited space by assuming a bright coloration similar to that of the males and fighting in a tournamentlike fashion. The winners get to lay their eggs in the favored spots. Afterward, they survey the sites from look-out positions on nearby rocks, descending occasionally to sniff and taste the sites and to scratch a little more earth on top (Eibl-Eibesfeldt, 1966).

Still other forms of evidence may be less direct but equally strong. In his study of *Pomacentrus flavicauda,* a fish of the Great Barrier Reef, Low (1971) noted that each territory covers a particular kind of interface of sand and coral in which sheltering crevices and an adequate supply of algae are located. The fish apparently never leave this spot. They challenge not just rival *P. flavicauda* but any intruder belonging to an alga-feeding species. Nonherbivores are ignored. When resident *P. flavicauda* were removed by Low, alga-feeders of several species moved into the vacated territories. Theoretically, the existence of feeding competitors of other species should reduce the density of food and force individuals to expand their territory size in order to harvest the same quantity of energy. The higher the number of competing species, the larger we should expect to find the average territory size—insofar as territory size is plastic and other factors, such as habitat differences, are accounted for. Precisely this result, in impressive detail, has recently been obtained in field studies of song sparrows by Yeaton and Cody (1974).

The functions served by territorial defense, like those of most other components of social behavior, are idiosyncratic and difficult to classify. We can nevertheless distinguish several major categories in which the known or probable function matches the size and location of the defended space. The following classification is an extension of one developed for birds in sequential contributions by Mayr (1935), Nice (1941), Armstrong (1947), and Hinde (1956). I have modified it slightly in order to cover other groups of animals as well.

Type A: a large defended area within which sheltering, courtship, mating, nesting, and most food gathering occur. This type of territory exists in especially high frequency among species of benthic fishes, arboreal lizards, insectivorous birds, and small mammals.

Type B: a large defended area within which all breeding activities occur but which is not the primary source of food. Examples of species utilizing this less common type include the nightjar *Caprimulgus europaeus* and the reed warbler *Acrocephalus scirpaceus.*

Type C: a small defended area around the nest. Most colonial birds utilize such a restricted form of territory, including a majority of seabirds, herons, ibises, flamingos, weaver finches, and oropendolas. Examples among the insects include sphecid wasps ("mud daubers") and bees that nest in aggregations.

Type D: pairing and/or mating territories. Examples include the leks of certain insects, including male damselflies and dragonflies, as well as those of birds and ungulates.

Type E: roosting positions and shelters. Many species of bats, from flying foxes to the cave-dwelling forms of *Myotis* and *Tadarida,* gather in roosting aggregations within which personal sleeping positions are defended. The same is true of such socially roosting birds as starlings, English sparrows, and domestic pigeons.

It is useful to recognize two additional classifications orthogonal to the one just given. For territories of types A and B, territorial defense can be absolute or spatiotemporal. That is, the resident can guard its entire territory all of the time, or it can defend only those portions of the territory within which it happens to encounter an intruder at close range. Spatiotemporal feeding territories are quite common in mammals, especially those that are frugivores or carnivores, because the area that must be covered to secure enough food is typically too large to be either monitored or advertised continuously. Absolute feeding territories are more frequently encountered in birds, which have excellent vision, access to vantage points, and the flight speed to scan relatively large foraging areas. This difference between the two major vertebrate groups is the reason why territoriality was originally elucidated in birds, and why its general significance in that group has never been in doubt, while in mammals the subject has always been plagued by seemingly inadequate data and semantic confusion.

The third classification is the following: the territory can be either fixed in space or floating. Animals are committed to floating territories when the substratum on which they depend is mobile. An example is the bitterling (*Rhodeus amarus*), a fish that lays its eggs in the mantle cavity of *Anodonta* mussels and other freshwater bivalves. Each bitterling limits sexual fighting to the vicinity of its mussel, and when the mussel moves around, the fish's territory moves with it (Tinbergen, 1951).

The results of 30 years of field research reveal territoriality to have a patchy phylogenetic distribution. It occurs widely through the vertebrates and is common, but far less general, among the arthropods, including especially the crustaceans and the insects. True territorial behavior has also been reported in one species of mollusk, the owl limpet *Lottia gigan-*

tea, by Stimson (1970) and in a few nereid polychaete worms (Evans, 1973).

The Theory of Territorial Evolution

The home range of an animal, whether defended as a territory or not, must be large enough to yield an adequate supply of energy. At the same time it should ideally be not much greater than this lower limit, because the animal will unnecessarily expose itself to predators by traversing excess terrain. This optimal-area hypothesis seems to be borne out by what little data we have bearing directly on the energy yield of home ranges. C. C. Smith (1968), for example, found that the territory size of *Tamiasciurus* tree squirrels appears to be adjusted to provide just enough energy to sustain an animal on a year-round basis. In 26 territories Smith measured, the ratio of energy available to energy consumed during one-year periods varied from just under 1 to 2.8, with a mean of 1.3. The poorer the energy yield per square unit of a given habitat, the larger the territory each squirrel occupied to compensate.

The optimum-area, or optimum-yield, hypothesis is further supported by data revealing a general correlation among terrestrial vertebrates between the size of the animal and the size of the home range it occupies. This relation, which is surprisingly consistent, was first demonstrated by McNab (1963) in the mammals and extended by later authors to other vertebrate groups. The relationship obtained by comparing many species fits roughly the following logarithmic function:

$$A = aW^b$$

where A is the home range area of a given species, W the weight of an animal belonging to it, and a and b fitted constants. It is also approximately true that the rate of energy utilization (E) is a linear function of the metabolic rate (M), that is,

$$E = cM$$

where c is another fitted constant. Finally, the metabolic rate, M, increases as a logarithmic function of the animal's weight, W:

$$M = \alpha W^B$$

where α and B are two more fitted constants. It follows that the area of the home range is a logarithmic function of energy needs.

The values of a, b, α, and B for three taxonomic groups of vertebrates are given in Table 12-1. It can be seen that each group has a distinctive set of values, reflecting peculiarities in locomotion and efficiency in the harvesting of energy. Schoener (1968a) has further demonstrated that the slope of the curve relating home range (or territory) to body weight in birds depends on their diet. As shown in Figure 12-2, the slope is greatest for predators, least for herbivores, and intermediate for species with mixed diets. This relationship convincingly supports the optimum-yield hypothesis. The hypothesis now reads that as predators grow larger, prey of suitable size grow scarcer and the predators must search over disproportionately larger areas to secure the minimum ration of energy. But why should the suitable prey grow scarcer? There are two reasons. Within any trophic level, say the herbivores or first-level carnivores, most organisms are concentrated at the small end of the size scale; hence disproportionately fewer items will be suitable for the bigger carnivore. Also, as a predator grows larger, it is more likely to feed on other predators, which are scarcer by virtue of the ecological efficiency rule. Schoener has provided evidence that the mammalian data can be decomposed in the same way, again yielding higher slopes for the predators.

One should keep in mind that these quantitative relationships pertain only to undefended home ranges and to feeding territories, which are a special kind of home range. Other forms of territories, for

Table 12-1 Regressions of metabolic rate (M) and area of home range (A) on body weight (W) in three groups of vertebrates. (M in kcal/day for mammals and birds, cm^3 O_2/hr for lizards. A in acres for mammals and birds, m^2 for lizards. W in kg for mammals and birds, g for lizards.)

Group	Relationship	Function	Authority
Mammals	Basal metabolism and body weight	$M = 70W^{0.75}$	Kleiber (1971)
	Home range and body weight	$A = 6.76W^{0.63}$	McNab (1963)
Birds	Basal metabolism and body weight	$M = kW^{0.69}$	Lasiewski and Dawson (1967)
	Home range and body weight	$A = kW^{1.16}$	Schoener (1968a)
Lizards	Standard (30°C) metabolism and body weight	$M = 0.82W^{0.62}$	Bartholomew and Tucker (1964)
	Home range and body weight	$A = 171.4W^{0.95}$	Turner et al. (1969)

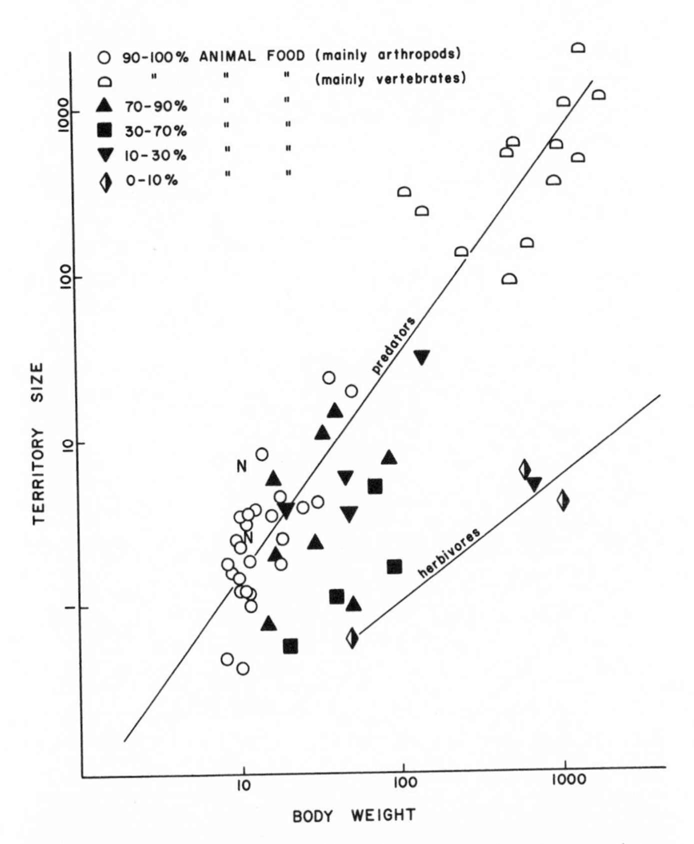

Figure 12-2 The relationship of territory size (in acres) to body weight (grams) for birds of various feeding categories. Each point represents a different species. Omnivores (10–90 percent animal food) are shaded, herbivores half-shaded, and predators clear. *N* = nuthatch species. (From Schoener, 1968a.)

example those deployed around shelters or display positions, are subject to wholly different sets of controls and are likely to be related to physiological properties of the animals in diverse ways. Even the feeding territories are sometimes demarcated by factors less complex or more complex than energy yield. Garden eels (*Gorgasia*), being plankton feeders, live in a superabundance of food. But being sedentary bottom dwellers evidently subject to heavy predation, they also do not leave their burrows. Therefore the radius of the feeding territory of each eel, which it defends from adjacent eels, is the exact length of its body (Clark, 1972). Troops of vervet monkeys (*Cercopithecus aethiops*), to take a more complex example, are highly variable in size. The largest troops dominate the smallest, which are forced into less favorable terrain and must defend larger areas in order to satisfy their energy needs (Struhsaker, 1967a). Such intricacies arising from higher social organization are probably responsible for the extreme variation in home range area among the primates generally, as revealed in the data recently collated by Bates (1970).

But why should animals bother to defend any part of their home range? MacArthur (1972) proved that pure contest competition for food is energetically less efficient than pure scramble competition. This is a paradox easily resolved. Territoriality is a very special form of contest competition, in which the animal need win only once or a relatively few times. Consequently, the resident expends far less energy than would be the case if it were forced into a confrontation each time it attempted to eat in the presence of a conspecific animal. Its energetic balance sheet is improved still more if it comes to recognize and to ignore neighboring territorial holders—the dear enemy phenomenon to be examined later in this chapter.

Clearly, then, a territory can be made energetically more efficient than a home range in which competition is of the pure contest or pure scramble form. But if this is the case, why are not all species with fixed home ranges also strictly territorial? The answer lies in what J. L. Brown (1964) has called economic defendability. Natural-selection theory predicts that an animal should protect only the amount of terrain for which the defense gains more energy than it expends. In other words, if an animal, a carnivore for example, occupies a much larger territory than it can monitor in one quick survey, it may find trotting from one end of its domain to the next just to oust intruders an energetically wasteful activity. Consequently, natural selection should favor the evolution of a spatiotemporal territory rather than an absolute territory. The carnivore will devote most of its energies to hunting prey, challenging only those intruders it encounters at close range. Or else it will deposit scent at strategic positions throughout the territory in an effort to warn off intruders. Arboreal lizards, such as iguanids, agamids, and chamaeleontids, which are able to scan large areas visually, also tend to maintain absolute territories. Terrestrial forms, such as many scincids, teiids, and varanids, generally have spatiotemporal territories or broadly overlapping home ranges (Judy Stamps, personal communication).

Horn (1968) utilized this same concept when investigating the conditions favoring colonial nesting in blackbirds. He proved that when resources are uniformly distributed and continuously renewing, there is an advantage to maintaining a complete defense of whatever portions of the area can be patrolled in reasonably short periods. But when food is patchily distributed and occurs unpredictably in time, it does not pay to defend fixed areas. The optimum strategy is then to nest colonially and to forage in groups. By this means the individual is able to utilize the knowledge of the entire group. Economic defensibility is actually only one important component of fitness that determines the evolution of territorial behavior. As Heller's work on chipmunks showed (Heller, 1971), territorial defense is curtailed if it exposes animals to too much predation. There is also the phenomenon of aggressive neglect: defense of a territory results in less time devoted to courtship, fewer copulations, and neglected and less fit offspring. In short, the territorial strategy evolved is the one that maximizes the increment of fitness due to extraction of energy from the defended area as compared with the loss of fitness due to the effort and perils of defense.

Schoener (1971) has taken the first step toward parameterizing this theory of territorial evolution. It is possible to estimate the permeability of a territory, measured by the density of intruders tolerated at any given moment of time, if we view the permeability as the balance struck when the rate of invasion by intruders becomes equal to the rate at which they are being expelled (Figure 12-3). In the simplest case, the invasion rate decreases linearly with increased density of invaders already on the territory. It may simply equal the following product:

Invasion Rate

The perimeter of the defended area × A constant determined by the probability an outsider will invade × $\left(1 - \frac{(N/A)}{H}\right)$

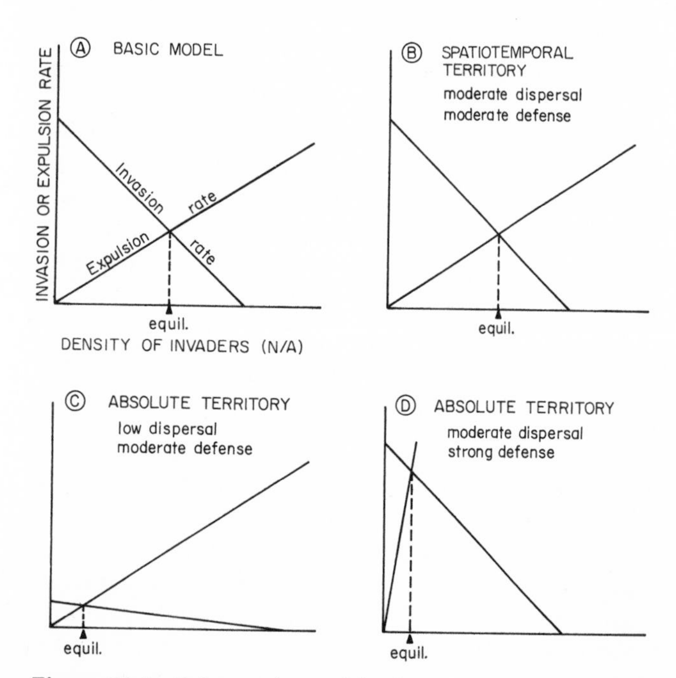

Figure 12-3 Schoener's model of territorial defense and three extensions that can be made from it.

where N/A is the density of invaders on the territory (number of invaders divided by the area of the territory), and H is the maximum density that can occur under any condition. The rate at which the territory holder chases out or destroys invaders might also be linear:

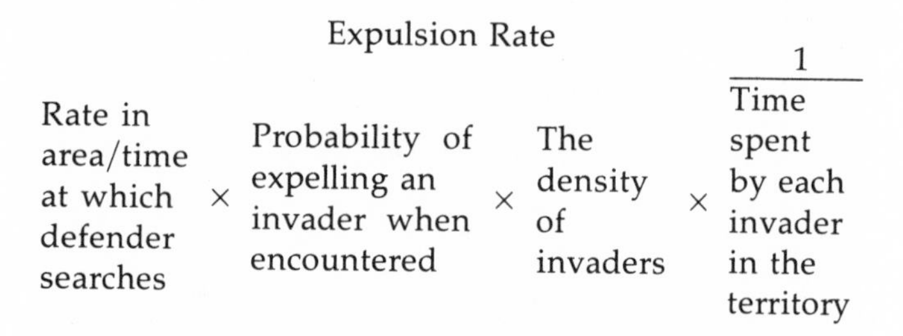

In Figure 12-3B-D are given three elementary extensions of Schoener's model. Note that a moderate to high invasion rate relative to the perimeter of the home range, combined with a moderate to low defensive response, produces a spatiotemporal territory. An absolute territory, in which all of a fixed area is defended all of the time, results when the dispersal rate of invaders is low, defense is strong, or both. The parameters envisaged by Schoener are probably correct, but we have no present way of estimating the form of the invasion and expulsion curves. Nor does the model incorporate the energy balance, which is all-important in natural selection, or other components of fitness.

Special Properties of Territory

Territorial behavior involves much more than the mere expulsion of intruders. And territories are more than defended areas: they possess both structure and dynamism and can be described as fields of variable intensity. Territories change in size and shape through the seasons and as the animal matures and ages. Field studies have revealed a rich set of territorial properties, some very general and others restricted to one or a small number of species. A brief survey of the most important follows.

The Elastic Disk

Territorial size in most animal species varies to a greater or lesser degree with population density. Julian Huxley (1934) compared the variable territory to an elastic disk with the resident animal at its center. When overall population density increases and pressure builds along its perimeter, the territory contracts. But there is a limit beyond which the animal cannot be pushed. It then stands and fights, or else the entire territorial system begins to disintegrate. When the surrounding population decreases, the territory expands. But, again, there is a limit beyond which the animal does not try to extend its control. In very sparse populations, either territories are not contiguous, or their boundaries simply become too vague to define.

An example of elastic territories is provided by the dunlin (*Calidris alpina*), a kind of sandpiper. In the subarctic locality of Kolomak, at 61°N in Alaska, food is relatively abundant and reliable, and populations of the sandpipers attain a density of 30 pairs per hectare. Farther north at 71°N, at the arctic site of Barrow, the food supply is unpredictable and the summers are shorter. Here the birds live at densities one-fifth those at Kolomak, or about 6 pairs per hectare. Since the territorial boundaries are contiguous at both localities, the average territorial size at Barrow is five times that at Kolomak.

The "Invincible Center"

Unless an adult male bird is grossly overmatched or ill, he is usually undefeatable by conspecific birds at the center of his territory (N. Tinbergen, 1939; Nice, 1941). This circumstance is only the extreme manifestation of the more general principle that the aggressive tendencies of animals, and the probability that any given intensity of aggression or display will result in the domination of rivals, increase toward the center (J. R. Krebs, 1971). What exactly is the "center"? In uniform terrain it is customarily the geometric center, but in a heterogeneous environment the behavioral center is more likely to be the location of either the animal's shelter or the most concentrated food supply within the territory, whichever is the more vital to the welfare of the animal. The territory holder spends most of its time near the center, routinely performing courtship displays and constructing shelters there. Male tree sparrows begin their day with a reaffirmation of the core area through song followed by more lengthy trips into the cortex (Weeden, 1965).

Changes with Season and Life History Stages

The values of the parameters that define the optimum territorial size shift during the life cycles of most kinds of animals. When a male bird begins to court, he needs to defend fewer positions than later, when his nestlings demand large amounts of food. But he needs to defend the positions more often, since the population as a whole is more mobile and floater males challenge more frequently. Such changes have been documented by Hinde (1952) in the great tit *Parus major* and by Marler (1956) in the chaffinch *Fringilla coelebs*. Males of these small European birds begin the breeding season by singing and fighting around selected display sites. Only later do they extend their defense to the entire territory. The boundaries of the territories of black-capped chickadees (*P. atricapillus*) strongly fluctuate through the breeding season, first expanding slightly as nests are built, then contracting drastically at the egg and nestling stages, and finally expanding again when the young reach the fledgling stage. In fact, the patterns of change vary greatly from species to species throughout the birds. The male of the mockingbird *Mimus polyglottos* stays on his territory the year round, expanding its size at the beginning of the breeding season in the spring (Hailman, 1960). The green heron *Butorides virescens*, in contrast, arrives at the breeding ground in the spring and immediately sets up a full-sized territory about 40 meters in greatest diameter. Thereafter, the defended area steadily shrinks until it is limited to the immediate vicinity of the nest, at which time the mated pair cooperate in defense (Meyerriecks, 1960). Some of the mystery presented by these differences vanishes when we consider the natural history of the species. The male mockingbird defends a feeding territory, which must be expanded and maintained while the young are growing. The male green heron, however, first defends a courtship display area. Later, he and his mate are no longer courting. They feed in shallow water outside the breeding area, and need only to defend their nest and young.

Seasonal variation of home range and territory is no less complicated and idiosyncratic in the mammals. Red squirrels (*Tamiasciurus hudsonicus*), for example, maintain two forms of territories in the mixed forests of Alberta. "Prime territories" are defended the year round by successful adults in mature coniferous stands, which provide a continuous supply of seeds that serve as food. Other habitats, especially those with a high proportion of deciduous trees, yield seeds only during the growing season. During the winter the resident red squirrels defend caches of seeds gathered in the warmer months. Wildebeest males, to take another, radically different example, defend display areas during the breeding season and travel with the nomadic herds the rest of the year (Estes, 1969).

Nested Territories

Social organizations are known in which an overlord male maintains a territory subdivided in turn by females. In the dwarf cichlid *Apistogramma trifasciatum* (Burchard, 1965) and some species of the lizard genus *Anolis* (Rand, 1967a,b), the females defend their domains against one another but not against the male. Such a nested territory is actually a combined territorial system and dominance order.

The Dear Enemy Phenomenon

A territorial neighbor is not ordinarily a threat. It should pay to recognize him as an individual, to agree mutually upon the joint boundary, and to waste as little energy as possible in hostile exchanges thereafter. James Fisher (1954) recognized this principle in the particular case of birds when he stated, "The effect is to create 'neighborhoods' of individuals who while masters on their own definite and limited properties are bound firmly and *socially* to their next-door neighbors by what in human terms would be described as a dear-enemy or rival-friend situation, but which in bird terms should more safely be described as mutual stimulation." The ability of birds to distinguish the songs of neighbors from those of strangers has since been proved in a variety of species by playing recordings of both kinds of individuals to the territorial males and observing their responses. In general, when a recording of a neighbor is played near a male, he shows no unusual reactions, but a recording of a stranger's song elicits an agitated aggressive response. But if the stranger is from such a remote area that its song belongs to a different dialect, the response is weaker.

Territories and Population Regulation

In a seminal paper on the population dynamics of titmice in Holland, H. N. Kluijver and L. Tinbergen (1953) concluded that territoriality plays a precise role in the regulation of populations. They recognized that the habitable environment is divided into areas that are optimal for breeding and others that are suboptimal. Kluijver and Tinbergen postulated that the optimal habitats support the densest, most stable populations—the endemic core of the species. The suboptimal habitats support sparser, less stable breeding populations. In the spring the birds that arrive first settle in the optimal habitats, spacing themselves out by territorial exclusion until the area is filled. Territoriality prevents overpopulation and thus guards the population from excessive fluctuation. Kluijver and Tinbergen referred to the stabilization as buffering. Late arrivals spill over into the suboptimal habitats, where they exist in more scattered territories or wander as floaters. These marginal populations are not buffered. They breed far less, their mortality is higher, especially in the fall and winter, and their numbers fluctuate more widely. The optimal habitat of the great tit *Parus major*, for example, is comprised of narrow strips of broad-leaved woodland, while the surrounding zones of pine woods serve as the suboptimal habitat.

In a later, detailed review of the subject, J. L. Brown (1969) envisaged the buffer effect as a three-stage development in the build-up of bird populations:

Level 1. At the lowest population density, territories are not circumscribed by competition. No individual is prevented from settling in the best habitat.

Level 2. As the population density rises, some individuals are excluded from the optimal habitats and are forced to establish territories in the poorer habitable areas.

Level 3. At the highest densities, some individuals are prevented from establishing territories altogether. They exist as a floating population that drifts back and forth, in and around the established territories. Brown inferred the floaters to be part of the buffering process for the optimal habitat and, to a lesser degree, for the less favorable habitats in which nesting to any extent occurs. When birds die on their territories, floaters move in to take their place and thereby maintain an approximately constant density in the habitable areas.

The existence of abundant suboptimal territories and floaters has been documented in so many species of birds and mammals, belonging to so many genera and higher taxa, as to suggest that they are a very general if not universal phenomenon. The readiness of floaters to fill vacated territories has been amply demonstrated by experiments in which territory holders were simply trapped or shot. The first such removal experiment was performed by Stewart and Aldrich (1951) and Hensley and Cope (1951). These investigators censused the territorial males of 50 species of birds in a 16-hectare plot of spruce-fir woodland, then shot as many birds as they could during a three-week period. The results were startling: the total number of territorial birds removed during the experiment was three times the number originally estimated to be present. A similar effect has been subsequently obtained for many other kinds of birds.

In general, it appears that territorial behavior in birds has evolved as an adaptation favoring individual survival and reproduction. Territoriality has the effect of secondarily regulating population density to some degree, but it cannot be said to represent a population-wide adaptation with a primary role of density regulation.

CHAPTER 13

Dominance Systems

Dominance behavior is the analog of territorial behavior, differing in that the members of an aggressively organized group of animals coexist within one territory. The dominance order, sometimes also called the dominance hierarchy or social hierarchy, is the set of sustained aggressive-submissive relations among these animals. The simplest possible version of a hierarchy is a *despotism:* the rule of one individual over all other members of the group, with no rank distinctions being made among the subordinates. More commonly, hierarchies contain multiple ranks in a more or less linear sequence: an alpha individual dominates all others, a beta individual dominates all but the alpha, and so on down to the omega individual at the bottom, whose existence may depend simply on staying out of the way of its superiors. The networks are sometimes complicated by triangular or other circular elements (Figure 13-1), but such arrangements seem a priori to be less stable than despotisms or linear orders. In fact, Tordoff (1954) found that triangular loops first established by a captive flock of red crossbills (*Loxia curvirostra*) were disruptive, so that changes in the order increasingly replaced them with straight chains. The dominance order of a flock of roosters assembled by Murchison (1935) was at first unstable and contained triangular elements, but it later settled into a slowly changing linear order (Figure 13-2). Ivan Chase (personal communication) obtained direct evidence that straight-chain hierarchies can result in higher group efficiency. When triads of hens formed a linear dominance order, a certain amount of food was eaten quickly by the alpha bird, sometimes assisted by the beta individual. When the dominance orders were circular, however, the hens fed warily, individuals frequently displaced one another, and the food was consumed more slowly.

Hierarchies are formed in the course of the initial encounters between animals by means of repeated threats and fighting. But after the issue has been settled, each individual gives way to its superiors with a minimum of hostile exchange. The life of the group may eventually become so pacific as to hide the existence of such ranking from the observer—until some minor crisis happens to force a confrontation. Troops of baboons, for example, often go for hours without displaying enough hostile exchanges to reveal their hierarchy. Then in a moment of tension—a quarrel over an item of food is sufficient—the ranking is suddenly revealed, appearing in graphic detail rather like an image on photographic paper dipped in developer fluid.

The societies of some species are organized into *absolute dominance hierarchies,* in which the rank order is the same wherever the group goes and whatever the circumstance. An absolute hierarchy changes only when individuals move up or down the ranks through further interactions with their rivals. Other societies, for example crowded groups of domestic cats, are arrayed in *relative dominance hierarchies,* in which even the highest-ranking individuals yield to subordinates when the latter are close to their personal sleeping places (Leyhausen, 1956, 1971). Relative hierarchies with a spatial bias are intermediate in character between absolute hierarchies and territoriality.

In its stable, more pacific state the hierarchy is sometimes supported by "status" signs. The identity of the leading male in a wolf pack is unmistakable from the way he holds his head, ears, and tail, and the confident, face-forward manner in which he approaches other members of his group. He controls his subordinates in the great majority of encounters without any display of overt hostility (Schenkel, 1947). Similarly, the dominant rhesus male maintains an elaborate posture signifying his rank: head and tail up, testicles lowered, body movements slow and deliberate and accompanied by unhesitating but measured scrutiny of other monkeys that cross his field of view (Altmann, 1962).

Figure 13-1 Three elementary forms of networks found in dominance orders. More complex networks are built up of combinations of such elements. (From Wilson et al., 1973.)

Finally, dominance behavior is mediated not only by visual signals but also by acoustic and chemical signals. Mykytowycz (1962) has found that in male European rabbits (*Oryctolagus cuniculus*) the degree of development of the submandibular gland increases with the rank of the individual. By means of "chinning," in which the lower surface of the head is rubbed against objects on the ground, dominant males mark the territory occupied by the group with their own submandibular gland secretions. Recent studies of similar behavior in flying phalangers and black-tailed deer indicate that territorial and other agonistic pheromones of these species are complex mixtures that vary greatly among members of the same population (Schultze-Westrum, 1965; Wilson, 1970). As a result individuals are able to distinguish their own scents from those of others.

Examples of Dominance Orders

Dominance hierarchies, like territories, are distributed in a highly irregular fashion through the animal kingdom. Among the invertebrates the hierarchies appear to be limited principally to evolutionarily more advanced forms characterized by large body size (Allee and Dickinson, 1954; Lowe, 1956). Among the insects, hierarchies are most clearly developed in species that are fully social yet still primitively organized, such as the bumblebees and paper wasps (Wilson, 1971a; Evans and Eberhard, 1970). The following case histories demonstrate some of the extreme variations in dominance relations in different species.

Domestic Fowl (*Gallus gallus*)

The common domestic fowl, sometimes referred to as *Gallus domesticus*, was the first vertebrate species in which dominance relations were systematically investigated (Schjelderup-Ebbe, 1922), and it has been the most intensively studied of all the animal species since that time. During the past 30 years A. M. Guhl and his associates at Kansas State University have concentrated on nearly every conceivable aspect of the subject; many of their key results and reviews are given in Craig and Guhl (1969) and Wood-Gush (1955).

The social behavior of chickens is relatively simple and is based to a large extent on the dominance

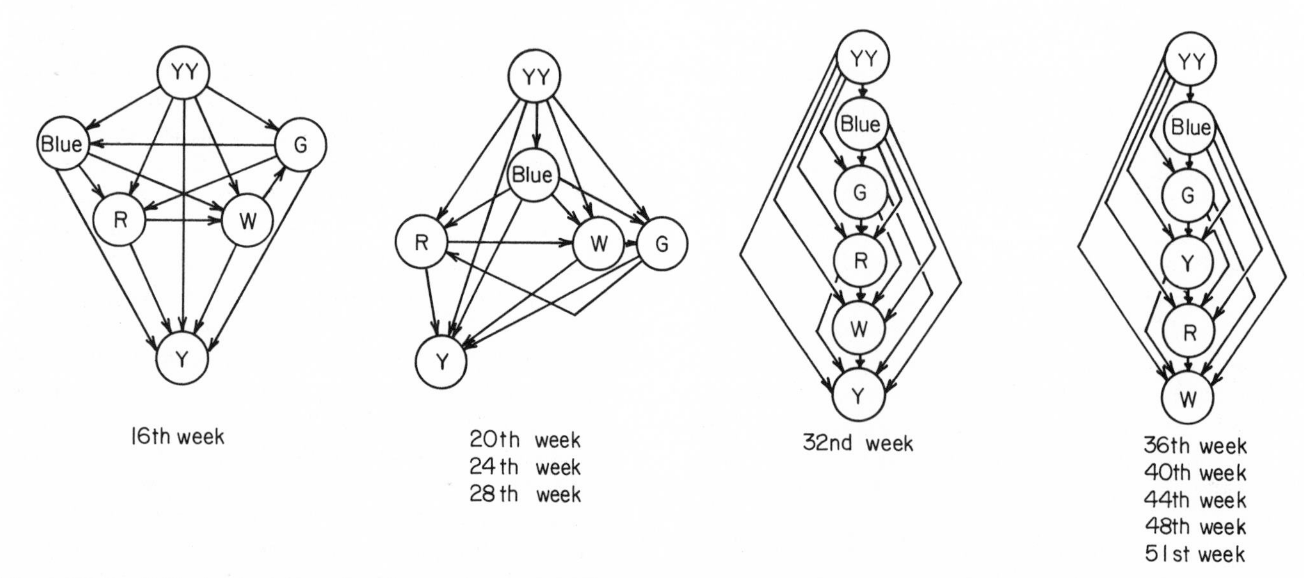

Figure 13-2 Shifts in the hierarchy of a newly formed flock of roosters. Triangular subnets give way to a more stable, linear order. The letters and the name "Blue" designate the individual roosters. (Redrawn from Murchison, 1935.)

order. As soon as a new flock is created by the experimenter, the power struggle begins. The hierarchy that quickly forms is in a literal sense a peck order: the chickens maintain their status by pecking or by a threatening movement toward an opponent with the evident intention of attacking in this manner. High-ranking birds are clearly rewarded with superior genetic fitness. They gain priority of access to food, nest sites, and roosting places, and they enjoy more freedom of movement. Dominant cocks mate far more frequently than subordinates. But dominant hens mate somewhat less frequently than others, because they display submissive and receptive postures to the cocks less consistently. Nevertheless, the fitness of dominant hens is probably greater because of the more than compensating advantages gained in access to food and nests. Cocks form a separate hierarchy above that of the hens. The adaptive explanation for this disjunction is that it facilitates mating: cocks subordinate to hens are unable to copulate with them. The heritability of fighting ability has been well documented. Strong genetic variation in this trait exists both between and within the various breeds of fowl. It is an interesting commentary on the evolution of dominance that when poultry breeders select for strains that lay more eggs, they also produce more aggressive chickens. In other words, the breeders have simply picked out the dominant birds, which incidentally have the most access to nests (McBride, 1958).

The critical size of a flock of hens is ten. In flocks below approximately that number, triangular and square elements straighten out and the resulting linear orders are stable for periods of months. In flocks above that number, looped elements stay common and the hierarchy continues to shift at a relatively high rate. However, revolts and rapid shifting can occur even in the small groups when one or more of the subordinate hens is injected with testosterone proprionate. It is to the advantage of a chicken to live in a stable hierarchy. Members of flocks kept in disorder by experimental replacements eat less food, lose more weight when their diet is restricted, and lay fewer eggs. Chickens remember one another well enough to maintain the hierarchy for periods of only up to two or three weeks. If separated for much longer periods, they reestablish dominance orders as if they were strangers. However, a chicken can be removed repeatedly for shorter intervals and reinserted without changing its rank.

Paper Wasps (*Polistes*)

In the primitive insect societies strife and competition prevail, leading to the emergence of a single female, the queen, who physically dominates the rest of the adult members of the colony. In the more advanced insect societies, particularly those in which strong differences between the queen and worker castes exist, the queen also exercises control over the workers but usually in a more subtle manner devoid of overt aggression. In many cases, as in the honeybees, hornets, and many ants and termites, the dominance is achieved by means of special pheromones that inhibit reproductive behavior and the development of immature members of the colony into the royal caste. The gradual change from what might be regarded as brutish dominance to the more refined modes of queen control is one of the few clear-cut evolutionary trends that extend across the social insects as a whole (Wilson, 1971a).

The existence of a primitive dominance system in the European paper wasp *Polistes gallicus* was intimated by the studies of Heldmann (1936a,b), who found that when two or more females start a nest together in the spring, one comes to function as the egg layer, while the others take up the role of workers. The functional queen feeds more on the eggs laid by her partners than the reverse, thus exercising a form of reproductive dominance. Pardi (1940, 1948) discovered that the queen establishes her position and controls the other wasps by direct aggressive behavior, and he went on to analyze this form of social organization in detail. Dominance behavior has since been documented in still greater and important detail in *P. canadensis* and *P. fuscatus* by Mary Jane West Eberhard (1969).

The relationship among the adult members of the *Polistes* colony is more than a simple despotism by the queen. There exists in most cases a linear order of ranking involving a principal egg layer, the queen, and the remainder of the associated females, called auxiliaries, who form a graded series in the relative frequencies of egg laying, foraging, and comb building. Dominant individuals receive more food whenever food is exchanged, they lay more eggs, and they contribute less work. They establish and maintain their rank by a series of frequently repeated aggressive encounters. At lowest intensity the exchange is simply a matter of posture; the dominant individual rises on its legs to a level higher than the subordinate, while the subordinate crouches and lowers its antennae. At higher intensities leg biting occurs, and at highest intensity the wasps grapple and attempt to sting each other. During the brief fights, the contestants sometimes lose their hold on the nest and fall to the ground. Injuries are rare, although Eberhard once saw a female of *P. fuscatus* killed in a fight. The severest conflicts occur between wasps of nearly equal rank and during the early days of the association when the nest is first being constructed. As time passes, the wasps fit more easily into their roles, and aggressive exchanges become more subdued, less frequent, and eventually purely postural in nature. When the first adult workers eclose from the pupal instar during later stages of colony development, their relations to the foundresses are invariably subordinate and nonviolent. The workers also form a hierarchy among themselves; those that emerge early tend to dominate their younger nestmates. If the alpha female is removed, the next highest auxiliaries intensify their hostility to one another until one of their number becomes unequivocally dominant; then the exchanges subside to the previous low level. Eberhard (1969) found that the tropical species *P. canadensis* differs from all the temperate species studied to date in that its contests are more violent and result in the losers leaving the nest to attempt nest founding on their own elsewhere. The queens are also less dependent on the indirect technique of egg eating to maintain reproductive control. Thus *P. canadensis* approaches a condition of true despotism, whereas the temperate species so far studied possess a colony organization that can be characterized as an uneasy oligarchy.

Spider Monkeys (*Ateles geoffroyi*)

The New World monkeys have not attained the degrees of social complexity found in the most advanced Old World monkeys and anthropoid apes. Spider monkeys can be taken as typical for the diffuse, muted form of dominance relations they display (Eisenberg and Kuehn, 1966). Aggression occurs: members of a group threaten one another by shaking branches, grinding their teeth, coughing, hissing, and even roaring. The animals slap and kick one another, and they sometimes slash at one another with their canines or nip with hard bites of the incisors. Dominants sometimes chase subordinates. However, such overt aggressive behavior occurs rather rarely. Males tend to be dominant over females and adults over juveniles, but the order is not linear and is difficult to define from the infrequent, often unpredictable exchanges of pairs. *Ateles geoffroyi* does not employ aggressive presenting and mounting, status posturing, or any of the other highly ritualized threat and conciliatory signals so prominently used by macaques and baboons among the Old World monkeys. Grooming is uncommon, and high-ranking animals groom more than they are groomed, the reverse of the grooming trend in most other monkeys. The adult male sometimes halts fighting between others, but he does not otherwise play the role of a control animal. In short, the relatively simple social organization of the spider monkey is reflected in its primitive and infrequently used dominance system.

Special Properties of Dominance Orders

The Xenophobia Principle

The relative calm of a stable dominance hierarchy conceals a potentially violent united front against strangers. The newcomer is a threat to the status of every animal in the group, and he is treated accordingly. Cooperative behavior reaches a peak among the insiders when repelling such an intruder. The sight of an alien bird, for example, energizes a flock of Canada geese, evoking the full panoply of threat displays accompanied by repeated mass approaches and retreats (Klopman, 1968). Chicken farmers are well aware of the practical implications of xenophobia. A new bird introduced into an organized flock will, unless it is unusually vigorous, suffer attacks for days on end while being forced down to the lowest status. In many cases it will simply expire with little show of resistance.

The Peace of Strong Leadership

Dominant animals of some primate societies utilize their power to terminate fighting among subordinates. The phenomenon has been described explicitly in rhesus and pig-tailed macaques and in spider monkeys. In squirrel monkeys this control function appears to operate in the absence of dominance behavior. Species organized by despotisms, such as bumblebees, paper wasps, hornets, and artificially crowded territorial fish and lizards, also live in relative peace owing to the generally acknowledged power of the tyrant. If the dominant animal is removed, aggression sharply increases as the previously equally-ranked subordinates contend for the top position.

The Will to Power

In a wide range of aggressively organized mammal species, from elephant seals, harem-keeping ungulates, and lions to langurs, macaques, and baboons, the young males are routinely excluded by their dominant elders. They leave the group and either wander as solitary nomads or join bachelor herds. At most they are tolerated uneasily around the fringes of the group. And, predictably, it is the young males who are also the most enterprising, aggressive, and troublesome elements. They contend among one another for in-group dominance and sometimes form separate bands and cliques that cooperate in reducing the power of the dominant males. Even the personalities of males in the two categories differ. The "establishment" males of a Japanese macaque troop remain calm and detached when shown a novel object, and thus do not risk the loss of their status. It is the females and young animals who explore new areas and experiment with new objects. The obvious parallels to human behavior have been noted by several writers, but most explicitly and persuasively by Tiger (1969) and Tiger and Fox (1971).

Nested Hierarchies

Societies that are partitioned into units can exhibit dominance both within and between the components. Thus flocks of white-fronted geese (*Anser albifrons*) develop a rank order of the several subgroups (parents, mated pairs without goslings, free juveniles) superimposed over rank ordering within each one of the subgroups (Boyd, 1953). Brotherhoods of wild turkeys contend for dominance, especially on the display grounds, and within each brotherhood the brothers establish a rank order (Watts and Stokes, 1971). Team play and competition between human tribes, businesses, and institutions are also based upon nested hierarchies, sometimes tightly organized through several more or less autonomous levels.

The Advantages of Being Dominant

In the language of sociobiology, to dominate is to possess priority of access to the necessities of life and reproduction. This is not a circular definition; it is a statement of a strong correlation observed in nature. With rare exceptions, the aggressively superior animal displaces the subordinate from food, from mates, and from nest sites. It only remains to be established that this power actually raises the genetic fitness of the animals possessing it. On this point the evidence is completely clear.

Consider the simple matter of getting food. Wood pigeons (*Columba palumbus*) are typical flock feeders. Solitary birds are attracted by the sight of a group feeding on the ground, and no doubt there is great advantage to following the lead of others in locating food. Dominant birds place themselves at the center of the flock. Murton et al. (1966) noted that these individuals feed more quickly than those on the edge of the flock, and especially those on the forward edge, who constantly interrupt their pecking to look back at the advancing center. By shooting pigeons at dusk just before they flew to the roosts, Murton and his coworkers established that the subordinate birds accumulate less food. In fact, they have only enough to last the night, and they are in danger of perishing if the temperature drops sharply during the night or bad weather prevents foraging the next day.

Without systematic studies that include an evaluation of this question, it is impossible to guess whether the relation between status and food-gathering ability is a crucial one. Studies of maternal care in sheep and reindeer have revealed that low-ranking females are among the most poorly fed animals and also among the poorest mothers (Fraser, 1968). The teat order of piglets is a feeding dominance hierarchy in microcosm with an apparently direct adaptive basis. During the first hour of their lives the piglets compete for teat positions that, once established, are maintained until weaning. The piglets struggle strenuously, using temporary incisors and tusks to scratch one another (McBride, 1963). Preference is for the anterior teats, which provide more milk than the posterior teats and keep the piglets attached to them farther away from the trampling of the hind legs of the mother. The more milk a young pig receives, the more it weighs at weaning. The gradient of milk yield in the teats is probably great enough to provide a selective pressure for the competition to evolve. Gill and Thomson (1956) found that the four anteriormost piglets studied in each of a series of eight litters obtained an average of 15.3 percent more milk than the four posteriormost piglets. Those who occupied the three anteriormost pairs of teats got 83.8 percent more milk than the small group relegated to the posteriormost three or four pairs. Not surprisingly, piglets able to shift teat preference during early lactation moved their position forward. The orienting stimulus by which piglets find their correct positions quickly, even when the teats are partly hidden from view and smeared with mud, has not been established with certainty, but by process of elimination it would seem to involve smell. Piglets are often seen to rub their noses on the udder around the teat, and McBride has made the intriguing suggestion that they are depositing a personal scent.

The evidence favoring the hypothesis of dominance advantage in reproductive competition is even more persuasive. An experiment by DeFries and McClearn (1970) on laboratory mice deserves to be cited for the cleanness of its design. Groups were assembled of three males, distinguishable by genetic markers, and three females. In each of the replications the males fought for a day or two and established rigid hierarchies. The relationship between dominance and genetic fitness, as detected by the genetic markers in the offspring, was striking. In 18 of 22 groups established, the dominant male sired all of the litters. In 3 of the triads a subordinate male sired one litter, and in only one case did a subordinate male succeed in siring two litters. Dominant males, constituting one-third of the population, were the fathers of 92 percent of the offspring.

The top-ranking animal in a hierarchy is also under less general stress. It therefore expends less energy coping with conflict, and it is less likely to suffer from endocrine hyperfunction. Erickson (1967) found, for example, that subordinate pumpkinseed sunfish (*Lepomis gibbosus*) initiated fewer aggressive acts than did the dominant fish; and to the extent that they were the targets of aggression, they developed larger interrenal glands, the source of corticosteroids in fish. Two classes of males in rhesus troops engage in the least aggressive behavior: the lowest-ranking individuals, who remain on the periphery of the troop and are systematically excluded from the best food and resting places, and the dominant males, who enjoy their privileges with a minimum of effort. Tension and antagonism are greatest in the middle-ranking males, who are continuously striving to move up in the dominance hierarchies (Kaufmann, 1967).

Finally, rank sometimes carries perquisites that might further enhance survival value. Among macaques, dominant females are the beneficiaries of aunting behavior. Grooming, which serves as a basic cleaning operation as well as a social signal, is received by dominant animals of both sexes more than it is given. In rhesus monkeys, rank order can be reliably measured by the directionality of the grooming.

The Compensations of Being Subordinate

Defeat does not leave an animal with a hopeless future. The behavioral ontogenies of species seem designed to give each loser a second chance, and in some of the more social forms the subordinate need only wait its turn to rise in the hierarchy. The most frequent recourse, from insects to monkeys, is emigration. A common principle running throughout the vertebrates is that juveniles and young adults are the ones most likely to be excluded from territories, most likely to start at the bottom on the dominance orders, and therefore most likely to be found wandering as floaters and subordinates on the fringes of the group. In the more nearly closed societies such wanderers are preponderantly males. Emigration is a common form of density-dependent control of populations. Natural-selection theory teaches that where the emigration behavior is programmed to occur at a certain life stage and at a certain population density, and involves a determined outward journey as opposed to mere aimless drifting, the chances of success on the part of the migrant at least equal those of otherwise equivalent animals who remain at home.

Kin selection might provide another means by which subordination pays out a genetic benefit. If an animal that has little chance of succeeding on its own chooses instead to serve a close relative, this strategy may raise its inclusive fitness. A concrete example is provided by the social insects. When fertile females of the paper wasp genus *Polistes* emerge from hibernation and begin searching for a nest site, they tend to settle in the neighborhood of the nest in which they were born the previous summer. Groups of these wasps, many of whom are sisters, commonly cooperate in founding a new nest, with one assuming the dominant, egg-laying role and the others turning into functional workers. This voluntary subordination is not easy to explain, for even if the associated females were full sisters, the subordinate female would be taking care of nieces with a coefficient of relationship of 3/8, whereas she could choose to care for her own daughters and share a bond of 1/2. The missing piece of the theory has been supplied by what might be termed the "spinster hypothesis," invented by Mary Jane West (1967). West points out that nest-founding females of *Polistes* vary greatly in ovarian development and that rank in the dominance hierarchy varies directly with the development. It is further true that most new *Polistes* nests fail. Consequently, the probability of a female with low fertility establishing and bringing a nest through to maturity may simply be so low that it is more profitable, as measured by inclusive fitness, for these high-risk individuals to subordinate themselves to female relatives in foundress associations.

In still other societies we encounter direct incentives for subordinates to stay with their group. Individual macaques and baboons cannot survive for very long on their own, especially away from the sleeping areas, and they have almost no chance to breed. As Stuart Altmann and others have shown, even a low-ranking male still eats well if he belongs to a troop, and he gets an occasional chance to copulate with estrous females. Furthermore, patience can turn half a loaf into a full one, because the dominant animals will eventually grow old and die. The European black grouse *Lyrurus tetrix* even observes a kind of seniority system on the display grounds: the yearling cocks occupy peripheral territories, which attract few females; at two years of age they move into second-ranking positions near the center; and at three years of age they have the chance to become dominant cocks (Johnsgard, 1967). The turnover of dominant males may be a general phenomenon. Fraser Darling observed that red deer stags do not eat while herding a harem. After about two weeks they are easily defeated by a fresher, often younger stag. They then retire and wander to higher ground to feed, to regain their strength, and perhaps to try again. Dominant male impalas also wear themselves out quickly, yielding to fresher rivals or falling victim to predators.

The Determinants of Dominance

What qualities determine the status of an individual? Surprisingly little critical work has been directed to this important question, and investigators with useful data often present the results tangentially while discussing other topics. Our current knowledge can be summarized in the form of the following loose principles:

1. Adults are dominant over juveniles, and males are usually dominant over females. In multimale societies, it is typical for the rank ordering of the males to lie entirely above that of the females, or at most to overlap it slightly. In such cases juvenile males sometimes work their way up through the female hierarchy before achieving greater than omega status with reference to the males. Exceptional species in which females are dominant over males include the brown booby *Sula leucogaster*, the hyena, the vervet (*Cercopithecus aethiops*), and Sykes' monkey (*C. mitis*).

2. The greater the size of the brain and the more flexible the behavior, the more numerous are the determinants of rank and the more nearly equal they are in influence. Also, the more complex and orderly are the dominance chains. These correlations are very loose, and they become apparent at our present state of knowledge only when species are compared over the greatest phylogenetic distances. Arthropods, including social insects, display relatively simple types of aggressive behavior that result in despotisms, short-chain hierarchies of elementary structure, or chaotic systems in which dominance is established anew with each contact (as in the wasp *Vespula*). Fish, amphibians, and reptiles also form despotisms and short-chain hierarchies. Birds and mammals commonly form long-chain hierarchies, the members of which defend territories communally. In some of the higher monkeys and apes, we see the emergence of coalitions of peers, protectorates by dominant individuals, and strong maternal influence in the early establishment of rank.

3. The greater the cohesiveness and durability of the social group, the more numerous and nearly coequal the correlates of rank and the more complex the dominance order. The male rank orders of antelopes, sheep, and other ungulates, especially those formed temporarily during the breeding season, are predominantly based on size, with age perhaps being a second, closely associated factor. In the more aggressively organized Old World monkeys, particularly the baboons and macaques, status is based more on childhood history as it relates to the mother's rank, to membership in coalitions, and to "luck"—whether the animal is a member of an old family, for example, or has just immigrated from a neighboring troop, or has been fortunate enough to catch a stronger opponent in a weak moment when it could be defeated. When a group is newly constituted, such as a group of hens or rhesus monkeys thrown together in an enclosure, the initial dominance orders tend to be established on the basis of size, strength, and aggressiveness. But later the other more personal and experiential factors assert themselves as well.

Few studies have been conducted with the explicit goal of assigning weights to the determinants of rank. The most instructive to date is N. E. Collias' (1943) analysis of domestic fowl. Collias measured the following intuitively promising qualities in a series of White Leghorn hens: their general health as indicated by weight and general vigor of movement, their age, their stage of molting, their level of androgen as indicated by the size of the comb, and their rank in the home flock from which they were drawn. The hens were then matched pairwise on neutral ground and the outcome of their aggressive interactions was recorded. Winning in these encounters was found to depend most on the absence of molt, followed in order by comb size, earlier social rank, and weight. Age did not seem to matter. All of these factors in combination accounted for only about half the variance. Collias suggested that the additional contributors to rank included differences in fighting skill, luck in landing blows, degree of wildness and aggressivity, slight differences in handling, and the physical resemblance of particular opponents to past despots. Of course, most or all of these components are heritable, so it is correct to say that to a degree not yet measured the status of hens is determined genetically.

A similar multiplicity of factors has been discovered in the more social mammals. Hormone levels are deeply implicated. An increase in androgen titer, and hence masculinization of anatomical and behavioral traits, tends to move individuals upward in the hierarchy. The adrenal hormones also appear to have a role. Candland and Leshner (1971) found that dominant males in a laboratory troop of squirrel monkeys had the highest levels of 17-hydroxycorticosteroids and the lowest levels of catecholamines (epinephrine plus norepinephrine). The 17-ketosteroids, however, were related to dominance by a J-shaped function: dominant males had medium titers, middle-ranking males low titers, and low-ranking males high titers with levels rising as rank fell. Candland and Leshner then turned the procedure around to see if the dominance order could be predicted from the hormone levels. Before forming a laboratory troop of five squirrel monkeys, they obtained baseline measures of urinary steroids and catecholamines in the separate animals. The concentration of 17-hydroxycorticosteroids was sufficient to predict the subsequent rank order, while the catecholamine titer was a J-shaped function of declining rank. These results, while very suggestive, do not constitute proof. The mere existence of the correlation between rank order and hormone level does not establish causation in either direction. Moreover, both could stem from other, prior determinants, such as age, health, and experiences unrelated to dominance interactions.

The status of parents can also matter. Among Japanese macaques the sons of high-ranking females are able ipso facto to spend more time in the center of the troop and to associate more closely with dominant males during their childhood. They tend to receive the cooperation of the leaders and to succeed them in position when they die. Sons of low-ranking mothers, in contrast, remain near the periphery of the group and are the first to emigrate. The existence of such a hereditary aristocracy results in greater group stability (Kawai, 1958; Kawamura, 1958, 1967; Imanishi, 1963). Kawai has made the useful distinction between *basic rank,* the outcome of the interaction of two monkeys unaffected by the influence of kin, and *dependent rank,* in which kinship plays a biasing role. Similar dependent succession has been studied in rhesus macaques by Koford (1963), Sade (1967), Marsden (1968), and Missakian (1972). Young rhesus monkeys of both sexes begin play-fighting with older infants or yearlings. The outcome is rather unfair: each animal defeats age peers whose mothers rank below its own, and each is defeated by age peers with higher-ranking mothers. As the monkeys mature they extend their dominance position into the existing hierarchy of adults, thus coming to rank just below their mothers. Females remain at approximately this level. Males, however, tend to change in rank upward or downward, possibly as a result of physiological variation.

It is not wholly imprecise to speak of much of the residual variance in dominance behavior as being due to "personality." The dominance system of the Nilgiri langur *Presbytis johnii* is weakly developed and highly variable from troop to troop. Alliances are present or absent; there is a single adult male or else several animals coexist uneasily; and the patterns of interaction differ from one troop to another. Much of this variation depends on idiosyncratic behavioral traits of individuals, especially of the dominant males (Poirier, 1970a,b).

Intergroup Dominance

Sometimes groups dominate groups in much the same way that group members dominate one another. Intergroup dominance is not often seen in nature, because contact between well-organized societies regularly occurs along territorial boundaries where power is more or less balanced. However, if the territories are spatiotemporal, dominance orders can appear when groups meet in overlapping portions of the home ranges. Phyllis Jay (1965) observed such a pattern in low-density populations of the common langur (*Presbytis entellus*) at Kaukori and Orcha in northern India. Because the langur troops possessed distinct core areas and followed their own routes while foraging, they seldom encountered one another. When contacts did occur, the larger group took precedence, with the smaller group simply remaining at a distance until the larger group moved away.

Intergroup hierarchies can also be created by confining societies in spaces smaller than the average territory occupied by a single group. When this is done to colonies of social insects, the result is almost invariably fatal for the weaker unit (Wilson, 1971a). While studying the phenomenon systematically in rhesus monkeys, Marsden (1971) discovered an interesting secondary effect. As the subordinate troops retreated into a smaller space, their members fought less among one another. But within the dominant group, which was in the process of acquiring new space, aggressive interactions increased. Marsden's effect, if it occurs at all generally, has important implications for the evolution of cooperative behavior.

Scaling in Aggressive Behavior

The general pattern of scaling in aggressive behavior among animals is summarized in Figure 13-3. This scheme is the culmination of a long history of inves-

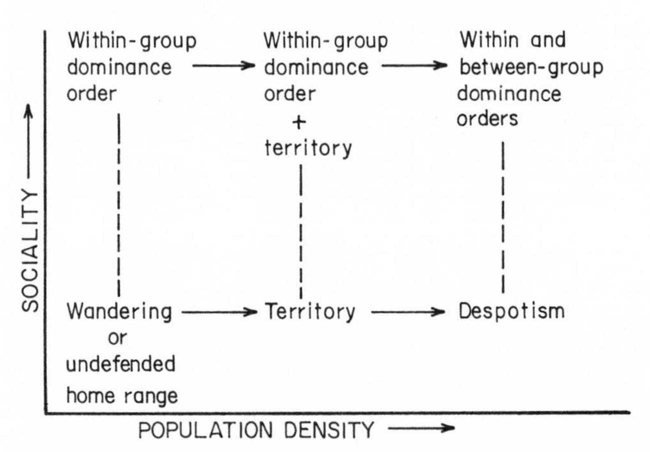

Figure 13-3 The prevailing patterns of scaling in aggressive behavior. The solid lines indicate true scaling: the transitions commonly observed to be part of the phenotypic variation of individuals. The dashed lines represent shifts unlikely to occur except by genetic evolution, which permits a species to substitute one pattern of scaling for another in the course of social evolution.

tigation by many zoologists. Perhaps the first explicit description of scaling was that of H. H. Shoemaker (1939), who found that canaries forced together in small spaces become organized into dominance orders. Given more space, they establish territories (the natural condition for *Serinus canaria* in the wild), even though low-ranking individuals continue to be dominated around bath bowls, feeding areas, and other nonterritorial public space. The phenomemon has been subsequently documented in other birds, sunfishes and char, iguanid lizards, house mice, Norway rats, *Neotoma* wood rats, woodchucks, and cats (review in Wilson, 1975). Kummer (1971) developed the concept with special reference to the social evolution of primates.

The existing data permit several generalizations about aggressive scaling. The clearest cases are found in species, such as certain lizards and rodents, in which the normal state is for solitary individuals or pairs to occupy territories. When forced together, groups of these individuals shift dramatically to despotisms or somewhat more complex dominance orders. In most such cases the shift from territoriality to a dominance system is really superficial in nature. In the case of despotism, one individual in effect retains its territory while merely tolerating the existence of the others.

Although some species display phenotypic variability that covers a substantial portion of the gradients, in other words utilize true behavioral scaling, many others are fixed at a single point. The males of sea lions, elephant seals, and other harem-forming pinnipeds maintain territories with about the same intensity regardless of population density. The adaptive significance of such rigidity is clear. Aggressive behavior in these animals serves the single function of acquiring harems. The means by which this goal is reached and its value to genetic fitness are unaffected by changes in the density of animals on the hauling grounds. In such cases, shifts from one point to another on the behavioral gradients occur by evolution, but probably only when changed environmental circumstances alter the optimum social strategy.

CHAPTER 14

Roles and Castes

Society, in the original,quasi-mystical vision created by Durkheim and Wheeler, is a superorganism that evolves to greater complexity through the complementary processes of differentiation and integration. As the society becomes increasingly efficient, larger, and geometrically more structured, its members become specialized into roles or castes and their relationships become more precisely defined through superior communication. Whole new ways of life—the practice of agriculture, industrialization, the storage of vast amounts of information, travel over fantastic distances, and more—await the society that can correctly engineer the division of labor of its members. Even the lowly ants have invented agriculture and slavery.

In considering the division of labor, three terms—role, caste, polyethism—must be carefully defined:

Role: a pattern of behavior that appears repeatedly in different societies belonging to the same species. The behavior has an effect on other members of the society, consisting either of communication or of activities that influence other individuals indirectly—or both. An animal, like a human being, can fill more than one role. For example, it might function as a control animal in terminating disputes and also as a leader when the group is on the move. Ideally, the full description of all roles together, insofar as they can be meaningfully distinguished, will fully define the society. In the broadest sense, male behavior during copulation constitutes a role, as does maternal care despite the fact that primatologists have not yet found it useful to speak of such behaviors in just that way. Idiosyncratic actions of individuals do not constitute roles; only regularly repeated categories fulfill the criterion. For example, the animal or set of animals that regularly watches for predators near the periphery of the group is playing a role, but a particular male who prefers to watch from a certain tree is not. Thus when Saayman (1971) spoke of the "roles" of three male chacma baboons in one particular troop as coincident with detailed differences in their behavior, he stretched the definition too far.

Caste: a set of individuals, smaller than the society itself, which is limited more or less strictly to one or more roles. Where the role is defined as a pattern of behaviors, which particular individuals may or may not display, the caste is defined inversely as a set of individuals characterized by their limitation to certain roles. In human societies a caste is a hereditary group, endogamously breeding, occupied by persons belonging to the same rank, economic position, or occupation, and defined by mores that differ from those of other castes. In social insects a caste is any set of individuals of a particular morphological type, or age group, or both, that performs specialized labor in the colony. It is often more narrowly defined as any set of individuals that are both morphologically distinct and specialized in behavior. A caste system may or may not be based in part on genetic differences. In the stingless bee genus *Melipona,* queens are determined as complete heterozygotes in multiple-locus systems, but in most or all other social insects the caste of individuals is fixed by purely environmental influences.

Polyethism: the differentiation of behavior among categories of individuals within the society, especially age and sex classes and castes. Both role playing and caste formation lead automatically to polyethism. In the social insects polyethism refers particularly to division of labor. A distinction is sometimes made, in compliance with the narrower usage of the word "caste," between caste polyethism, in which morphological castes are specialized to serve different functions, and age polyethism, in which the same individual passes through different forms of specialization as it grows older (Wilson, 1971a). Age polyethism is generally less rigid and sharply delimited than caste polyethism; individual workers may sometimes revert to earlier age specialization if there is a sufficient need on the part of the colony.

Table 14-1 Division of labor among workers of the ant *Daceton armigerum* by head width. (From Wilson, 1962b.)

Type of labor	Head width (mm) 1	2	3	4	Total number of observations
Surinam colony					
Total population (in artificial nest, April 5)[a]	13	60	20	9	102
Disposing of corpses and refuse[b]	0	19	12	2	33
Dismembering and feeding on fresh prey in nest[b]	0	14	25	5	44
Feeding larvae by regurgitation[b]	8	15[c]	3[c]	1[c]	27
Attending egg-microlarva pile[b]	24	3	0	0	27
Foraging in the field[a]	0	0	4	10	14
Trinidad colony					
Foraging in the field[a]	1	91	77	12	181
Resting in way-station[a]	0	8	19	10	37
Carrying prey[a]	0	1	1	10	12

[a] Numbers refer to separate, individual workers.

[b] Numbers refer to separate behavioral acts, without regard to the number of workers engaged.

[c] Consisting mostly of callows.

The Adaptive Significance of Roles

The differentiation of behavior within a society can best be measured by the judicious selection of sets of individuals and the comparison of their behavior patterns. Tables 14–1 and 14–2 present the results of such analyses in a colony of ants and in troops of primates, respectively. Notice that the two matrices closely resemble each other. The idea can be fleetingly entertained that they provide the means of comparing such different societies in a quantitative manner. The independent category of ants is the caste and the dependent one division of labor, whereas the monkey troop is partitioned into age-sex classes and "role profiles." But the distinction at this level is trivial. The castes of insects are based on age and sex in addition to size, while their places in the division of labor could equally well be labeled role profiles.

The deeper difference between the two patterns lies in the nature of the adaptiveness of the differentiation of behavior. We ask: At what level has natural selection acted to shape these varying profiles? The reader will recognize one or more version of the central problem of group selection in social evolution. The problem must be solved with reference to polyethism before the full significance of behavioral differentiation can be disclosed. For social insects the problem appears to be essentially solved. Selection is largely at the level of the colony. Castes are generated altruistically—they perform for the good of the colony. Caste systems and division of labor can therefore be treated by optimization theory. Vertebrates, however, are as usual mired in ambiguity. Kin selection is undoubtedly strong in small, closed societies such as primate troops. Hence an adult male of a single-male group can play the role of sentinel and defender in an altruistic manner. He risks injury or death for the good of the society. But his role is not quite the same as the role of defender in the insect society, for the male vertebrate is defending his own offspring. Much of role playing in vertebrate societies is patently selfish. The forager who discovers food gets the first share; the male who visits new troops improves his chances of rising in status by finding weaker opponents. Each behavior must be interpreted unto itself. Only when the contribution

Table 14-2 Differentiation of behavior in troops of the vervet monkey *Cercopithecus aethiops*, given as frequencies of contributions by age-sex classes to several categories of behavior. (Based on Gartlan, 1968.)

Behavior ("roles")	Age-sex class: Adult males	Adult females	Juvenile males	Subadult females	Infants
Territorial display	.66	0	.33	0	0
Vigilance; look-out behavior	.35	.38	.03	.12	.12
Receiving friendly approaches	.12	.46	.04	.27	.12
Friendly approach to others	.03	.32	0	.47	.15
Chasing of territorial intruders	.67	0	.33	0	0
Punishing intragroup aggression	1.00	0	0	0	0
Leading in group movement	.32	.49	0	.16	0

of the behavior to individual as opposed to group fitness is assayed will it become feasible to distinguish roles that are the secondary outcome of individual adaptations from those that are "designed" with reference to the optimum organization of the society. Meanwhile the concept of the vertebrate role must be regarded as loose and even potentially misleading. We shall explore this matter further, but first it will be useful to examine the less ambiguous paradigm of castes in insect societies and lower invertebrates. Here the basic theory has been initiated to which vertebrate societies can eventually be referred.

The Optimization of Caste Systems

Caste in the social insects is a large and complicated subject (see reviews by Wilson, 1971a, Lüscher, ed., 1977, and Oster and Wilson, 1978). Here we shall consider only two topics of general interest: the defensive castes of ants and termites, which illustrate the extremes of specialization and altruism found in the social insects as a whole, and the theory of caste ergonomics, through which the problem of optimization can be approached.

In the case of advanced polymorphism in ant colonies, especially complete dimorphism where intermediates have dropped out and the two remaining size classes are strikingly different in morphology, members of the larger class usually serve as soldiers. Often they play other roles as well. Soldiers of some species of *Camponotus* and *Pheidole* assist in food collection, and their abdomens swell with liquid food. Recent work has revealed that their per-gram capacity is much greater than that of their smaller nestmates, and they therefore serve as living storage casks (Wilson, 1974a). But it is apparent that the extensive changes in the head and mandibles that make the soldiers so deviant are directed primarily toward a defensive function. One of three fighting techniques is employed, depending on the form of the soldier. In one form the soldier may use the mandibles as shears or pliers: the mandibles are large but otherwise typical, the head is massive and cordate, and the soldiers are adept at cutting or tearing the integument and clipping off the appendages of enemy arthropods. Examples are found in *Solenopsis, Oligomyrmex, Pheidole, Atta* (Figure 14–1), *Camponotus, Zatapinoma,* and other genera of diverse taxonomic relationships. W. M. Wheeler, in his essay "The Physiognomy of Insects" (1927), pointed out that the peculiar head shape of this kind of soldier is due simply to an enlargement of the adductor muscles of the mandibles, which imparts to the mandibles greater cutting or crushing power. A second form of soldier has pointed, sickle-shaped or hook-shaped mandibles that are used to pierce the bodies of enemies. Some formidable examples are the major workers of the army ants (*Eciton*) and driver ants (*Dorylus*), which are able to drive off large vertebrates with their simultaneous bites and stings. The third basic type of soldier is less aggressive, using its head instead to block the nest entrance—thus serving literally as a living door. The head may be shield-shaped (many members of the tribe Cephalotini) or plug-shaped (*Pheidole lamia* and several subgenera of *Camponotus*). The colonies possessing such forms usually nest in cavities in dead and living plants and cut nest entrances with diameters just a little greater than the width of the head of an individual soldier.

The soldier is also the most specialized caste found in the termites. The soldier castes of ants and termites display many remarkable convergences in anatomy and behavior. The three basic forms found in ants—the shearer-crusher, the piercer, and the blocker—also occur in various termite species. In addition there are bizarre "snapping" soldiers in *Capritermes, Neocapritermes,* and *Pericapritermes* (Kaiser, 1954; Deligne, 1965). Their mandibles are asymmetrical and so arranged that the flat inner surfaces press against each other as the adductor muscles contract. When the muscles pull strongly enough, the mandibles slip past each other with a convulsive snap, in the same way that we snap our fingers by pulling the middle finger past the thumb with just enough pressure to make it slide off with sudden force. If the mandibles strike a hard surface, the force is enough to throw the soldier backward through the air. If they strike another insect, which seems to be the primary purpose of the adaptation, a stunning blow is delivered. Even vertebrates receive a painful flick.

The extreme soldier castes of some ant and termite species are so specialized that they function as scarcely more than organs in the body of the colony superorganism. Their existence supports the concept that in the case of insects the colony rather than the individual is the unit of organization of most importance in evolution. If natural selection is indeed mostly at the colony level, and workers are mostly or wholly altruistic with respect to the remainder of the colony, their numbers and behavior can be closely regulated through evolution to approach maximum colony fitness. In the ergonomics theory I developed earlier (Wilson, 1968a; see extensions in Oster and Wilson, 1978), I postulated that the mature colony, on reaching its predetermined size, can therefore be expected to contain caste ratios that approximate the *optimal mix.* This mix is simply the ratio of castes that can achieve the maximum rate of production of virgin queens and males while the colony is at or near its maximum size. It is helpful to think of a colony of

Figure 14-1 A soldier of the leaf-cutting ant *Atta cephalotes* is surrounded by smaller nestmates. The middle-sized workers shown here are most active in foraging outside the nest, while the smallest individuals specialize more in the care of the brood. Soldiers weigh as much as 90 milligrams and the smallest workers as little as 0.42 milligrams. (Photograph by courtesy of C. W. Rettenmeyer.)

social insects as operating in somewhat the same way as would a factory constructed inside a fortress. Entrenched in the nest site and harassed by enemies and capricious changes in the physical environment, the colony must send foragers out to gather food while converting the secured food inside the nest into virgin queens and males as rapidly and as efficiently as possible. The rate of production of the sexual forms is an important but not an exclusive component of colony fitness.

Suppose we are comparing two genotypes belonging to the same species. The relative fitness of the genotypes could be calculated if we had the following complete information: the survival rates of queens and males belonging to the two genotypes from the moment they leave the nest on the nuptial flights; their mating success; the survival rate of the fecundated queens; and the growth rates and survivorship of the colonies founded by the queens. Such complete data would, of course, be extremely difficult to obtain. In order to develop an initial theory of ergonomics, however, it is possible to get away with restricting the comparisons to the mature colonies. In order to do this and still retain precision, it would be necessary to take the difference in survivorship between the two genotypes outside the period of colony maturity and reduce it to a single weighting factor. But we can sacrifice precision without losing the potential for general qualitative results by taking the difference as zero. Now we are concerned only with the mature colony, and, given the artificiality of our convention, the production of sexual forms becomes the exact measure of colony fitness. The role of colony-level selection in shaping population characteristics within the colony can now be clearly visualized. If, for example, colonies belonging to one genotype contain, on the average, 1,000 sterile workers and produce 10 new virgin queens in their entire life span, and colonies belonging to the second genotype contain, on the average, only 100 workers but produce 20 new virgin queens in their life span, the second genotype has twice the fitness of the first, despite its smaller colony size. As a result, selection would reduce colony size. The lower fitness of the first genotype could be due to a lower survival rate of mature colonies, or to a smaller average production of sexual forms for each surviving mature colony, or to both. The important point is that the rate of production can be expected to shape mature colony size and organization to maximize this rate.

The production of sexual forms is determined in large part by the number of "mistakes" made by the mature colony as a whole in the course of its fortress-factory operations. A mistake is made when some potentially harmful contingency is not successfully met—a predator invades the nest interior, a breach in the nest wall is tolerated long enough to desiccate a brood chamber, a hungry larva is left unattended, and so forth. The cost of the mistakes for a given category of contingencies is the product of the number of times a mistake is made times the reduction in queen production per mistake. With this formal definition, it is possible to derive in a straightforward way a set of basic theorems on caste. In the basic model I developed, the average output of queens is viewed as the difference between the ideal number made possible by the productivity of the foraging area of the colony and the number lost by failure to meet some of the contingencies. (The model can be modified to incorporate other components of fitness without altering the results.) The evolutionary problem which I postulate to have been faced by social insects can be solved as follows: the colony produces the mixture of castes that maximizes the output of queens. In order to describe the solution in terms of simple linear programming, it is necessary to restate the solution in terms of the mirror equivalent of the first statement: the colony evolves the mixture of castes that allows it to produce a given number of queens with a minimum quantity of workers. In other words, the objective is to minimize the energy cost.

The simplest case involves two contingencies whose costs would exceed a postulated "tolerable cost" (above which, selection takes place), together with two castes whose efficiencies at dealing with the two contingencies differ. The inferences to be made from this simplest situation can be extended to any number of contingencies and castes.

The most important step is to relate the total weights, W_1 and W_2, of the two castes in a colony at a given instant to the frequency and importance of the two contingencies and the relative efficiencies of the castes at performing the necessary tasks. By stating the problem as the minimization of energy cost (see Wilson, 1968a), the relation can be given in linear form.

The optimal mix of castes is the one that gives the minimum summed weights of the different castes while keeping the combined cost of the contingencies at the maximum tolerable level. The manner in which the optimal mix is approached in evolution is envisaged as follows. Any new genotype that produces a mix falling closer to the optimum is also one that can increase its average net output of queens and males. In terms of energetics, the average number of queens and males produced per unit of energy expended by the colony is increased. Even though colonies bearing the new genotype will contain about the same adult biomass as other colonies, their average net output will be greater. Consequently, the new genotype will be favored in colony-level selection, and the species as a whole will evolve closer to the optimal mix.

The general form of the solution to the optimal-mix problem is given in Figure 14–2. It has been postulated that behavior can be classified into sets of responses in a one-to-one correspondence to a set of kinds of contingencies. Even if this conception only roughly fits the truth, it is enough to develop a first theory of ergonomics. For example, continuation of the graphical analysis (Figures 14–3 and 14–4) shows that so long as the contingencies occur with relatively constant frequencies, it is an advantage for the species to evolve so that in each mature colony there is one caste specialized to respond to each kind of contingency. In other words, one caste should come into being that perfects the appropriate response, even at the expense of losing proficiency in other tasks.

Roles in Vertebrate Societies

We can now consider the key question about roles in vertebrate societies, which is the following: To what extent are age-sex classes and other categories of individuals defined by behavioral profiles comparable to the castes of insects? In other words, can there be an ergonomics of vertebrate societies? The answer, as suggested earlier, lies in the intensity of group selection with reference to behavioral differentiation.

The best way to attack the problem may be to partition the behavioral differences into *direct* and *indirect roles*. A direct role is a particular behavior or set of behaviors displayed by a subgroup that benefits other subgroups and therefore the group as a whole. An indirect role is behavior that benefits only the individuals that display it and is neutral or even destructive to other subgroups. The direct role is fa-

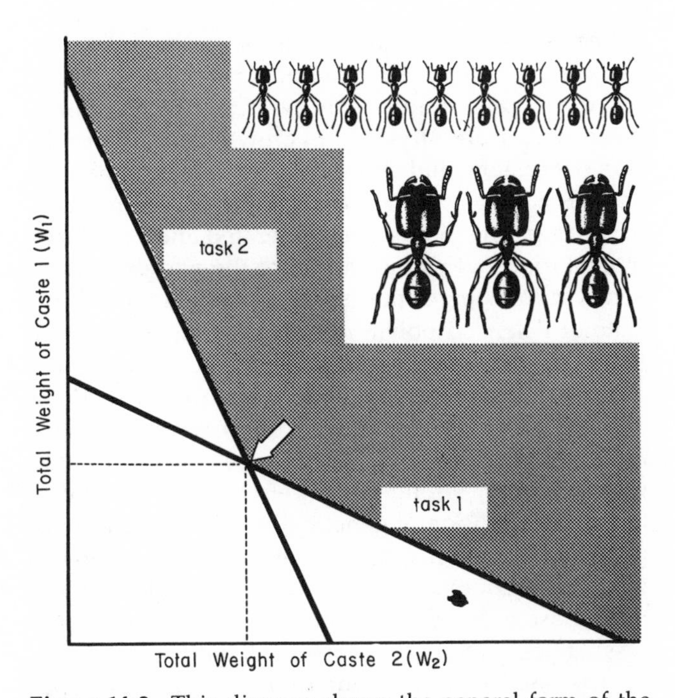

Figure 14-2 This diagram shows the general form of the solution to the optimal-mix problem in evolution. In this simplest possible case, two kinds of contingencies ("tasks") are dealt with by two castes. The optimal mix for the colony, measured in terms of the respective total weights of all the individuals in each caste, is given by the intersection of the two curves. Contingency curve 1, labeled "task 1," gives the combination of weights (W_1 and W_2) of the two castes required to hold losses in queen production to the threshold level due to contingencies of type 1; contingency curve 2, labeled "task 2," gives the combination with reference to contingencies of type 2. The intersection of the two contingency curves determines the minimum value of $W_1 + W_2$ that can hold the losses due to both kinds of contingencies to the threshold level. The basic model can now be modified to make predictions about the effects on the evolution of caste ratios of various kinds of environmental changes. (From Wilson, 1968a.)

vored by group selection or at least does not run counter to it. It can be detrimental to the individual and the individual's progeny, as in the actions of castes of ants and termites. In this case favorable group selection almost certainly occurs. Or the direct role can add to individual fitness while at the same time reinforcing group survival or at least not diminishing it. Direct roles favored by group selection are subject to ergonomic optimization with respect to the numbers of individuals playing the role and the intensity with which it is expressed. This is evidently what Gartlan (1968) had in mind when describing primate societies in terms of roles: "The group is an adaptive unit, the actual form of which is determined by ecological pressures. Different roles of relevance to particular ecological conditions are

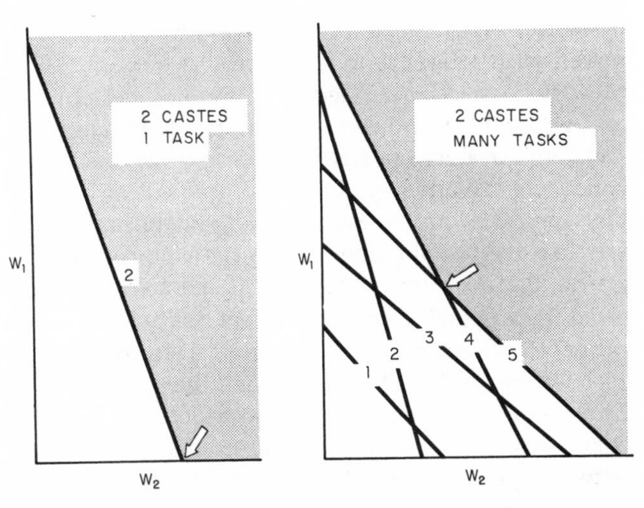

Figure 14-3 The diagram on the left shows that, when there are more castes than tasks, the number of castes will be reduced in evolution to equal the number of tasks. The surplus castes removed will be the least efficient ones (in this case, caste 1). The diagram on the right shows that if there are more tasks than castes, the optimal mix of castes will be determined entirely by those tasks, equal or less in number to the number of castes, which deal with the contingencies of greatest importance to the colony (in this case, tasks 4 and 5). (From Wilson, 1968a.)

performed by different animals." An indirect role, in contrast, is simply the outcome of selfish behavior that can be manifested by some but not all members of the society. If the magnitude of individual genetic selection is at least comparable to the rate of group extinction opposing it, the role will be maintained in a balanced polymorphic state (see Chapter 5). But in

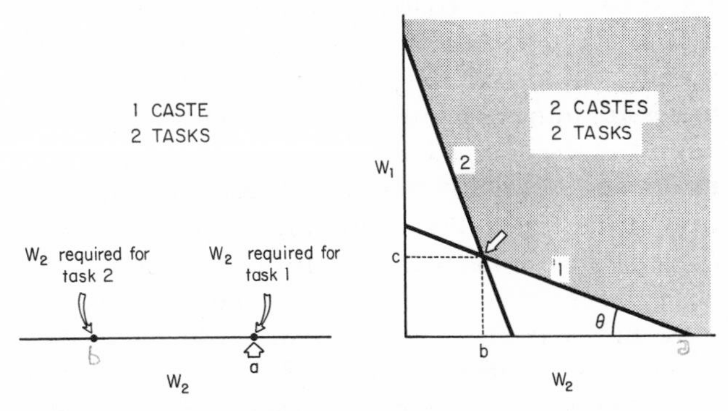

Figure 14-4 It is always to the advantage of the species to evolve new castes until there are as many castes as contingencies, and each caste is specialized uniquely on a single contingency. This theorem can be substantiated readily by comparing the two graphs in this figure. With the addition of caste 1 in the right-hand figure, the total weight of workers is changed from a to $b + c$. Since caste 1 specializes in task 1, θ is ~~acute;~~ therefore, $a - b > c$ and $a > b + c$ for all a, b, and c. (From Wilson, 1968a.)

less than 45°

no sense can the indirect role be ergonomically optimized with reference to the society as a whole.

Most roles so far defined in nonhuman vertebrates are apparently indirect in nature. Consider the "leaders" in flocks of European wood pigeons. They constitute the advancing front of the feeding assemblies, but they are in that position only because they are displaced by the dominant birds who control the center. Because they constantly glance backward toward the advancing group, they eat less and are more prone to starvation in hard times (Murton et al., 1966). Some authors have spoken of the role of young birds and mammals as dispersants of the species, colonizing new terrain and exchanging genes between populations. While it is true that individuals journey farther while they are young, the difference is generally the outcome of their subordinate position in their place of birth. Its adaptive basis is the greater chance it gives young animals to gain territory or to rise to dominance in new places. The role the young play with respect to population dynamics and gene flow is probably wholly indirect. Fruit bats (*Pteropus giganteus*) form large daytime resting aggregations in certain trees in the Asiatic forests. Each male has his own resting position, with subordinate individuals occupying the lower limbs and hence suffering the most exposure to ground-dwelling predators. The subordinate males usually see danger first and alert the remainder of the colony by their excited movements. They serve as very effective sentinels for the group as a whole (Neuweiler, 1969), but their role is clearly indirect in nature. Similar examples can be multipled indefinitely.

Cases of direct roles are much harder to find among the vertebrates. Adults of the African wild dog appear to divide labor in a way that benefits the pack as a whole. Adults, including the mother bitch, remain behind with the pups during a chase, and the successful hunters regurgitate meat to them upon returning to the den. Adult male olive baboons cooperate to an impressive degree in policing the area around the troop. When juveniles register alarm or excitement, the nearest adult male investigates the cause. If his own reaction is strong enough, the other adult males rush to his assistance (Rowell, 1967). Silver-backed males of the mountain gorilla play multiple roles in the troops they lead. When a group loses the leader-male, it appears to search for a new one (Schaller, 1963). Perhaps no evidence more strongly suggests the direct nature of a role than the effort by the group to recruit another animal to fill it.

Do castes occur in vertebrate societies in addition to direct roles? If group selection is strong enough, there is no reason why caste systems cannot have evolved. They might even have a purely physiological basis, as in most social insects. In that case individuals at inception would possess equal potential for development into any caste. Once an animal crossed a certain threshold in growth and differentiation, its caste would be fixed for some period of time. Although physiological caste systems seem intuitively to be the most easily generated and optimized in evolution, we cannot discount the possibility that genes influencing some aspects of behavior exist in a state of balanced polymorphism, for the reason that their carriers benefit noncarriers and therefore the group as a whole. Such genes might be altruistic or nonaltruistic, that is, either counteracting or reinforcing individual-level selection. To put it another way, relatively strong group selection might tend to favor the evolution of genetic diversity within as opposed to between societies. A society with genes near the ergonomic mix would have higher fitness than those away from the mix, including those possessing less diversity. However, specific gene *frequencies* are harder to maintain by selection than specific *genes*, and such genes can individually program physiological caste systems.

Castes in vertebrates, if they exist, should take the form of distinctive physiological or psychological types that recur repeatedly at predictable frequencies within societies. Some would probably be altruistic in behavior—homosexuals who perform distinctive services, celibate "maiden aunts" who substitute as nurses, self-sacrificing and reproductively less efficient soldiers, and the like. The most direct and practicable test of the caste hypothesis is whether phenotypic variance within societies exceeds that of comparable samples from closely related nonsocial or at least less social species. If it does not, the hypothesis is negated; if it does, the hypothesis is supported but still not proved. In cases where greater variance is further associated with higher genetic diversity, the possibility of a genetic caste system is indicated. Vertebrate zoologists appear not to have consciously investigated these possibilities, although Trivers' recent theoretical work on parent-offspring conflict envisages celibate and other self-sacrificing behavioral types as one possible outcome of vertebrate kin selection (see Chapter 16). The evidence is sparse and equivocal. Jolicoeur (1959) reports that populations of wolves are highly polymorphic in size and color, while Fox (1972) has detected strong differences among members of the same litter in reactivity, exploratory behavior, and prey-killing ability. Other authors have commented on the existence of striking variation within packs of African wild dogs and of differences in facial features of baboons and chimpanzees that permit human observers to recognize individuals at a glance. Such variation is subject to several competing explanations, but at least it is consistent with the hypothesis of hereditary castes in advanced mammalian societies.

The existence of direct roles and castes in vertebrate societies is thus indicated only by marginal evidence limited to the most social of the mammals. This limitation puts the utility of the role as a scientific concept in considerable doubt. The word can be used in a metaphorical sense, intuitively and changeably defined, but it is not likely to acquire a firm operational definition in the immediate future. The classification of indirect roles remains a formidably difficult—and perhaps useless—task. After a brave start, the primate literature has foundered in a simple listing of categories. Some authors, for example Bernstein and Sharpe (1966) and Crook (1971), virtually equated roles with role profiles. The category was defined by sex, age, and perhaps also some diagnostic social trait, and then its other statistically distinctive qualities were explored. Thus reference was made to the "roles" of the control male, the secondary male, the isolate male, the central female, the peripheral female, and others. Gartlan (1968), in contrast, equated roles with acts of social behavior: territorial vigilance, approaching other troop members in a friendly manner, and so forth. Such orthogonal classifications are multiplicative when taken together and hence increase confusion at an exponential rate. To make this criticism is not to doubt the value of lists of social behaviors and analyses of behavioral profiles. All that is suggested here is that they be called by their correct names and not obscured by unnecessary reference to roles. If this is true, can it be said that the concept of the role has any useful place at all in vertebrate sociobiology? The answer is a qualified yes. There exist a few patterns of social behavior that can be conveniently labeled roles and treated as separate elements in the analysis of certain vertebrate societies. Two of them, leadership and control, will now be briefly reviewed. It is only necessary to bear in mind that each is a heterogeneous collection of behaviors defined loosely by function across species, and referred to as a role because of its employment by a subgroup of the society in affecting the behavior and welfare of the group as a whole.

Leadership

When zoologists speak of leadership as a social role, they usually mean the simple act of leading other group members during movement from one place to another. In many instances such a role is filled casually, even accidentally. Schooling fish, such as mullet and silversides, are "led" from moment to moment by whatever fish happen to be brought to the forward edge by the movement of the school as a whole. Individuals frequently try to turn inward toward the center of the group, so that the second ranks are brought to the front. When the school encounters a predator or impassable object, the members turn away individually. As a result, the entire school wheels to the side or reverses direction, and fish along the flank or rear become the new leaders (Shaw, 1962). The least-organized bird flocks, for example the feeding groups of starlings, move in a similar fashion, with the leaders often being simply the fastest fliers (Allee, 1942). In flocks of ring doves and jackdaws the most experienced birds in a particular situation take the initiative, and others follow (Lorenz, 1935; Collias, 1950). Leadership in large ungulate groups is also casual and shifting in character. The vanguard of reindeer herds consists chiefly of the most timid and restless individuals, who are first to stop eating, first to rest and chew their cud, and first to get up again (V. M. Sdobnikov, in Allee et al., 1949).

A few mammalian species possess stronger forms of leadership, more nearly consonant with the role as played by human beings. When members of a wolf pack travel in single file, any one of several individuals can take the lead. But during chases the dominant male assumes command. He directs the attack on prey and sometimes the pursuit after others give up. In herds of red deer a fertile hind consistently leads the group, and her followers sometimes include even young stags. Herds of the African elephant are organized in a nearly identical fashion. Clans of mountain sheep are also highly structured. The adult males and females usually stay apart except during the rutting season, and leadership is assumed in each group by the largest and oldest individuals (Geist, 1971a).

Control

Since the role concept was first introduced into the primate literature, the key paradigm has been the control animal. The point stressed by most writers is that dominance and the control function are separable. The two forms of interaction are generally correlated but nevertheless distinct. In some animals, for example the squirrel monkey *Saimiri sciureus*, a control animal exists but there is no overt dominance order. This is an important generalization, but unfortunately it has been semantically obscured. There has been a failure to distinguish between one or more control behaviors, which can be defined if need be down to the neuromuscular mechanisms, and the behavior profile of control animals. The elemental behavior pattern constituting the role is the intervention in aggressive episodes with the result of reducing or halting them. In monkey groups control is almost always achieved by threat or punishment. Kawamura (1967) describes it in the Japanese macaque as follows: "When one monkey of the troop is being attacked by another and emits an exaggerated cry for help, the leader males quickly rush in to attack and punish the aggressor. When the leaders arrive on the scene, many other monkeys flatter them while the aggressor attacks still another monkey as a new enemy, thereby adding to the confusion. Because the monkeys create such a furor, observers wonder at times whether the real purpose of the display is to punish the aggressor. Usually, however, the leaders do eventually find the original offender and punish it, even though it appears as though they are no longer angry with it." If an animal performing control behavior is characterized further, he is usually found to be prominent in leading the group in defense against intruders and to serve as an attention focus for other members of the group. But it must be admitted that these are additional roles and not part of control behavior per se, unless we care to broaden the definition of control to the point of uselessness. The correct way to analyze roles is to define them as discrete behavior patterns in particular species, to establish their degree of correlation within the group members, and finally to identify categories of individuals according to the roles usually invested in them. The correspondence of role profiles to certain age and sex groups, or even to castes, is an important but separable issue.

Roles in Human Societies

The very poverty and vagueness of roles in nonhuman primate societies underscores their richness and importance in human behavior. Human existence, as Erving Goffman and his fellow microsociologists have argued, is to a large extent an elaborate performance of roles in the presence of others. Each occupation—the physician, the judge, the waiter, and so forth—is played just so, regardless of the true workings of the mind behind the persona. Significant deviations are interpreted by others as signs of mental incapacity and unreliability.

Role playing in human beings differs from that in other primates, including even the chimpanzee, in several ways intimately connected to high intelligence and language. The roles are self-conscious: the actor knows that he is performing for the sake of others to some degree, and he continuously reassesses his persona and the impact his behavior is having on others. Models from his own social class and occupation are chosen and imitated. Role playing is thorough. The individual may change his clothing and personality and even his manner of speech while off the job, but while on it his performance must be consistent or others will suspect him of insincerity or incompetence. Human roles are very numerous. In advanced societies each individual is familiar with the behavioral norms of scores or hundreds of occupations and social positions. Division of labor is based on these memorized distinctions, in a fashion analogous to the physiological determination of castes in social insects. But whereas social organization in the insect colonies depends on programmed, altruistic behavior by an ergonomically optimal mix of castes, the welfare of human societies is based on trade-offs among individuals playing roles. When too many human beings enter one occupation, their personal cost-to-benefit ratios rise, and some individuals transfer to less crowded fields for selfish reasons. When too many members of an insect colony belong to one caste, various forms of physiological inhibition arise, for example the underproduction or overproduction of pheromones, which shunt developing individuals into other castes.

Nonhuman vertebrates lack the basic machinery to achieve advanced division of labor by either the insect or the human methods. Human societies are therefore unique in a qualitative sense. They have equaled and in many cultures far exceeded insect societies in the amount of division of labor they contain. We can speculate that if the evolutionary trajectory of higher nonhuman primates were now to be continued beyond the chimpanzee, it would reach a role system similar to the human model. With an increase in intelligence would come the capacity for language, the consciousness of personae, the long memories of personal relationships, and the explicit recognition of "reciprocal altruism" through equal, long-term trade-offs. Did in fact such qualities emerge as a consequence of higher intelligence during human evolution? Or was it the other way around—intelligence constructed piece by piece as an enabling device to create the qualities? This distinction, which is not trivial, will be explored further in the more extended discussion of man in Chapter 26.

CHAPTER 15

Sex and Society

Sex is an antisocial force in evolution. Bonds are formed between individuals in spite of sex and not because of it. Perfect societies, if we can be so bold as to define them as societies that lack conflict and possess the highest degree of altruism and coordination, are most likely to evolve where all of the members are genetically identical. When sexual reproduction is introduced, members of the group become genetically dissimilar. Parents and offspring are separated by at least a one-half reduction of the genes shared through common descent and mates by even more. The inevitable result is a conflict of interest. The male will profit more if he can inseminate additional females, even at the risk of losing that portion of inclusive fitness invested in the offspring of his first mate. Conversely, the female will profit if she can retain the full-time aid of the male, regardless of the genetic cost imposed on him by denying him extra mates. The offspring may increase their personal genetic fitness by continuing to demand the services of the parents when raising a second brood would be more profitable for the parents. The adults will oppose these demands by enforcing the weaning process, using aggression if necessary. The outcomes of these conflicts of interest are tension and strict limits on the extent of altruism and division of labor.

The strong tendency of polygamous species to evolve toward sexual dimorphism reinforces this canonical genetic constraint. When sexual selection operates among males, adults become larger and showier, and their behavior patterns and ecological requirements tend to diverge from those of the females. One consequence is a partitioning of the incipient society, not into castes designed to promote the efficiency of the society, but into secondary sex roles that enhance individual as opposed to group genetic fitness. In other words, males and females tend to diverge in the kinds of activities and in the ways they find most profitable.

The inverse relation between sex and social evolution becomes clear when a phylogenetic survey of the animal kingdom is conducted. The vertebrates are all but universally sexual in their mode of reproduction. To judge from Uzzell's review (1970), the relatively few cases of parthenogenetic populations recorded in fishes, amphibians, and lizards are local derivatives that do not evolve far on their own. And with the exception of man, vertebrates have assembled societies that are only crudely and loosely organized in comparison with those of insects and other invertebrates. Sex is a constraint overcome only with difficulty within the vertebrates. Sexual bonds are formed by a courtship process typically marked in its early stages by a mixture of aggression and attraction. Monogamy, and especially monogamy outside the breeding season, is the rare exception. Parent-offspring bonds usually last only to the weaning period and are then often terminated by a period of conflict. Social ties beyond the immediate family are mostly limited to a few mammal groups, such as canids and higher primates, that have sufficient intelligence to remember detailed relationships and thereby to form alliances and cliques. Even these are relatively unstable and in most species mixed with elements of aggression and overt self-serving.

The highest forms of invertebrate sociality are based on nonsexual reproduction. The phylogenetic groups possessing the highest degrees of caste differentiation, namely, the sponges, coelenterates, ectoprocts, and tunicates, are also the ones that create new colony members by simple budding. The social insects reproduce primarily by sexual means, and the limited amounts of conflict that occur within the colonies can be traced to genetic differentiation based on sexual reproduction. The Hymenoptera, the order in which advanced social life has most frequently originated, is also characterized by haplodiploidy, a mode of sex determination that causes sisters to be more closely related genetically to each other than parents and offspring. According to prevailing theory (see Chapter 19), this peculiarity accounts for the fact that the worker castes of ants, bees, and wasps are exclusively female. Thus increased sociality in insects appears to be based on a moderation of the shearing force of sexuality. In the invertebrates as a whole, sociality is also loosely associated with hermaphroditism. Groups in which the two conditions coexist include the sponges (Porifera), corals (Anthozoa), ectoprocts, and sessile tunicates. However, a few colonial groups are not hermaphroditic, while many hermaphrodites are noncolonial. (A thorough general review of hermaphroditism, with an investigation of its adaptive significance, is provided by Ghiselin, 1969).

In short, social evolution is constrained and shaped by the necessities of sexual reproduction and not promoted by it. Courtship and sexual bonding are devices for overriding the antagonism that arises automatically from genetic differences induced by sexual reproduction. Because an antagonistic force is just as important as a promotional one, the remainder of this chapter will present a review of current knowledge about the evolution of sex and its multifaceted relationships to social behavior.

The Meaning of Sex

Sexual reproduction is in every sense a consuming biological activity. Reproductive organs tend to be elaborate in structure, courtship activities lengthy and energetically expensive, and genetic sex-determination mechanisms finely tuned and easily disturbed. Furthermore, an organism that reproduces by sex cuts its genetic investment in each gamete by one-half. If an egg develops parthenogenetically, all of the genes in the resulting offspring will be identical with those of the parent. In sexual reproduction only half are identical; the organism, in other words, has thrown away half its investment. There is no intrinsic reason why gametes cannot develop into organisms parthenogenetically instead of sexually and save all of the investment. Why, then, has sex evolved?

It has always been accepted by biologists that the advantage of sexual reproduction lies in the much greater speed with which new genotypes are assembled. During the first meiotic division, homologous chromosomes typically engage in crossover, during which segments of DNA are exchanged and new genotypic combinations created. The division is concluded by the separation of the homologous chromosomes into different haploid cells, creating still more genetic diversification. When the resulting gamete is fused with a sex cell from another organism, the result is a new diploid organism even more different than the gamete from the original gametic precursor. Each step peculiar to the process of gametogenesis and syngamy serves to increase genetic diversity. To diversify is to adapt; sexually reproducing populations are more likely than asexual ones to create new genetic combinations better adjusted to changed conditions in the environment. Asexual forms are permanently committed to their particular combinations and are more likely to become extinct when the environment fluctuates. Their departure leaves the field clear for their sexual counterparts, so that sexual reproduction becomes increasingly the mode. (For detailed information on the theory of the evolution of sex, see Maynard Smith, 1978).

Evolution of the Sex Ratio

Why are there usually just two sexes? The answer seems to be that two are enough to generate the maximum potential genetic recombination, because virtually every healthy individual is assured of mating with a member of another (that is, the "opposite") sex. And why are these two sexes anatomically different? Of course, in many microorganisms, fungi, and algae, they are not; gametes identical in appearance are produced (isogamy). But in the majority of organisms, including virtually all animals, anisogamy is the rule. Moreover, the difference is usually strong: one gamete, the egg, is relatively very large and sessile; the other, the sperm, is small and motile. The adaptive basis of the differentiation is division of labor enhancing individual fitness. The egg possesses the yolk required to launch the embryo into an advanced state of development. Because it represents a considerable energetic investment on the part of the mother, the embryo is often sequestered and protected, and sometimes its care is extended into the postnatal period. This is the reason why parental care is normally provided by the female, and why most animal societies are matrifocal. The spermatozoan is specialized for searching out the egg, and to this end it is stripped down to the minimal DNA-protein package powered by a locomotory flagellum. Scudo (1967), entirely on the basis of an analysis of the searching role of the sperm, concluded that anisogamy must be developed to a high degree before its advantages outweigh those of the ancestral state of isogamy.

It is also generally profitable for parents to produce equal numbers of offspring belonging to each sex. Such mechanisms as XY and XO sex determination (where X and Y represent sex chromosomes and O denotes the absence of a chromosome) are to be viewed not as some inevitable result of chromosome mechanics but rather as specialized devices favored by natural selection because they generate 50/50 sex ratios with a minimum of complication. The evolutionary process thought to underlie the 50/50 ratio was first modeled by R. A. Fisher (1930). In barest outline "Fisher's principle" can be stated as follows. If male births in a population are less frequent than female births, each male has a better chance to mate than each female. All other things being equal, the male is more likely to find multiple partners. It follows that parents genetically predisposed to produce a higher proportion of males will ultimately have more grandchildren. But the tendency is self-negating in the population as a whole, since the advantage will be lost as the male-producing gene spreads and males become commoner. As a result, the sex ratio will converge toward 50/50. An exactly symmetric argument holds with reference to the production of females. Subsequent authors have refined and extended this model to the point where the following more precise statement can be made. Ideally a parent will not produce equal *numbers* of each sex; it should instead make equal *investments* in them. If one sex costs more than the other, the parent should produce a correspondingly smaller proportion of offspring belonging to it. Ordinarily, cost can be assessed in amounts of energy expended. Thus if a newborn female weighs twice as much on the average as a male, and no further parental investment is made after birth, the optimal sex ratio at birth should be in the vicinity of 2 males/1 female. Probably an even more precise assessment than energy expenditure is reproductive effort, the decrement in future reproductive potential as a consequence of the present effort (see Chapter 4). When parental care is added, differences in the amount of care devoted to the two sexes must be added to the deficit side of the ledger. If a daughter, for example, proves twice as costly to raise to independence as a son, the optimum representation of females among the offspring is cut by one-half. Once parental care ends, differential mortality between the sexes has no effect on the optimum sex ratio.

Other selection pressures can intervene to shift the ratio away from numerical parity. Parasitic species that found populations with small numbers of inseminated females are not bound by Fisher's principle (Hamilton, 1967). Because a large percentage of matings are between sibs, many of the males seeking mates will be in competition with other males who share sex-determining genes by common descent. In the parasitic life style it is advantageous to produce as many inseminated females as possible, even at the expense of unbalancing the initial sex ratio in favor of females. This advantage will override the selection working to restore male representation to parity, since the Fisher effect is weakened by inbreeding. Hamilton proved that under such conditions the "unbeatable" sex ratio will be $(n - 1)/kn$, where k is either 1 or 2, depending on the mode of sex inheritance, and n is the number of females founding the population. (Sex ratios are conventionally given as male-to-female.) When $n = 1$, the ideal ensemble is all-female, but the practical solution is either gynandromorphism or the production of a single male capable of fertilizing all of his sisters. The parasitic Hymenoptera appear to have solved this problem by haploidiploidy, in which males originate from unfertilized eggs and females from fertilized ones. A female has the capacity to control the sex of each offspring simply by "choosing" whether to release sperm from her spermatheca, the sperm-storage organ, just before the egg is laid. This control is used by some hymenopterous species to yield other sex ratios appropriate to special circumstances. The social bees, wasps, and ants ordinarily produce males only prior to the breeding season, reverting to all-female broods during the remainder of the year. A common pattern seen in parasitic wasps is the production of all-male broods on small or young hosts and of an increasing proportion of females on hosts capable of supporting a larger biomass (Flanders, 1956; van den Assem, 1971).

With physiological control of sex determination so prominently developed in the insects, the possibility should not be overlooked that it also occurs at least to a limited extent in the vertebrates. Trivers and Willard (1973) have constructed an ingenious argument to reveal which pecularities can be expected to result from such an adaptive distortion of the sex ratio. Their reasoning proceeds syllogistically as follows:

1. In many vertebrate species, large, healthy males mate at a disproportionately high frequency, while many smaller, weaker males do not mate at all. Yet nearly all females mate successfully.

2. Females in the best physical condition produce the healthiest infants, and these offspring tend to grow up to be the largest, healthiest adults.

3. Therefore, females should produce a higher proportion of males when they are healthiest, because these offspring will mate most successfully and produce the maximum number of grandchildren. As the females' condition declines, they should shift increasingly to the production of daughters, since female offspring will now represent the safer investment.

The first two propositions have been documented in rats, sheep, and human beings. The rather surprising conclusion of the argument (no. 3) is also consistent with the evidence. It provides a novel explanation for some previously unexplained data from mink, pigs, sheep, deer, seals, and human beings. For example, in deer and human beings environmental conditions adverse for pregnant females are associated with a reduced sex ratio, favoring the birth of daughters. The most likely mechanism is differential mortality of the young in utero. It is known that stress induces higher male fetal mortality in some mammals, especially during the early stages of pregnancy. The ultimate cause of the mortality could be natural selection in accordance with the Trivers-Willard principle.

Sexual Selection

The final question in our basic series about the nature of sex is: Why do the sexes differ so much? The traits of interest are the secondary sexual characteristics, which occur in addition to the purely functional differences in the gonads and reproductive organs. The males of many species are larger, showier in appearance, and more aggressive than the females. Often the two sexes differ so much as to seem to belong to different species. Among the ants and members of such aculeate wasp families as the Mutillidae, Rhopalosomatidae, and Thynnidae, males and females are so strikingly distinct in appearance that they can be matched with certainty to species only by discovering them *in copula*. Otherwise experienced taxonomists have erred to the point of placing them in separate genera or even families. The ultimate vertebrate case is encountered in four families of deep-sea angler fishes (Ceratiidae, Caulophrynidae, Linophrynidae, Neoceratiidae) in which the males are reduced to parasitic appendages attached to the bodies of the females.

Part of the solution to the mystery of sexual divergence was supplied by Charles Darwin in his concept of *sexual selection*, first developed at length in *The Descent of Man and Selection in Relation to Sex* (1871). According to Darwin, sexual selection is a special process that shapes the anatomical, physiological, and behavioral mechanisms that function shortly before or at the time of mating and serve in the process of obtaining mates. He excluded selection that leads to the evolution of such primary reproductive traits as the form of the male gonads or the egg-laying behavior of females. Darwin reasoned that competition for mates among the members of one sex leads to the evolution of traits peculiar to that sex. Two distinct processes were judged to be of about equal importance in the competition. They are, in Julian Huxley's (1938) phraseology, *epigamic selection*, which consists of the choices made between males and females, and *intrasexual selection*, which comprises the interactions between males or, less commonly, between females. To use Darwin's own words, the distinction is between "the power to charm the females" and "the power to conquer other males in battle." As early as 1859, when he first used the expression "sexual selection," Darwin envisaged it as basically different from most forms of natural selection in that the outcome is not life or death but the production or nonproduction of offspring.

Pure epigamic selection is not easy to document in the field. The displays of male birds are ordinarily directed at both males and females, and sexual selection is based as much on the territorial exclusion of rival males as on competition for the attention of potential mates. Epigamic selection unalloyed by intermale aggression can be seen during part of the courtship rituals of the ruff *Philomachus pugnax*, a European shore bird. The males are highly variable in color and display frenetically on individual territories that are grouped tightly together in a communal arena (see Figure 15-1). The rivals scuttle about with their ruffs expanded and wings spread and quivering. Sometimes they pause to touch their bills to the ground or shudder their entire bodies. Females wander singly or in groups from territory to territory, expressing their willingness to mate by crouching. The possession of a territory is not essential in all cases. Females have been observed to follow individual satellite males as they wandered from the territory of one dominant male to the territory of another (Hogan-Warburg, 1966). True epigamic selection also occurs in *Drosophila*. The *yellow* mutant of *D. melanogaster* is characterized not only by the altered body color from which it draws its name but also by subtle alterations in male courtship activity. One step in the display is wing vibration, a ritualized flight movement which is perceived by the female's antennae. The vibration bouts of *yellow* males

Figure 15-1 Males of the ruff occupy small territories and display competitively to females. (From Lack, 1968. Reproduced with permission from David Lack, *Ecological Adaptations for Breeding in Birds*, Methuen & Co., Ltd.; drawing by Robert Gillmor.)

are shorter in duration and spaced further apart than those of the normal genotypes, and they are less successful in obtaining the appropriate response from the female (Bastock, 1956).

Pure epigamic display can be envisioned as a contest between salemanship and sales resistance. The sex that courts, ordinarily the male, plans to invest less reproductive effort in the offspring. What it offers to the female is chiefly evidence that it is fully normal and physiologically fit. But this warranty consists of only a brief performance, so that strong selective pressures exist for less fit individuals to present a false image. The courted sex, usually the female, will therefore find it strongly advantageous to distinguish the really fit from the pretended fit. Consequently, there will be a strong tendency for the courted sex to develop coyness. That is, its responses will be hesitant and cautious in a way that evokes still more displays and makes correct discrimination easier.

In intrasexual selection, which is based on aggressive exclusion among members of the courting sex, the matter is settled in a more direct way. A member of the passive sex simply chooses the winner, or, to put the matter more realistically, it chooses from among a group of winners who represent a small subset of the potential mates. By picking a winner, the individual not only acquires a more vigorous partner but shares in the resources guarded by it. The latter consideration can be overriding. Males of the long-billed marsh wren (*Telmatodytes palustris*) attempt to stake out territories in stands of cattail, which provide the richest harvest of the aquatic invertebrates on which the birds feed. There they build as many nests and attract as many females as they can. Strong indirect evidence compiled by Verner and Engelsen (1970) suggests that the females choose territories according to the richness of the food and that they somehow are able to assess this quality without reference to the displays of the males. The richer the territory, the easier it is for the male to secure food, and the more time it has to build and maintain nests. Verner and Engelsen believe that the number of visible nests serves as a primary visual index.

When resources are not part of the bargain, intrasexual conflict often evolves onward to acquire a style and intensity impressive even to the most hardened human observer. Males of the highly polygamous prairie sharp-tailed grouse, *Pedioecetes phasianellus,* for instance, compete on communal display grounds, where only a tiny fraction will succeed in inseminating females. Since the young are raised exclusively by the females, for the male everything turns on prowess on the display grounds. The cocks respond to the challenge by viciously fighting other males, pecking them savagely, beating adversaries with their wings, and pulling out their feathers. Even the master cock is not immune from attack. As he tries to mate, he is assaulted by other cocks, who sometimes hit him hard enough to displace him before he can copulate (Scott, 1950).

Rampant *machismo* has also evolved in some insects with similar mating patterns. The horns of male rhinoceros beetles and their relatives and the mandibles of male stag beetles are among the weapons used. Beebe (1947) has described fighting in the hercules beetle (*Dynastes hercules*), a gigantic member of the Scarabaeidae from South America (see Figure 15.2). The battle follows a highly predictable sequence from the moment it is joined:

> The projecting horns touch and click, spread wide and close, the whole object of this opening phase being to get a grip outside the opponent's horns. When the four horns are closed together, there is a dead-lock. All force is now given over to pinching, with the apparent desire to crush and injure some part of head or thorax . . . Again and again, both opponents back away, freeing their weapons, and then rush in for a fresh grip. When a favorable hold is secured outside the other's horns, a new effort, exercised with all possible force, is initiated. This is a series of lateral jerks, either to the right or left, with intent to shift the pincer grip farther along the thorax as far as the abdomen and if possible on to mid-elytra . . . Once this hold is attained and a firm grip secured the beetle rears up and up to an unbelievably vertical stance. At the zenith of this pose it rests upon the tip of the abdomen and the tarsi of the hind legs, the remaining four legs outstretched in mid-air, and the opponent held sideways, kicking impotently. This posture is sustained for from two to as many as eight seconds, when the victim is either slammed down, or is carried away in some indefinite direction to some indeterminate distance, at the end of which the banging to earth will take place. After this climax, if the fallen beetle is neither injured nor helpless on its back, it may either renew the battle, or more usually make its escape.

The displays of species showing extreme intrasexual selection function both to attract females and to intimidate other males. Precopulatory displays are short or absent. The male hercules beetle, for example, evidently engages in none whatever. Occasionally he picks a female up and carries her aimlessly about for a short while, but the significance of the behavior is unknown. During both transportation and copulation the female remains outwardly passive.

Preoccupation with the more dramatic vertebrate examples leads to the impression that intrasexual competition is exclusively precopulatory in timing, ending with the act of insemination. But, as the classification of modes of sexual selection in Table 15-1 suggests, numerous postcopulatory devices exist, some of which are refined and ingenious in nature.

Figure 15-2 On the floor of a Venezuelan rain forest, two males of the hercules beetle fight for dominance and access to a nearby female. The struggle consists to a large degree of grappling and lifting with the huge horns that sprout from the head and prothorax. The orchid shown in this illustration is *Teuscheria venezuela;* mosses, liverworts, and lichens cover other parts of the ground and litter. (Original drawing by Sarah Landry.)

Male mice are capable of inducing the Bruce effect: when they are introduced to a pregnant female, their odor alone is enough to cause her to abort and to become available for reinsemination. Nomadic male langurs routinely kill all of the infants of a troop after they drive off the resident males; the usurpers then quickly inseminate the females. A similar form of infanticide is perpetrated by male lions. By far the greatest diversity of postcopulatory techniques occurs in the insects (Parker, 1970). The reason for this phylogenetic peculiarity appears simple. Female insects generally need to fertilize a great many eggs, often during a prolonged period; at the same time they must economize on the weight of sperm carried in their spermatheca. In the extreme case, exemplified by parasitic wasps, honeybees, ants, and at least some *Drosophila,* the spermatozoans are paid out one to an egg. As a consequence, males still find some profit in trying to inseminate females that have already mated. Their sperm can displace at least some of those inserted by their predecessors.

The threat posed by sperm displacement has provoked the evolution of a series of countermeasures making up much of the list of devices in Table 15-1. Mating plugs, commonly added to the female's genital tract by the coagulation of secretions from the male accessory gland, occur through a very wide array of insect groups. Some authors have concluded that the plugs serve chiefly to prevent sperm leakage, but in at least some of the Lepidoptera and in water beetles of the genera *Dytiscus* and *Cybister* the principal function appears to be prevention of subsequent matings. Also, competitive sperm blockage has not been ruled out as at least a secondary function in the great majority of remaining cases. Copulatory plugs also occur in some mammals, including marsupials, bats, hedgehogs, and rats. The coagulation of seminal fluid is induced by the enzyme vesiculase, which in rodents is secreted by a "coagulating gland" adjacent to the seminal vesicle (Mann, 1964).

During copulation the male may transmit substances that reduce the receptivity of the female. Craig (1967) has postulated that such a pheromone, which he calls "matrone," is secreted by the accessory gland in mosquitoes of the genus *Aedes.* A similar gland is produced by male house flies (*Musca domestica*) from secretory cells lining the male ejaculatory duct (Riemann et al., 1967). An even more effective means of thwarting sperm displacement is prolonged copulation. Male house flies remain *in copula* about an hour in spite of the fact that virtually all of the sperm are transferred during the first 15 minutes. Finally, Parker (1970a,b) has distinguished a "passive phase" in the courtship of many insect species, during which the male attaches himself physically to the female for more or less prolonged intervals without sexual contact. The attachment, which according to species occurs before or after copulation has taken place, prevents rival males from mounting

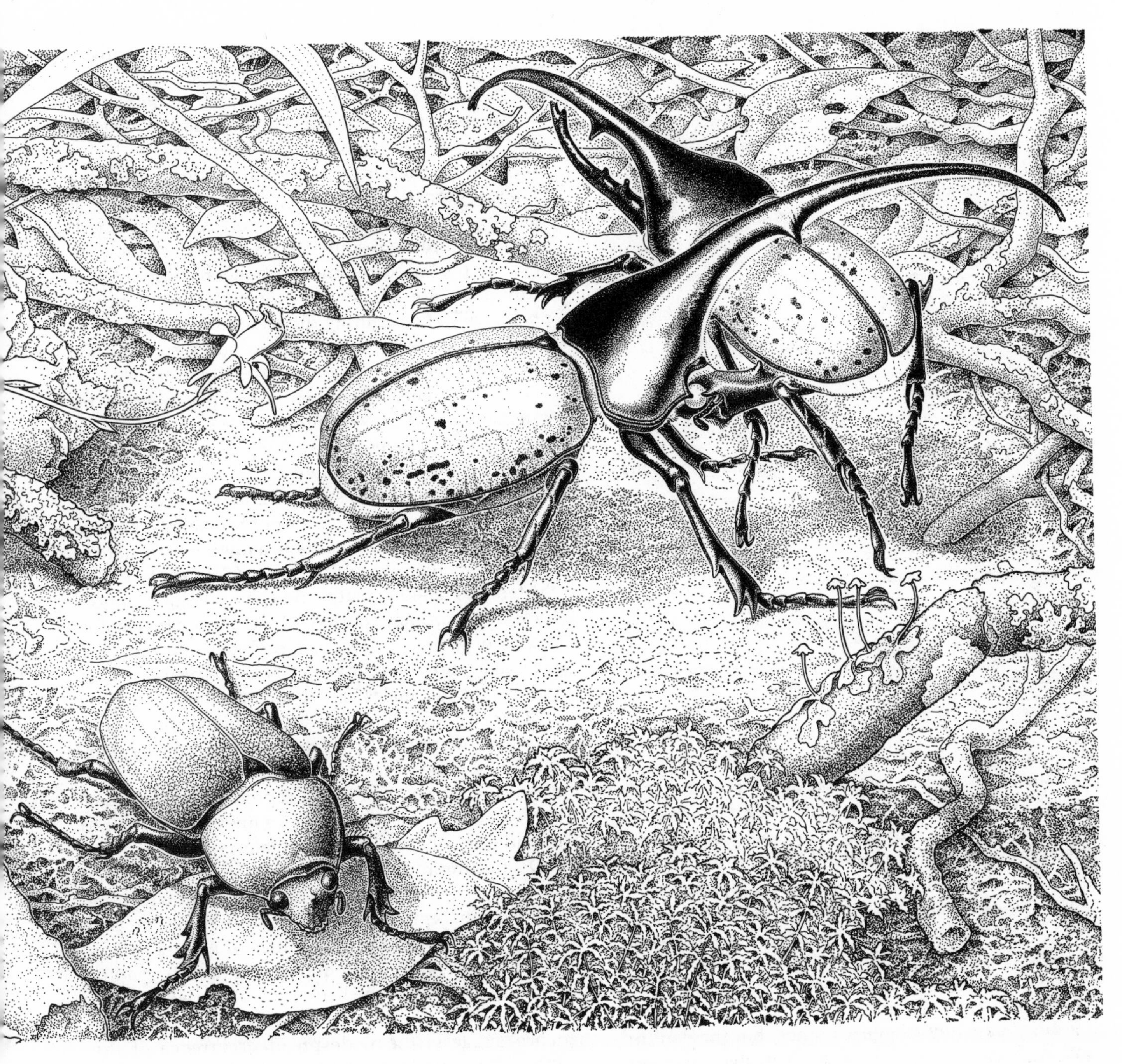

the female. The tandem position of dragonflies is the most familiar example. The male holds on to the abdomen of the female, and the two fly about together while the female lays eggs on the water surface.

It must be kept in mind that the aggression displayed during intrasexual competition is of a special kind. In Chapter 11 I argued that most forms of animal aggression are techniques that evolve when shortages of resources chronically limit population growth. The aggressive behavior thus becomes part of the density-dependent controls. In the case of intrasexual selection there is also competition for a limiting resource. But the shortage, usually of females available for insemination but sometimes of males available to care for the females' offspring, does not limit population growth, and the aggression does not contribute to the density-dependent controls. Indeed, intrasexual selection is likely to become most intense when other resources, such as land and food, are in greatest supply and population growth is most rapid. At that time females are able to reproduce at higher rates, which places a premium on fertility per se, and the abundance of other resources frees the males for pursuit of the females. The principle of allocation comes into play: the male behavior evolves so as to carry intrasexual competition to its greatest

Table 15-1 The modes of sexual selection.

I. Epigamic Selection
   A. *Based on choices made among courting partners*
      1. The choice among the different types of suitors is dependent on their relative frequencies
      2. The choice is not frequency-dependent
   B. *Based on differences in breeding time: superior suitors offer to breed more at certain times than at others*
II. Intrasexual Selection
   C. *Precopulatory competition*
      1. Differential ability in finding mates
      2. Territorial exclusion
      3. Dominance within permanent social groups
      4. Dominance during group courtship displays
   D. *Postcopulatory competition*
      1. Sperm displacement
      2. Induced abortion and reinsemination by the winning suitor
      3. Infanticide of loser's offspring and reinsemination by the winning suitor
      4. Mating plugs and repellents
      5. Prolonged copulation
      6. In "passive phase" of courtship, suitor remains attached to partner during a period before or after copulation
      7. Suitor guards partner but without physical contact
      8. Mated pair leaves vicinity of competing suitors

heights. The most elaborate forms of courtship display and intrasexual aggression develop under conditions in which males have the fewer problems with food and predators. The lek systems of insects, birds, and African grassland antelopes such as the Uganda kob are located away from the feeding grounds. The violent dominance hierarchies of the elephant seal and some other pinnipeds have evolved on island hauling grounds where both time devoted to feeding and mortality from predators are minimal. Thus not only do ordinary competition and intrasexual selection differ basically from each other, but they are in conflict. Insofar as social behavior evolves as a response to resource shortages and predation, the principle of allocation reinforces the antagonism between sexual reproduction and social evolution.

The Theory of Parental Investment

The ultimate basis of sexual selection is greater variance in mating success within one sex. Because of anisogamy, females—defined as the sex producing the larger gametes—are virtually assured of finding a mate. The eggs are the limiting resource. Females therefore have more to offer in terms of energetic investment with each act of mating, and they are correspondingly more likely to find a mate. Males invest relatively little with each mating effort, and it is to their advantage to tie up as many of the female investments as they can. This circumstance is reversed only in the exceptional cases where males devote more effort to rearing offspring after birth. Then the females compete for males, in spite of the initial advantage accruing from anisogamy. Active competition for a limiting resource tends to increase the variance in the apportionment of the resource. Some individuals are likely to get multiple shares, others none at all. The resulting differential in reproductive success leads to evolution in secondary sexual characteristics within the more competitive sex.

This difference in variance was documented by Bateman's classic experiment (1948) on *Drosophila melanogaster*. The technique consisted simply of introducing five males to five virgin females, so that each female could choose among five males and each male had to compete with four other males. The flies carried chromosomal markers allowing Bateman to distinguish them as individuals. Only 4 percent of the females failed to mate, and even this small minority were vigorously courted. Most of those that mated did so only once or twice, by which time they had received more than sufficient sperm. In contrast, 21 percent of the males failed to mate, and the most successful individuals produced almost three times as many offspring as the most fertile females. Furthermore, most of the males repeatedly attempted to mate, and, in contrast to the reproductive success of the females, theirs increased in a linear proportion to the number of times they copulated.

Data on reproductive success in wild populations are few. Other than *Drosophila*, animals in which the variance in reproductive success of males exceeds that of females include dragonflies, the dung fly *Scatophaga stercoraria*, the common frog *Rana temporaria* of Europe, prairie chickens and other lek-forming grouse, elephant seals, and baboons. Indirect evidence suggests the widespread occurrence of this difference in variance in other vertebrates. Building on this principle, Trivers (1972) has constructed the outlines of a general theory of parental investment intended to account for a wide range of differing patterns of sexual and parental behavior. His arguments are based on the graphical analysis of *parental investment*, which is defined as any behavior toward offspring that increases the chances of the offspring's survival at the cost of the parent's ability to invest in other offspring. A second variable analyzed is *reproductive success*, measured by the numbers of surviving offspring. The central principle of sexual selection is reformulated in the graph presented in Figure 15-3. Here we see that one sex, usually the female,

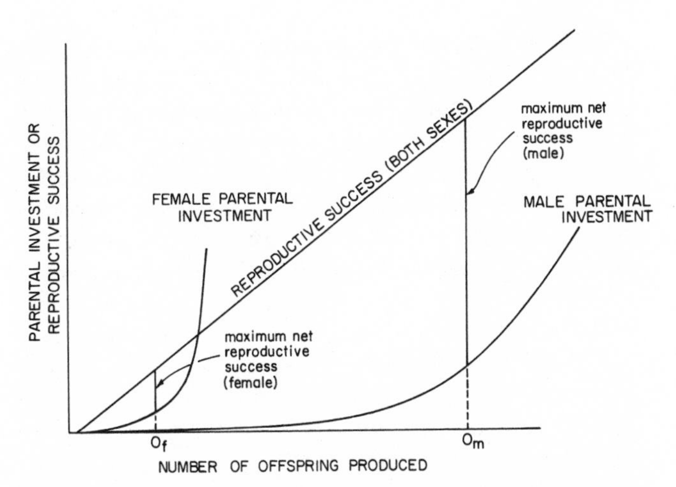

Figure 15-3 A conflict of the sexes arises when the optimum number of offspring (O_f for the female, O_m for the male) differs between them. In the imaginary case depicted here, the usual situation is exemplified: the female must expend a greater effort to create offspring, and her greatest net production of offspring comes at a lower number than in the case of the male. Under these conditions the male is likely to turn to polygamy in order to attain his optimum number. In a few species the situation is reversed, and the female is polygamous. (Modified slightly from Trivers, 1972.)

invests more heavily in each offspring. An egg costs more than a spermatozoan in the sense that it more drastically reduces the number of additional eggs that can be produced at that time or later. The parent that commits itself to the greater part of parental care, again usually the female, will find it difficult or impossible to begin reproducing again until the first offspring are fledged. Hence for that parent investment rises more quickly as a function of the number of offspring produced per reproductive episode. The reproductive success curve, however, will be the same for both sexes: a one-to-one linear increase with the number of offspring produced, by which it is in fact defined. Consequently, the parent making the greater investment per offspring will want to stop at fewer offspring than its mate. From this disparity flows Bateman's principle, that variance in net reproductive success will be greater in the sex with the smaller per-offspring investment. Furthermore, this sex will experience the more intense degree of sexual selection and be prone to evolve the more extreme epigamic displays and techniques of intrasexual selection.

Although the basic theory is built on parameters difficult to measure in practice, there is an indirect way it can be decisively tested. In the exceptional cases where males have taken on more than their share of parental care, we should also find the exceptional circumstance that females are the competitive sex, using the more conspicuous displays and perhaps contending directly for possession of the males. This prediction is easy to confirm in full. Species in which such a reversal of sex role exists and is associated with extended male parental care include the following: pipefishes and seahorses of the family Syngnathidae; Neotropical "poison-arrow" frogs of the family Dendrobatidae; jacanas, which are gallinulelike wading birds; four species of tinamous in the genera *Crypturellus* and *Nothocercus;* phalaropes; the painted snipe *Rostratula benghalensis;* the button quail *Turnix sylvatica;* and the Tasmanian native hen *Tribonyx mortierii.*

The Trivers mode of analysis can be extended to a consideration of parental investment through time in order to interpret patterned changes in sexual interaction. Figure 15-4 presents the cumulative investment curves of the male and female of an imaginary bird species. The principles it illustrates can be broadened with little effort to include any kind of animal as well as human beings. At each point in time there will be a temptation for the partner with the least accumulated investment to desert the other. This is particularly true of the male immediately following insemination. The female's investment has surged upward, while that of the male remains small.

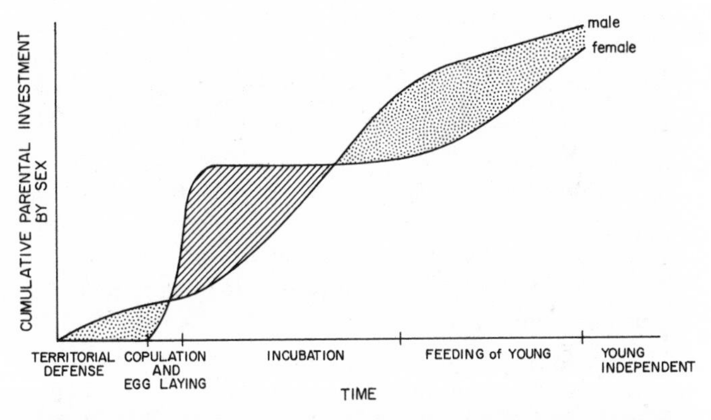

Figure 15-4 The cumulative parental investment of two mated animals can change through time, causing shifts in their attitudes and relationships. In this imaginary example modeled on the bird life cycle, the male has more to lose at certain stages (stippled) and the female at others (cross-hatched). *Territorial defense:* the male defends the area to protect food supply and nest sites. *Copulation and egg laying:* the female commits her eggs to the male while the male commits his defended nest to the female. *Incubation:* the male incubates eggs while the female does nothing relevant to the offspring; consequently, the cumulative female investment remains constant while that of the male rises, and the two investments attain parity a second time. *Feeding of young:* each parent feeds the young but the female does so at a more rapid rate, causing the investments to converge a third time. (Modified slightly from Trivers, 1972.)

As parental care by each sex accumulates, the tendency to desert will depend not only on the difference in the amount of investment but on the ability of the partner to rear the offspring alone. If one partner is deserted it will no doubt try to finish the job, since so much has been committed already. But if a substantial risk exists that a solitary parent will fail because the task is overwhelming, desertion carries the risk to the potential deserter of a loss in genetic fitness. When the expected loss in fitness is not likely to be compensated by success in future matings with other partners, desertion is likely to be rare at this stage of the cycle. As Trivers has pointed out, there may come a time when the investments of both partners are so great that natural selection will favor desertion by either partner even if the investment of one is proportionately less. This is so because desertion places the faithful mate in a cruel bind: it has invested so much that it cannot abandon the investment regardless of the difficulties that lie ahead. Under these circumstances the relationship between the partners may develop into a game as to which can desert first. The outcome might be determined not so much by wiliness as by the opportunities that exist for the possession of a second mate.

Out of this unsentimental calculus of marital conflict and deceit can be drawn a new perception of cuckoldry. When fertilization occurs by insemination, the universal mode of reptiles, birds, and mammals, the male cannot always be completely sure that his mate's eggs have been fertilized by his own sperm. To the degree that he invests in the care of the offspring, it is genetically advantageous to him to make sure that he has exclusive access to the female's unfertilized eggs. Frequently the preemption comes about as a bonus resulting from other kinds of behavior. Males that exclude others from territories or control them within dominance systems are avoiding sperm competition. The same effect is achieved through the time lag that normally occurs between bonding and copulation in monogamous birds, an interval that serves as a de facto quarantine period for the detection of alien sperm. The theory suggests that a particularly severe form of aggressiveness should be reserved for actual or suspected adultery. In many human societies, where sexual bonding is close and personal knowledge of the behavior of others detailed, adulterers are harshly treated. The sin is regarded to be even worse when offspring are produced. Although fighting is uncommon in hunter-gatherer peoples such as the Eskimos, Australian aborigines, and !Kung Bushman, murder or fatal fighting appears to be frequent in these groups in comparison with other societies and is usually a result of retaliation for actual or suspected adultery.

Until recently biologists and social scientists have viewed courtship in a limited way, regarding it as a device for choosing the correct species and sex and for overcoming aggression while arousing sexual responsiveness in the partner. The principle significance of Trivers' analysis lies in the demonstration that many details of courtship can also be interpreted with reference to the several possibilities of maltreatment at the hands of the mate. The assessment made by an individual is based on rules and strategies designed by natural selection. Social scientists might find such an interpretation rather too genetic for their tastes, yet the implications for the study of human behavior are potentially very great.

The Origins of Polygamy, Monogamy, and Pair Bonding

Because of the Bateman effect, animals are fundamentally polygamous. At the start, virtually all are anisogamous. In those species in which parental care is also lacking, variance in reproductive success is most likely to be greater in males than in females. Under many circumstances the addition of parental care will reinforce this inequality, because parental investment in postnatal care is seldom quantitatively the same in both sexes. Monogamy is generally an evolutionarily derived condition. It occurs when exceptional selection pressures operate and equalize total parental investment and literally force pairs to establish sexual bonds. This principle is not compromised by the fact that the great majority of bird species are monogamous. Although polygamy in birds is in most cases phylogenetically derived, the condition represents a tertiary shift back to the primitive vertebrate state. Monogamy in modern birds was almost certainly derived from polygamy in some distant avian or reptilian ancestor.

Before going any further, let me define the essential terminology pertaining to mating systems. *Monogamy* is the condition in which one male and one female join to rear at least a single brood. It lasts for a season and sometimes, in a small minority of species, extends for a lifetime. *Polygamy* in the broad sense covers any form of multiple mating. The special case in which a single male mates with more than one female is called polygyny, while the mating of one female with more than one male is called polyandry. Polygamy can be simultaneous, in which case the matings take place more or less at the same time, or it can be serial. Simultaneous polygyny is sometimes referred to as harem polygyny. In the narrower sense preferred by zoologists, polygamy also implies the formation of at least a temporary *pair bond*. Otherwise, multiple matings are commonly defined as *promiscuous*.

Entirely by itself, anisogamy favors polygamy as broadly defined. There also exists several general conditions that promote polygamy still further. They include (1) local or seasonal superabundance of food, which permits the female to raise the young on her own and the male to go off in search of additional females; (2) the risk of heavy predation, which makes it advantageous for the family to divide; (3) and the existence of precocial young, which requires less parental care. All of these factors were discovered in birds, where polygamous and monogamous species commonly coexist and provide the opportunity for evolutionary comparisons; but the same biases probably operate with equal force on other, less well studied groups.

Monogamy can therefore be viewed as a more specialized, derived condition, the origin of which invites closer analysis. In general, fidelity is a special condition that evolves when the Darwinian advantage of cooperation in rearing offspring outweighs the advantage to either partner of seeking extra males. Three biasing ecological conditions are known that seem to account for all of the known cases of monogamy: (1) the territory contains such a scarce and valuable resource that two adults are required to defend it against other animals; (2) the physical environment is so difficult that two adults are needed to cope with it; and (3) early breeding is so advantageous that the head start allowed by monogamous pairing is decisive.

Defense of a scarce and valuable resource. An estimated 91 percent of all bird species are monogamous during at least the breeding season. This adaptation provides superior defense for scarce nest sites, or territories containing scattered renewable food sources, or both (Lack, 1966, 1968). A few species even form life-long bonds. One well-analyzed example is the oilbird (*Steatornis caripensis*) of Trinidad and northern South America. According to Snow (1961), the permanence of the bond stems ultimately from the combined circumstances that the birds nest in caves, are long-lived, and breed very slowly. Appropriate nesting sites along the cave walls are extremely few, and their scarcity appears to be the principal factor limiting the size of the population. Cooperative defense of the sites is needed to maintain them during the long reproductive lives of the mated couple.

Adaptation to a difficult physical environment. I know of no case in which monogamy serves exclusively as a device for overcoming challenges offered by the physical environment. However, at least one species exists in which generally severe conditions have intensified the need for defense of the territory to the point of favoring permanent monogamy. The sowbug *Hemilepistus reaumuri* is an isopod crustacean that lives in the dry steppes of Arabia and North Africa. During the hottest, driest time of the year the crustaceans are forced to retreat deep into burrows in order to survive. Linsenmair and Linsenmair (1971) discovered that each burrow is occupied by an adult pair that remain mated for life. The highly territorial behavior of the mated sowbugs prevents overcrowding of the burrows and depletion of the scarce, unpredictable food supply in the surrounding area. This example may be regarded as a special case of the defense of a limiting resource. It deserves particular attention because it identifies the possibility that parameters in the physical environment affect the entire population simultaneously.

An early start in breeding. When the timing of breeding is important, cooperation between the mated pair can provide a decisive edge. In his excellent study of the kittiwake gull (*Rissa tridactyla*) in Britain, Coulson (1966) found that about 64 percent of the breeding birds retained their mate from the previous season. Females in this category began to lay eggs three to seven days earlier, produced a higher seasonal total, and reared a larger number of chicks than did comparable individuals that had taken a new mate. The difference stemmed from the ability of "married" pairs to cooperate more smoothly from the beginning of courtship and nesting. Yet divorces were common. Over the 12-year period of the study, Coulson found that two-thirds of the birds that changed partners did so while the first mate was still living in the colony. Also, birds that failed to hatch eggs the previous season were three times more likely to change mates than those that had enjoyed success. The latter correlation suggests that "divorce" is adaptively advantageous to birds originally bound to a reproductively incompatible partner.

Communal Displays

Communal sexual displays provide some of the great spectacles of the living world. In southeast Asia thousands of male fireflies sit in certain trees in the forest and flash synchronously and rhythmically through the night. The light is so strong and the location of the trees so consistent that mangrove trees inhabited close to the shoreline are used as navigational aids (Buck, 1938; Lloyd, 1966, 1973). Millions of 13-year and 17-year locusts gather to mate in one or another forest locality in the eastern United States; the singing of the males can be literally deafening to the human ear (Alexander and Moore, 1962). These insect performances are rivaled by arena mating birds, the battles royal of male mountain sheep and elk, and others among the more dramatic vertebrate displays.

The primary role of communal displaying appears to be enhancement of attractiveness through the increase in volume and reach of the signal. In simplest terms, a group of males is more likely to attract a single female than is a solitary male, and a male is more likely to encounter a receptive female when he is in a group. The effect can be strengthened further when the display grounds are situated in an open space, atop a prominent high point, or in some other distinctive location that makes orientation easy. Such features characterize the swarm areas of parasitic wasps, ants, nematoceran flies, and other communally breeding insects. Birds also commonly rely on landmarks. Moreover, the display grounds of many species are traditional in nature, remembered by older individuals from one season to the next.

When an area is consistently used for communal displays, it is referred to as a *lek* or *arena*. The animals are said to be engaged in lek displays or arena displays, and the entire breeding system is called a lek or arena system. The most complicated and spectacular lek system occur in birds. The phenomenon has arisen independently in lines belonging to ten families: the ruff *Philomachus pugnax* and great snipe *Capella media* (Scolopacidae); many grouse species, including capercaillie and blackcock (Tetraonidae); a few hummingbirds (Trochilidae); the argus pheasant *Argusianus argus* (Phasianidae); most manakins (Pipridae); the cock of the rock *Rupicola rupicola* (Cotingidae); the bustard *Otis tarda* (Otidae); some bowerbirds (Ptilonorhynchidae); two species of birds of paradise (Paradisaeidae) and Jackson's dancing whydah *Drepanoplectes jacksoni* (Ploceidae, a large Old World family that also includes the weaver finches and *Passer* sparrows). The males belonging to species on this list are among the most colorful of the bird world. The brilliant red cock of the rock, for example, is easily the most spectacular cotingid, and the birds of paradise are justly considered the most beautiful of all birds. The basis of the correlation is that lek systems in birds are universally associated with extreme polygyny and sexual dimorphism, both of which promote secondary sexual evolution in males.

Birdlike lek systems occur in many of the open-country antelopes of Africa, including the common and defassa waterbucks (*Kobus ellipsiprymnus, K. defassa*), Uganda kob (*K. kob thomasi*), puku (*K. vardoni*), springbuck (*Antidorcas marsupialis*), Grant's and Thomson's gazelles (*Gazella granti, G. thomsoni*), and wildebeest (*Connochaetes taurinus*). In the Uganda kob, an antelope that has carried this trend unusually far, successful males cram small territories next to one another in sites well removed from the feeding and watering areas. Receptive females wander through the networks as part of the nursery herds and are mated by those territorial males able to detain them. Bachelor herds roam the periphery of the lek, sometimes joining the nursery herds there, but their members are seldom if ever able to copulate (Buechner and Roth, 1974).

Full-scale leks are also formed by the fruit-eating bat *Hypsignathus monstrosus* of Africa (Bradbury, 1975). The adults display the greatest sexual dimorphism of all the 875 bat species of the world. The males have grotesque muzzles and enlarged larynxes. During the breeding season they gather in nocturnal aggregations at traditional sites in the forest canopies. Each male stakes out a small territory that it defends from rivals with harsh cries and gasps. From the moment of arrival at his post the male sings, emitting metallic notes at 80–120 times a minute while beating his partially unfolded wings at twice this rate. Females visit the lek and fly along its axis, causing sudden increases in the rates of display as they pass by.

Other Ultimate Causes of Sexual Dimorphism

The reader will recognize the following thread of reasoning that runs strongly through the theory of sexual evolution: polygamy, enhanced by one or more forces in the environment such as the very unequal apportionment of territorially defended resources, leads to increased sexual selection, which leads in turn to increased sexual dimorphism. But as is true with other sociobiological phenomena, the final effect—in this case the enhancement of sexual dimorphism—can be approached along evolutionary pathways. These alternate chains of causation, as understood principally from studies on birds, are represented diagrammatically in Figure 15-5.

The relation between strong sexual dimorphism and unstable environments was first examined in depth by Moreau (1960) in the ploceine weaver finches and by Hamilton (in Hamilton and Barth, 1962) in the parulid warblers and other passerine birds of the New World. Most of the species of *Ploceus* that breed in the dry habitats of Africa move in large, itinerant flocks and wear dull plumage when out of the breeding season. In a parallel fashion, the passerine birds that migrate the farthest in the New World assume the dullest plumage and often associate in feeding flocks during the off season. Hamilton and Barth, following Moynihan (1960), concluded that the convergence toward dull plumage is a device for reducing hostile interaction during flock forma-

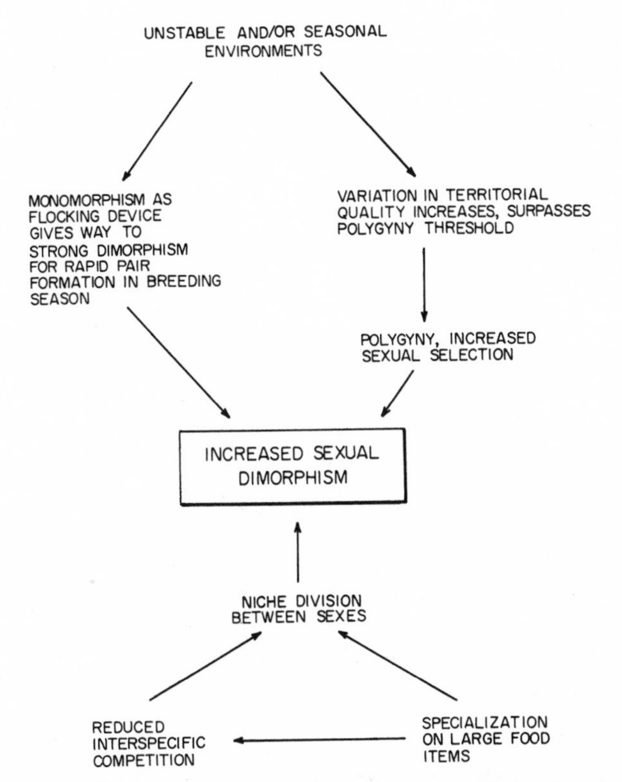

Figure 15-5 The alternate chains of events leading to increased sexual dimorphism in birds.

tion. The strongest seasonal dimorphism is assumed by the species that have the longest migration routes and the shortest periods in which to breed. It seems to follow that dimorphism has been heightened in such cases by the need to form pair bonds as quickly as possible, a requirement that amplifies the usual process of sexual selection.

Selander (1966, 1972) has observed that sexual dimorphism is commonly based on a difference in food preference in species of birds that specialize on large food items. The relationship has been documented in woodpeckers, hawks and their allies, owls, frigate birds, jaegers and skuas, and the extinct huia (*Heteralocha acutirostris*) of New Zealand. Its basis appears to be the relative scarcity of the items, which places a premium on a division of the niche between mated birds that must cooperate in utilizing the same resources in order to rear young. The hypothesis gains strength from the discovery by Schoener (1965) that one category of birds showing interspecific character displacement in bill size is comprised of species that feed on scarce food items, especially those unusually large in size. A third, independent line of evidence comes from the data collected by Schoener (1967, 1968b) on West Indian *Anolis*. When small and medium-sized species of this insectivorous lizard occur alone on small islands, the sexes diverge in size. To a startling degree, the head length of the male approaches a mean of 17 millimeters, and the head length of the female approaches a mean of 13 millimeters, regardless of the species. The implication appears to be that there exists an optimum division of labor between the males and the females living inside the territories of the males. Such ecological partitioning does not require group selection; it can stem entirely from selection at the individual level. Specifically, that female best survives who is able to eat well within the territory of a male, while that male breeds the most who is best able to accommodate females. Most ecologists agree with this more straightforward explanation.

CHAPTER 16

Parental Care

The pattern of parental care, being a biological trait like any other, is genetically programmed and varies from one species to the next. Whether any care is given in the first place, and of what kind and for how long, are details that can distinguish species as surely as the diagnostic anatomical traits used by taxonomists. The females of most hemipterous bugs, for example, simply deposit their eggs on the host plant and depart. In a few cases one parent—whether the female or the male depends on the species—stands guard over the egg mass until the nymphs emerge. Adults of a subset of these species protect the nymphs as well, standing near or over them and warding off predatory insects. In a still smaller group, which includes the tingid *Gargaphia solani* and scutellerid *Pachycoris fabricii*, the young orient to the mother and follow her from place to place. Some arachnids abandon their eggs or guard them to the time of hatching; others carry their newly hatched young around in brood pouches on the abdomen.

Parental care in vertebrate species is even more diversified. Birds are constrained by their own warm-bloodedness, which requires that the eggs and young be kept within a narrow range of temperature. But the 8,700 living bird species use virtually every conceivable device for accomplishing this (Kendeigh, 1952). Many species, from ostriches to pheasants, have precocial young that are able to run and feed within hours after emergence from the egg. The megapodes of Australia and southeastern Asia not only possess precocial young but have given up nearly every trace of postnatal parental care. The female simply buries the eggs in sand, volcanic ash, or mounds of rotting vegetation and allows the sun and heat of decomposition to provide the heat for incubation. At the opposite extreme are species in which one of the parents sits on the eggs without food until the young birds hatch; these spartan types include the emus, the eider duck, the argus pheasant, and the golden pheasant. Altricial bird species, those with helpless young that require protection and nursing within a nest, also vary greatly in the amount and kind of aid they provide. Lesser but still impressive amounts of diversity exist within the fishes, amphibians, reptiles, and mammals. Such variation is evidently due to the sensitivity of parental behavior to natural selection.

The Ecology of Parental Care

The theory of parental care postulates a web of causation leading from a limited set of primary environmental adaptations through alterations in the demographic parameters to the evolution of parental care as a set of enabling devices. Figure 16-1 gives the essence of the theory. When species adapt to stable, predictable environments, *K* selection tends to prevail over *r* selection, with the following series of demographic consequences that favor the evolution of parental care: the animal will tend to live longer, grow larger, and reproduce at intervals instead of all at once (iteroparity). Further, if the habitat is structured, say a coral reef as opposed to the open sea, the animal will tend to occupy a home range or territory, or at least return to particular places for feeding and refuge (philopatry). Each of these modifications is best served by the production of a relatively small number of offspring whose survivorship is improved by special attention during their early development. At the opposite extreme, species sometimes penetrate new, physically stressful environments by developing idiosyncratic protective devices that include care of offspring through the most vulnerable period of their development. Specialization on food sources that are difficult to find, to exploit, or to hold against competitors is occasionally augmented by territorial behavior and the strengthened defense of the food sources when offspring are present. A few species of vertebrates even train their offspring in foraging techniques. Finally, the activity of predators

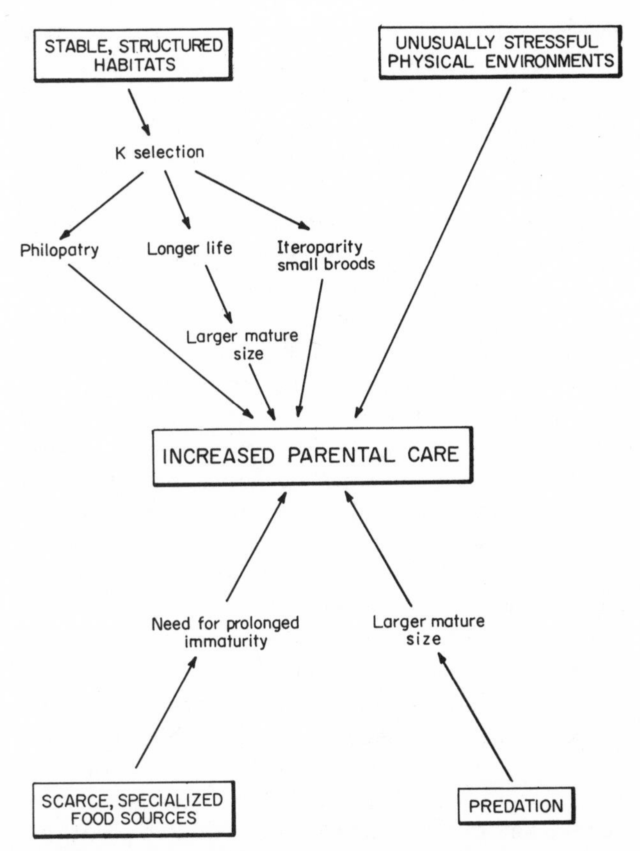

Figure 16-1 The prime environmental movers and intermediate biological adaptations that lead to increased parental care.

can prolong parental investment to protect the lives of the offspring. All four of these environmental prime movers—stable, structured environments leading to K selection, physical environments that are unusually difficult, opportunities for certain types of food specialization, and predator pressure—can act singly or in combination to generate the evolution of parental care. Let us now examine some of the logic and evidence behind principal pieces of the theory.

Reduced brood size. The smaller the brood, the more likely the iteroparous adult is to care for it. Also, the more effort the parent puts into rearing young, the more precisely controlled will be the brood size. The idea was first developed by David Lack (1954, 1966), who showed that clutches of songbirds deviating by as few as one or two eggs from the average number of the species produce fewer fledglings than do those at the average size. Lack argued that fewer eggs fell below the parents' potential to raise them, while too many eggs resulted in undernourishment and high mortality for the growing brood as a whole. Wynne-Edwards (1962) offered a competing hypothesis, that clutch size is adjusted altruistically by parents to prevent overpopulation. Both logic and evidence have subsequently favored Lack's view. In particular, Cody (1966) extended the theory of clutch size in a way that permits an independent test of Lack's hypothesis. Cody recognized three adaptive "goals" between which some compromise is essential: large clutch size, efficient food search, and effective predator escape. He postulated that large clutch size increases r, efficient food search increases K, and predator avoidance increases both, so that the absence of predators from a particular area will not alter the balance between clutch size and feeding efficiency. Cody's argument yields several nonobvious predictions. On the seasonal north temperate mainland, where r selection is generally more important than K selection, clutch size will be larger and feeding efficiency somewhat less. This effect should be lessened on offshore islands at the same latitudes, which enjoy a generally milder and less fluctuating climate. Toward the mainland tropics, where predation and K selection are considered to be more important, the compromise should lean toward feeding efficiency and predator avoidance, and clutch size should be diminished accordingly. On tropical islands yet another trend can be expected to appear: predators are less important and selection already leans toward feeding efficiency, so that the reduction in clutch size should be less than on the nearby mainland. All of these predictions are consistent with the evidence (see also MacArthur, 1972). Similar theory, appropriately modified to take into account special biological properties, will apply to the brood size of other kinds of animals that supply postnatal care.

Longevity and delayed maturity. The more the parental investment measured by the length of the gestation period, the larger the size of the offspring at birth, and the greater the amount of care devoted to the neonatal infant, the earlier the mortality will be programmed. When a heavy early investment is made, prolonged postnatal care, extended immaturity, and long life are likely to emerge as coadaptations. Furthermore, the older the parent, the more personal risks it is likely to take on behalf of the offspring. Long life and low fertility can be mutually reinforcing in still another way. Suppose that long life has been favored by circumstances wholly independent of reproductive effort—say, a rich, stable environment relatively free of predators. Suppose further that conditions are not favorable for the emigration of offspring. Then progeny of this K-selected species are likely to become direct competitors of the parents. If the parents have lived only part of their life span, each unit of genetic fitness gained by an offspring at their expense is compensated for by as little as one-half a unit of inclusive fitness. On this basis alone it pays not to reproduce frequently. In practice, competition from offspring offers no separate theoretical problem, since it can be computed as part of the reproductive effort.

The positive correlations between low reproductive effort, lateness of maturity, and large investment in individual offspring generally hold through the major vertebrate groups. In lizards, for example, greater reproductive effort is reflected in the proportionately larger weight and size of the clutches and in the greater effort devoted to courtship, which in turn is measured by increases in the degrees of sexual dimorphism and elaborateness of courtship behavior (see Tinkle, 1969).

Large size. Longer-lived animals not only mature later, but also are generally larger. The expected correlation between size and parental care holds in the birds and mammals, but it is weak or absent in the fishes and reptiles. Williams (1966) reviewed the evidence in the latter group in some depth and concluded that the lack of correlation is due to a compromise with extraneous factors. A small fish may have a greater need for defending its eggs in nests, while its size makes it less effective in providing protection against predators. Oral incubation and viviparity are alternatives available to it, but they impose a reduction in fertility that make them less valuable. The social insects appear to conform to the a priori size rule. For example, the cryptocercid cockroaches, which are closely related to the primitive termites and show strongly developed parental care, are large in size, long-lived, and breed very slowly.

Unusually stressful physical environments. Next to a stable, predictable environment, the condition most likely to promote the evolution of parental care is very nearly the direct opposite. When species penetrate new habitats in which one or more physical parameters are exceptionally stressful, they sometimes add parental care as the only means of advancing the young to a developmental stage in which they are able to cope with new conditions. *Bledius spectabilis,* for example, is a beetle that lives in an extreme environment for an insect—the intertidal mud of the northern coast of Europe, where it constantly faces the hazards of high salinity and oxygen shortage. The species is also exceptional within the large taxonomic group to which it belongs (family Staphylinidae) in the amount of care given by the female to her brood. She keeps the larvae in a burrow, protects them from intruders, and brings them fresh algae at frequent intervals (Bro Larsen, 1952).

Scarce or difficult food sources. The slowest breeding of all birds are the eagles, condors, and albatrosses. Only one young is fledged at a time, and the full breeding cycle occupies more than a year. Maturity requires at least several years; condors and royal albatrosses do not begin to breed until they are about nine years old. The ecological trait that all these species have in common is dependence on food that is sparse and difficult to obtain (Amadon, 1964). Foraging consists of long, skillful searches. Homing occurs over long distances, and resourcefulness in transport is often required. Some eagles, for example, wander over thousands of square kilometers in search of prey. During the breeding season, however, movement must be severely restricted. The male normally does all of the hunting for himself, his mate, and the single chick. When the young bird is nearly grown, the female begins to hunt also. The prey of the largest eagles are moderate-sized mammals such as tree sloths, monkeys, and small antelopes. Bringing them to the nest after the kill has been made often requires exceptional strength and skill. Little wonder, then, that the young must attain a large size themselves before attempting an independent existence.

A comparable degree of prolonged immaturity characterizes large mammalian carnivores such as wolves, African wild dogs, and the great cats. Lions engage in training sessions during which the adult females initiate their young to the hunting of prey. According to Schenkel (1966) these exercises resemble real stalking of nearby animals up to a point, but are not carried through to the kill. When the female lions leave on "real" hunting trips, they walk off from the cubs in a determined gait and the youngsters do not even try to follow. Schenkel's wild cubs began to hunt on their own when they were about 20 months old, while they were still under the care of the females. Their first victims were warthogs, but they also stalked wildebeest and zebras frequently. When some of the young lions busied themselves at these activities, the other cubs watched intently from a short distance.

Parent-Offspring Conflict

The traditional view of the relationship between parent and offspring has always assumed unilateral parental investment. The offspring was considered to represent so many units of genetic fitness to the parent, a more or less passive vessel into which a certain amount of care is poured to enlarge the investment. Until recently, behaviorists have not come to grips with the phenomenon of parent-offspring conflict during the weaning period. As the juvenile grows older, the mother discourages it with increasing firmness. For example, the female macaque removes the juvenile's lips from her nipple by pushing its head with the back of her hand; she holds its head beneath her arm, or strips the infant away from her body altogether and deposits it on the ground. The juvenile, sometimes screaming in protest, struggles to get back into a favorable clinging position (Rosenblum, 1971). Among ungulates the discouragement often shades into open hostility. A young moose passes through two crises during the period of declining dependence on its mother. The first is in the spring, when it is one year old and its mother has just given birth to a new calf. The dam suddenly turns hostile and drives the yearling from her territory. The young moose lingers in the immediate vicinity and repeatedly attempts to return to the dam. In the fall, at the onset of the rutting season, the territorial barriers relax and the yearling is able to draw close to its mother again. But this new proximity precipitates the second crisis. Dams now treat their daughters as rivals, while bulls chase away the young males as if they were adults. At this stage the young animal finally becomes independent of its mother (Margaret Altmann, 1958).

Mammalogists have commonly dealt with conflict as if it were a nonadaptive consequence of the rupture of the parent-offspring bond. Or, in the case of macaques, it has been interpreted as a *mechanism* by which the female forces the offspring into independence, a step designed ultimately to benefit both generations. Hansen (1966), writing on the rhesus, expressed this second hypothesis as follows: "One of the primary functions that the mother monkey served was seen in her contribution to the gradual, but definite, emancipation of her infant. Although this process was aided and abetted by the developing curiosity of the infants to the outer world, this release from maternal bondage was achieved in considerable part by responses of punishment and rejection."

A wholly different approach to the subject has been taken by Trivers (1974). Rather than viewing conflict as the rupture of a relationship, or as a device promoting the independence of the young animal, Trivers interprets it as the outcome of natural selection operating in opposite directions on the two generations. How is it possible for a mother and her child to be in conflict and both remain adaptive? We must remember that the two share only one-half of their genes by common descent. There comes a time when it is more profitable for the mother to send the older juvenile on its way and to devote her efforts exclusively to the production of a new one. To the extent that the first offspring stands a chance to achieve an independent life, the mother is likely to increase (and at most, double) her genetic representation in the next breeding generation by such an act. But the youngster cannot be expected to view the matter in this way at all. So long as the continued protection of its mother increases its own inclusive genetic fitness, the young animal should try to remain dependent.

If the mother's inclusive fitness suffers first from the relationship, conflict will ensue. More precisely, selection will favor rejection behavior on the part of the mother when the cost in units of fitness to herself exceeds the benefit in the same units, while the offspring will try to hang on until the cost to its mother exceeds twice the benefits to itself. At that point the offspring's inclusive fitness is reduced and independence becomes profitable. We can expect that when the offspring is very small in size, the ratio of cost-to-mother/benefit-to-offspring is also very small and the mother and offspring will "agree" to continue the dependent relationship. As the youngster grows it becomes increasingly more expensive to maintain in units of inclusive fitness, so that the following two thresholds are crossed in sequence:

Cost-to-mother/benefit-to-offspring exceeds 1: the conflict begins as the mother's fitness declines but the inclusive fitness of the offspring is not yet diminished by the relationship.

Cost-to-mother/benefit-to-offspring exceeds 2: the conflict ends and the offspring willingly leaves, because the inclusive fitness of both participants is now diminished.

The hypothetical time course of the relationships is represented in Figure 16-2. The principal conflict can be expected to begin during the period of weaning, when the young animal becomes independent of milk or other food provided directly by the parent. Weaning conflict has been documented in a variety of mammals, including rats, dogs, cats, langurs, vervets, baboons, and the rhesus and other species of macaque. Among birds it has been recorded in the

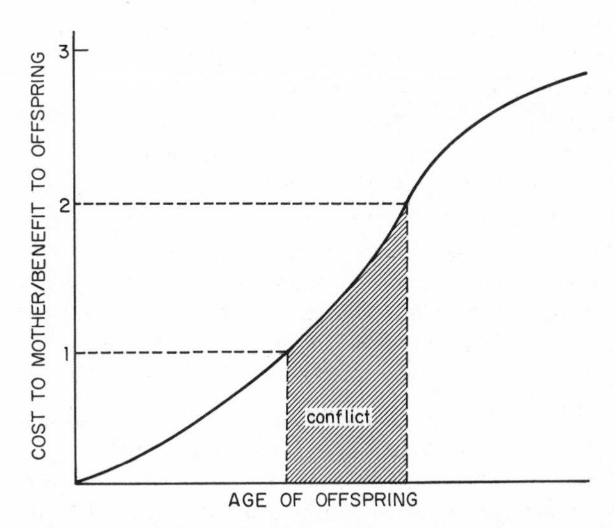

Figure 16-2 Trivers' model of the timing of parent-offspring conflict.

herring gull, red warbler, Verreaux's eagle, and white pelican, and is perhaps widespread in altricial birds generally. It even appears to occur in mouth-brooding fish (Reid and Atz, 1958).

The period of weaning conflict is in fact one extreme case, entailing disagreement over whether any aid at all will be given the offspring. It is equally easy to conceive of circumstances earlier in the life of the young animal when the interests of both individuals are served when the parent helps, but disagreement will exist over *how much* help is to be given. This lesser conflict results from the fact that whereas the adult is selected to bestow the amount of investment that maximizes the difference between benefit and cost, the offspring is selected to try to secure an amount of investment that maximizes the difference between the benefits to itself and the cost to its mother devalued by the relevant coefficient of relationship, which is normally ½. The two functions are visualized graphically in Figure 16-3.

Triver's hypothesis is consistent with the time course of conflict observed in cats, dogs, sheep, and rhesus monkeys. In each of these species the conflict begins well before the onset of weaning and tends to increase progressively thereafter. In dogs (Rheingold, 1963a) and cats (Schneirla et al., 1963) the period of maternal care has been explicitly divided into three successive intervals characterized by increasing conflict. In the first stage, most nursing is initiatied by the mother, and she seldom if ever resists the infant's advances. During the second interval, approaches are initiated by both individuals with about equal frequency; the mother occasionally rejects the offspring and may even treat it with hostility. The third interval is the period of weaning. The young animal initiates most or all nursing episodes and is usually rejected. Paradoxically, but in a manner consistent with the theory, the more wide-ranging and independent the juvenile becomes, the more frequently it seeks to renew contact with the mother.

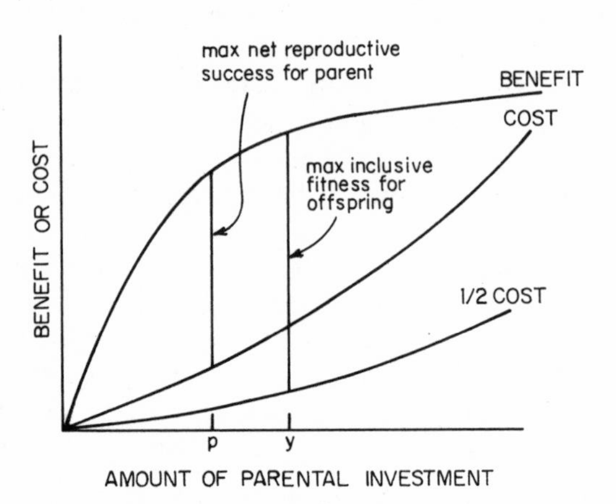

Figure 16-3 Parent-offspring conflict of varying degrees throughout the period of parental care is possible under the conditions envisaged here. The benefit, cost, and half the cost of a parental act toward an offspring at one moment in time are shown as functions of the amount of the parental investment in the act. An example of an investment in mammals would be the quantity of milk provided during one day of nursing. At p the parent's inclusive fitness (benefit minus cost) is maximized; at y the offspring's inclusive fitness (benefit minus ½ cost) is maximized. The parent and offspring are consequently selected to disagree over whether p or y should be invested. (Modified from Trivers, 1974.)

Trivers' model can be fitted closely in nonobvious ways to the detailed results of experiments conducted on rhesus infant development by Robert Hinde and his associates (Hinde and Spencer-Booth, 1971; Hinde and Davies, 1972a,b). When an infant was separated from its mother for a few days and then reunited with her, it sought contact more frequently than it did before the separation. In contrast, control infants left with their mothers decreased their rate of contact during the same interval. Separated infants also displayed more signs of distress, such as calls and immobility, after they were reunited with their mothers. The more the mothers rejected the infants before the separation, the more distress they displayed afterward. Even more significant, infants separated by having their mothers removed were more distressed than those who were taken from their mothers. All of these results are consistent with the view that the infant strives to increase the amount of maternal investment and is alert to signs that the mother is reducing the investment. The data are not consistent with the two principal competing hypotheses, namely, that maternal rejection has been selected as a device for promoting independence in the young, or that independence is attained primarily by attraction to other members of the society.

Trivers has elaborated his model to account for inclusive selection through a wide array of relatives and nonrelatives. However, the refinement may eventually prove applicable only to human beings and a few of the other most intelligent vertebrates. Consider the offspring that behaves altruistically toward a full sibling. If it were the only active agent, its behavior would be selected when the benefit to the sibling exceeded 2 times the cost to itself. From the mother's point of view, however, inclusive fitness is gained whenever the benefit to the sibling simply exceeds the cost to the altruist. Consequently, there is likely to evolve a conflict between parents and offspring in the attitudes toward siblings: the parent will encourage more altruism than the youngster is prepared to give. The converse argument also holds:

the parent will tolerate less selfishness and spite among siblings than they have a tendency to display, since its inclusive fitness will begin to suffer at a lower intensity. The same inequalities hold through an indefinitely widening circle of relatives and nonrelatives. Altruistic acts toward a first cousin are ordinarily selected if the benefit to the cousin exceeds 8 times the cost to the altruist, since the coefficient of relationship of first cousins is $1/8$. However, the parent is related to its nieces and nephews by $r = 1/4$, and it should prefer to see altruistic acts by its children toward their cousins whenever the benefit-to-cost ratio exceeds 2. Parental conscientiousness will also extend to interactions with unrelated individuals. From a child's point of view, an act of selfishness or spite can provide a gain so long as its own inclusive fitness is enhanced. The exploited individual (or society as a whole) may retaliate against the individual and one or more members of its family. But the act will be favored in selection if the benefit it brings is greater than the loss inflicted by the retaliation, where the loss is summed over the individual and its relatives and devalued by the appropriate coefficients of relationship. The parents can be expected to view the matter according to approximately the same calculus. However, since they lose more inclusive fitness by costs inflicted on the offender's siblings and other relatives, they are likely to be less tolerant of the act. In human terms, the asymmetries in relationship and the differences in responses they imply will lead in evolution to an array of conflicts between parents and their children. In general, offspring will try to push their own socialization in a more egoistic fashion, while the parents will repeatedly attempt to discipline the children back to a higher level of altruism. There is a limit to the amount of altruism the parents wish to see; the difference is in the levels that selection causes the two generations to view as optimum. Trivers has summarized the argument as follows: "Conflict during socialization need not be viewed solely as conflict between the culture of the parent and the biology of the child; it can also be viewed as conflict between the biology of the parent and the biology of the child." Above all, the young individual is not simply a malleable organism being molded by its parents, as psychologists have conventionally viewed it. On the contrary, the youngster can be expected to be receptive to some of the actions of its parents, neutral to others, and hostile to still others.

The implications of the conflict theory do not stop here. Under some circumstances parents can be expected to influence the offspring's behavior into its adult life. Altruistic behavior might be induced when it results in an increase in inclusive fitness through benefits bestowed on the parents and other relatives. The celibate monk, the maiden aunt, or the homosexual need not suffer genetically. In certain societies their behavior can redound to improved fitness of parents, siblings, and other relatives to an extent that selects for the genes that predisposed them to enter their way of life. Moreover, their relatives, and especially their parents, will respond in a way that reinforces the choice. The social pressure need not be conscious, at least not to the extent of explicitly promoting the welfare of the family. Instead, it is likely to be couched in the sanctions of custom and religion. In case the benefit-to-cost ratio is less than 1 for the individual but greater than 1 for members of the family, a conflict will arise that makes selection for the trait far less likely. Even so, such an evolutionary trend can be sustained where the greater experience and initial advantage of the relatives prove overwhelming.

Parental Care and Social Evolution in the Insects and Mammals

One of the few really fundamental differences between insect and vertebrate societies exists in the realm of parental care. The species of social insects display diversity in the form and intensity of attention paid the young, and this variation is only weakly correlated with the degree of complexity of social organization. In some of the most advanced species virtually no contact occurs between the adults and immature forms. Moreover, it is probable that parent-offspring interaction has little effect on the development of social behavior. For example, the stingless bees (Meliponini) seal their eggs in waxen brood chambers and have no interaction with the developing young until they emerge as full adults, ready to go to work. In spite of this circumstance, the stingless bees have some of the most complex societies found in all the insects. Vertebrate species, in contrast, display a strong correlation between the amount of parental care and the complexity of social organization. Furthermore, the behavior of the parent strongly influences the social development of the offspring. Both of these relations are well marked within the mammals, especially the primates.

To illustrate the more orderly relation that exists between brood care and social organization in the mammals, we can do no better than examine the extremes within the primates. The tree shrews, constituting the family Tupaiidae, are so primitive that their placement as primates has been disputed, some authors preferring to place them within the Insectivora close to the elephant shrews. However, the weight of evidence, chiefly osteological and serological in nature, indicates that they fit close to if not directly in line with the origin of the primates. Tree shrews turn out to have a simple but quite peculiar form of parental care, characterized by absenteeism on the part of the mother. Martin (1968) was able to follow the complete breeding cycle in six pairs of *Tupaia glis belangeri* under seminatural conditions. Social organization is minimal. Sexual pair bonding is strong, and the male marks both the cage and his mate with scent. The male and female sleep together in a nest that is constructed mostly by the male. When the female becomes pregnant she builds a second nest, in which she later bears her young. The litter size is typically two. The mother soon leaves the helpless infants and returns to the first nest to rejoin the male. Thereafter, she visits the nursery only every 48 hours. The infants move from one of her six nipples to another in an unsystematic way. After a few minutes the mother shakes them off and runs away. The young are left to groom themselves and, presumably, to clean out their own feces. The adults do not retrieve them when they are left outside the nest, and the infants do not utter distress calls. Parents forced to remain too close to their young often kill and eat them. When they are about 33 days old, the young tree shrews start emerging from their nest for short foraging trips, and the rhythm of maternal visits begins gradually to break down. At first the young return to their own nest at night and whenever they are frightened, but after 3 days they shift to the parental nest. Sexual maturity is reached at about 90 days of age. Afterward the young evidently scatter to find mates and territories of their own.

Among the nonhuman primates, chimpanzees are at the opposite extreme from the tree shrews in both parental care and social organization. The development of the young chimpanzee has been studied in captive groups by many investigators, but our chief understanding of the process comes from the naturalistic studies conducted by Jane van Lawick-Goodall (1967; 1968a,b; 1971) on the wild population of the Gombe Stream National Park in Tanzania. The main significance of her research has been to show how exceptionally subtle, complicated, and even manlike is the social development of the young chimpanzee. The process unfolds over a period of a little more than ten years. It consists in essence of the slow gaining of locomotory competence, during which the infant leaves the mother for lengthening intervals to explore the environment, manipulate objects, and play with other members of the troop. At first the youngster is broadly tolerated by the adults it encounters, and their friendly responses contribute an important part of its socialization. But as it approaches maturity, it begins to receive rebuffs from adults. Aggressive interactions intensify as the chimpanzee next finds its way into the adult hierarchy. Although the mother mildly rejects her youngster's attempts to nurse during the weaning period, she remains an ally and solace throughout its adolescence.

Alloparental Care

When other members of the society assist the parent in the care of offspring, the potential for social evolution is enormously enlarged. The socialization of individuals can be shaped in new ways, dominance systems can be altered, and alliances can be forged. The term "aunt" was used by Rowell et al. (1964) to denote any female primate other than the mother who cares for a young animal. No genetic relationship was implied; the inspiration is the British "auntie," or close woman friend of the family. The parallel term "uncle" was suggested by Itani (1959) for the equivalent male associates in macaque societies. Since the use of both expressions must always be accompanied by a disavowal of any necessary hereditary relationship, it seems more useful to employ the neutral terms *alloparent* (or helper) and alloparental care. "Allomaternal" and "allopaternal" can then be used as adjectives to distinguish the sex of the helper.

Alloparental care is mostly limited to advanced animal societies. It is, for example, the essential behavioral trait of the higher social insects, the form of altruism displayed by the sterile worker castes. In at least 60 species of birds, including certain babblers, jays, wrens, and others, young adults assist the parents in raising their younger siblings. When eider ducks (*Somateria mollissima*) migrate, mothers and their offspring are often joined by barren females whose behavior closely resembles that of the mothers. Fraser Darling (1938), who discovered the phenomenon, even called these individuals "aunties" well before the term was introduced to primate studies. Among mammals, the phenomenon has been reported in porpoises (*Tursiops*) and both the African and Asiatic elephants. But alloparental care is most richly expressed in the primates. It has been recorded in lemurs (*Lemur catta, Propithecus verreauxi*), New World monkeys (*Lagothrix lagothricha, Saimiri sciureus*), Old World monkeys (species of *Cercopithecus, Colobus, Macaca, Papio,* and *Presbytis*), and chimpanzees (see the general review by Hrdy, 1976).

Within individual primate species, important differences exist between female and male alloparents. Females generally restrict their efforts to the fondling of infants, play, and "baby sitting"; males not only perform these roles but under various conditions also affiliate with infants as future sexual consorts and exploit them as conciliatory objects during encounters with other males. The evidence suggests that the function of the behavior generally differs in males and females, and in the case of male helpers it varies greatly from species to species.

Why should alloparents care for the infants of others, and why should mothers tolerate such behavior? In the case of females, plausible explanations exist for the actions of both participants. First, young females who handle infants gain experience as mothers before committing themselves to motherhood. Gartlan (1969) and Lancaster (1971) have argued that although maternal care may possess innate basic components, it is a sufficiently complex and physically difficult activity to require practice. In this view, play-mothering is one of the final episodes of the socialization process. The relevant evidence is somewhat equivocal but as a whole seems to support the hypothesis with reference to at least a few primate species. Of the seven langur females that Phyllis Jay observed to drop infants through awkwardness, all were very young and in four cases known to be nulliparous. Similarly, female vervets seen by Gartlan to be carrying infants upside down or in other unusual positions were all less than fully grown. The crucial experience appears to be contact with other animals, including infants, rather than just primiparity. When female monkeys and chimpanzees are reared in the wild, their competence at caring for their firstborn is evidently as high as it will ever be. In the rhesus, for example, primiparous mothers are more anxious in manner and reject their infants less firmly than multiparous mothers but are equivalent in the way they retrieve, constrain, cradle, and nurse them (Seay, 1966). When females are reared in varying degrees of social isolation, however, their initial responses are far less adequate and only at later births do they approach normalcy. The rhesus monkeys reared with artificial mothers by Harlow and his coworkers rejected and mistreated their firstborn in a way that would have been fatal to many. But of six such individuals, five gave adequate care to their second infants (Harlow et al., 1966). A captive gorilla who had killed her first infant cared for her second (Schaller, 1963). Similar improvement with experience has been reported in chimpanzees (van Lawick-Goodall, 1969).

A second line of evidence supporting the learning-to-mother hypothesis is the fact that in most species in which allomaternal care is prominent, and in at least some where it occurs infrequently, the behavior is displayed principally by juvenile and subadult females. Of 347 allomaternal contacts recorded in vervets by Lancaster, 295 were initiated by nulliparous females between one and three years of age; the remaining 52 involved females three years or older who were experienced mothers. In caged rhesus populations, two-year-old females are the most active allomaternal class. Nulliparous individuals are more hesitant than experienced mothers in approaching infants, but they nevertheless initiate contacts at a higher rate (Spencer-Booth, 1968).

Thus the evidence conforms to the learning-to-mother hypothesis, although it cannot yet be said to prove it. But even if we grant for the moment that allomothers are benefiting in this way, why should the real mothers tolerate it? The mothers can be expected to lose fitness by turning their children over to the ministrations of incompetents. One reason for taking the risk could be kin selection. By permitting daughters, nieces, and other close female relatives to practice with their children, mothers can improve their inclusive fitness through the proliferation of additional kin when the relatives bear their own first offspring. Such selection will not work if the infants used in practice are damaged as much as the first babies of females who lack practice, for in this case the mother will be trading damage to an infant with $r = 1/2$ for potential benefit to an infant (yet to be born) with $r = 1/4$ or less. In practice, however, there is no reason to expect an even trade. In most species of primates, mothers do not release their infants to allomaternal care until the behavior of the babies has developed somewhat. Even then the mother is alert, sometimes to the point of aggressiveness, and she permits the infant to be kept only for short periods of time. In other words, the allomaternal helpers can practice with the babies without endangering them nearly as much as if they were given sole responsibility. The kin-selection hypothesis might be tested by establishing the degrees of relationship of mothers and helpers, but so far the data are insufficient (Sarah Blaffer Hrdy, personal communication).

Adoption

Although allomaternal care is widespread in the primates, only under special circumstances does it lead to the full adoption of strange infants. In particular, mothers who are nursing young of their own (and are therefore best able to rear orphans) are typically hostile to strange infants who try to approach them. A possible exception is provided by the langur *Presbytis johnii*. Lactating females were seen by Poirier (1968) to respond permissively toward other infants. When more than one infant struggled to gain access to a nipple, mothers showed no preference for their own young. Even in species with aggressive females, orphans are probably rarely left to starve. Female macaques who have lost their own babies readily accept other infants, and they may even go so far as to kidnap them (Itani, 1959; Rowell, 1963). Since mothers are more likely to lose their infants than the reverse, it is probable that orphans will find a willing foster mother. Even if none is available, other females might be able to assume the role fully. When caged rhesus females were induced to adopt infants under experimental conditions, they began to produce apparently normal milk (Hansen, 1966). The foster mothers observed in captive groups of rhesus monkeys usually served first as helpers. This circumstance makes it more likely that in nature a relative will adopt an orphaned infant. Van Lawick-Goodall (1968a) recorded three instances of adoption in wild chimpanzees of the Gombe Stream National Park, two by older juvenile sisters and one by an older brother. Similarly, foster mothers observed by Sade (1965) in the feral rhesus population of Cayo Santiago were older sisters.

PART III

The Social Species

CHAPTER 17

The Four Pinnacles of Social Evolution

To visualize the main features of social behavior in all organisms at once, from colonial jellyfish to man, is to encounter a paradox. We should first note that social systems have originated repeatedly in one major group of organisms after another, achieving widely different degrees of specialization and complexity. Four groups occupy pinnacles high above the others: the colonial invertebrates, the social insects, the nonhuman mammals, and man. Each has basic qualities of social life unique to itself. Here, then, is the paradox. Although the sequence just given proceeds from unquestionably more primitive and older forms of life to more advanced and recent ones, the key properties of social existence, including cohesiveness, altruism, and cooperativeness, decline. It seems that social evolution slowed as the body plan of the individual organism became more elaborate.

The colonial invertebrates, including the corals, the jellyfishlike siphonophores, and the bryozoans, have come close to producing perfect societies. The individual members, or zooids as they are called, are in many cases fully subordinated to the colony as a whole—not just in function but more literally, through close and fully interdependent physical union. So extreme is the specialization of the members, and so thorough their assembly into physical wholes, that the colony can equally well be called an organism. It is possible to array the species of colonial invertebrates into an imperceptibly graded evolutionary series from those forming clusters of mostly free and self-sufficient zooids to others with colonies that are functionally indistinguishable from multicellular organisms.

The higher social insects, comprised of the ants, termites, and certain wasps and bees, form societies that are much less than perfect. To be sure, they are characterized by sterile castes that are self-sacrificing in the service of the mother queen. Also, altruistic behavior is prominent and varied. It includes the regurgitation of stomach contents to hungry nestmates, suicidal weapons such as detachable stings and exploding abdomens used in defense of the colony, and other specialized responses. The castes are physically modified to perform particular functions and are bound to one another by tight, intricate forms of communication. Furthermore, individuals cannot live apart from the colony for more than short periods. They can recognize castes but not individual nestmates. In a word, the insect society is based upon impersonal intimacy. But these similarities to the colonies of lower invertebrates are balanced by some interesting qualities of independence. Social insects are physically separate entities. The secret of their success is in fact the ability of a colony to dispatch separately mobile foragers that return periodically to the home base. Also, the queens are not always the exclusive egg layers. Female workers sometimes insert eggs of their own into the brood cells. Because these eggs are unfertilized, they develop into males. The evidence is strong that in some species of ants, bees, and wasps a low-keyed struggle continually takes place between queens and workers for the opportunity to produce sons. Conflict is sometimes overt in more primitive forms. Groups of female wasps starting a colony together contend for dominance and the egg-laying rights that go with the alpha position. Losers perform as workers, once in a while stealthily inserting eggs of their own into empty brood cells. Similarly, bumblebee queens control their daughters by aggression, attacking them whenever they attempt to lay eggs. If the queen is removed from the relatively simple wasp and bumblebee societies, certain of the workers fight among themselves for the right to replace her.

Aggressiveness and discord are carried much further in vertebrate societies, including those of mammals. Selfishness rules the relationships between members. Sterile castes are unknown, and acts of altruism are infrequent and ordinarily directed only toward offspring. Each member of the society is a potentially independent, reproducing unit. Although an animal's chances of survival are reduced if it is forced into a solitary existence, group membership is not mandatory on a day-to-day basis as it is in colonial invertebrates and social insects. Each member of a society is on its own, exploiting the group to gain food and shelter for itself and to rear as many offspring as possible. Cooperation is usually rudimentary. It represents a concession whereby members are able to raise their personal survival and reproductive rates above those that would accrue from a solitary life. By human standards, life in a fish school or a baboon troop is tense and brutal. The sick and injured are ordinarily left where they fall, without so much as a pause in the routine business of feeding, resting, and mating. The death of a dominant male is usually followed by nothing more than a shift in the dominance hierarchy, perhaps accompanied, as in the case of langurs and lions, by the murder of the leader's youngest offspring.

Human beings remain essentially vertebrate in their social structure. But they have carried it to a level of complexity so high as to constitute a distinct, fourth pinnacle of social evolution. They have broken the old vertebrate restraints not by reducing selfishness, but rather by acquiring the intelligence to consult the past and to plan the future. Human beings establish long-remembered contracts and profitably engage in acts of reciprocal altruism that can be spaced over long periods of time, indeed over generations. Human beings intuitively introduce kin selection into the calculus of these relationships; they are preoccupied with kinship ties to a degree inconceivable in other social species. Their transactions are made still more efficient by a unique syntactical language. Human societies approach the insect societies in cooperativeness and far exceed them in powers of communication. They have reversed the downward trend in social evolution that prevailed over one billion years of the previous history of life. When the human form of social organization is placed in this perspective, it perhaps seems less surprising that it has arisen only once, whereas the other three peaks of evolution have been scaled repeatedly by independently evolving lines of animals.

Why has the overall trend been downward? The reason must have something to do with the greater physical malleability of the lower invertebrates. Because their body plan is so elementary, such colonial animals as coral polyps and bryozoans can be grossly modified to permit an actual physical union with one another. Compared with insects and vertebrates, they require less "rewiring" of nerve cells, rerouting of circulatory systems, and other adjustments of organ systems needed to coordinate colonial physiology. The generally sedentary habits of zooids also make it easier for them to be fused. However, this advantage is not the decisive one; some of the most elaborate invertebrate colonies, including those of siphonophores and thaliaceans, are also the most motile. The simple body construction of lower invertebrates also makes it possible for them to reproduce by directly budding off new individuals from old. Colonies therefore consist of genetically identical individuals. And this, in the final analysis, is the most important feature of all. Absolute genetic identity makes possible the evolution of unlimited altruism. It is already the basis for the extreme specialization and coordination of somatic cells and organs within the metazoan body. The most advanced colonial invertebrates have followed essentially the same road, leading to superorganisms whose organs are created by the extreme modification of zooids (see Chapter 18).

The social insects have none of the preadaptations of the lower invertebrates. Their bodies are in many ways as elaborately constructed as those of the vertebrates, and they are fully motile. Physical union is impossible. Yet they have produced altruistic castes and degrees of colony integration almost as extreme as it is possible to imagine. I would like to suggest that part of the explanation of this achievement is the sheer enormity of the sample size. Over 800,000 species of insects have been described, constituting more than three-quarters of all the known kinds of animals in the world. In the late Paleozoic Era the ancestors of the living insects were among the first invaders of the land, and they took full advantage of this major ecological opportunity. Whereas the ocean and fresh water were already crammed with the representatives of many animal phyla, most of which had originated in Precambrian times, the land was like a new planet, filling with plant life and nearly devoid of animal competitors. The result was an adaptive radiation of species unparalleled before or since. The purely statistical argument states that out of this immense array of new types, it was more probable for at least a few extreme social forms to arise—in comparison, say, with the mere 7,000 species of annelid worms or 5,300 species of starfish and other echinoderms. The argument can perhaps be made clearer by imagining a concrete example: if the rate of invention of advanced sociality were 10^{-12} per species per year for animals generally, 800,000 insect species would certainly have achieved it many times by chance alone, whereas 10,000 species belonging to another phylum might never do so.

The statistical argument gains strength if we calculate on the basis of numbers of genera, families, and higher taxa instead of species, because these higher taxonomic categories reflect stronger degrees of ecological difference. A wolf, for example, representing the family Canidae, differs more in ecology from a deer (family Cervidae) than it does from other canids, such as foxes and wild dogs. The very extensive insect radiation has made it more probable that at least one entire group of species has arisen that is especially predisposed toward sociality. Such a group can in fact be identified; it is the order Hymenoptera, composed of the ants, bees, and wasps. Although constituting only about 12 percent of the living insect species, the hymenopterans hold a near monopoly on higher social existence. Eusociality, the condition marked by the possession of sterile castes, has originated within this order on at least 11 separate occasions, within the roachlike ancestors of the termites once, and in no other known insect group. This remarkable fact brings us back to the overriding factor of kinship. Because of the haplodiploid mode of sex inheritance (to be explained in Chapter 19), female hymenopterans are more closely related to their sisters than they are to their daughters. Thus, all other circumstances being equal, it is genetically more advantageous to join a sterile caste and to rear sisters than it is to function as an independent reproductive. Hymenopterans have other preadaptive traits that make the origin of social life easier, including a tendency to build nests, a long life, and the ability to home; but the haplodiploid bias remains the one feature that they possess uniquely. Even so, the maximum degree of relationship among hymenopteran sisters (measured by the coefficient of relationship, r) is $3/4$, which is substantially less than that in the colonial invertebrates ($r = 1$). The 25 percent or more of genetic difference is enough to explain the amount of discord observed within the hymenopteran societies.

In vertebrates the maximum r between siblings is $1/2$, meaning 50 percent identity of genes by common descent, the same degree that exists between parents and their offpsring. As a result, no special genetic advantage accrues to members of a sterile caste, and with the remotely possible exception of homosexuals in human beings (see Chapter 26), no sterile caste is known to have originated. The nonhuman vertebrates as a whole are more social than the insects in the one special sense that a larger percentage have achieved some level of sociality. But their most advanced societies are not nearly so extreme as those of the insects. In other words, a strong impelling force appears to generate social behavior in vertebrate evolution, but it is brought to a halt by the equally strong countervailing force of lower genetic relationship among closest relatives. Consequently, it seems best to dwell not on the matter of genetic relationship, which is simple and canonical, but on the nature of the impelling force. This force, I would like to suggest, is greater intelligence. The concomitants of intelligence are more complex and adaptable behavior and a refinement in social organization that are based on personalized individual relationships. Each member of the vertebrate society can continue to behave selfishly, as dictated by the lower degrees of kinship. But it can also afford to cooperate more, by deftly picking its way through the conflicts and hierarchies of the society with a minimum expenditure of personal genetic altruism. We must bear in mind that whereas the primary "goal" of individual colonial invertebrates and social insects is the optimization of group structure, the primary "goal" of a social vertebrate is the best arrangement it can make for itself and its closest kin within the society. The social behavior of the lower invertebrates and insects has been evolved mostly through group selection, whereas the social behavior of the vertebrates has been evolved mostly through individual selection. The requisite refinement and personalization in vertebrate relationships are achieved by (1) enriched communication systems; (2) more precise recognition of and tailored responses to groupmates as individuals; (3) a greater role for learning, idiosyncratic personal behavior, and tradition; and (4) the formation of bonds and cliques within the society. Let us examine each of these qualities briefly.

The majority of vertebrate species utilize at least two or three times more basic displays than do most insect species, including even the social insects. But the actual number of messages that can be transmitted is far greater, for two reasons. First, context is more important to the meaning of each vertebrate display. A distinct message can be associated with the place the display occurs, the time of year, or even the sex and rank of the animal. The signal is also more likely to be part of a composite display. For example, a movement of the head may accompany one or the other of several vocalizations, each of which lends it a different meaning. Scaling is also more prominently developed in vertebrates than in insects. Variations in the intensity of the signal, often very slight, are used to convey subtle changes in mood. All of these improvements together enlarge vertebrate repertories to such an extent that they are able to transmit perhaps an order of magnitude more bits of information per second than insect repertories. We cannot be sure of the exact amount, owing to the severe technical difficulties encountered in measuring the information content of more complex communication systems.

The recognition of individuals as such is mostly a

vertebrate trait. Tunicate colonies "recognize" those of differing genotypes by failing to coalesce with them on contact (Burnet, 1971). *Drosophila* adults can identify the odors of different genetic strains when choosing mates (Hay, 1972), while social insects generally discriminate nestmates from all other members of the species through colony odors that adhere to the surface of the body (Wilson, 1971a). These responses are to classes of individuals and not to separate organisms, however. Only a few cases of truly personal recognition have been documented in the invertebrates. When females of the social wasp *Polistes* found colonies together, they organize themselves into dominance hierarchies that appear to be based on knowledge of one another as individuals (Pardi, 1948; Mary Jane Eberhard, 1969). Sexual pair bonds are formed by individual starfish-eating shrimp *Hymenocera picta* (Wickler and Seibt, 1970) and desert sowbugs *Hemilepistus reaumuri* (K. E. and C. Linsenmair, 1971; K. E. Linsenmair, 1972). Both of these species use bonding as a device to cope with specialized ecological requirements. Other examples among invertebrates will no doubt be discovered, but they will certainly continue to constitute a very small minority. Vertebrates, in contrast, generally have the power of personal recognition. It is probably lacking in schooling fishes, in amphibians, and in at least the more solitary reptiles. But personal recognition is a widespread and possibly universal phenomenon in the birds and mammals, the two vertebrate groups containing the most advanced forms of social organization.

Vertebrates are also capable of quick forms of learning that fit them to the rapidly changing nexus of relationships within which they live. When an ant colony faces an emergency, its members need only respond to alarm pheromones and assess the general stimuli they encounter. But a rhesus monkey must judge whether the excitement is created by an internal fight, and if it is, learn who is involved, remember its own past relation to the participants, and judge its immediate actions according to whether it will personally benefit or lose by taking action of its own. The social vertebrate also has the advantage of being able to modify its behavior according to observations of success or failure on the part of the group as a whole. In this manner traditions are born that endure for generations within the same society. Play became increasingly important as vertebrate social evolution advanced, facilitating the invention and transmission of traditions and helping to establish the personalized relationships that endure into adulthood. Socialization, the process of acquiring these traits, is not the cause of social behavior in the ultimate, genetic sense. Rather, it is the set of devices by which social life can be personalized and genetic individual fitness enhanced in a social context (see Chapter 7).

Finally, the typically vertebrate qualities of improved communication, personal recognition, and increased behavioral modification make possible still another property of great importance: the formation of selfish subgroups within the society. It is possible for mated pairs, parent-offspring groups, clusters of siblings and other close kin, and even cliques of unrelated individuals to exist within societies without losing their separate identities. Each pursues its own ends, imposing severe limits on the degree to which the society as a whole can operate as a unit. The typical vertebrate society, in short, favors individual and in-group survival at the expense of societal integrity.

Man has intensified these vertebrate traits while adding unique qualities of his own. In so doing he has achieved an extraordinary degree of cooperation with little or no sacrifice of personal survival and reproduction. Exactly how he alone has been able to cross to this fourth pinnacle, reversing the downward trend of social evolution in general, is the culminating mystery of all biology. We will return to it at the end of the survey of social organisms composing the remainder of this book.

CHAPTER 18

The Colonial Invertebrates

For years the study of colonial organization in microorganisms and the lower animals has been confused by a dilemma. On the basis of several criteria many of the species can be considered to belong to the highest social grade ever attained in three billion years of evolution. The very term *colony* implies that the members are physically united, or differentiated into reproductive and sterile castes, or both. When the two conditions coexist in an advanced stage, the "society" can be viewed equally well as a superorganism or even as an organism. Many invertebrate zoologists have pondered and debated this philosophical distinction. The dilemma can be restated simply as follows: At what point does a society become so nearly perfect that it is no longer a society? On what basis do we distinguish the extremely modified zooid of an invertebrate colony from the organ of a metazoan animal?

These questions are not trivial. They address a theoretical issue seldom made explicit in biology: the conception of all possible ways by which complex metazoan organisms can be created in evolution. To make the issue wholly clear let us go directly to the *ne plus ultra* of invertebrate social forms, and of animal societies generally, the colonial hydrozoans of the order Siphonophora. Approximately 300 species of these bizarre creatures have been described. Vaguely resembling jellyfish, all live in the open ocean where they use their stinging tentacles to capture fish and other small prey. The most familiar genus is *Physalia*, the Portuguese man-of-war. Another example is *Nanomia*, illustrated in Figure 18–1. These creatures resemble organisms. To the uninitiated they appear basically similar to scyphozoans, the "true" jellyfish of the ocean, which are unequivocally discrete organisms. Nevertheless, each siphonophoran is a colony. The zooids are extremely specialized. At the top of each *Nanomia* sits an individual modified into a gas-filled float, which gives buoyancy to the rest of the colony strung out below it. Nectophores act like little bellows, squirting out jets of water to propel the colony through the water. By altering the shape of their openings, they are able to alter the direction of the jets and hence the path followed by the colony. Through their coordinated action the *Nanomia* colony is able to dart about vigorously, moving at any angle and in any plane, and even executing loop-the-loop curves. Lower on the stem sprout saclike zooids called palpons and gastrozooids, which are specialized for the ingestion and distribution of nutrients to the remainder of the colony. Long branched tentacles arise as organs from both the palpons and gastrozooids. They are used to capture prey and perhaps to defend the colony as well. The roster of specialists is completed by the sexual medusoids, which are responsible for the production of new colonies by conventional gamete formation and fertilization, and the bracts, which are inert, scalelike zooids that fit over the stem like shingles and evidently help protect it from physical damage. New zooids are generated by budding in one or the other of two growth zones located at each end of the nectophore region.

The behavior and coordination of *Nanomia* colonies has been the object of excellent studies by Mackie (1964, 1973). The zooids behave to some degree as independent units, but they are also subject to considerable control from the remaining colony members. Each nectophore, for example, has its own nervous system, which determines the frequency of contraction and the direction in which the water is squirted. However, the nectophore remains quiescent except when aroused by excitation arriving from the rest of the colony. When the rear portion of the colony is touched the forward nectophores begin to contract and then others join in. Experiments have shown that the coordination is due to conduction through nerve tracts that connect the nectophores. When the float is touched, the nectophores reverse the direction of their jet propulsion and drive the colony backward. The conduction that coordinates this

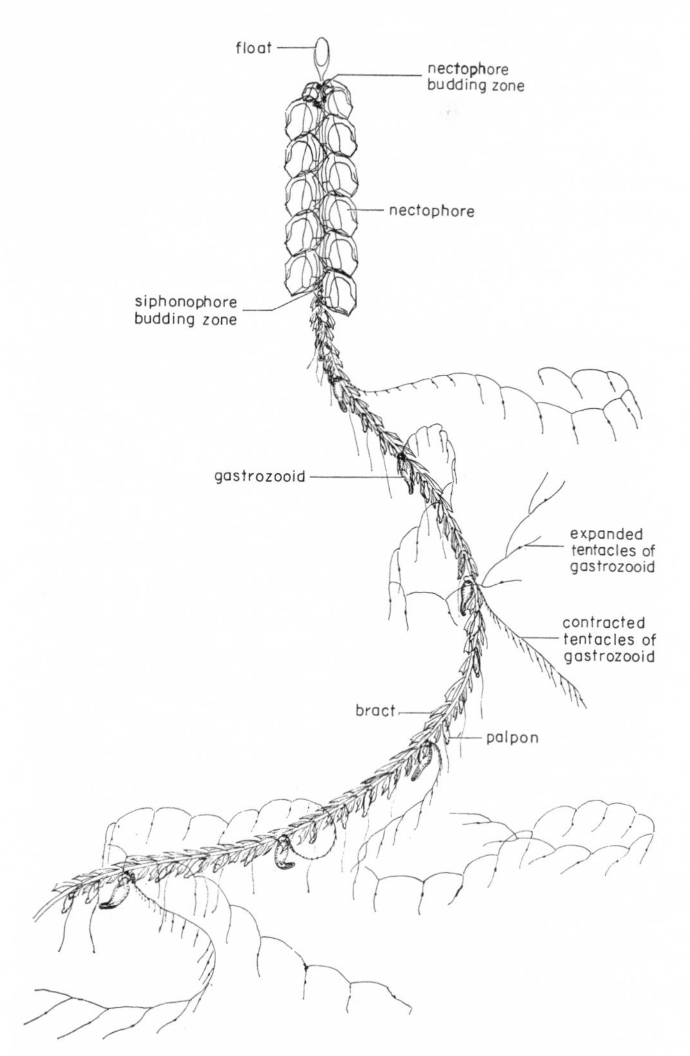

Figure 18-1 A colony of the siphonophoran *Nanomia cara*. The float that provides buoyancy, the nectophores that propel the entire ensemble, the gastrozooids that capture and digest prey, and other individual colony members, such as the bracts and palpons, are modified to such an extreme as to be comparable to the organs of single metazoan animals. (Modified from Mackie, 1964.)

second movement is not through nerves but through sensitive cells in the epithelium. The gastrozooids of *Nanomia* are more independent in action. Several gastrozooids may collaborate in the capture and ingestion of a prey, but their movements and nervous activity are wholly separate. They and the palpons, which are auxiliary digestive organs, pump digested food out along the stem to the rest of the colony. Even the empty zooids participate in the peristaltic movements, with the result that the digested material is flushed more quickly back and forth along the stem. In other respects, however, their behavior remains independent.

Much of the difficulty in conceptualizing *Nanomia* and other siphonophores as colonies rather than organisms stems from the fact that each of the entities originates from a single fertilized egg. This zygote undergoes multiple divisions to form a ciliated planula larva. Later, the ectoderm thickens and begins to bud off the rudiments of the float, the nectophores, and other zooids. Fundamentally, the process, which specialists in colonial invertebrates call "astogeny," does not differ from the ontogeny of scyphozoan medusas or other true individual organisms among the Coelenterata. The resolution of the paradox is that siphonophores are both organisms *and* colonies. Structurally and embryonically they qualify as organisms. Phylogenetically they originated as colonies. Other hydrozoans, including anthomedusans and leptomedusans as well as milleporine and stylasterine "corals," display every stage in the evolution of coloniality up to the vicinity of the siphonophore level. Some species form an elementary grouping of fully formed, independently generated polypoids that are nevertheless connected by a stolon that either runs along the substrate or rises stemlike up into the water. In others there is a physical union of the zooid body wall along with varying degrees of specialization. As castes differentiate among the zooids, as for example in the familiar leptomedusan genus *Obelia*, some lose their reproductive capacity, while the reproductive individuals (gonozooids) lose the capacity to feed and to defend themselves. In the end, among the evolutionarily most advanced species, the ensemble collapses into a highly integrated unit. It assumes a distinctive and relatively invariable form that subordinates the zooids to mere coverings, floats, tentacle bearers, and other organlike units. This last stage is the one attained by the order Siphonophora and to a less spectacular degree by *Velella* and other pelagic hydroids of the order Chondrophora.

The achievement of the siphonophores and chondrophores must be regarded as one of the greatest in the history of evolution. They have created a complex metazoan body by making organs out of individual organisms. Other higher animal lines originated from ancestors that created organs from mesoderm, without passing through a colonial stage. The end result is essentially the same: both kinds of organisms escaped from the limitations of the diploblastic (two-layered) body plan and were free to invent large masses of complicated organ systems. But the evolutionary pathways they followed were fundamentally different.

The Adaptive Basis of Coloniality

The Darwinian advantage of membership in colonies is not immediately obvious. At the highest levels of integration, a majority of the zooids do not reproduce, and many are freely autotomized when injury or overgrowth causes them to encumber the colony as a whole. Even fully formed zooids may suffer some disadvantages from colony membership. Bishop and Bahr (1973) demonstrated that as colony size in the freshwater bryozoan *Lophopodella carteri* increases, the clearance rate of water per zooid (for example, microliter/zooid/minute) decreases. Hence larger colony size results in less food for each of the member individuals. Arrayed against these negative factors are a variety of advantages that have been documented in one or more species of colonial organisms:

1. *Resistance to physical stress in the neritic benthos.* Coloniality is most common among the invertebrates that inhabit the bottom of the ocean in the shallow water along the seacoast. There wave action is strongest and sessile organisms are most likely to be choked by sedimentation. Coral reefs and the fouling communities on pilings and rocks are made up principally of colonial coelenterates, bryozoans, and tunicates. Careful studies of zoantharian corals have revealed that the massed calcareous skeletons of the zooids, when constructed in certain ways, anchor the colonies more securely to the sea floor and increase the survival time of the organisms they protect. The colony raises individuals from the bottom, away from the densest concentrations of suspended soil particles. The orientation of the zooids in corals and other colonial forms allows them to generate faster currents than any attainable by isolated single organisms of similar construction.

2. *Liberation of otherwise sessile forms for a free-swimming, pelagic existence.* The zooids of siphonophore and chondrophore colonies are polypoids, which are basically hydralike individuals adapted for attachment to the sea floor and a sedentary existence. By modifying some of the polypoids into floats and swimming bells, the colonies have been able to swim free in the open ocean. Some of the members, such as the gastrozooids, gonozooids, and bracts, are still polypoid in construction, but they are easily carried along by the swimming specialists.

3. *Superior colonizing and competitive abilities.* As Bonner (1970) has emphasized, the clear advantage gained by aggregation in the single-celled myxobacteria and cellular slime molds is the capacity of masses of these cells to shape their overall form and to elevate fruiting bodies on stalks. The spores liberated from the fruiting bodies consequently travel farther than would have been the case had they been formed by the individual bacteria or myxamebas still in the soil. Dispersal is the "aim" of sporulation, because aggregation is induced when local environmental conditions deteriorate. Coloniality in the sessile invertebrates is associated not with the enhancement of dispersal but with the improvement of colony growth and survival following dispersal. Asexual budding is the fastest form of growth, especially when performed laterally to create an encrusting assemblage. It also enables a colony to overgrow and to choke out competing forms. Corals, for example, compete with one another like plants, by cutting out the light of those beneath or by covering and suffocating competitors occupying the same surface. In both instances the capacities to produce large masses and to continue growing at high densities are decisive.

4. *Defense against predators.* Most of the heterozooids of the polymorphic Ectoprocta for which functions have been established specialize in the defense of the colony, either by adding to the strength of the colony wall or by actively repelling invaders.

General Evolutionary Trends in Coloniality

The most nearly complete systematic account of colonial life in the invertebrates is that by the Russian zoologist W. N. Beklemishev (1969). After surveying most of the colonial taxa and providing a simple morphological classification of the assemblage, Beklemishev formulated the major evolutionary rules that he believed apply broadly across the invertebrates. His thinking was influenced by two venerable ideas: the concept of the superorganism and the view that biological complexity evolves by the dual processes of the differentiation and integration of individuals. He accordingly identified three complementary trends as the basis of increasing coloniality: (1) the weakening of the individuality of the zooids, by physical continuity, sharing of organs, and decrease in size and life span, as well as by specialization into simplified, highly dependent heterozooids; (2) the intensification of the individuality of the colony, by means of more elaborate, stereotyped body form and closer physiological and behavioral integration of the zooids; and (3) the development of cormidia, or "colonies within colonies." Within at least the cheilostome ectoprocts, Banta (1973) has concluded that coloniality first increased by division of labor, presumably in association with the delimitation of cormidia, then by physiological interdependence of the polymorphic zooids, and finally by structural interdependence of the zooids. In later stages of evolution all three processes proceeded concurrently.

The cormidia are particularly interesting, because they correspond to the organ systems and appendages of metazoan individuals. Examples of cormidia

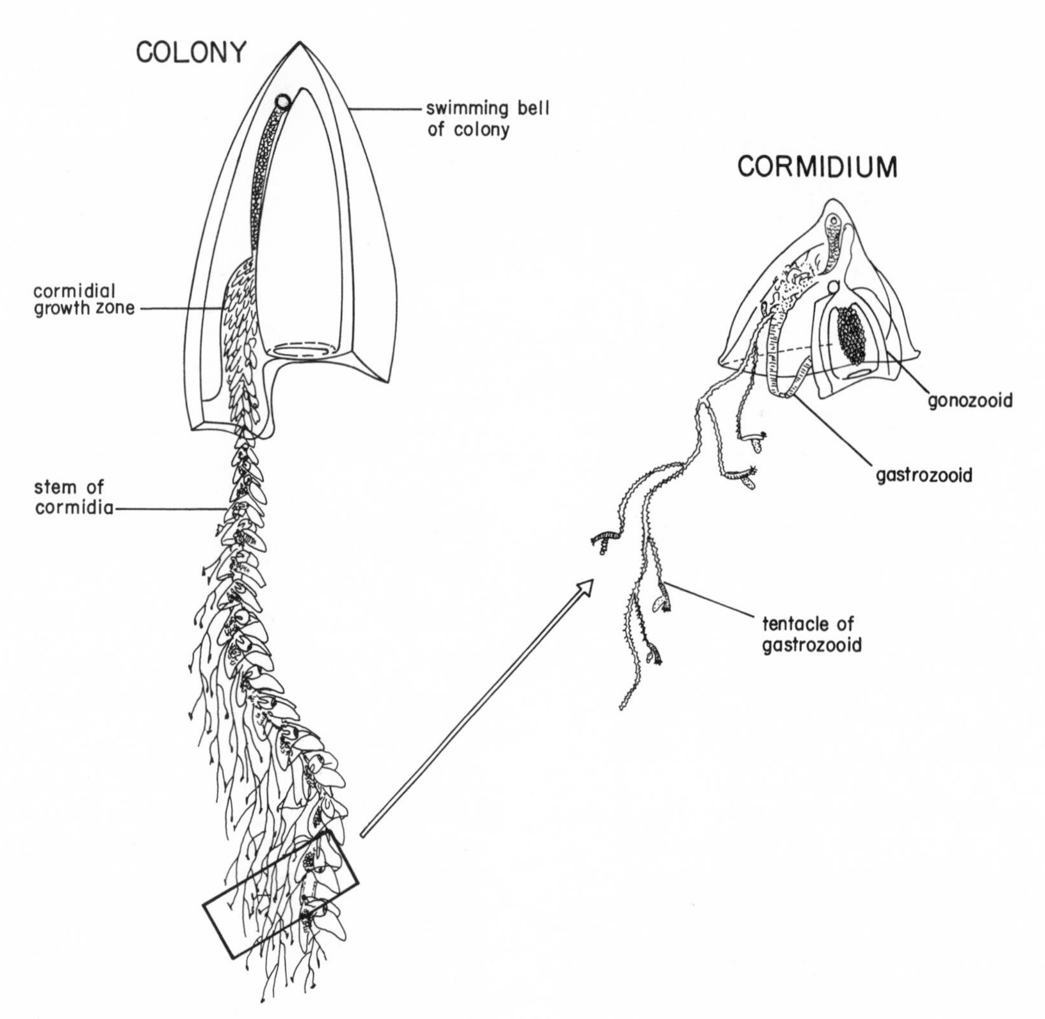

Figure 18-2 The full hierarchy of organizational units in colonial invertebrates is displayed by the siphonophore *Muggiaea*. The cormidium comprises a group of zooids (individuals) that are coordinated among themselves strongly enough to be recognized as a distinct element within the colony. In the case of *Muggiaea*, the cormidium nearly ranks as a full colony, since it can break away and lead a separate existence for a while. (Modified from Hyman, 1940.)

include the nectosome, or region of swimming bells (nectophores), in *Nanomia* and other siphonophores; the "leaves," "petals," and branches of sea pens and certain other octocorallian corals; the shoots or internodes in ectoproct colonies; and others. In *Muggiaea* and related calycophoran siphonophores, one kind of cormidium is so nearly independent as to exist on the borderline between the cormidium as an organizational unit and a full colony. It consists typically of a helmet-shaped bract, a gastrozooid with a tentacle, and one or more gonophores of one sex, which double as swimming bells. When fully developed, these units break loose and lead a temporarily free existence. Known as eudoxomes, they were considered to be distinct species of Siphonophora until their true relationship to the larger colony unit was discovered (see Figure 18–2).

The Coelenterates

The fullest demonstrable range of colonial evolution has occurred within the phylum Coelenterata. Although the individual organisms themselves all retain the basic diploblastic body plan, the associations show a huge amount of diversity, from the solitary scyphozoan jellyfish, hydras, and sea anemones through virtually every conceivable gradation to colonies so fully integrated as to be nearly indistinguishable from organisms. Some of the colonies are sessile, mosslike forms, others complex motile assemblages resembling jellyfish.

The living species of corals present a tableau of the evolution of sessile colonies (Bayer, 1973). In the order Stolonifera are found colonies of the most basic type, consisting of nearly independent zooids connected by a living stolon. Growth is plantlike, with new zooids sprouting up from the stolon as it creeps over the substrate (see Figure 18–3). In other species of Stolonifera, new zooids originate from stolonic outgrowths on the body walls of older zooids. When they reach a certain height, they generate still other new individuals in the same fashion. The result is a dendritic colony that increases in density from bottom to top. Within the order Alcyonacea further integration is achieved by the formation of a common jellylike mesogloea in which the gastrovascular cavities are closely packed. The zooids of species producing the largest colonies are differentiated into two forms: the more elementary autozooids, which eat, digest, and distribute nutrients to the remainder of the colony; and the siphonozooids, which possess the reproductive organs and circulate water by means of large ciliated grooves running down the side of the pharynx (see Figure 18–3). The apparent significance of the differentiation lies in this last function, which prevents water stagnation at high population densities. The ultimate colonial development among the corals has been attained by the alcyonacean *Bathyalcyon robustum*. Each mature colony consists of a single giant autozooid, in the body wall of which are embedded large numbers of daughter siphonozooids. In effect, the siphonozooids have become organs of the parental autozooid.

Figure 18-3 Two grades in the colonial evolution of corals. Top: a simple form (*Clavularia hamra*), in which largely independent zooids grow from near the terminus of the ribbon-shaped stolon. Bottom: a more advanced species (*Heteroxenia fuscescens*) in which the zooids collaborate to construct a massive calcareous base. This specimen is cut in half to reveal the two types of zooids: the siphonozooids, which have a ciliated groove along their sides and gonads that penetrate deeply into the base, and the autozooids, which lack these organs. The siphonozooids specialize in reproduction and the creation of water currents, while the autozooids eat and digest the food. (From Bayer, 1973; after H. A. F. Gohar. Reprinted with permission from *Animal Colonies: Development and Function through Time*, ed. R. S. Boardman, A. H. Cheetham, and W. A. Oliver, Jr. Copyright © 1973 by Dowden, Hutchinson and Ross, Inc., Publishers, Stroudsburg, Pennsylvania.)

A wholly different colonial strategy appears in the order Gorgonacea. Here growth is treelike, and integration consists of little more than regularity in the patterns of branching. The colonies of some species resemble fans, others palm leaves, while still others rise in delicate exploding whorls of zooid-bearing branches. Among the Coralliidae, which are undoubted gorgonians, the zooids are as fully dimorphic as in the large alcyonaceans. But the similarity is probably due to convergence. In the coralliids the autozooids, not the siphonozooids, contain the gonads.

The Ectoprocts

The phylum Ectoprocta, or Bryozoa, containing the bulk of the "bryozoans" of older zoological classifications, displays the most advanced colonial organization of any of the coelomate groups. The specialization of the zooids is extreme, rivaling that found in the acoelomate siphonophores. The vast majority of ectoproct species are sessile, forming encrusting or arborescent colonies on almost any available firm surface in both marine and freshwater environments. To the naked eye some of the aggregations resemble sheets of fine lacework, others miniature moss or seaweed. The colonies of one genus, *Cristatella*, are ribbon-shaped bodies that creep over the substrate at rates not exceeding 3 centimeters a day (see Figure 18–4). Ectoprocts feed on plankton, which they capture with lophophores, hollow organs crowned with ciliated tentacles. All of the species are colonial. The zooids communicate by pores through the skeletal walls. The pores are usually plugged by

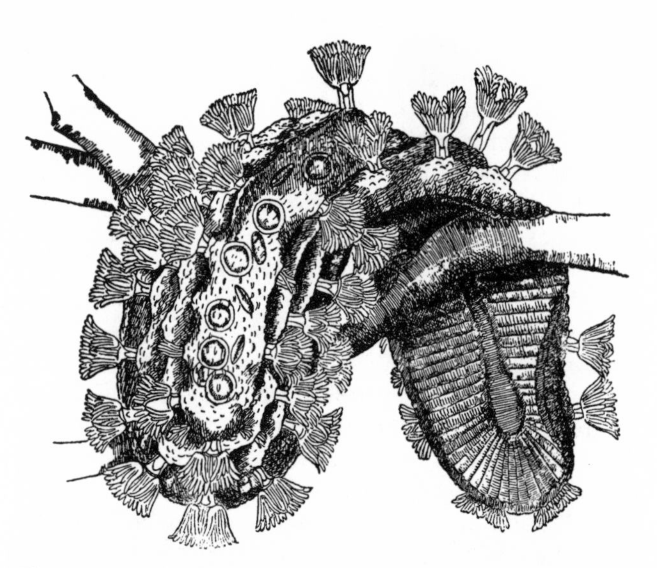

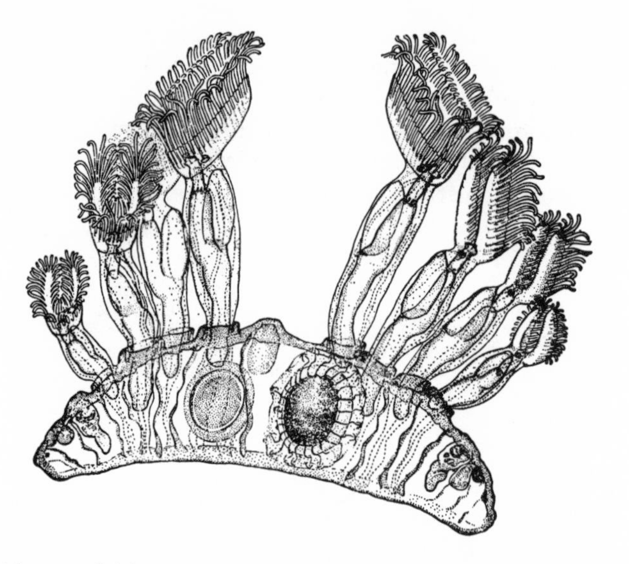

Figure 18-4 The motile colony of the fresh water ectoproct *Cristatella mucedo*. Left: view of an entire colony creeping over plant stems; individual zooids project their brushlike lophophores upward while beneath them, round statoblasts, which are asexually produced reproductive bodies, can be seen embedded in the gelatinous supporting structure of the colony (from J. Jullien, 1885). Right: transverse section of a *Cristatella* colony (from Brien, 1953.)

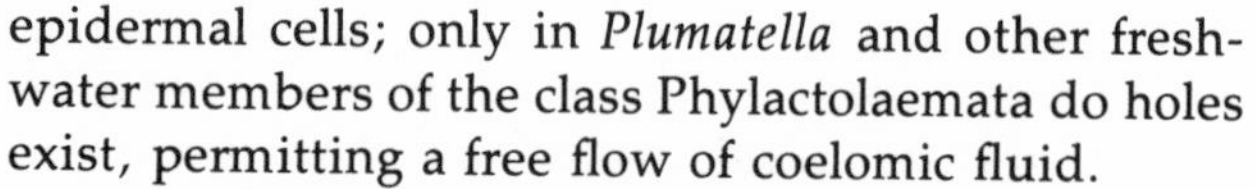

epidermal cells; only in *Plumatella* and other freshwater members of the class Phylactolaemata do holes exist, permitting a free flow of coelomic fluid.

Polymorphism of zooids of an extreme form is found in the primarily marine class Gymnolaemata. The diversity of specialists among these individuals is enormous, and bryozoologists have scarcely begun to study them systematically. Even the classification of the basis morphological types is in a state of flux. Autozooids, which are individuals with independent reproductive and feeding organs, are distinguished as a major category from heterozooids, the class of all specialists together. One of the most distinctive types of specialists is the avicularium, found in some species of the order Cheilostomata. The operculum of this zooid has been modified into a sharp-edged lid that opens and snaps shut by means of opposing sets of muscles. The pedunculate avicularia of *Bugula* and *Synnotum* bear a remarkable resemblance to the head of a bird and are in fact capable of turning and biting at intruding objects. The difference between such an individual and the presumed ancestral autozooid is much greater than the difference between any two castes of the social insects and is matched elsewhere in the animal kingdom only by the diversification of the siphonophore zooids. Other ectoproct specialists include the vibraculum, in which the operculum is modified into a flexible bristle that can be lashed back and forth; spinozooids, characterized by spines that project from the body wall; gonozooids, which are specialized for sexual reproduction; and interzooids, highly reduced forms that serve as pore-plates or pore-chambers fitted between completely formed neighboring zooids. Kenozooids consist of a wide variety of supporting and anchoring elements, such as the rhizoids and other root attachments, the tubular elements of the stolons from which other zooids sprout, and attachment disks. Kenozooids, together with interzooids, are often so simplified in structure that their identity as individuals rather than organs can sometimes be established only by comparing them with more completely formed, intermediate evolutionary stages. Their existence led Silén (1942) to provide the ultimate structural definition of a zooid (that is, an individual organism) in the Ectoprocta as a cavity enclosed by walls. They illustrate convincingly the principle that when the unit of selection is the colony instead of the individual, the individual can be reduced in the course of evolution until it is virtually annihilated as an entity.

CHAPTER 19

The Social Insects

The social insects challenge the mind by the sheer magnitude of their numbers and variety. There are more species of ants in a square kilometer of Brazilian forest than all the species of primates in the world, more workers in a single colony of driver ants than all the lions and elephants in Africa. The biomass and energy consumption of social insects exceed those of vertebrates in most terrestrial habitats. The ants in particular surpass birds and spiders as the chief predators of invertebrates. In the temperate zones ants and termites compete with earthworms as the chief movers of the soil and leaf litter; in the tropics they far surpass them.

Insects present the biologist with a rich array of social organizations for study and comparison. The full sweep of social evolution is displayed—repeatedly—by such groups as the halictid bees and sphecid and vespid wasps. There are so many species in each evolutionary grade that one can sample them as a statistician, measuring variance and partialing out correlates. And the subject is in its infancy. The great majority of social genera are unexplored behaviorally, so that one can only guess the outlines of their colonial organization. The ants are a prime example. Somewhat fewer than 8,000 species have been described in the most limited sense of being assigned a name in the binomial system. W. L. Brown, the leading authority on ant classification, estimates that at least another 4,000 remain to be discovered, and the rate at which new species appear in the literature seems more than ample to bear him out. Of the perhaps 12,000 living species, less than 100, or 1 percent, have been studied with any care, and less than 10 thoroughly and systematically. The species comprise about 270 genera (Brown, 1973), which represent the grosser units that can be compared with greatest profit when making comparative studies. I judge only 49 to have been subjected to careful sociobiological inquiry, and in most cases these studies have been rather narrowly conceived and executed. The remaining 220 genera are known mostly from natural history notes having little to do with their social behavior. Perhaps half of these can be said to be virtually unknown in every respect except habitat and nest site. The other major groups of social insects, the termites, the social bees, and the social wasps, have been even less well explored.

Hence, despite the fact that there have been about 3,000 publications on social wasps, 12,000 on termites, 35,000 on ants, and 50,000 on social bees (Spradbery, 1973), the development of insect sociobiology lies largely in the future. The number of investigators, the rate of publication, and the accretion of knowledge are all in what appears to be the early exponential phase of growth. Entomologists have also only begun to address the central questions of the subject, which can be stated in the following logical sequence:

——What are the unique qualities of social life in insects?

——How are insect societies organized?

——What were the evolutionary steps that led to the more advanced forms of social organization?

——What were the prime movers of social evolution?

These questions have been considered systematically in my earlier book *The Insect Societies,* to which the reader is referred for fuller explanation. The remainder of this chapter consists of a précis of insect sociobiology, which starts with partial answers for the questions just given and proceeds to a more systematic account of the behavior of key groups.

What Is a Social Insect?

The "truly" social insects, or eusocial insects, as they are often more technically labeled, include all of the ants, all of the termites, and the more highly organized bees and wasps. These insects can be distinguished as a group by their common possession of three traits: (1) individuals of the same species cooperate in caring for the young; (2) there is a reproductive division of labor, with more or less sterile individuals working on behalf of fecund nestmates; and (3) there is an overlap of at least two generations in life stages capable of contributing to colony labor, so that offspring assist parents during some period of their life. These are the three qualities by which the majority of entomologists intuitively define eusociality. If we bear in mind that it is possible for the traits to occur independently of one another, we can proceed with a minimum of ambiguity to define *presocial* levels on the basis of one or two of the three traits. Presocial refers to the expression of any degree of social behavior beyond sexual behavior yet short of eusociality. Within this broad category there can be recognized a series of lower social stages, which are defined in matrix form in Table 19-1. The full logic of the reconstruction of social evolution, worked out over many years by such veteran entomologists as Wheeler (1923, 1928), Evans (1958), and Michener (1969), can be perceived by examining the matrices closely. In the *parasocial* sequence, adults belonging to the same generation assist one another to varying degrees. At the lowest level, they may be merely communal, which means that they cooperate in constructing a nest but rear their brood separately. At the next level of involvement, *quasisociality*, the brood are attended cooperatively, but each female still lays eggs at some time of her life. In the *semisocial* state, quasisocial cooperation is enhanced by the addition of a true worker caste; in other words, some members of the colony never attempt to reproduce. Finally, when semisocial colonies persist long enough for members of two or more generations to overlap and to cooperate, the list of three basic qualities is complete, and we refer to the species (or the colony) as being eusocial. Precisely this sequence has been envisioned by Michener and his coworkers as one possible evolutionary pathway taken by bees.

The alternate sequence is comprised of the *subsocial* states. In this case there is an increasingly close association between the mother and her offspring. At the most primitive level, the female provides direct care for a time but departs before the young eclose as adults. It is possible then for the care to be extended to the point where the mother is still present when her offspring mature, and they might next assist her in the rearing of additional brood. It remains only for some of the group to serve as permanent workers, and the last of the three qualities of eusociality has been attained. The subsocial route is the one believed by Wheeler and most subsequent investigators to have been followed by ants, termites, social wasps, and at least a few groups of the social bees.

It is fair to say that as an ecological strategy eusociality has been overwhelmingly successful. It is useful to think of an insect colony as a diffuse organism, weighing anywhere from less than a gram to as much as a kilogram and possessing from about a hundred to a million or more tiny mouths. It is an animal that forages amebalike over fixed territories a few square meters in extent. A colony of the common pavement ant *Tetramorium caespitum*, for example, contains an average of about 10,000 workers that weigh 6.5 grams in the aggregate and control 40 square meters of ground. The average colony of the American harvester ant *Pogonomyrmex badius*, a larger species, contains 5,000 workers who together weigh 40 grams

Table 19-1 The degrees of sociality in the insects, showing intermediate parasocial and subsocial states that can lead to the highest (eusocial) form of organization.

	Qualities of sociality		
Degrees of sociality	Cooperative brood care	Reproductive castes	Overlap between generations
Parasocial sequence			
Solitary	–	–	–
Communal	–	–	–
Quasisocial	+	–	–
Semisocial	+	+	–
Eusocial	+	+	+
Subsocial sequence			
Solitary	–	–	–
Primitively subsocial	–	–	–
Intermediate subsocial I	–	–	+
Intermediate subsocial II	+	–	+
Eusocial	+	+	+

and patrol tens of square meters. The giant of all such "superorganisms" is a colony of the African driver ant *Dorylus wilverthi*, which may contain as many as 22 million workers weighing a total of over 20 kilograms. Its columns regularly patrol an area between 40,000 and 50,000 square meters in extent. The solitary tiphioid wasps, the nearest living relatives of ants, are by comparison no more than minuscule components of the insect fauna. Termites similarly outrank cockroaches. The cryptocercid cockroaches, the advanced subsocial forms closest to the ancestry of the termites, are in particular an insignificant cluster of species limited to several localities in North America and northern Asia. Only in the social bees and wasps have species in intermediate evolutionary grades held their own in competition with eusocial forms. They provide an array from which the full evolutionary pageant can now be deduced.

The Organization of Insect Societies

Once a species has crossed the threshold of eusociality, there are two complementary means by which it can advance in colonial organization: through the increase in numbers and degree of specialization of the worker castes, and through the enlargement of the communication code by which the colony members coordinate their activities. This statement is the insectan version of the venerable prescription that a society, like an organism and indeed like any cybernetic system, progresses through the differentiation and integration of its parts. In Chapter 14 I derived the less obvious theorem that castes tend to proliferate in evolution until there is one for each task. The theoretical limit has probably not been attained by many species of social insects, but it has been pursued by the most advanced forms to the extent that the number of discernible, functional types within the worker caste is often five or more and perhaps often exceeds ten. The reason for the vagueness of the estimate is simple. Castes can be physical, meaning that they are based upon permanent anatomical differences between individuals. Or they can be temporal, which means that individuals pass through developmental stages during which they serve the colony in various ways. The individual, in other words, belongs to more than one caste in its lifetime. Purists may hesitate to call a developmental stage a caste, but examination of ergonomic theory will show why it must be so defined.

Three basic physical castes are found in the ants, all members of the female sex: the worker, the soldier, and the queen. I refer to them as basic because they exist usually, but not always, as sharply distinctive forms unconnected to any other castes by intermediates. The males constitute an additional "caste" only in the loosest sense. No certain case of true caste polymorphism *within* the male sex has yet been discovered. Two forms of the male occur in some species of *Hypoponera*, but even in these cases they are not known to coexist in the same colony. Soldiers are often referred to as major workers, and the smaller coexisting worker forms as minor workers. Where soldiers exist in a species, minor workers are also invariably found. The latter caste is the more versatile of the two, typically attending to food gathering, nest excavation, brood care, and other quotidian tasks. In many species the soldiers assist to some degree, but in most cases they serve as defenders of the nest and living vessels for the storage of liquid food. The entire worker caste has been lost in many socially parasitic species, while in a few free-living species, especially in the primitive subfamily Ponerinae, the queen has been completely supplanted by workers or workerlike forms. In only a minority of species are all three females castes found together. All ant species, however, produce males in abundance as part of the normal colony life cycle.

Although termites are phylogenetically unrelated to ants, they have evolved a caste system that is remarkably similar to the ant system in several major respects. Their colonies are generated by one or at most a small number of queens. Like the ants, termites have produced a soldier caste that is highly specialized in both head structure and behavior for colony defense, and a minor worker caste that is numerically dominant in the colonies, morphologically similar from species to species, and behaviorally versatile. The number of physical castes in the phylogenetically most advanced termite species is somewhat greater than in the most highly evolved ant species, but the average degree of specialization of individual castes is about the same. Finally, the higher termites have developed temporal polyethism resembling that of the ants in broad outline.

Differences also exist. The neuter castes of termites consist of both sexes, rather than females alone as in the ants, and there are no termite "drones" that live solely for the act of mating and that are programmed for an early postreproductive death. Where ant larvae are grublike and incapable of contributing to the labor of the colony other than through the biosynthesis of nutrients, immature termites are active nymphs not radically different in form and behavior from the mature stages. In the more primitive termites, the nymphs contribute to the work of the colony, but in the higher termites (the family Termitidae), the immature forms are wholly dependent on a well-differentiated worker caste. Finally, the termites generally have a wide array of "supplementary reproductives," fertile but wingless individuals of both sexes that develop in colonies whenever the primary reproductives are removed. The nearly universal occurrence of these substitute castes provides termite colonies with a degree of resiliency leading in extreme cases to a potential immortality seldom encountered in ants and other social hymenopterans.

Caste finds a wide range of expression among the species of social bees and wasps. In the primitively eusocial halictine bees, it emerges as a mere psychological difference among morphologically similar adults, but goes on to include, in a few species, several striking forms of queen-worker dimorphism. In the honeybee species *Apis mellifera* strong morphological and physiological differences exist between queens and workers, and the caste of individuals is determined by a complex interaction between pheromone-mediated behavior on the part of nurse workers and specialized diets fed to the larvae. Next, at least one group of species of stingless bees, the genus *Melipona,* has superimposed a genetic control of caste upon the conventional physiological device employed by related groups. Most of these phylogenetic advances, with the most conspicuous exception being the invention of genetic control, have been paralleled in the evolution of the social wasps. Together the social bees and wasps differ from the ants and termites in one major respect: for some reason none of them has fashioned well-defined worker subcastes. It is true that the species with very large colonies display a division of labor comparable to that of the most advanced ants and termites. But where the division in the latter insects is based in part on physical subcastes and in part on programmed, temporal polyethism, in most bees and wasps it is based almost entirely on temporal polyethism.

The ways by which social insects communicate, like their caste systems, are impressively diverse. They include tappings, stridulations, strokings, graspings, antennations, tastings, and puffings and streakings of chemicals that evoke various responses from simple recognition to recruitment and alarm. We must add to this list other, often subtle and sometimes even bizarre, effects: the exchange of pheromones in liquid food that inhibit caste development, the soliciting and exchange of special "trophic" eggs that exist only to be eaten, the acceleration or inhibition of work performance by the presence of other colony members nearby, various forms of dominance and submission relationships, programmed execution and cannibalism, and still others.

Three generalizations are useful in gaining perspective on this subject. First, most communication systems in the social insects appear to be based on chemical signals. The known visual signals are sparse and simple. In some groups, particularly the termites and subterranean ants, they play no role in the day-to-day life of the colony. Airborne sound is only weakly perceived by social insects and has not been definitely implicated in any important communication system. Many species, however, are extremely sensitive to sound carried by the substrate, but they evidently employ it only in limited fashion, chiefly during aggressive encounters and alarm signaling. Modulated sound signals appear to play a role in recruitment in the advanced stingless bees of the genus *Melipona* and in the honeybees, which have incorporated them into the waggle dance. Touch is universally employed by insect colonies, but, with the possible exception of dominance and trophallaxis control in the vespine wasps, it has not been molded into a Morse-like system capable of transmitting higher loads of information.

In contrast, chemical signals, evoking the sensations of either odors or smells, have been implicated in almost every category of communication. In 1958 I suggested that the separation of these substances by the dissection of their glandular sources could provide the means of analyzing much social behavior that had previously seemed intractable: "The complex social behavior of ants appears to be mediated in large part by chemoreceptors. If it can be assumed that 'instinctive' behavior of these insects is organized in a fashion similar to that demonstrated for the better known invertebrates, a useful hypothesis would seem to be that there exists a series of behavioral 'releasers,' in this case chemical substances voided by individual ants that evoke specific responses in other members of the same species. It is further useful for purposes of investigation to suppose that the releasers are produced at least in part as glandular secretions and tend to be accumulated and stored in glandular reservoirs" (Wilson, 1958). With each improvement in organic microanalysis permitting the separation and bioassay of secretory substances, new evidence has been added to support this conjecture.

Pheromones, as the chemical releasers were first called by Karlson and Butenandt (1959), may be classified as olfactory or oral according to the site of their reception. Also, their various actions can be distinguished as releaser effects, comprising the classical stimulus-response mediated wholly by the nervous system (the stimulus being thus by definition a chemical "releaser" in the terminology of animal behaviorists), or primer effects, in which endocrine and reproductive systems are altered physiologically. In the latter case, the body is in a sense "primed" for new biological activity, and it responds afterward with an altered behavioral repertory when presented with appropriate stimuli. Examples of releaser pheromones include the alarm and trail substances of workers and the attractive scents of queens, while the best-understood primer pheromones include the substances secreted by queen and king termites that inhibit the development of nymphs into their own castes. A pheromone may have both releaser and primer effects: 9-keto-decenoic acid, the principal "queen substance" produced by honeybee queens, attracts males and inhibits the building of royal cells on the part of workers (releaser effect); it also inhibits the development of worker ovaries (primer effect). The sum of current evidence indicates that pheromones play the central role in the organization of insect societies.

The second generalization is that most of the communication systems have parallels in behavior patterns already present in some form or other in solitary and presocial insects. Nest building is a case in point. The primitive ants, termites, and social wasps build nests that are scarcely more complicated than those of many of their solitary relatives. The nests of primitively social bees are frequently simpler than those of their solitary relatives. Elaboration of nest structure occurred in certain phyletic lines after the eusocial state was attained, and its evolution can be easily traced. The dominance hierarchies that play a key role in bumblebee and wasp societies have a precedent in the territorial behavior of many solitary insect species, including at least a few hymenopterans. Elaborate brood care, a hallmark of higher sociality, has its precursor in progressive larval feeding in a multitude of subsocial species belonging to several insect orders.

This brings us finally to the third generalization about communication in insect societies. The remarkable qualities of social life are mass phenomena that emerge from the integration of much simpler individual patterns by means of communication. If communication itself is first treated as a discrete phenomenon, the entire subject is much more readily analyzed. To date it has been found convenient to recognize about nine categories of responses in social insects, as given in the following list:

1. Alarm
2. Simple attraction (multiple attraction = "assembly")
3. Recruitment, as to a new food source or nest site
4. Grooming, including assistance at molting
5. Trophallaxis (the exchange of oral and anal liquid)
6. Exchange of solid food particles
7. Group effect: either increasing a given activity (facilitation) or inhibiting it
8. Recognition, of both nestmates and members of particular castes
9. Caste determination, either by inhibition or by stimulation

Most of these categories have been examined elsewhere in the present book (see especially Chapters 3 and 8–10), as well as in the monographs cited earlier.

The Prime Movers of Higher Social Evolution in Insects

The single most notable fact concerning eusociality in insects is its near monopoly by the single order Hymenoptera. Eusociality has arisen at least 11 times within the Hymenoptera: at least twice in the wasps, more precisely at least once each in the stenogastrine and vespine-polybiine vespids and probably a third time in the sphecid genus *Microstigmus;* 8 or more times in the bees; and at least once or perhaps twice in the ants. Yet throughout the entire remainder of the Arthropoda, true sociality is known to have originated in only one other living group, the termites. This dominance of the social condition by the Hymenoptera cannot be a coincidence. Throughout at least the Cenozoic Era, less than 20 percent of all insect species have belonged to the order. Furthermore, eusociality is limited within the Hymenoptera to the aculeate wasps and to their immediate descendants, the ants and the bees, which together constitute no more than 50,000 estimated living species, or perhaps 6 percent of the total number of insect species in the world. This overwhelming phylogenetic bias is the most important clue we have to go on in searching for the prime movers of higher social evolution.

The tendency of aculeate Hymenoptera to evolve eusocial species can probably be ascribed in part to their mandibulate (chewing) mouthparts, which lend themselves so well to the manipulation of objects, or to the penchant of aculeate females for building nests to which they return repeatedly, or to the frequent close relationship between mother and young. These and perhaps some other biological features are prerequisites for the evolution of eusociality. But they are shared in full by many other, species-rich groups of arthropods, including the spiders, earwigs, orthopterans, and beetles, none of which, with the exception of the cockroaches that gave rise to termites, achieved full sociality. Time and again phyletic lines have pressed most of the way to eusociality, in some cases to the very threshold, and then unaccountably stopped.

At the present time the key to hymenopteran success appears to be haplodiploidy, the mode of sex determination by which unfertilized eggs typically develop into males (hence, haploid) and fertilized eggs into females (hence, diploid). Haplodiploidy is a characteristic of the Hymenoptera shared by only a few other arthropod groups (certain mites, thrips, and whiteflies; the iceryine scale insects; and the beetle genera *Micromalthus, Xylosandrus,* and, perhaps, *Xyleborus*). Two authors have independently suggested a connection between haplodiploidy and the frequent occurrence of eusociality. Richards (1965) suggested that the control which haplodiploidy grants the female over the sex of her own offspring has eased the way to colonial organization. This is undoubtedly true. The postponement of male production until late in the season, by the simple expedient of passing sperm through the spermathecal duct to meet all eggs, is a characteristic of advanced sociality, for example, in the annual halictid bees. At the same time, it is not a characteristic of many other Halictidae that are primitively eusocial—but eusocial nonetheless. In other words, sex control by the mother is a general feature of higher social evolution but not a prerequisite for the attainment of full sociality.

Hamilton (1964) created an audacious genetic theory of the origin of sociality that assigns a wholly different central role to haplodiploidy. Working from traditional axioms of population genetics, he first deduced the following principle that applies to any genotype: in order for an altruistic trait to evolve, the sacrifice of fitness by an individual must be compensated for by an increase in fitness in some group of relatives by a factor greater than the reciprocal of the coefficient of relationship (r) to that group. As explained in Chapter 4, the coefficient of relationship (also called the degree of relatedness) is the equivalent of the average fraction of genes shared by common descent; thus, in sisters r is ½; in half-sisters, ¼; in first cousins, ⅛; and so on. The following example should make the relation intuitively clearer: if an individual sacrifices its life or is sterilized by some inherited trait, in order for that trait to be fixed in evolution it must cause the reproductive rate of sisters to be more than doubled, or that of half-sisters to be more than quadrupled, and so on. The full effects of the individual on its own fitness and on the fitness of all its relatives, weighted by the degree of relationship to the relatives, are referred to as the "inclusive fitness." This measure can be treated as the equivalent of the classical measure of fitness, which takes no account of effects on relatives. Hamilton's theorem on altruism consists merely of a more general restatement of the basic axiom that genotypes increase in frequency if their relative fitness is greater.

Next Hamilton pointed out that owing to the haplodiploid mode of sex determination in Hymenoptera, the coefficient of relationship among sisters is ¾; whereas, between mother and daughter, it remains ½. This is the case because sisters share all of the genes they receive from their father (since their father is homozygous), and they share on the average ½ of the genes they receive from their mother. Each sister receives ½ of all her genes from the father and ½ from the mother, so that the average fraction (r) of genes shared through common descent between two sisters is equal to

$$\left(1 \times \frac{1}{2}\right) + \left(\frac{1}{2} \times \frac{1}{2}\right) = \frac{3}{4}$$

Therefore, in cases where the mother lives as long as the eclosion of her female offspring, those offspring may increase their inclusive fitness more by care of their younger sisters than by an equal amount of care given to their own offspring. In other words, hymenopteran species should tend to become social, all other things being equal.

This strange calculus, when extended to other kin (see Table 19-2), leads to even stranger conclusions. Consider, for example, the prediction that males should be more consistently selfish than females toward everyone else in the colony. This is expected to be the case because under all conditions except complete queen domination of the workers, a male's expected reproductive success is greater than that of a similar-sized female (see below). In order for selection to favor male altruism, such altruism would have to confer greater benefits than similar altruism by a female—an unlikely situation. Not only is this prediction met in nature; its fulfillment seems explicable only by this particular theory. The selfishness of male behavior is well known but has never before been adequately explained—in our language, the

Table 19-2 The degrees of relatedness (r) among close kin in hymenopteran groups. (Based on Trivers and Hare, 1976; modified from Hamilton, 1964.)

	Mother	Father	Sister	Brother	Son	Daughter	Nephew or niece
Female	½	½	¾	¼	½	½	⅜
Male	1	0	½	½	0	1	¼

word "drone" has come to designate any lazy, parasitic person. Not only do hymenopteran males contribute virtually nothing to the labor of the colony, but they are also highly competitive in begging food from female members of the colony and become quite aggressive in contending with other males for access to females during the nuptial flights. Nature has even provided a control experiment: termites are not haplodiploid and yet have equaled the hymenopterans in social evolution, for different reasons that will be discussed below. According to the theory termite males should not be drones. And they are not. Males constitute approximately half of the worker force, contribute an equal share of the labor, and are as altruistic to nestmates as are their sisters.

A second, nonobvious prediction of the theory is that workers of hymenopteran colonies should favor their own sons over their brothers. In other words, workers should lay unfertilized eggs and try to rear them to the exclusion of the queen's unfertilized eggs. This bias follows in part from the simple fact that females are related to their sons by a degree of $1/2$ but to their brothers by a degree of only $1/4$. It is enhanced by the relations between sister workers, in a manner to be explained shortly. Although the result seems odd, it can be reasonably well documented. Males are commonly derived from worker-laid eggs in nests of paper wasps, bumblebees, stingless bees of the genus *Trigona*, and ants of the genera *Oecophylla* and *Myrmica*. The origin of males from workers appears to be a widespread phenomenon in the social Hymenoptera. But it is not universal; in the ant genera *Pheidole* and *Solenopsis*, for example, ovaries are completely lacking in the worker caste.

A still more detailed and rigorous test of the kin-selection hypothesis can be made by examining the asymmetries within the haplodiploid system (Trivers and Hare, 1976). The test can be made objective by challenging it with a competing hypothesis. In particular, Brothers and Michener (1974) and Michener and Brothers (1974) have proposed that eusocial behavior in halictid bees evolved by the successful domination and control of some female bees over others—as opposed to "voluntary" submission of the dominated bees due to kin selection. They have noted that queens of the primitively eusocial bee *Lasioglossum zephyrum* control other adult females by a pair of simple behaviors. Other adult females are systematically nudged, an action that appears to be aggressive in nature and may have the effect of inhibiting ovarian development. The individuals most frequently nudged are the ones with the largest ovaries, and hence the greatest potential as rivals to the queens. Nudging is followed by backing, in which the nudger retreats down the nest galleries, apparently attempting to draw the other bee after it. The effect is to maneuver the follower closer to the brood cells, where it can assist in the construction and provisioning of the cells used by the queens. The bees that follow the most consistently are the ones with the smallest ovaries. It is not difficult to imagine, along with Michener and Brothers, that sterile castes can evolve if certain genotypes arise that are very powerful in controlling nestmates. Alexander (1974) has independently advocated the influence of exploitation, especially of parents over offspring, as a general factor in the social evolution of insects.

Trivers and Hare (1976) have shown how to discriminate between the kinship and exploitation hypotheses, by making use of the asymmetries in the haplodiploid system. According to the exploitation hypothesis, we expect a queen in full control of a colony to produce an equal dry weight of reproductive females (new, virgin queens) and males. This would be in accordance with the original Fisher model that predicts a maximum benefit/cost ratio when the energetic investments in the two sexes are equal, that is, when the dry weight of the queens produced is equal to the dry weight of the males produced (Chapter 15). On the other hand, kin selection in haplodiploid systems will lead to strong deviations from the 1:1 ratio. Two circumstances involving kin selection are possible:

1. *Denying the queen the production of males.* If a worker is able to assist her mother in raising the queen's daughters (and her own sisters), but lays unfertilized eggs and succeeds in having the colony raise only her own sons, she trades an average r to her own offspring of $1/2$ for an r (to sisters and sons) of $5/8$ (average of $3/4$ and $1/2$). If the other workers collaborate with the egg-laying worker, they will raise sisters and nephews and thereby trade an r to their own offspring of $1/2$ for an average of $9/16$. Finally, the queen also gains by the arrangement, because she now has daughters and grandsons at an average $r = 3/8$; whereas if the workers left and had their own offspring exclusively, the queen would have only granddaughters and grandsons at $r = 1/4$. However, the arrangement is still inferior to the one in which the workers let her have all the daughters and sons. If workers do manage to produce the males, then most of the females in the colony—the queen and the nonlaying workers—will prefer to invest the same in new queens as in males. For example, the queen will be related by $r = 1/2$ to the new queens (her daughters) and by $r = 1/4$ to the males (grandsons via laying workers); but a male is in turn twice as valuable, per unit investment, as a new queen, because he will father females related by

$r = 1$ and males (via laying workers) related by $r = 1/2$, while a new queen will (like her mother) produce females related by $r = 1/2$ and grandsons related by $r = 1/4$. Nonlaying workers also prefer equal investment: they are related to new queens by $3/4$ and to males by $3/8$, but (as just shown) a male is twice as valuable, per unit investment, as a new queen. When laying workers only produce some of the males, the situation is complicated, but Trivers and Hare (1976) have shown that the queen still prefers nearly equal investment, while nonlaying workers begin to prefer more investment in the females. The more males that come from the queen, the sharper the conflict over the ratio of investment.

2. *Allowing the queen to produce males but controlling the ratio in other ways.* Even if the queen is permitted to be the mother of all the males, the workers can still adjust the ratio to their optimum as opposed to the queen's optimum. The methods at their disposal are differential destruction according to sex of the eggs, larvae, and pupae. The evidence already exists that the rate of colony growth, in *Leptothorax* ants at least, is determined almost entirely by the workers and not the queens (see Wilson, 1974b). In the case where the queen lays all of the eggs, the workers trade $r = 1/4$ for $r = 3/4$ if they invest in a sister instead of a brother. The equilibrium ratio should be 3:1 in favor of queens (sisters) as opposed to males (brothers), since the expected reproductive success of the males will then be three times that of the queens on a per-gram basis, balancing the one-third initial investment.

In summary, the kin-selection hypothesis predicts that to the extent that workers control the reproduction of the colony—one might even say to the extent that they "exploit" the queen—the ratio of investment will fall between 1:1 and 3:1 in favor of queen production. If the mother queen is in control, that is "exploiting" the workers, the ratio should be the usual Fisherian 1:1. For various species of ants thus far measured, the ratio is significantly greater than 1:1 and in many cases it falls very close to 3:1 (Trivers and Hare, 1976). This remarkable result appears to confirm the operation of kin selection in ants as the controlling force, as opposed to individual selection leading to domination and exploitation by the queen. But a great deal more research is required before the matter can be regarded as settled.

It remains to be pointed out that although termites are not haplodiploid, they possess one remarkable feature that may provide the clue to their social beginnings: along with the closely related cryptocercid cockroaches, they are the only wood-eating insects that depend on symbiotic intestinal protozoans. As first pointed out by L. R. Cleveland (in Cleveland et al., 1934), the protozoans are passed from old to young individuals by anal feeding, an arrangement that necessitates at least a low order of social behavior. Cleveland postulated that termite societies started as feeding communities bound by the necessity of exchanging protozoans and, in a sequence that is the reverse of hymenopteran social evolution, only later evolved social care of the brood. It is not theoretically necessary to the origin of eusociality for sibs to be unusually closely bound by kinship in the hymenopteran manner. Williams and Williams (1957), in an extension of the Wright theory of group selection (1945), demonstrated that eusocial behavior, including the formation of sterile, altruistic castes, can evolve in insects if competition between groups of sibs is intense enough. The point is that the termites have gone this far. The achievement is remarkable, and biologists should continue to reflect on the conditions that made it possible.

The Social Wasps

Although only about 725 species of truly social wasps are known (see Richards, 1971), the study of their behavior has repeatedly yielded results of major interest. Four of the basic discoveries of insect sociobiology—nutritional control of caste (P. Marchal, 1897), the use of behavioral characters in studies of taxonomy and phylogeny (A. Ducke, 1910, 1914), trophallaxis (E. Roubaud, 1916), and dominance behavior (G. Heldman, 1936a,b; L. Pardi, 1940)—either originated in wasp studies or were based primarily on them. Even more important, the living species of wasps exhibit in clearest detail the finely divided steps that lead from solitary life to the advanced eusocial states (Wheeler, 1923; Evans, 1958; Evans and Eberhard, 1970).

Eusocial behavior in wasps is limited almost entirely to the family Vespidae. The only known exception is an apparently primitive eusocial organization recently discovered in the sphecid *Microstigmus comes* (Matthews, 1968a,b). In order to put these and other social hymenopterans in perspective consider the phylogenetic arrangement given in Figure 19-1 of the seven superfamilies of the aculeate Hymenoptera. The aculeates, as they are familiarly called by entomologists, include the insects referred to as "wasps" in the strict sense. Also placed in this phylogenetic category are the ants (Formicoidea), which are considered to have been derived from the scolioid wasp family Tiphiidae, and the bees (Apoidea), which are considered to have originated from the wasp superfamily Sphecoidea. The Vespoidea is comprised of three families, the Masaridae, Eumenidae, and Vespidae. These wasps are often called the

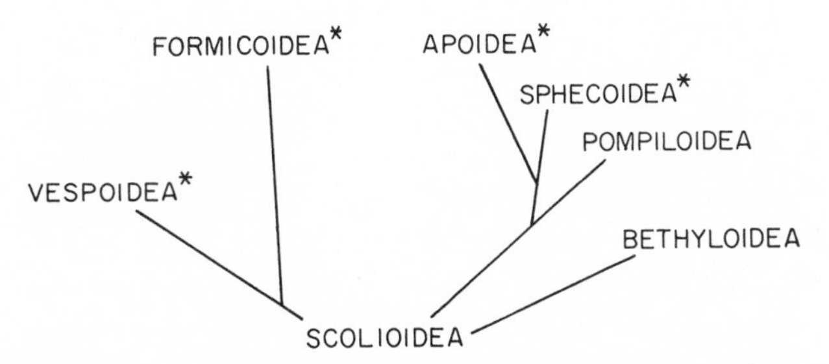

Figure 19-1 Evolution of the aculeate Hymenoptera, which include the "wasps" in the strict sense. An asterisk identifies the superfamilies in which eusocial behavior occurs, having evolved two or more times in certain cases. The Vespoidea and Sphecoidea are superfamilies of wasps; the Formicoidea are the ants and the Apoidea are the bees. (Modified from Evans, 1958.)

Diploptera because of the extraordinary ability of the adults to fold their wings longitudinally. The trait does not occur in the stenogastrine vespids or in the great majority of Masaridae, but its absence there may be a derived rather than a primitive characteristic. Vespoids are further distinguished from other wasps by the manner in which the combined median vein and radial sector slant obliquely upward and outward from the basal portion of the fore wing. Most can also be recognized at a glance by the presence of a notch on the inner margin of each eye.

Among the more primitively eusocial vespid wasps are the paper wasps of the genus *Polistes*. Various of the 150 species are found throughout the world with the exception of New Zealand and the polar regions; and in Europe and North America *Polistes* colonies outnumber those of all other social wasps combined. *P. fuscatus*, the familiar brown paper wasp of temperate North America, has been the subject of an excellent study by Mary Jane West Eberhard (1969). This species has an annual life cycle, with each colony lasting only for a single warm season. In the colder parts of the United States, the only individuals to overwinter are the queens. After being inseminated by the short-lived males in late summer and fall, they take refuge in protected places such as the spaces between the inner and outer walls of houses and beneath the loose bark of trees. In the spring the ovaries begin to develop several weeks before nest initiation, and during this time queens often aggregate in sunny places. Then, presumably when their ovaries reach an advanced stage of development, the queens begin to sit alone on old nests and future nest sites, where they react aggressively to other females who come close.

Eberhard found that nests in Michigan are usually started by a single female. Of 38 nests observed during May when they contained only from one to ten cells, 37 were attended by a single female. A single nest had two foundresses when it was less than 24 hours old. However, by the time the first brood appears in late June, the majority of the foundresses have been joined by from two to six auxiliaries—overwintered queens who for some reason have not managed to start a nest of their own. These wasps are usually subordinate in status and reproductive capacity to the foundresses. Their subordinacy is expressed behaviorally in overt ways: the auxiliaries assume submissive postures, undertake food-gathering flights and regurgitate to the dominant foundress, and defer to the foundress in egg laying. The foundress not only attempts to prevent her associates from laying eggs; she also eats their eggs when they occasionally sneak them into unoccupied cells. In time the ovaries of the subordinates regress. Marking experiments have revealed that such auxiliaries prefer to associate with foundresses who are sisters. But they move rather readily from nest to nest during the period of colony founding, and a few even attempt to start their own nests while serving as subordinates in established nests.

Through the summer, and on to the onset of the colony's decline and dissolution in early fall, the adult population grows rapidly (Figure 19-2). The complete development from egg to adult takes an average of 48 days, so that roughly three widely overlapping, complete brood sequences can be completed in a season. By the end of summer as many as 200 or more adult individuals may have been reared in a single nest, but their mortality is consistently high, and only a fraction are to be found together at a given time. The first individuals to appear are all workers, that is, females whose wings are generally less than 14 millimeters in length and whose ovaries are undeveloped. Together with the foundress, and possibly the original auxiliaries, they make up the entire adult population until the end of July. They carry on all the work of the colony: foraging for insect prey, nectar, and wood pulp for nest construction, building new cells onto the edge of the nest, and caring for the brood and nonworking adults of the colony. In early August males and "queens" (larger females capable of overwintering) begin to emerge; these purely reproductive forms come to replace the workers entirely by fall. The reproductives are essentially parasites, and as they grow in number they exert an increasingly disruptive influence on the life of the colony. The males are treated aggressively by the workers, and during the peak of male abundance in mid-August the chasing of males is a conspicuous feature of behavior on the nest.

Around the middle of August the *Polistes fuscatus* males begin to leave the nests and to cluster in cracks

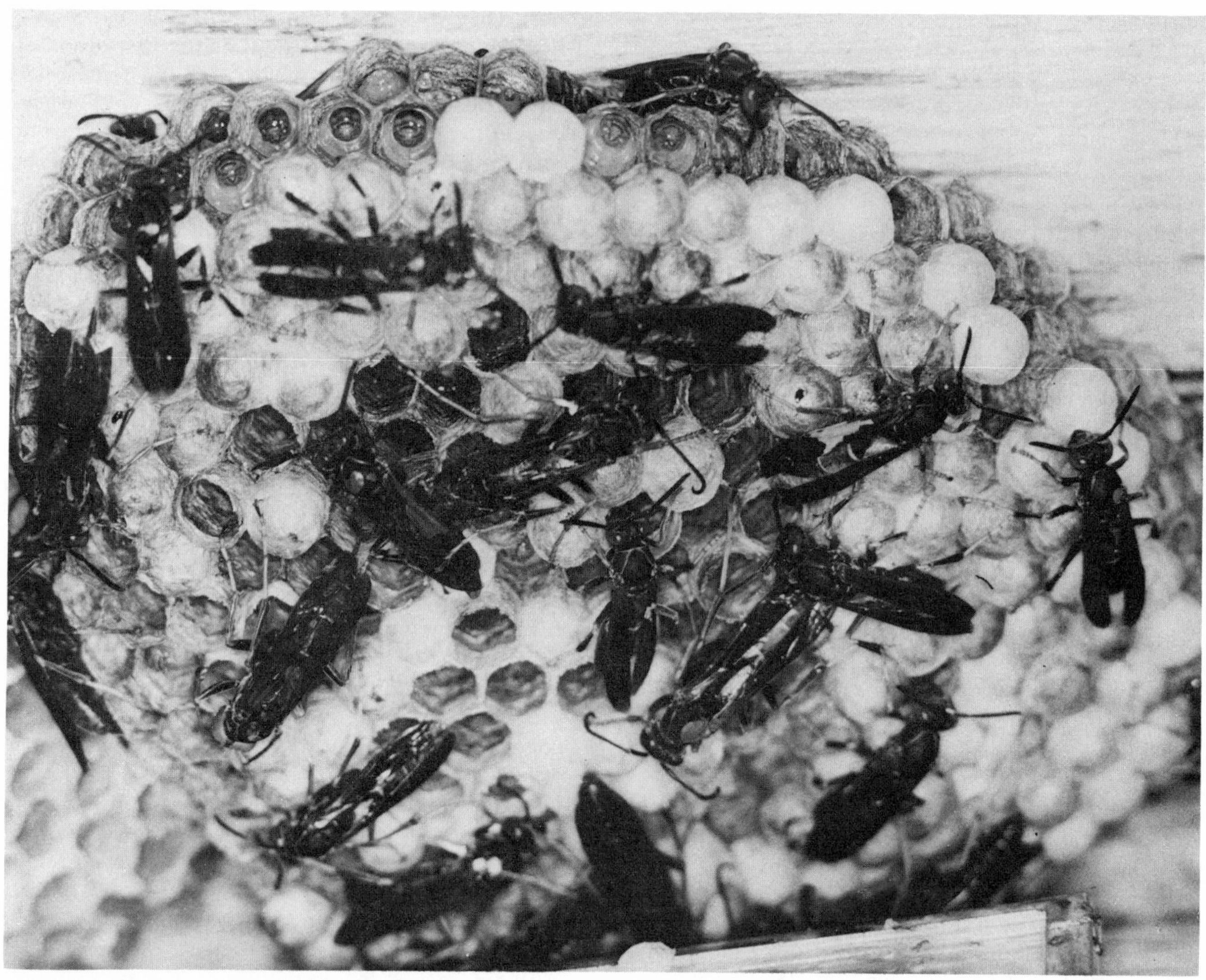

Figure 19-2 A colony of the paper wasp *Polistes fuscatus* in Michigan. The nest, which is viewed from below, consists of a single comb of brood cells fashioned from chewed vegetable fibers. Some wasps bear paint marks used by the investigator, as an aid in recognizing individuals. New cells are added on the periphery, with the result that the youngest members of the brood are located initially at this position. The heads and thoraces of mature larvae can be seen in the cells at the top of the photograph. Somewhat older pupae are located in the capped cells near the center of the comb. Finally, the center is occupied by uncapped cells from which fully adult worker wasps have emerged. Eggs have already been laid in some of the cells, initiating a new generation. (Photograph by courtesy of Mary Jane West Eberhard.)

and on old, abandoned nests. Later, females begin to join these groups. Mating takes place on or close to sunlit structures nearby or within the cavities destined to serve as hibernacula. With the onset of winter the males die off and the inseminated females hibernate singly to await the coming of spring and the renewal of the colony life cycle.

The *Polistes* life cycle illustrates one of the most important generalizations concerning the sociobiology of wasps. Since the time of von Ihering (1896) it has been noted repeatedly that the nests of tropical species tend to be founded by multiple foundresses, while those of temperate species tend to be founded by single females that overwinter in solitude. The extreme development of the first type is seen in some of the tropical Polybiini, close relatives of *Polistes,* in which new colonies are started by swarms of morphologically similar individuals who leave the old nest at about the same time. The extreme development of the second type is shown by the temperate species of Vespinae, the hornets or yellowjackets, in which new colonies are always begun in the spring by a single fecundated individual belonging to the morphologically very distinct queen caste. An extension of the generalization, but not an essential part of it, is that colonies of the tropical swarming species tend to have multiple functional queens but those of temperate species have only one queen.

The Ants

Ants are in every sense of the word the dominant social insects. They are geographically the most widely distributed of the major eusocial groups, ranging over virtually all the land outside the polar regions. They are also numerically the most abundant. At any given moment there are at least 10^{15} living ants on the earth, if we assume that C. B. Williams (1964) is correct in estimating a total of 10^{18} individual insects—and take 0.1 percent as a conservative estimate of the proportion made up of ants. The ants contain a greater number of known genera and species than all other eusocial groups combined.

The reason for the success of these insects is a matter for conjecture. Surely it has something to do with the innovation, as far back as the mid-Cretaceous period 100 million years ago, of a wingless worker caste able to forage deeply into soil and plant crevices. It must also stem partly from the fact that primitive ants began as predators on other arthropods and were not bound, as were the termites, to a cellulose diet and to the restricted nesting sites that place colonies within reach of sources of cellulose. Finally, the success of ants might be explained in part by the ability of all of the primitive species and most of their descendants to nest in the soil and leaf mold, a location that gave them an initial advantage in the exploitation of these most energy-rich terrestrial microhabitats. And perhaps this behavioral adaptation was made possible in turn by the origin of the metapleural gland, the acid secretion of which inhibits growth of microorganisms. It may be significant that the metapleural gland (or its vestige) is the one diagnostic anatomical trait that distinguishes all ants from the remainder of the Hymenoptera.

The "bulldog ants" of the genus *Myrmecia* (see Figure 19-3) are important in several respects for the study of sociobiology. They are among the largest ants, workers ranging in various species from 10 to 36 millimeters in length, but are nevertheless easy to culture in the laboratory. They are, next to *Nothomyrmecia* and *Amblyopone,* the most primitive of the living ants. The first encounter with foraging *Myrmecia* workers in the field in Australia is a memorable experience for an entomologist. One gains the strange impression of a wingless wasp just on its way to becoming an ant: "In their incessant restless activity, in their extreme agility and rapidity of motion, in their keen vision and predominant dependence on that sense, in their aggressiveness and proneness to use the powerful sting upon slight provocation, the workers of many species of *Myrmecia* and *Promyrmecia* show more striking superficial resemblance to certain of the Myrmosidae or Mutillidae than they do to higher ants" (Haskins and Haskins, 1950).

Yet so little do ants vary in the broad features of their social life cycles that *Myrmecia* can be taken as an adequate paradigm for most of this group of insects. The colonies are moderate in size, containing from a few hundred to somewhat over a thousand workers. They capture a wide variety of living insect prey, which they cut up and feed directly to the larvae. The ants are formidable predators, being able to haul down and to paralyze honeybee workers. They also collect nectar from flowers and extrafloral nectaries, which appears to be the main article in their diets when the nest is without larvae. In most species the queens are winged when they emerge from the pupae, whereas the workers are smaller and wingless—the universal condition of ants. Intermediates between the two castes normally occur in some species, and occasionally the usual queen caste has been replaced either by ergatogynes with reduced thoraces and no wings or by mixtures of ergatogynes and short-winged queens. However, these exceptions represent secondary evolutionary derivations and not the primitive states left over from the ancestral wasps. In some of the larger species, such as *M. gulosa,* the worker caste is differentiated into two overlapping subcastes. The larger workers do most or all of the foraging, while the smaller ones devote themselves principally to brood care.

Many species of *Myrmecia* engage in a spectacular mass nuptial flight. The winged queens and males fly from the nests and gather in swarms on hilltops or other prominent landmarks. As the females fly within reach they are mobbed by males, who form solid balls around them in violent attempts to copulate. After being inseminated, the queen sheds her wings, excavates a well-formed cell in the soil beneath a log or stone, and commences rearing the first brood of workers. In 1925 John Clark made the discovery, later confirmed and extended by Wheeler (1933) and Haskins and Haskins (1950), that the queens do not follow the typical "claustral" pattern of colony founding seen in higher ants. That is, they do not remain in the initial cell and nourish the young entirely from their own metabolized fat bodies and alary muscle tissues. Instead, they periodically emerge from the cells through an easily opened exit shaft and forage in the open for insect prey. This "partially claustral" mode of colony foun-

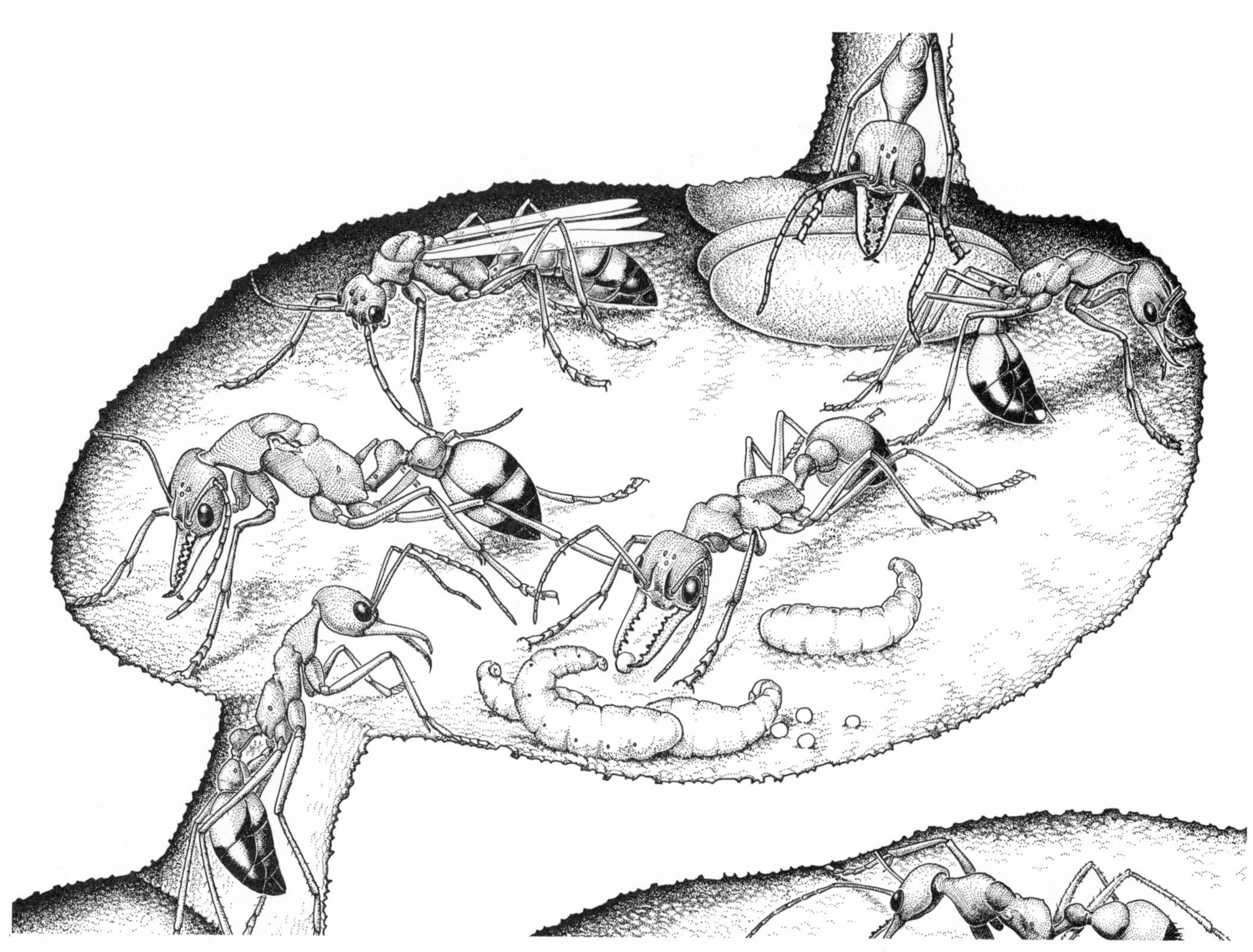

Figure 19-3 A segment of the earthen nest of a colony of primitive Australian bulldog ants (*Mymecia gulosa*). To the extreme left is the mother queen, distinguished by her larger size, heavier thorax, and three ocelli in the center of her head. Behind her is a winged male, who is her son. The other adults are workers, all daughters of the queen. To the right, one lays a trophic egg while another offers one of its trophic eggs to a wormlike larva. Spherical queen-laid eggs, which will be permitted to hatch into larvae, are scattered singly over the nest floor. To the rear of the chamber are three cocoons containing pupae of the ants. (Drawing by Sarah Landry; from Wilson, 1971a.)

dation, which is now known to be shared with most of the Ponerinae, is regarded as a holdover from a more primitive form of progressive provisioning practiced by the nonsocial tiphiid wasp ancestors.

The adaptive radiation of ants from something like the *Myrmecia* prototype has been extraordinarily full. Food specialization in many species is extreme, exemplified by the species of the ponerine genus *Leptogenys* that prey only on isopod crustaceans; by certain *Amblyopone* that feed exclusively on centipedes; by species of the ponerine genera *Discothyrea* and *Proceratium* that feed only on arthropod eggs, especially those of spiders; by certain members of the myrmicine tribe Dacetini that prey only on springtails; and by ponerines in the genus *Simopelta* and in the tribe Cerapachyini, all of which, so far as we know, prey exclusively on other ants. The majority of ant groups exhibit a high degree of variability in prey choice while a few have come to subsist on seeds. Still others rely primarily or exclusively on the anal "honeydew" excretions of aphids, mealybugs, and other homopterous insects reared in their nests. Unquestionably the most remarkable group of all are the fungus-growing ants of the myrmicine tribe Attini. The 11 genera and 200 attine species are limited entirely to the New World. They are extremely successful in the tropics—in Brazil *Atta* is the most destructive insect pest of agriculture—and a few species range as far north as New Jersey in the United States (Weber, 1972). These ants rear specialized symbiotic yeasts or fungi on organic material that they gather and carry into their nests. The substratum varies according to species: in *Cyphomyrmex rimosus*, for example, it is chiefly or entirely caterpillar feces; in *Myrmicocrypta buenzlii*, dead vegetable matter and insect corpses; and in the famous leaf-cutter ants in *Atta* and *Acromyrmex*, fresh leaves, stems, and flowers.

Nesting habits have been no less diversified. A few ant species, such as members of the genus *Atta* and the extreme desert dwellers *Monomorium salomonis* and *Myrmecocystus melliger*, excavate deep galleries and shafts down into the soil, sometimes to depths of 6 meters or more. In contrast, some arboreal members of the subfamily Pseudomyrmecinae and dolichoderine genus *Azteca* are limited to cavities of one or a very few species of plants. Some of the host plants in turn are highly specialized to house and nourish ant colonies. Experiments have shown that these plants are probably unable to survive without their insect guests. The tiny myrmicine *Cardiocondyla wroughtoni* sometimes nest in cavities left in dead leaves by leaf-mining caterpillars, while a few formicine species, *Oecophylla longinoda* and *O. smaragdina*, *Camponotus senex*, and certain species of *Polyrhachis*, have evolved the habit of using silk drawn from their own larvae to construct tentlike arboreal nests. During the nomadic phase of their activity cycles, colonies of army ants move from one site to another in search of new populations of insects and other small animals as prey. In the extreme case, exemplified by *Eciton* of the New World tropics, the ants form shelters for the queen and brood from the tangled masses of their own bodies.

The Social Bees

All the bees together constitute the superfamily Apoidea. On morphological grounds they fall closest to the sphecoid wasps, although the lack of an adequate fossil record has made it impossible to pinpoint the exact ancestral phyletic line. In a word, the Apoidea can be loosely characterized as sphecoid wasps that have specialized on collecting pollen instead of insect prey as larval food. The adults are still wasplike in that they eat nectar (and sometimes store it, in the form of honey), but, unlike the vast majority of true wasps, including all of the sphecoids, they feed their larvae on pollen or pollen-honey mixtures. Some of the eusocial species feed their larvae on specialized glandular products derived ultimately from pollen and nectar.

Eusociality has arisen at least eight times within the Apoidea by both the parasocial and subsocial routes, and presociality of nearly every conceivable degree has emerged on an uncounted number of other occasions. This prevalence and great variability of social behavior in bees provides an opportunity to study the evolution of social behavior paralleled only in the wasps, an opportunity that has only begun to be exploited.

Among the more primitively eusocial forms are the bumblebees. Comprising about 200 species of the genus *Bombus*, they are notable as social insects primarily adapted to colder climates. Most are restricted to the temperate zones of North America and Eurasia.

In these zones the life cycle of *Bombus* is annual. Only the fertilized bumblebee queens hibernate. The history of a colony unfolds in the following way. In early spring the solitary queen leaves her hibernaculum and searches on wing until she finds an abandoned nest of a field mouse or some other similarly shaped cavity, and an open but relatively undisturbed habitat such as a fallow field or abandoned garden. She pushes her way into the nest and then modifies it for her own use by constructing an entrance tunnel and lining the inner cavity with fine material teased out of the nest walls. While in the nest the queen begins to secrete wax in the form of thin plates from intersegmental glands on the abdomen. From this material she fashions the first egg shell in the form of a shallow cup set onto the floor of the nest cavity. Next she places a pollen ball into the egg cell and lays 8–14 eggs onto the surface of the ball. Finally, she constructs a dome-shaped roof of wax and other materials over the cell, so that the entire brood cell is sealed and spherical in shape. About the time the first eggs are laid, the queen also constructs a wax honeypot just inside the entrance of the nest cavity and begins to fill it with some of the nectar gathered in the field. When the first workers emerge, they assist the queen in expanding the nest and caring for additional brood as depicted in Figure 19-4.

By the end of summer the colony contains, depending on the species, from around 100 to 400 workers. As fall approaches the annual colonies produce males and queens and begin to break up. The demise of the bumblebee colonies seems to be controlled by endogenous factors. In the mild climate of northern New Zealand, species of *Bombus* introduced from Europe fly at all times of the year, and solitary queens can start nests during at least nine months of the year. Colonies sometimes overwinter and attain unusual size. In spite of this opportunity for perennial growth, however, the New Zealand colonies never return to the production of workers after they have reared queens.

Mating behavior varies greatly among the species of *Bombus*. In some, the males hover around the nest entrances and wait for the young queens to emerge.

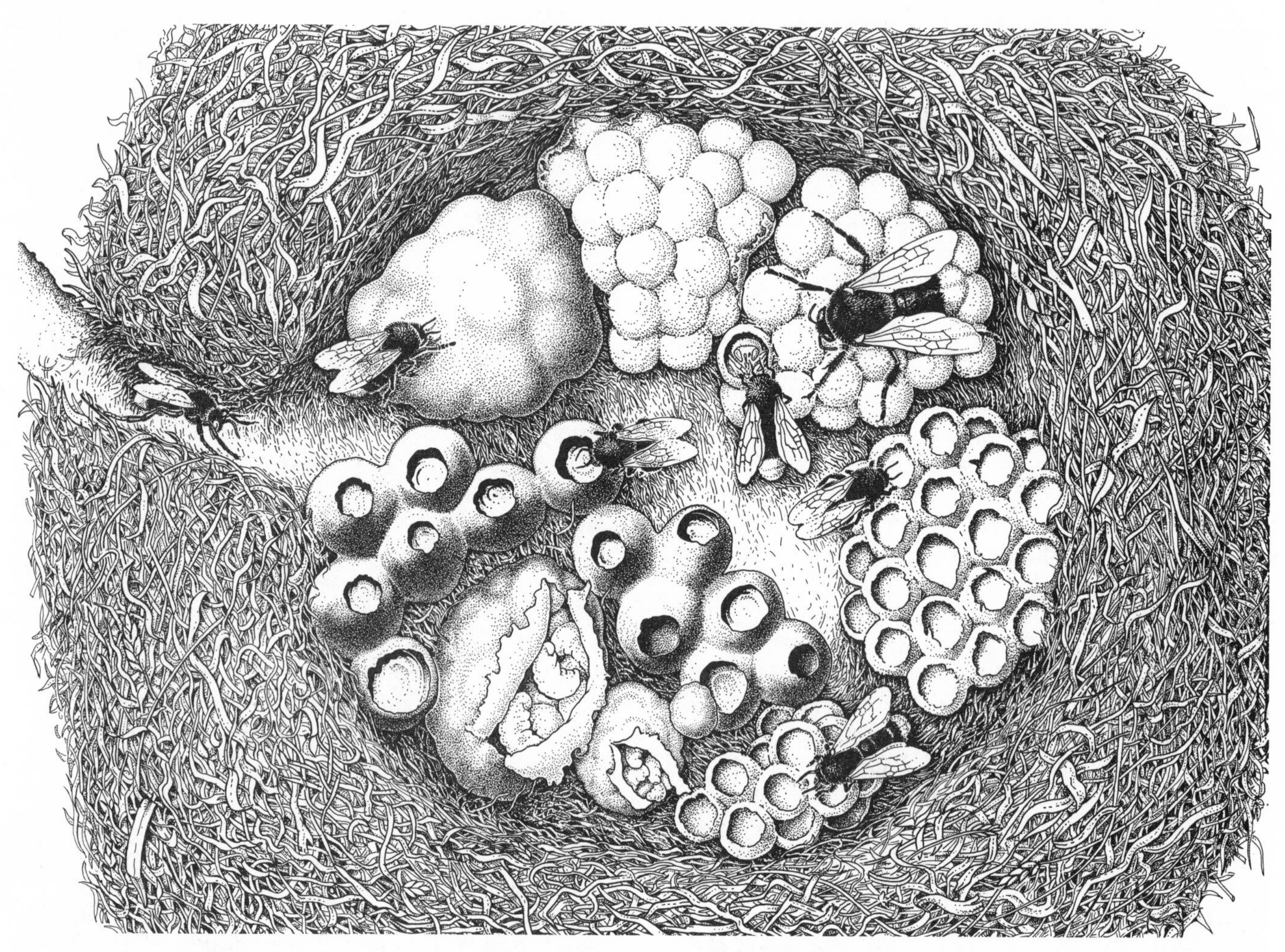

Figure 19-4 A colony of the European bumblebee *Bombus lapidarius.* The nest has been fashioned out of the center of an abandoned mouse nest in an old cultivated field. The large queen sits atop a cluster of cocoons inside which are worker pupae (one pupa has been exposed to show its position). At the upper and lower left are three communal larval cells: the waxen envelopes of the bottom two have been torn open to reveal the larvae inside. Large waxen honeypots occupy the left and center of the ensemble. At the lower right are clusters of abandoned cocoons, which are now used to store pollen. (Drawing by Sarah Landry; from Wilson, 1971a.)

In others, the male selects a prominent object, such as a flower or fence post, and alternately stands on it and hovers over it, ready to dart at any passing object that resembles a queen in flight. In a third group of species, the males establish flight paths that they mark at intervals by dabbing spots of scent from the mandibular gland onto objects along the route. The males fly around the path hour after hour, day after day, waiting for the approach of the females. After mating, the queens hibernate in specially excavated chambers in the soil, and the following spring they initiate new colonies.

Queen bumblebees differ from workers only in their larger size and the greater extent of their ovarian development, and intermediates between the two castes are common. There is also great variation in size within the worker caste. The larger workers tend to forage more, and the smaller workers spend more time in nest work. In a few species, the smallest workers do not fly and are thus bound to the nest permanently. Nest guarding occurs in some species and is usually undertaken by workers who possess better-developed ovaries.

Within the Apidae, whose species constitute the *haut monde* of the social bees, *Bombus* occupies a relatively lowly position. Its solutions to the problems of social organization are as a rule crude, and it has not achieved many of the more spectacular control mechanisms that distinguish honeybees and the meliponine stingless bees.

The common honeybee *Apis mellifera* can be taken as representative of the most advanced social bees. By the general intuitive criteria of social complexity —colony size, magnitude of queen-worker difference, altruistic behavior among colony members, periodicity of male production, complexity of chemical communication, regulation of the nest temperature and other evidences of homeostatic behavior—the honeybee is at about the level of the other highest eusocial insects, that is to say, the stingless bees, the ants, the higher polybiine and vespine wasps, and the higher termites. In one feature, the waggle dance, the species comes close to standing truly apart from all other insects. The really remarkable aspect of the waggle dance is that it is a ritualized reenactment of the outward flight to food or new nest sites; it is performed within the nest and somehow understood by other workers in the colony, who are then, and this must be counted the remarkable part, able to translate it back into an actual, unrehearsed flight of their own. A similar ability to interpret modulated symbols is evidently shared by certain meliponine bees, who transmit sound signals correlated in duration and frequency with the distance of food finds. But other cases of symbolical communication have yet to be demonstrated in the social insects.

At the risk of oversimplification, it can be said that the key to understanding the biology of the honeybee lies in its ultimately tropical origin. It seems very likely that the species originated somewhere in the African tropics or subtropics and penetrated colder climates before it came under human cultivation. Thus, unlike the vast majority of social bees endemic to the cold temperate zones, the honeybee is perennial, and, being perennial, it is able to grow and to sustain large colonies. Having large colonies, it must forage widely and exploit efficiently the flowers within the flight range of its nests; the waggle dance and the release of scent from the Nasanov gland of the abdomen are clearly adaptations to this end. Also, being ultimately tropical in origin, its colonies multiply by swarming; there is no need to have a hibernation episode in the colony life cycle as in the temperate paper wasps and bumblebees. And finally, since the queen is relieved of the necessity to overwinter and initiate colonies in solitude, she has regressed in evolution toward the role of a simple egg-laying machine, with the result that the queen and worker castes differ drastically from each other in both morphology and physiology. Within the scope of these interlocking effects are to be found just about all of the phenomena that distinguish *Apis mellifera* from the exclusively cold temperate bee species (see Figure 19-5). When we turn to the tropical faunas and consider what else has evolved to eusocial levels within the Apoidea, the contrasts are not nearly so sharp. The prevailing group of tropical eusocial bees, the Meliponini, not only resemble *Apis* in their life cycle, but are comparable to it in complexity of social organization. Of course, a great many, perhaps most, of the primitively eusocial bees exist in the tropics, but this does not affect the important generalization that the most advanced bee societies are tropical in origin.

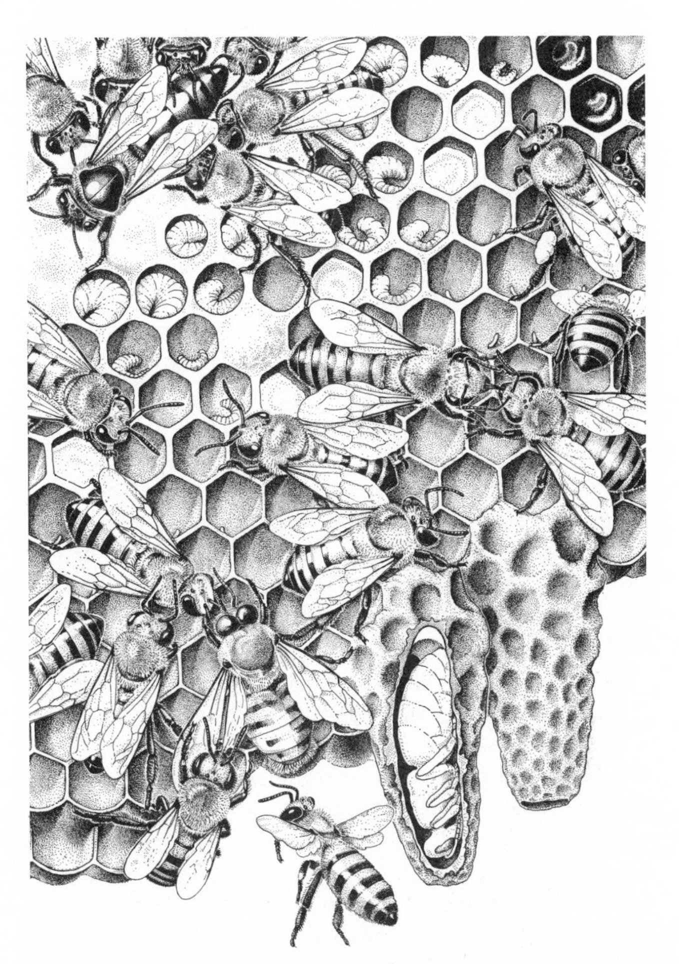

Figure 19-5 A portion of a colony of honeybees. In the upper left-hand corner the mother queen is surrounded by a typical retinue of attendants. She rests on a group of capped cells, each of which encloses a developing worker pupa. Many of the open cells contain eggs and larvae in various stages of development, while others are partly filled by pollen masses or honey (extreme upper right). Near the center a worker extrudes its tongue to sip regurgitated nectar and pollen from a sister. At the lower left another worker begins to drag a drone away by its wings; the drone will soon be killed or driven from the nest. At the lower margin of the comb are two royal cells, one of which has been cut open to reveal the queen pupa inside. (Drawing by Sarah Landry; from Wilson, 1971a.)

The Termites

Termites are almost literally social cockroaches. Detailed similarities exist in anatomy between the most primitive termite family, the Mastotermitidae, and the relatively primitive wood-eating cockroaches constituting the family Cryptocercidae. Even the intestinal microorganisms that digest cellulose are similar. Of the 25 species of hypermastigote and polymastigote flagellate protozoans found in the gut of the cockroach *Cryptocercus punctulatus,* all belong to families also found in the more primitive termites. Even one genus, *Trichonympha,* is shared. These intestinal protozoans can be successfully "transfaunated" from cockroach to termite and vice versa. It is of course too much to hope that any of the living cockroaches are really the ancestors of the termites. All known cockroaches have horny fore wings; the clear, membranous wings of the termites are more primitive. Other differences indicate that the two groups of insects originated from a common, cockroachlike ancestor. But they are not cardinal distinctions, and some entomologists have gone so far as to place termites in the same order (Dictyoptera) as the cockroaches and mantids.

Because the termites have climbed the heights of eusociality from a base extremely remote in evolution from the Hymenoptera, it is of great interest to know whether their social organization differs from hymenopteran organization in any fundamental way. Although value judgments of the degree of convergence of two radically differing stocks are difficult to make, much less to justify quantitatively, I believe the following assessment can reasonably be made. The termites have adopted mechanisms that are mostly but not entirely similar to those in the ants and other social Hymenoptera. Also, the level of complexity of termite societies is approximately the same as that in the more advanced hymenopteran societies. This remarkable fact seems to tell us that there are constraints in the machinery of the insect brain that limit not only the options of social organization but also the upper limit that the degree of organization can attain. These limits appear to have been reached between 50 and 100 million years ago in both the termites and the social Hymenoptera.

The Kalotermitidae, known as the dry wood termites, have a relatively primitive social organization. Their colonies, which rarely contain more than a few hundred individuals, live in ill-defined galleries inside the wood on which they feed. The termites rely on an intestinal flagellate fauna to digest the wood and do not utilize symbiotic fungi or store food. When the primary queens and males are lost, they are quickly replaced by secondary "neoteinics" that transform in one molt from a labile, workerlike caste called pseudergates. When present, the primary reproductives prevent the transformation of pseudergates by means of inhibitory pheromones passed out of their anuses. Soldier inhibition also occurs, but the physiological basis is not yet known. The exchange of oral and anal liquids, as well as integumentary exudates, occurs very frequently among all members of the colony. Anal exchange is essential to the transmission of flagellates to young nymphs and newly molted individuals of all ages.

It is a curious fact that most kalotermitids, as well as most other relatively primitive termite groups, are concentrated in the temperate zones. The tropics, constituting the true headquarters of the world fauna, are dominated by the "higher" termites of the family Termitidae. The majority of the termitids are soil dwellers and are responsible for most of the elaborately structured mounds that are such a conspicuous feature of the tropical landscape. Various of their species have specialized on virtually every conceivable cellulose source. To reach this food, workers extend galleries through the soil, or construct covered trailways over the surface of the ground, or even march in columns along exposed odor trails.

As an example of a relatively advanced termitid, we can take *Amitermes hastatus,* which has been studied in detail by Skaife (1954a,b; 1955). The species occurs in South Africa, in the mountains of the southwest Cape at elevations from about 100 to 1,000 meters above sea level. It nests in the sandy soil of the natural veld, throwing up conspicuous hemispherical or conical mounds constructed of a black mixture of soil and excrement. In the late summer months of February and March large numbers of white nymphs with wing pads are to be found in the larger nests. By the end of March, or April at the lat-

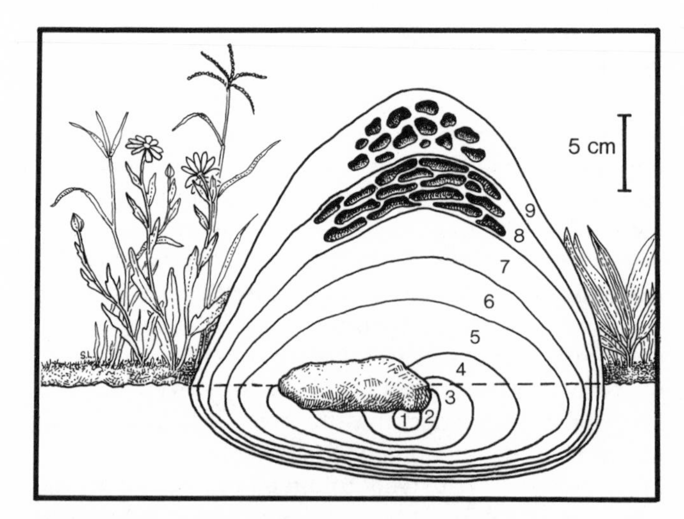

Figure 19-6 The growth of a typical mound of the South African termitid *Amitermes hastatus,* over a period of nine years. Each successive year's growth is indicated by a number. Representative outer and inner cells are shown at the top of the mound. There is no royal cell. (Based on Skaife, 1954a.)

est, these individuals have transformed into winged reproductives. For several weeks the alates wander slowly through the nest. Then, soon after the onset of the autumn rains, the nuptial flight occurs. One day between 11 o'clock in the morning and 4 o'clock in the afternoon, immediately after a ground-soaking rain and with the temperature rising, the exodus begins. The workers first excavate large numbers of tightly grouped exit holes, each about 2 millimeters in diameter, giving the apex of the mound the appearance of a coarse sieve. True to the pattern of most termite species, this is the only time the workers breach the walls of their nest and expose themselves to the outside air. Workers, soldiers, and alates boil out of the holes in a state of intense excitement, the alates fly off almost immediately, and within three or four minutes the others retreat back into the nest, plugging the exit holes after them. Most, but not all, of the alates leave in this first flight. As soon as they land, they break off their wings at the basal fracture line by swiftly pressing the wing tips to the ground.

The subsequent pairing and nest-founding behavior follows the basic sequence of termites generally. The construction of an initial nest chamber is conducted principally by the queen; sometimes the king does not assist at all. The pair remain in the incipient nest through the winter and apparently do not copulate until the arrival of warmer weather. In the spring months of October and November the queen lays the first five or six eggs. The individuals of the first brood develop into stunted workers. Soldiers make their appearance in later broods, and finally

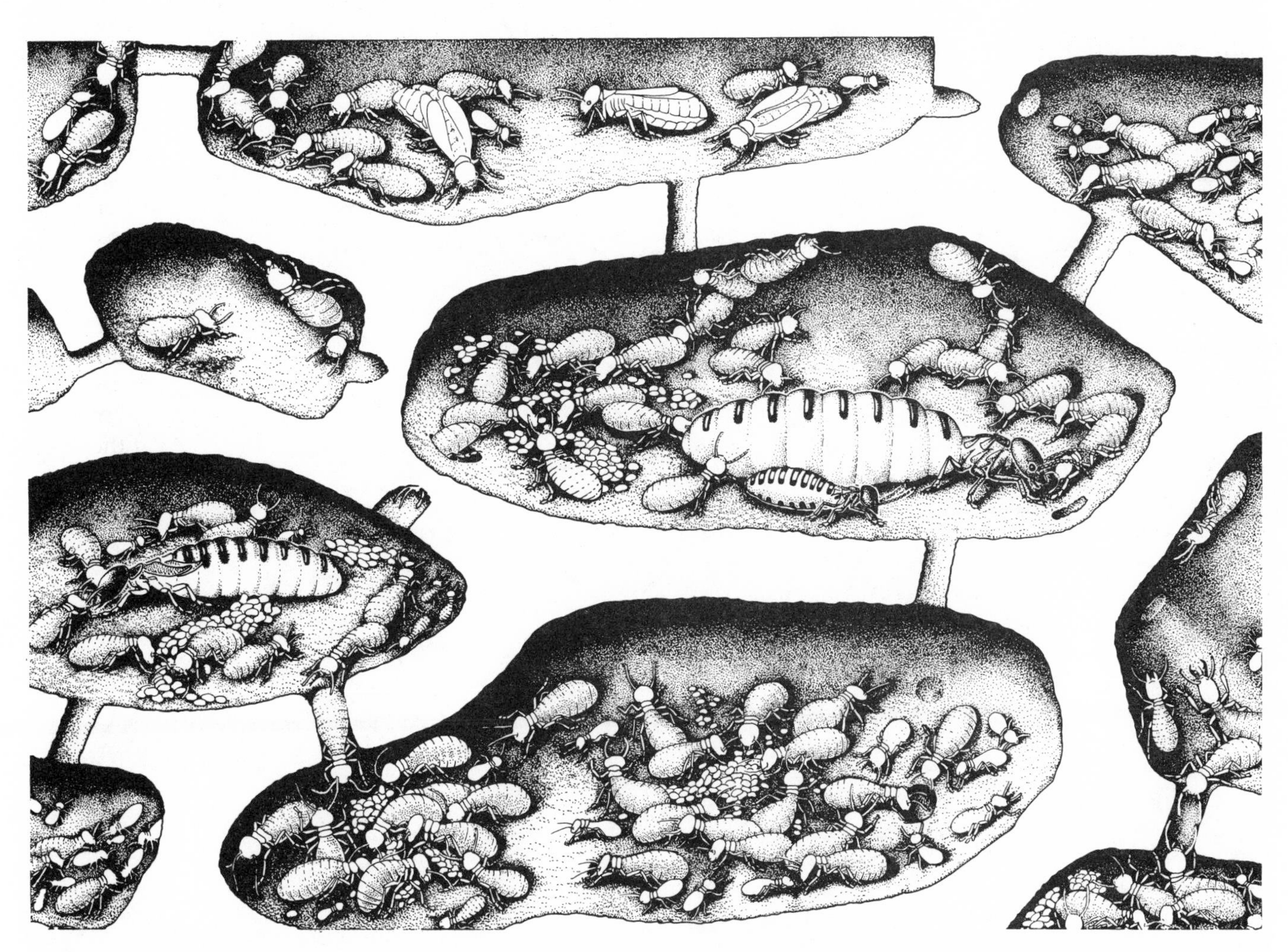

Figure 19-7 The interior of a typical nest of the higher termite *Amitermes hastatus* of South Africa. The primary queen and much smaller primary male sit side by side in the middle cell. To the lower left can be seen a secondary queen, who is also functional in this case. In the chamber at the top are reproductive nymphs, characterized by their partially developed wings. Workers attend the queens and are especially attracted by their heads, to which they offer regurgitated food at frequent intervals. Other workers care for the numerous eggs. A soldier and presoldier (nymphal soldier stage) are seen in the lower right chamber, while worker larvae in various stages of development are found scattered through most of the chambers. (Drawing by Sarah Landry; from Wilson, 1971a.)

after four years alate reproductives are produced. The growth of a typical nest is displayed in Figure 19-6. Skaife has estimated the age of some mounds of *Amitermes hastatus* to be greater than 15 years, but judging strictly from the size of the mounds, he did not believe any to be more than 25 years old. This mortal stage of individual colonies, if true, is an unexpected feature, because presumably the colonies are capable of producing secondary reproductives when the queen dies. When the primary queen does fail, the workers put her to death, apparently by licking her abrasively. As Skaife describes it, "She is surrounded by a crowd of workers, all with their mouthparts applied to her skin, and this goes on for three or four days, her body slowly shrinking until no more than the shrivelled skin is left." Secondary and tertiary queens do appear in the presence of the queen—at least sometimes (see Figure 19-7). Skaife, however, was unable to rear them in queenless colonies kept in artificial nests, and he found that only about 20 percent of the natural mounds contained them. Clearly, then, either the supplementary reproductives are rare, or appear only under special conditions, or the colonies that possess them are relatively short-lived.

CHAPTER 20

The Cold-Blooded Vertebrates

The fishes, amphibians, and reptiles are sophisticated in some of the elements of social organization but not in the ways the elements are assembled. In territoriality, courtship, and parental care, these cold-blooded vertebrates are the equal of mammals and birds, and various of their species have served as key paradigms in field and laboratory investigations. But for some reason, possibly lack of intelligence, they have not evolved cooperative nursery groups of the kind that constitute the building blocks of mammalian societies. For other reasons, possibly the lack of haplodiploid sex determination or the presence of the right ecological imperatives, they have not become altruistic enough to generate insectlike societies. Even so, the cold-blooded vertebrates offer special attractions in the study of sociobiology. As the remainder of this chapter will show, schooling in fishes has unique features which are just now beginning to be appreciated. In a sense, schooling is sociality in a new physical medium, making three-dimensional geometry important for the first time in social organization (all other societies consist of individuals arrayed on a plane). The amphibians are no less interesting but for wholly different reasons. Recent research has shown that frogs possess well-developed, highly diversified social systems parallel with those possessed by birds. Since they are phylogenetically far removed from the birds, and the traits under consideration are labile at the level of the genus and species, frogs provide us with an independent evolutionary experiment just beginning to be examined. Much the same is true of the reptiles, particularly the territorial species of lizards.

Fish Schools

In 1927 Albert E. Parr published an article that opened the subject of schooling to objective biological research. Rejecting vague earlier notions of a "social instinct," he postulated that fish schools result from the balance struck between the programmed mutual attraction and repulsion of individual fish based on the visual perception of one another. Species differ in the degree to which they are committed to schooling and in the form of the groupings. Parr identified schooling by implication as an adaptive biological phenomenon, to be analyzed like any other at both the physiological and the evolutionary level. The past 50 years have seen the accumulation of a very large amount of information on the behavioral basis of schooling and its ecological significance that confirms the validity of Parr's approach. The best recent reviews are those of Shaw (1970), who covers the large English and German literature well, and Radakov (1973), who deals with the equally large Russian literature. The Soviet studies, hitherto mostly unknown to Western zoologists, are notable for the attention they pay to the ecological significance of the schools, in line with the more modern aspects of sociobiological research being conducted in other countries.

A fish school, to cite Radakov, is "a temporary group of individuals, usually of the same species, all or most of which are in the same phase of the life cycle, actively maintain mutual contact, and manifest, or may manifest at any moment, organized actions which are as a rule biologically useful for all the members of the group." One can quarrel with this characterization, adding, deleting, or modifying the separate qualifications; but intuitive semantic argumentation has already clouded the "theory" of this subject for too long. Radakov's definition, which is close to the consensus, is more than adequate for a description of the current substantive issues.

At a distance a fish school resembles a large organism. Its members, numbering anywhere from two or three into the millions, swim in tight formations, wheeling and reversing in near unison. Either dominance systems do not exist or they are so weak as to have little or no influence on the dynamics of the school as a whole. There is, moreover, no consistent leadership. When the school turns to the right or left, individuals formerly on the flank assume the lead (see Figure 20-1). The average school size varies according to the species, as does the spacing of its members, its average velocity, and its three-dimensional shape (Breder, 1959; Pitcher, 1973). Although the fish are usually aligned with military precision while the group is on the move, they assume a more nearly random orientation while resting or feeding. Their alignments also shift in particular ways when the fish are attacked by predators. Spacing within the moving school is evidently determined to a large extent by hydrodynamic force. Individual fish tend to seek positions in which they can be as close as possible to their neighbors without suffering serious loss of efficiency due to turbulence created by the other fish. Each individual generates a trail of dying vortices behind it. In most schools the side-to-side spacing is slightly more than twice the distance from the flank of one fish to the outer edge of the trail of vortices close to the zone of their production. It is even possible for fish to coast along the edges of vortices for short distances, utilizing the energy expended by the schoolmates in front of them. But energy expenditure is not the sole consideration. Schools sometimes condense into what Breder (1965) has called "pods," in which the bodies of the members actually touch. Under some circumstances such formations can help protect individual fish from predators. Young catfish in the genus *Plotosus*, for example, mass together in a solid ball when disturbed, their sharp pectoral fins projecting out in all directions like thorns on a cactus. In general, fish tend to form the most compact schools when well fed and to thin out and become less aligned when hungry. The shift can be interpreted as the surrender of some of the advantage of predator avoidance in exchange for an increased probability of finding food.

Extensive experimentation by Shaw and others has shown that the orientation of individual fish to their school is primarily visual. Minnows, in particular *Menidia menidia* and *Atherina mochon*, display the appropriate optomotor reactions in the first few days of life and achieve parallel alignment soon afterward. *Menidia* reared in isolation still form schools but far less smoothly than those raised in groups. Jack mackerel (*Trachurus symmetricus*) adjust their velocity to match that of their schoolmates, paying closest attention to those directly off their flank (Hunter, 1969).

Why do fishes school in the first place? They are obviously able to do so only when they are not bound to a permanent territory. Species that spend part or all of their lives feeding in open water, moving opportunistically from one site to another, are the ones with the potential to evolve schooling behavior. It is possible to infer the ecological factors that "free" species from a territorial existence from comparative analyses of species varying strongly in territorial behavior. These studies show that the conditions required are those that sever wandering individuals from their home bases and make it more advantageous for the fish to migrate continuously from place to place. The reverse evolution, from nomadic schooling to the solitary occupancy of territory, is equally possible. Some species, such as the sticklebacks, alternate the two behaviors in the same life cycle, departing from the feeding schools at the onset of the breeding season to set up territories.

Nomadism is a necessary condition for the evolu-

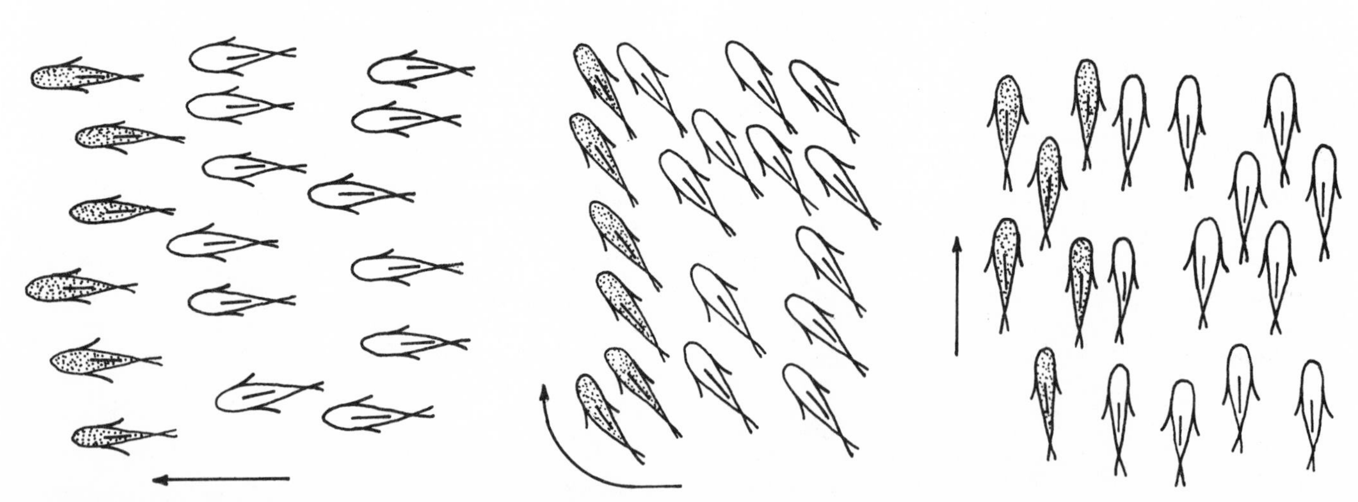

Figure 20-1 A school of fish changes its leadership when it changes direction. The leaders at the left (stippled) are shifted to the flank when the school makes a 90% turn, as shown in sequence in the center and at the right. (Modified from Shaw, 1962.)

tion of schooling behavior without being a sufficient one. Nor can any other single ecological imperative be assigned as *the* prime factor. Schooling is a highly eclectic phenomenon that originated independently in numerous phylogenetically distinct groups (Shaw, 1962). Perhaps 2,000 marine species school. Most belong to three orders that include the most abundant fish of the sea: the Clupeiformes, or herrings; the Mugiliformes, comprising the mullets, atherinid "minnows," and related forms; and the Perciformes, which include the schooling jacks, pompanos, bluefishes, mackerels, tuna, and occasional schooling snappers and grunts. A single freshwater order, the Cypriniformes, contains another 2,000 schooling species. These include the freshwater minnows and characins. The evidence is now overwhelming that a variety of advantages accrue from the behavior, and that these apply singly or in various combinations according to the species:

1. *Protection from predators.* The strongest and most distinctive changes in schooling behavior occur when the fish are confronted by a predator. Some species, such as sticklebacks and catfishes, close ranks. Most spring away as a school, often taking off at a sharp angle from the original course. Still others, such as the sand eels of the genus *Ammodytes,* flee only a short distance before regrouping to form a circle around the predator. If the larger fish charges, the *Ammodytes* wheel away to either side, then close ranks to surround it once again (Kühlmann and Karst, 1967).

2. *Improved feeding ability.* In theory at least, individual members of the school can profit from the discoveries and previous experience of all other members of the school during the search for food. This advantage can become decisive, outweighing the disadvantages of competition for food items, whenever the resource is unpredictably distributed in patches. Thus larger fish that prey on schools of smaller fish or cephalopods might be expected to hunt in groups for this reason alone. In fact, many of the largest predators, which have the least reason to fear predation themselves, do run in schools.

3. *Energy conservation.* Schooling fish can ride the edges of vortices made by other school members in front of them, thus utilizing energy that would otherwise be lost while conserving their own. It is also possible that heat is retained by the crowding, an important consideration for cold-water species.

4. *Reproductive facilitation.* Fish species that range widely through open water exist in population densities far below the densities of species that remain in special habitats on the sea bottom. Membership in schools almost certainly makes it easier for individuals to find mates or to spawn near others, but whether this advantage has been sufficient by itself to cause the evolution of schooling cannot be decided on the basis of existing evidence.

The Social Behavior of Frogs

The popular image of frogs and other anurans, held even by many zoologists, is one of simple creatures that lead a monotonous, solitary existence, interrupted only by brief bouts of courtship and spawning. In fact, the life histories of the hundreds of anuran species are enormously diverse. Although a great many do follow the basic aquatic egg-tadpole-adult sequence, the events often entail elaborate communication and even temporary social organization of breeding groups. Furthermore, profound changes in the life cycle have occurred, especially among tropical forms. Some species carry the tadpoles on the back or in the vocal pouch of the male, and others build nests in vegetation above streams so that the tadpoles can drop into the water when they hatch. Still others omit the tadpole stage altogether.

Territoriality is the rule in the families Dendrobatidae, Hylidae, Leptodactylidae, Pipidae, and Ranidae. At dusk, male bullfrogs (*Rana catesbeiana*) leave their retreats and take up calling stations in open water, where they adopt a characteristic high floating position by inflating the lungs completely with air. This exposes the brilliant yellow gular area, which may serve as a supplementary visual signal when the frogs emit their deep-throated calls. If one male approaches another closer than by about 6 meters, the

Figure 20-2 Males of a tropical frog (*Dendrobates galindoi*) wrestle for possession of a territory. In most cases spacing is maintained by repetitious calling. (From Duellman, 1966.)

Figure 20-3 A speculation concerning the social life of dinosaurs. The reconstructed habitat is a dry floodplain in Wyoming in Late Jurassic times. The large sauropod dinosaurs are *Diplodocus*. Because they were the closest ecological equivalents of modern plains ungulates and elephants, they have been arbitrarily assigned the same social organization as elephants. A herd of females and young moves in from the left, led by an old matriarch. In the foreground two males fight for dominance, neck-wrestling like giraffes and clawing at one another with their elongated middle toenails. The *Diplodocus* were among the largest of all dinosaurs; adults reached 30 meters in length, stood about 4 meters at the shoulder and could extend their heads 10 meters into the air when they reared up on their hind legs.

Here they are represented as agile open-country animals and not sluggish aquatic forms of the kind popularized in the older literature. A "pack" of flesh-eating dinosaurs, *Allosaurus*, is seen in the right background. To the left a small "flock" of bipedal dinosaurs scurries through a stand of horsetails. Other characteristic plants are the cycadeoid *Williamsonia*, the palmlike plant to the right, a true cycad just in front of it, and araucarian pines in the background. (Drawing by Sarah Landry; based on Robert T. Bakker, 1968, 1971, and personal communication, and John H. Ostrum, 1972.)

resident individual gives a sharp, staccato "hiccup" vocalization and advances a short distance toward the intruder. In most cases the intruder withdraws. If he does not, the two frogs join in battle. One may leap at or on top of the other, forcing him away. More commonly, however, the two males wrestle face to face with their arms locked around each other, kicking violently with their hind legs, until one is forced over onto his back (S. T. Emlen, 1968). Similar encounters occur in dendrobatid frogs, which defend territories on land (see Figure 20-2).

When males gather to call in choruses, they are in reality forming leks similar to those of birds. The sounds of the group carry much farther and can be sustained more continuously than those of a lone male. A member of a chorus presumably has a better chance of mating than if he were singing elsewhere, alone and in competition with the group. Choruses are typically formed by species that breed in rain pools and bodies of fresh water temporarily swollen by rain. They produce some of the most spectacular sounds of nature. The wailing of thousands of spadefoot toads (*Scaphiopus*) in a Florida roadside ditch, in the pitch-black darkness of a hot summer night, brings to mind the lower levels of the Inferno. It might be counterpointed a short distance away by the soft trilling of *Hyla avivoca* or the sharp, metallic ringing of *Pseudacris ornata*. Choruses of South American frogs sometimes consist of bedlamlike mixtures of ten or more species.

The Social Behavior of Reptiles

The behavior of reptiles has been poorly explored in comparison with that of birds and mammals, largely because reptilian behavior tends to be reduced markedly in captivity. Yet a picture of reptilian social life is beginning to emerge that indicates considerable diversity among species, with a few flashes of sophistication. The average complexity of social behavior is probably below that of the birds and mammals. That is, many more species are strictly solitary, while very few possess social systems even approaching the middle evolutionary grades of these two other vertebrate groups. Nevertheless, among the reptiles as a whole are to be found a surprising array of adaptations, some of them advanced even by mammalian standards.

Consider home range and territoriality. As in the remainder of the vertebrates, these phenomena are highly labile. Within the lizards a broad ecological basis underlying the form of land tenure can be detected. Most members of the families Agamidae, Chamaeleontidae, Gekkonidae, and Iguanidae sit and wait for their prey, often in exposed situations, and they rely heavily on optical cues. They also tend to be territorial, watching their domain constantly and warning off invaders of the same species with visual signals. In contrast, members of the Lacertidae, Scincidae, Teiidae, and Varanidae typically search for their food in places where vision is obstructed. Many root through the soil and leaf litter, depending strongly on olfactory cues. Probably as a consequence of this behavior, their home ranges overlap broadly. If territories exist, they are spatiotemporal. Considerable variation in land usage also exists within species. In both the land and marine iguanas of the Galápagos, territorial defense is limited to the breeding season. In *Uta stansburiana* it varies in form and intensity between localities. Many cases have been documented of a density-dependent shift between strict territoriality at one extreme and coexistence of adults organized into dominance hierarchies at the other. When black iguanas (*Ctenosaura pectinata*) occur in less disturbed habitats, so that individuals are able to spread out, each solitary adult male defends a well-defined territory.

Most reptilian dominance systems appear to be little more than transmuted forms of territorial hegemony, with a tyrant permitting a few subordinates to exist within his domain. The subordinates themselves are seldom organized. One exception exists in *Anolis aeneus*. Multiple females live within the territories of single males and are themselves arrayed into hierarchies consisting of at least three levels (Stamps, 1973).

Parental care is generally poorly developed in reptiles. It has been observed in both wild and captive king cobras (*Ophiophagus hannah*) by Oliver (1956). The females build nests and defend them against all intruders—making these large snakes especially dangerous to man. Since snakes are otherwise the least social of all the reptiles, this unique behavior pattern is quite remarkable and makes the king cobra one of the most promising reptile species for future field investigations. It may also be surprising to learn that the most advanced forms of parental care are practiced by the crocodilians—the alligators, crocodiles, caimans, and related forms. The females of all of the 21 living species lay their eggs in nests and defend them against intruders (Greer, 1971). The more primitive behavior is hole nesting, employed by the gharial and 7 species of crocodiles. The remaining crocodilians, including alligators, caimans, the tomistoma, and the other crocodile species, build mound nests of leaves, sticks, and other debris. The mounds serve to raise the eggs above rising water, and they probably also generate extra heat by decomposition. Just before they hatch, the young emit high-pitched croaks, particularly when disturbed by nearby movements. The response of the mother is to start tearing material off the top of the nest. Her assistance is probably essential for the escape of the young in many cases, since the outer shell of the nest is baked into a hard crust by the sun after the eggs are buried. In some species at least, the mother also leads the young to the edge of the water and protects them for varying periods afterward.

Crocodilians are archosaurs, the last surviving members of the group of ruling reptiles that dominated the land vertebrate fauna of Mesozoic times. Since they practice a relatively sophisticated form of maternal care, it is entirely reasonable to inquire whether dinosaurs, their distant relatives, lived in social groupings. A few scraps of evidence exist to indicate that this could have been the case for at least some of the species. The celebrated egg clutch of *Protoceratops* discovered by the 1922 American Museum of Natural History expedition to Mongolia appears to have been buried in a sand nest, perhaps not much different from the hole nest of modern crocodilians. More significant, however, are the dinosaur footprints and trackways that have been discovered in Texas and Massachusetts (Bakker, 1968; Ostrom, 1972). The animals that made them appear to have passed in groups, laying down tight rows of foot tracks. At Davenport Ranch, Texas, 30 brontosaurlike animals evidently progressed as an organized herd. The largest footprints occur only at the periphery of the trackway and the smallest near the center. Furthermore, the largest plant-eating dinosaurs may not have been the sluggish, stupid creatures envisioned in popular accounts of the past. Bakker (1968, 1971) has argued on the basis of very general physiological principles and new anatomical reconstructions that many of the species were erect in carriage, homoiothermal, and swiftly moving. Herds of brontosaurs and ornithiscians might have roamed the dry plains and open forests much like the antelopes, rhinoceroses, and elephants of the present time. In Figure 20-3 Sarah Landry and I have taken the maximum amount of liberty in reconstructing this scene. The animals shown are *Diplodocus*. Since they were among the largest of the dinosaurs, we have assigned them the same social organization as African elephants.

CHAPTER 21

The Birds

Birds are the most insectlike of the vertebrates in the details of their social lives. A few species, including the African weaverlike birds *Bubalornis albirostris* and *Philetairus socius,* the wattled starling *Creatophora cinerea,* the West Indian palmchat *Dulus dominicus,* and the Argentinian parrot *Myiopsitta monachus,* build communal nests in which each pair occupies a private chamber and rears its own brood. The advantage of collaborating to this extent appears to be the improvement in defense against predators (Lack, 1968). In the language of entomology, such birds form communal groups. They are closely paralleled by certain bee species. Insects in the communal stage are considered to be on the "parasocial" route of evolution that can eventually lead to full-fledged colonies with sterile castes. Communal nesting is distinguished from cooperative breeding, in which more than one pair of adults join at the same nest to rear young together. In many bird species certain of the individuals, known as helpers, assist in raising the young of others and do not lay eggs on their own. This, too, is notably insectlike. When helpers attach themselves to breeders at the very start, as in the long-tailed tit *Aegithalos caudatus,* the species resembles the "semisocial" bee and wasp species, which are also on the parasocial route. When helpers consist of offspring from former broods who remain with the parents at the nest site, a condition exemplified by the social jays, the entomologists would classify the species as "advanced subsocial," equally well along the alternate, subsocial route of evolution. Whether or not the distinction between the parasocial and subsocial states will prove to be as useful in the study of birds as it has been in entomology, it is undeniable that the presence of helpers is an advanced social trait by insectan standards. To attain the level of ants and termites all that would be needed is for a helper "caste" to evolve, whose members remain permanently in the role. So far as is known this final step has never been taken by any bird species. Bird helpers are potentially fully reproductive and ready to start their own nests whenever the opportunity arises.

The reason for the resemblances, I believe, lies in the mode of parental care shared by both groups. Birds, like the presocial and social insects, provide extended parental care requiring repeated expeditions to gather food for the young. In the great majority of cooperatively breeding bird species, as well as in those that are hosts for brood parasites, the young are altricial—helpless at birth—and must be kept in specially constructed nests. These two factors together appear to be the basis for the widespread occurrence of bonding between the two parents, a condition that is relatively infrequent in other vertebrate groups. The stage is set, first, for older siblings and other kin to improve their inclusive genetic fitness by assisting their parents, and, second, for parasitic forms to exploit the process by inserting their eggs into the nests. Parasitism may be promoted further by the relative anonymity of altricial young and the stereotypy of the communication between them and the parents.

The reader is by this time aware that the *elements* of social behavior in birds have played a large role in the development of the general principles of sociobiology. In particular, the adaptive significance of aggregations has been analyzed with special reference to bird flocks (see Chapter 3), while the study of communication has been based to a large extent on birds (Chapters 8–10). Birds provide much of the documentation of territoriality and dominance (Chapters 12–13), the endocrine control of reproductive and aggressive behavior (Chapters 7 and 11), sexual behavior with special reference to colonial nesting and polygamy (Chapter 15), and parental care (Chapter 16). Most of these components are shared by most other vetebrates. What is needed now, and the remainder of this chapter will provide, is a closer examination of the most advanced *patterns* of avian social organization, particularly those based upon cooperative breeding.

The Evolution of Cooperative Breeding

Ornithologists have gained some understanding of the ecological basis of cooperative breeding (Lack, 1968; J. L. Brown, 1974). Brown in particular has evaluated the demographic factors involved and thereby aligned this aspect of bird sociality for the first time with the theory of population biology. In Figure 21-1 I have presented a simple scheme that attempts to link together causal and intermediate factors to account for all of the known cases of cooperative breeding. Note that there appear to be two major pathways leading to the phenomenon. One has been taken by species with precocial young (able to leave the nest soon after birth) and the other by species with altricial young (helpless at birth). The form of the communal nesting also differs in an important way. In the first group of species, which includes the ostrich *Struthio camelus*, the rhea *Rhea americana*, and several species of the primitive tropical American birds called tinamous, two to four hens lay in one nest guarded by a single male. The male generally takes exclusive charge of the nest, although in the ostrich he is sometimes assisted by the dominant female. Females of the tinamou *Nothocercus bonapartei* remain in the male's territory and are prepared to lay another clutch if the first is destroyed. But in *Crypturellus boucardi* and *Nothoprocta cinerascens* they move on to lay for other males. The environmental prime movers of this peculiar form of mutual tolerance among females are unknown, but certain conditions that predispose species to acquire it in evolution are clear enough. First, the precocial nature of the young means that a single parent can look after all their needs. It is then advantageous for a male to control a territory into which he can entice multiple females. This is basic polygyny, and it is conceivable that a large variance in the quality of the male territories exists. What is peculiar, however, is the fact that individual females do not attempt to preempt access to the male and the single nest within each territory. One would expect them at least to follow the pattern common in other bird species of subdividing the male's territory and constructing private nests of their own. The reason they do not do so may well be that they are closely related. A little band of sisters could gain maximum inclusive fitness if it performed as a unit, especially if it could make use of more than one male, as in the case of the *Crypturellus* and *Nothoprocta* tinamous. The study of kinship within this small group of birds will no doubt prove instructive.

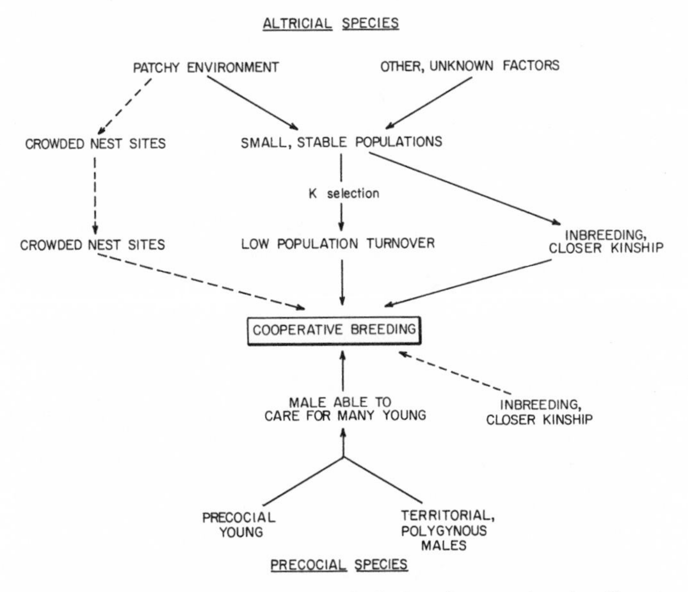

Figure 21-1 The hypothesized chain of causation leading to the evolution of cooperative breeding, the most advanced form of social behavior known in birds The solid lines indicate relationships that are documented and considered to be of crucial importance; dashed lines suggest those that are still undocumented but may play at least auxiliary roles.

The second class of cooperatively nesting birds is much larger, containing over 90 percent of the known species. As suggested in Figure 21-1, there appear to be several causative factors, which are complexly linked with one another. These factors have been elucidated separately in studies by several authors. Pulliam et al. (1972), for example, suggested that gregariousness in the yellow-faced grassquit *Tiaris olivacea* intensifies as population size is reduced and inbreeding thereby increased. In the Jamaican population, which is nearly continuous and hence relatively large, individuals are strongly territorial. The populations of Costa Rica are small and semiisolated, and the members gather in relatively large flocks. On Cayman Brac, both the population and the flocks are intermediate in size. The implication of this finding is that the smaller the effective population size, the greater the degree of kinship among interacting individuals, and the less likely they are to respond aggressively. Several zoologists who have investigated cooperatively breeding species have similarly commented on the small size and stability of the populations.

The division of species into small, semiisolated populations is itself an effect of other, more purely environmental factors. The factors have not been identified conclusively in birds, but their general nature can be guessed. First, it is evident that the spe-

Figure 21-2 The helper phenomenon in the Florida scrub jay *Aphelocoma coerulescens.* This drawing depicts a typical scene at the Archbold Biological Station in central Florida. At the nest the two parents and a yearling feed the nestlings, which are the siblings of the helpers. To the right two other helpers have spotted an indigo snake (*Drymarchon corais*), one of the dangerous predators of jay nestlings. One crouches on the ground in a threat posture. The other perches nearby in the "hiccup stance," an alarm signal that will soon alert the birds at the nest. The habitat is the floris-

tically peculiar Florida "scrub" to which *A. coerulescens* is restricted. The nest is constructed of dead twigs in a low myrtle oak (*Quercus myrtifolia*). Other typical plants include wire grass (*Aristida oligantha*), seen in the lower right-hand corner, and saw palmetto (*Serenoa repens*) and sand pine (*Pinus clausa*) in the right background. (Drawing by Sarah Landry; based on G. E. Woolfenden, 1974a,b, and personal communication.)

cies preadapted for sociality have become specialized on patchily distributed resources. The form of the patchiness has a profound influence on the kind of sociality evolved. Where the resources are fine-grained, meaning that individual birds search from one patch to another during the course of single foraging trips, the result is likely to be the formation of flocks. Food and water are the resources most likely to be fine-grained, whereas nesting and roosting sites tend to remain fixed and stable. As a result, the birds maintain individual territories where they breed, but form flocks to search for food and water. The more unpredictable these resources are in space and time, the more pronounced the optimum flocking behavior. This causal relationship appears to be the most plausible explanation for flocking by terns and some other colonial seabirds (Ashmole, 1963), starlings (Hamilton and Gilbert, 1969), and Australian desert-dwelling parrots (Brereton, 1971). If the principal resources are more nearly coarse-grained (widely distributed or large enough to require careful exploration one by one), the result is likely to be radically different. Now individuals wander less widely. Populations are restricted to limited habitats and are more prone to be both genetically isolated from one another and smaller in size. The possible result is the entrainment of events represented in Figure 21-1. Small, isolated populations tend to be stable and K-selected. K selection favors longer life, less potential fecundity, and a more prolonged parent-offspring relationship (see Chapters 4 and 16). All of these alterations in the life cycle promote cooperation and altruism during the reproductive period.

In essence, then, the evolutionary origin of cooperative breeding is viewed as depending most upon small effective breeding size. Species that practice flocking behavior might also evolve cooperative beeding if the nest sites are so restricted that population size is reduced and kinship significantly increased. However, the two processes can be entirely decoupled. Many flocking species form large breeding colonies in which average kinship is low, and intrasexual competition and aggression at the breeding sites are consequently intense. Conversely, species that live and feed in special habitats—the coarse "grains" on the environment—may utilize habitat patches so extensive, or exist in such high densities, that their population size is relatively large. They too will evolve intrasexual competition and aggression at the breeding sites. Cooperative breeding, according to the current hypothesis, depends upon the existence of a limiting resource—food, nest site, or whatever—that keeps populations small, philopatric, and isolated.

The Jays

We will turn now to a consideration of the jays, in which the evolution of cooperative breeding has been unusually well worked out by a comparison of closely related species (J. L. Brown, 1972, 1974; Woolfenden, 1974a,b). The social systems of these birds range from the basic avian pattern of pair bonding with defended territories to some of the more extreme forms of colonial nesting and cooperative breeding known in the birds.

Brown (1974) points out that social evolution within the jays has followed two alternate pathways. One culminates in the only colonially nesting species, the piñon jay. Up to several hundred adult pairs build nests in clusters and forage together in closely packed flocks that "roll" through the open woodland like groups of starlings or wood pigeons. Only the immediate vicinity of the nest is defended by the resident pair, and the colony as a whole does not protect its home range from other piñon jays. Some adults serve as helpers, but the phenomenon is not nearly so well developed as in the scrub jay and Mexican jay. A possible early intermediate stage is represented by Steller's jay, *Cyanocitta stelleri*. This species is not truly colonial, since the nests are evenly spaced owing to aggressive behavior on the part of the resident pairs. But the home ranges are left mostly undefended, and as a result they overlap widely. Steller's jay can be interpreted as a species whose territorial defense has begun to diminish, setting the stage for the clumping of the nests into a colonial system.

The second pathway culminates in cooperatively breeding species, such as the Mexican jay, the tufted jay, and the Florida scrub jay. Cooperative breeding is especially well advanced in the Florida scrub jay (*Aphelocoma coerulescens*), the behavior of which has been painstakingly studied over a period of five years by G. E. Woolfenden (1973, 1974a,b, and personal communication). This handsome blue-and-white bird is limited to the "scrub" of peninsular Florida, a highly discontinuous, sandy habitat with a distinctive flora. The eastern North American form of *A. coerulescens* is so attached to the scrub that it is the most distinctive of the Floridian birds, having never been recorded beyond the borders of the state. Its populations are very stable and bear the expected marks of prolonged K selection. Individuals are long-lived for wild birds, often surviving for eight years or more. They do not begin to breed before they are two years old. Pairs are bonded for life and occupy permanent territories. Approximately half of the breeding pairs studied by Woolfenden were assisted by helpers; the number actually fluctuated from year to year, varying from 36 to 71 percent. The helpers did not participate in nest construction or incubation, but they were active in every other activity, including defense of the territory and nest from other jays, attacks on predators, and feeding the young (see Figure 21-2).

By marking large numbers of jays and following them through the first several years of their lives, Woolfenden was able to determine the relationships and ultimate fates of the helpers. In 74 seasonal breedings (complete seasons of breeding by individual pairs), helpers assisted both parents 48 times, a father and stepmother 16 times, a mother and stepfather twice, a brother and his mate 7 times, and an unrelated pair only once. Thus the closest kin are strongly preferred—and a basis for the evolution of the altruistic trait by kin selection exists. Woolfenden was also able to demonstrate that the presence of the helpers actually increases the rate of reproduction of the breeders and hence their own inclusive fitness. Among 47 seasonal breedings by unassisted pairs observed over a period of several years, the average number of fledglings produced per pair was 1.1, while the average number of offspring still alive three months after fledging was 0.5. In contrast, 59 seasonal breedings by pairs accompanied by helpers produced an average of 2.1 fledglings per pair, and 1.3 of these were still alive three months after fledging. Hence the presence of helpers increased the replacement rate of the jay family by a factor of two to three. Woolfenden was aware that breeding pairs lacking helpers are also the youngest and least experienced and that this factor alone might account for the difference. But when experience was partialed out, by eliminating inexperienced birds, the role of helpers remained equally strong. Finally, the analysis was made still more rigorous by comparing the success of the same pairs of birds during years in which they had helpers and years in which they were alone. Again, the advantage of being helped proved clear-cut.

Surprisingly, the enhancement of reproduction does not appear to be a result of the increased feeding rate of the young. The number of helpers had no influence on the number of offspring fledged, and the weight of the fledglings had no discernible effect on their subsequent survival rate. The most likely remaining hypothesis is that helpers increase survival rates by improving communal defense against predators, notably the large snakes that are especially dangerous to the nestlings. The helpers add to the vigilance system of the family, and they assist in the mobbing of snakes that approach too closely to the nest. But whether their presence actually reduces mortality of the young birds remains to be established.

The data on the Florida scrub jay are important because little other evidence exists to indicate whether cooperative breeding really improves reproductive success, in other words, whether helpers really help. In only one additional species, the superb blue wren *Malurus cyaneus,* has such an enhancement been documented (Rowley, 1965). Gaston's study of the long-tailed tit *Aegithalos caudatus* in England indicates that helpers have no effect, while helpers of the Arabian babbler *Turdoides squamiceps* may even hinder reproduction (Amotz Zahavi, personal communication).

If in some species helpers do not help the breeders, the implication is that they themselves benefit in some way from the relationship. Woolfenden has found that this is probably the case even in the "altruistic" scrub jays. A strict dominance order exists among the nonbreeders in each family group, with males above females. If the breeder male dies or leaves, he is most likely to be replaced by the dominant helper male. It is also true that the presence of helpers results in some expansion of the territory, which may ultimately grow in area by one-third or more. When this occurs, the dominant helper male sometimes sets up a personal territory within that of the group, pairs, and begins to breed on his own. In short, the population grows to some extent by budding, with helpers being the beneficiaries. Thus the helper phenomenom could be due at least in part to individual selection. The relative contributions of individual and kin selection to the evolution of cooperative breeding in this and other bird species remain to be measured.

CHAPTER 22

Evolutionary Trends within the Mammals

The key to the sociobiology of mammals is milk. Because young animals depend on their mothers during a substantial part of their early development, the mother-offspring group is the universal nuclear unit of mammalian societies. Even the so-called solitary species, which display no social behavior beyond courtship and maternal care, are characterized by elaborate and relatively prolonged interactions between the mother and offspring. From this single conservative feature flow the main general features of the more advanced societies, including such otherwise diverse assemblages as the prides of lions and the troops of chimpanzees:

——When bonding occurs across generations beyond the time of weaning, it is usually matrilineal.

——Since the adult females are committed to an expenditure of substantial amounts of time and energy, they are the limiting resource in sexual selection. Hence polygyny is the rule in mammalian systems, and harem formation is common. Monogamous bonding is relatively rare, having arisen in such scattered forms as beavers, foxes, marmosets, titis, gibbons, and nycterid bats. In this regard the mammals depart from the largely monogamous birds. They are also distinguished by the absence of any species that shows reversal of sex roles, wherein females court the males and then leave them to care for the young.

Although these very broad generalizations can be safely made, most of the sociobiology of mammals is in an early stage of exploration, well behind that of the insects and birds. Most accounts of natural history touch on the subject only in an anecdotal fashion, especially in the case of burrowing and nocturnal species. Authors often erroneously label dense populations and breeding aggregations as "colonies" and mothers accompanied by larger offspring as "bands." The sociobiology of the majority of the families and genera of two of the greatest mammalian orders, the bats and rodents, is virtually unknown. The same is true of the marsupials, which represent a remarkable experiment in social evolution comparable to that of the eutherians.

It is difficult if not impossible to put existing information on mammalian social systems into one grand evolutionary scheme. In the first place the data are still too fragmentary. But more fundamentally, most social traits in mammals are very labile. Beyond the universal occurrence of maternal care and the most obvious immediate consequences just listed, particular features of social organization occur in a highly patchy manner within taxonomic units as small as the family and genus. The bats are an interesting case in point. Various species within the same family and even within the same genus sometimes occupy three or more "grades" of social evolution. In a given taxon some may be solitary, others monogamous or harem-forming or living in permanent groups of mixed sexes. The combination of such systems displayed by related species varies from family to family and is not easily predicted from existing knowledge of other aspects of natural history. Bradbury (1975), whose excellent review is the basis for this conclusion, cites an example from the genus *Saccopteryx* to illustrate how subtle the environmental factors can be that control social evolution. On Trinidad, groups of *S. bilineata* rest principally on the buttresses of large trees. When disturbed by a bird or mammal, the bats drop to safety in the dark recesses between the buttresses and remain motionless. This habit allows the formation of moderately large, stable aggregations and, from that, a more elaborate social system. The males keep year-round harems while competing with one another by means of complex singing, barking, gland shaking, and hovering. The related species *S. leptura* occurs in the same localities but forms groups of five individuals or less on the exposed boles of trees. When disturbed they immediately fly off to some other, usually well-known site. Evidently as a result of this escape strategy, and more particularly the small group size it necessitates, the *S. leptura* males do not form harems, and their signal repertory is smaller than that of *S. bilineata*.

A few other trends are visible within the bats as a whole. Smaller species, which have the most difficulty with thermoregulation, tend to nest more in protected sites, such as caves and the hollows of large trees. Consequently, they form larger aggregations and as a rule cluster while resting, traits that set the stage for the evolution of the more advanced forms of social organization. But the correlation is weak. One of the most spectacular lek systems occurs in the large, sexually dimorphic *Hypsignathus monstrosus*, an African bat that also rests in the open in the forest canopy. Huge permanent aggregations are formed in trees by some of the large fruit-eating bats of the family Pteropidae, evidently as a protective device against predators. Overall correlations between diet and social systems are still weaker and perhaps even nonexistent.

A relative intractability to quick evolutionary analysis also characterizes the other mammalian orders. This is very much the case in the largest and most interesting eutherian groups, including the rodents, artiodactyls, and primates. It is also true of the marsupials, which provide us with the one great evolutionary experiment outside the eutherians. In the case of artiodactyls and primates, the analysis has begun to reach sufficient depth to establish correlations at the level of the genus and species. These mammal groups will be the subjects of later special chapters. Also, it is now possible to assess to some extent the relative degree of evolutionary lability in individual social traits. In Chapter 26, the procedure will be used to help reconstruct the early evolution of man.

General Patterns

The details of mammalian social evolution are best summarized not by a general phylogenetic tree but by the Venn diagram displayed in Figure 22-1. This arrangement recognizes that the close mother-offspring relationship is universal and that the other social traits are added or subtracted at the genus or species level with relative ease. The square encloses the set of all mammalian species at a given moment in time. Evolutionary changes in individual species are depicted as tracks through time across boundaries of the subsets. Additional, smaller subsets can be delimited. Details vary, for example, in the mode of intrasexual cooperation, the degree of cohesion, and the openness of the societies. Also, most of the forms of interaction change seasonally in one species or another, and the patterns of these changes differ at the species level.

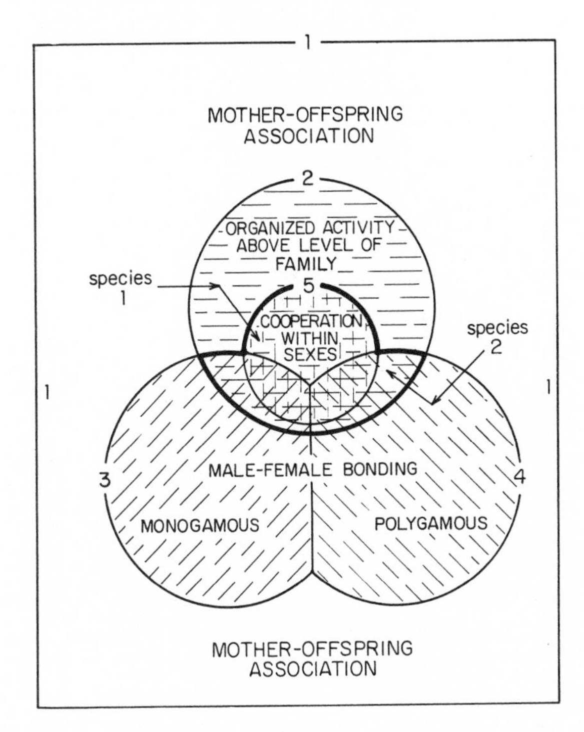

"SOLITARY"

1 only: mother-offspring groups, with males visiting only for purposes of mating.

SOCIAL

2 only: unorganized herds, schools, and other motion groups.

3 only: mated pairs, often territorial.

4 only: harems, often territorial.

5: packs, bands, troops

2+3, 2+4, and 5 (heavy outline): structured societies above the level of the family.

Figure 22-1 The diversity of mammalian social systems is represented by a Venn diagram delimiting species that possess combinations of particular social traits. The square encloses all mammalian species at a given moment of time and the circles the subsets of species that possess individual social traits. The heavy line in the center encloses the mammal species that are considered to have the most advanced social organizations. Phylogenetic trees and evolutionary grades are not used because of the great complexity of pattern created by lability in most social traits at the level of the genus and species, making generalized diagrams of this nature impracticable. However, the inferred evolution of individual species can be represented as tracks through the subsets, as illustrated by the cases of the imaginary species *1* and *2*.

Yet despite the patchy distributions of particular social systems among the species, certain broad phylogenetic trends can be detected within the Mamma-

Figure 22-2 A mob of the whiptail wallaby (*Macropus parryi*), considered to be the most social of all the marsupials. The scene is Gorge Creek, New South Wales, in the early morning. The entire group is still assembled, but the wallabies will soon begin to break up into smaller subgroups that move into more open areas to feed. There is virtually no coordination of the mob as a whole. Individuals and small groups carry on diverse activities within close proximity of one another. In the foreground various females and joeys (young animals) rest, groom, and go through the first motions of feeding. To the left two females can be seen sniffing each other for identification. To the right of center two males spar in the ritual combat that determines rank in the dominance hierarchy. A third male watches the en-

counter. To the rear of this group a male inspects the cloacal area of a female, a frequent procedure used to "check" whether females are in estrus. In the left background a courting male bends toward an estrous female while pawing up earth and grass. Three subordinate males surround the pair, each ready to commence courtship displays of his own should the dominant male leave the vicinity. The habitat is open woodland. The ground cover consists principally of grass and clover with a sprinkling of bracken ferns and thistles. (Drawing by Sarah Landry; based on J. H. Kaufmann, 1974a.)

lia as a whole and within a few of its larger orders (Eisenberg, 1966). Stem groups such as the more primitive living marsupials and insectivores tend, as expected, to be solitary. Species that forage nocturnally or underground are also predominantly solitary. As a rule the most complex social systems within each order occur in the physically largest members. This is true, for example, of the marsupials, rodents, ungulates, carnivores, and primates. Perhaps the trend partially reflects the simple fact that the largest animals forage above ground and during the day. But another significant correlate must be their increased intelligence. The biggest forms in each taxonomic group, regardless of their way of life, ordinarily possess larger, more complexly structured brains and are capable of greater feats of learning. Finally, species adapted to life in open environments are more likely to be social. For example, the most social of all marsupials are the species of wallabies and kangaroos that graze in the grasslands and open woodlands of Australia. The few rodent species known to form coteries of mixed sexes are all inhabitants of grasslands. Among the ungulates, the great herds are formed predominantly by species limited to grasslands and savannas. Although the herds are very loosely structured in most cases, those of horses, mountain sheep, elephants, and a few other forms comprise cohesive, highly organized societies.

The remainder of this chapter will be devoted to two mammalian species possessing the most advanced form of social behavior in their own groups. The whiptail wallaby is located at the apex of the marsupials. The bottle-nosed dolphin is a promising but still enigmatic species that will represent the cetaceans (orders Mysticeti and Odontoceti, including all the whales and dolphins), the least understood of all major groups of mammals. In the subsequent four chapters, which conclude the book, reviews will be presented of the ungulates, carnivores, and primates.

The Whiptail Wallaby (*Macropus parryi*)

Whiptail wallabies, which are probably the most social of all living marsupials, range from northern Queensland to northeastern New South Wales. Their preferred habitat is open *Eucalyptus* woodland with an abundance of grass. These attractive little macropods are diurnal grazers, feeding exclusively on grass and some other herbaceous plants, including ferns. A free-living population was studied by John H. Kaufmann (1974) at Gorge Creek, in the Richmond Range of New South Wales, for a period of 13 months. The animals were found to be grouped into three loosely organized "mobs" which remained stable throughout the year. Each mob contained 30 to 50 members.

Social organization was weak and individualistic. The wallaby mob was little more than a loosely structured aggregation, with individuals and small groups carrying on differing activities in close proximity to one another (see Figure 22-2). Dominance hierarchies existed among the subadults and adults. They were diffuse and infrequently expressed by the females but strongly marked, linear, and reinforced at frequent intervals in the case of the males. The aggressive behavior was highly ritualized. Its mildest form was physical displacement, in which one wallaby caused another to move aside. The first animal sometimes simply approached and sniffed the other or touched its nose, inducing it to step away. Sometimes it leaped at its opponent from behind and seized it around the middle. Displacement occurred most frequently when males contended for access to a female or when females were trying to ward off amorous males. In the case of male conflict, displacement often led to chasing and fighting. Kaufmann was impressed by the "gentlemanly" nature of fighting in the Gorge Creek population. One male usually challenged another by standing upright—the fighting position—and perhaps also by placing his paws gently on the opponent's neck or upper body. When the challenge was accepted, the fight proceeded in a predictable manner. The combatants faced each other erect, rearing to the fullest possible height by standing on their toes. They then pawed with open hands at each other's head, shoulders, chest, and throat. Sometimes pawing gave way to wrestling, in which the two males seized each other around the neck or shoulders and tried to throw each other over. In a small percentage of fights one animal kicked his opponent in the abdomen with his hind legs. This was done with far less than maximum force and usually indicated that the kicker was about to give up. Fighting clearly served to reinforce the dominance relationships among males. It was initiated in most cases by the higher-ranking animal, and was most vigorous among males of nearly equal rank. It was never observed to result in visible injury.

Superior rank paid off in access to estrous females. In the few hours in which a female remained in this condition as many as a half dozen or more males trailed her. But usually only the alpha male copulated with her. When this individual was occupied with another estrous female, the second-ranking male took his place. The shortness of duration and the unpredictability of timing of estrus resulted in a great deal of sexual searching on the part of the males. In fact, the commonest overt social interaction seen among the whiptail wallabies was the sexual "checking" of females by males. At Gorge Creek the checking procedure was initially olfactory. Typically, the male approached the female from behind and quickly sniffed at her tail, perhaps going so far as to lift the tail and to paw and lick the female's cloaca. Next the male stood in front of the female, pushing his head toward her or waving it back and forth and up and down. Sometimes he crossed his arms over his chest or placed them gently on her head or shoulders. When the female was not in estrus, her usual response was to move away or to hit at him with her paws until he retreated. As the female entered estrus, the approaches became more prolonged and persistent. At first low-ranking males trailed her, but at the peak of estrus they were invariably forced away by the highest-ranking male in the neighborhood. An exclusive consort relationship was then established that lasted from one to four days.

Although complex, ritualized behaviors have emerged in the evolution of courtship and aggression, resulting in a well-developed dominance system among the males, the wallabies apparently have not produced other modes of internal organization. Their aggregations are stable and the group home ranges are persistent and nearly exclusive. Tolerance between groups is remarkably high and may be facilitated by recognition of individuals across the groups. In this one special way whiptails resemble chimpanzees. But in other aspects their behavior is strongly individualistic, and the total social pattern over short periods of time tends to be chaotic. Although aggression plays an important role in whiptail social life, allogrooming has not evolved to a compensatory level as it has in most eutherians. Finally, the relationships between the mother and her offspring are as complex as in the social eutherians, but relationships between young age peers remain rudimentary. Social play among the peers is virtually nonexistent despite the fact that the particular adult interactions to which play normally relates are as complex and personalized in the whiptails as in other mammals.

Dolphins

Are bottle-nosed dolphins more intelligent than other animals and perhaps the equals of human beings? Do they communicate with one another by a highly sophisticated but alien language not yet decoded by human observers? These are notions widely held by the public and even among scientists, thanks to popular books such as John C. Lilly's *Man and Dolphin* (1961) and *The Mind of the Dolphin: A Nonhuman Intelligence* (1967). Given the inflated claims that authors like Lilly make about dolphins' high intelligence and "language," it is important to emphasize that there is no evidence whatever that dolphins and other delphinids are more advanced in intelligence and social behavior than other animals. In intelligence the bottle-nosed dolphin probably lies somewhere between the dog and the rhesus monkey (Andrew, 1962). The communication and social organization of delphinids generally appear to be of a conventional mammalian type.

The factual basis on which the dolphin myth was created is the undeniably large size of the animal's brain and its exceptional ability to imitate. As pointed out by McBride and Hebb (1948), the brain of the Atlantic bottle-nosed dolphin *Tursiops truncatus* is about as large as a human being's, weighing approximately 1,600 to 1,700 grams, and it is also comparable in the degree of cortical convolution. But the brain size and cortical area alone are not precise measures of intelligence. The mass of the brain tends to increase in relation to body size, so that the sperm whale, a gigantic distant relative of the dolphin, possesses a brain weighing as much as 9,200 grams. Perhaps the sperm whale is really a genius in disguise; the possibility cannot be totally discounted. But consider the brain of the elephant, which weighs approximately 6,000 grams, or four times as much as that of a human being. The behavior of this largest of land animals is now well enough known for us to be reasonably sure that its intelligence is far below the human level and probably comparable to that of the more intelligent cercopithecoid monkeys and apes. Furthermore, in signal repertory and social organization the elephant does not differ radically from other ungulates. Thus brain size, while being roughly correlated with intelligence, is not a precise measure of it.

The significant question remains, however, as to why the dolphin brain is so large. The answer may lie in the dolphin's truly remarkable imitative powers. Not only are these animals as easily trained as seals and chimpanzees to perform circus tricks; they show a strong tendency to imitate the actions of other species in the absence of reinforcement. Lilly reported that some captive dolphins answered laughter, whistles, and Bronx cheers with similar sounds. Phrases such as "One, two, three," "TRR," and "It's six o'clock" were also mimicked, albeit poorly. When Brown et al. (1966) placed an Atlantic bottle-nosed dolphin in the same tank with a Pacific *Stenella* dolphin, it made a spinning leap like that of the *Stenella* after seeing this distinctive maneuver only once. In the wild, bottle-nosed dolphins do not leap in this manner and the Atlantic specimen had

Figure 22-3 Altruistic cooperative behavior in the dolphin *Delphinus delphis.* On the left a group assists an individual that has just been struck by an electroharpoon. With blood pouring from its side, the injured animal is unable to rise to the surface to breathe and would soon drown if others did not push it upward, as shown here. Other members of the school mill nearby; to the far right can be seen two youngsters crowding close to their mothers. (Drawing by Sarah Landry; based on a written account by Pilleri and Knuckey, 1969.)

never previously had an opportunity to see a spinning leap. Tayler and Saayman (1973) have provided a remarkable series of additional examples involving captive Indian Ocean bottle-nosed dolphins (*Tursiops aduncus*). When placed in the same tank as Cape fur seals, they imitated the seals' sleeping postures and various swimming, comfort, and sexual movements. One dolphin observed a diver cleaning algae from an observation window, then proceeded to repeat the movements while making sounds similar to those made by the air-demand valve and emitting streams of bubbles resembling the diver's exhaust air. Another watched a diver remove algae from the flow of a tank with a mechanical scraper, then manipulated the tool itself well enough to loosen some of the algae, which it proceeded to eat. In this final case the dolphin displayed a capacity comparable to the learning of the use of tools by chimpanzees.

Why has the dolphin become such a superb imitator? Andrew offered a plausible hypothesis for the vocal mimicry. As in the mimicking birds and primates, the behavior might cause a convergence of signals among group members and permit individuals to recognize their own group at a distance. This faculty would seem to be especially valuable to animals that cruise the open sea at high speeds, repeatedly joining and breaking away from schools of their own species. This factor alone could account for the hypertrophy of the capacity for vocal mimicry and the enlargement of the brain. Moreover, the dependence of delphinids on echolocation for orientation and the detection of prey has preadapted them for a strongly developed system of auditory communication. The tendency to imitate movements is less easily explained. Our knowledge of the behavior of free-ranging dolphin schools is still fragmentary, although studies are currently under way (see Saayman et al., 1973). There is a possibility that the members of schools adapt quickly to special challenges from the environment, profiting from the maneuvers of the most successful individuals during escapes from predators or the pursuit of fish. Such flexibility could also lead to coordinated behavior under particular circumstances. Hoese (1971) witnessed two bottle-nosed dolphins cooperate to strand small fish by pushing waves onto the muddy shore of a salt marsh. The dolphins then rushed up onto the bank for a short distance, seized the fish, and slid back down into the water.

Another form of cooperative behavior occurs during the rescue of disabled animals. When a member of a delphinid school is harpooned or otherwise injured, the usual response of the remainder of the school is to desert the area, leaving the injured member to its fate. But occasionally the school clusters around the animal and lifts it to the surface of the water, where it can continue to breathe. Such a scene is depicted in Figure 22-3. Similar behavior has been observed in both free-ranging and captive bottle-nosed dolphins. It represents a form of altruistic behavior comparable to acts of rescue observed in wild dogs, African elephants, and baboons. However, it does not necessarily reflect a higher order of intelligence. By itself the behavior is not as complicated as say, nest building by weaverbirds or the waggle dance of honeybees. It could well represent an innate, stereotyped response to the distress of companions. Drowning that results from an incapacitating injury must be one of the chief causes of mortality among cetaceans. The automatic rescue of offspring and other relatives contributes greatly to inclusive fitness and is likely to have been fixed in the innate behavioral repertory of the species.

Allomaternal behavior is also well developed in *Tursiops truncatus* (Tavolga and Essapian, 1957). In captivity at least, older nonpregnant females associate with pregnant females and help to tend the newborn calves by swimming next to them. They sometimes lift stillborn calves to the surface in what can be interpreted as a rescue attempt.

The schools of social delphinids are highly variable in size, ranging usually from 10 to 100 members, although occasionally schools of hundreds or even thousands have been seen in some species. Free-swimming dolphins of the genus *Stenella* form three kinds of schools distinguishable on a demographic basis (Evans and Bastian, 1969). The first consists of a lone male, sometimes accompanied by a female; the second, four to eight subadult males; and the third, five to nine adult females and young. This triple array is strongly reminiscent of the herd organization of many ungulate species, in which males remain apart from nursery groups except during the breeding season. The impression is strengthened by the fact that captive Atlantic bottle-nosed dolphins (*Tursiops*) form dominance hierarchies, with a senior bull ruling over subordinate males and females. The bull is especially aggressive during the breeding season, when he bites and rakes other adults with his teeth. He controls juveniles by ramming them with his head, striking them with his flukes, and threatening them with loud percussive jaw claps. Adult females sometimes dominate both lower-ranking males and other females, although the relationships are loose and imprecise (Tavolga, 1966). The resemblance of these features to ungulate social behavior may have a basis in ecology. Like the ungulates of the savannas and semideserts, delphinids "graze" and "browse" over wide areas. Their food consists of fish rather than vegetation, but the resource is similar in being patchily distributed in space and time. Under these circumstances it is generally advantageous to move in herds of variable size, with male and nursery groups capable of independent travel.

The communication systems of delphinids appear to be of approximately the same size and complexity as those of most other species of birds and mammals. Dreher and Evans (1964) were able to distinguish 16 distinct whistles in the Atlantic bottle-nosed dolphin, 16 in the Pacific bottle-nosed dolphin, and 19 in *Delphinus bairdi*. When any two of these three species are compared in detail, 60–70 percent of the signals are found to be held in common. To these can be added several percussive sounds produced by slapping the water with the flukes and snapping the jaws (Caldwell and Caldwell, 1972; see also Busnel and Dziedzic, 1966). Thus a reasonable first estimate of the total number of signals would lie between 20 and 30, well below the total systems of the rhesus monkey, the chimpanzee, and other higher nonhuman primates but comparable to those of most other vertebrates. However, this approximation could easily be too low. Because of the difficulty of studying free-living schools, investigations of the sociobiology of dolphins and other cetaceans are still in an early stage. It is extremely difficult to mark individual animals and to follow them during their lengthy travels in the open water of the sea. Moreover, the auditory signals employed in communication may be difficult to distinguish from the ultrasonic emissions used for purposes of echolocating prey and orienting under conditions of poor visibility. Finally, the challenge of communication in a featureless space may include unique problems that have been solved by cetaceans in ways unattainable by other marine animals. In particular, there is a strong prospect that these mammals have evolved long-distance contact signaling to hold the families and schools together. Such a function has already been suggested for the elaborate songs of the humpback whale (see Chapter 9).

CHAPTER 23

The Ungulates and Elephants

The ungulates, or hoofed mammals, are a heterogeneous assemblage once placed under the single order Ungulata but now recognized to comprise two phylogenetically distinct orders: the Perissodactyla, or odd-toed ungulates, including the horses, rhinoceroses, and tapirs; and the Artiodactyla, or even-toed ungulates, including the camels, pigs, deer, giraffes, antelopes, cattle, goats, sheep, and related forms. Ungulates are vegetarians whose limbs are largely specialized for running to escape big cats and other mammalian carnivores. Hooves in effect place the feet on points, permitting a faster striding rate and greater speed when the animals are running in the open. Elephants are called subungulates, an allusion to the fact that they originated from the same ancestral stock as the ungulates. They, too, are vegetarians but rely more on sheer bulk and strength to defeat predators.

Throughout most of the Cenozoic Era, for roughly 50 million years, the perissodactyls declined while the artiodactyls and elephants expanded. Since Pleistocene times, during the past 3 million years, artiodactyls and elephants have also declined. But the artiodactyls suffered the least of the three groups, so that today they are by a wide margin the dominant large herbivores throughout the world. And the premier artiodactyls are the ruminants, the suborder Ruminantia, comprised of the deer, antelopes, cattle, sheep, and their allies. Ruminants are distinguished by their peculiar mode of digestion. Food is swallowed with a minimum of mastication and later brought up from the four-chambered stomach as a cud which is then chewed and reswallowed. A huge population of symbiotic protozoans and bacteria living in the stomach breaks down the cellulose and is then itself partially digested and absorbed. The technique of rumination, combined with the use of microorganisms, allows the animals to feed on rough forage more efficiently and has undoubtedly contributed to their general ecological success.

Two characteristics of ungulates make them especially favorable for studies of social evolution: their strong tendency toward herd formation and the relatively large number of species (187 worldwide). In the past ten years there has been a dramatic upsurge of studies of both captive and free-ranging populations. The social systems considered together present a relatively simple pattern that can be transformed with minor distortion onto a single axis, or "sociocline." At one end is the primitive state shared with most other mammalian groups, including undoubtedly the Paleocene condylarths that gave rise to the ungulates and elephants: adults live alone except to pair, and the young animals remain closely associated with their mother until they are partially or fully grown. Some ungulates, for example the moose, retain this elementary organization while forming temporary aggregations at the best feeding grounds. Other species, such as the horses, pigs, and many antelopes, have taken a major additional step. Multiple female-offspring units are allied for prolonged periods of time, during which the members recognize one another and may or may not exclude strangers. Finally, the elephants have carried this tendency to its extreme, with tight kinship groups persisting across generations. The adult cows assist others altruistically in times of stress, young are nursed indiscriminately by whichever members happen to be lactating, and a single old matriarch leads the group in every progression and formation.

In short, ungulate and elephant societies are matrifocal assemblages capable of considerable sophistication. The role of males varies greatly among species in a manner that can be viewed as orthogonal to the evolution of the female-offspring units. In all known species the males compete in some manner for access to the females. Some do so simply by territorial defense, servicing females whose home ranges overlap their own or, as in migratory populations of wildebeest, whichever females pass through their domains while in estrus. The males of at least one antelope species, the Uganda kob of Africa, concentrate their territories into leks, which are traditional sites visited by the females for the primary purpose of mating. The males of other species, in such diverse groups as the horses, camels, and bush-pigs, contend for dominance over nursery herds. The most successful individuals then enjoy unlimited access to the estrous females. In still other species, including the elk and pronghorn "antelope" of North America, males control harems only during the season of rut.

The array of social states displayed by ungulate and elephant species can also be viewed as ensembles of points in three-dimensional space, the axes of which are herd size, intensity of alliance among the adult females, and the form of attachment of males to the female herds. The correlations among these variables are weak. The ecological imperatives that determine the position of each species have been considered in a preliminary manner by Eisenberg (1966) in his review of mammalian sociobiology and investigated at greater depth in special studies of sheep, deer, and bison by Geist (1971a,b), Asiatic ungulates and elephants by Eisenberg and Lockhart (1972), and African bovids by Estes (1974), Jarman (1974), and Leuthold (1974). What the combined data of these writers reveal is the elaboration of social behavior as a consequence of the shift from the cover of closed forests to more open habitats such as savannas, grasslands, and meadows. Estes has argued that the extraordinary speciation of the African antelopes, constituting 37 percent of the entire ungulate fauna of the world, was made possible by this change. A majority of the species remaining in the forests are small and solitary, while most of those on the plains are to some extent social. The great herds of the Serengeti and other savanna reserves are the visible evidence of this correlation. It is no coincidence that the herds are pursued by lions, the most social of the cats, and by wild dogs, the most social of the canids. In a word, the wildlife spectacles of Africa are to a large degree based on social organization.

The remainder of this chapter is devoted to a series of natural history sketches of species that in aggregate range across the entire spectrum of social stages. These examples have also been selected to provide the greatest possible phylogenetic array, from the morphologically primitive tragulids to the advanced and specialized wildebeest and African elephant.

Chevrotains (Tragulidae)

The behavior of the chevrotains, or mouse deer, is of extraordinary interest because of their primitive po-

Figure 23-1 Fighting between males of the mouse deer *Tragulus napu*. (Photograph by Karen Minkowski; by courtesy of Katherine Ralls.)

Figure 23-2 Societies of the vicuña, a small member of the camel family found on the barren plains of the high Andes. A territorial family group is arrayed in the foreground. The single dominant male faces the observer in a hostile pose, making himself appear as large as possible by standing erect on a rock with his head and tail held high. Behind him his harem, composed of ten females and three young, rests and feeds. In repose the vicuña pulls the legs under the body to conserve heat; the effect is enhanced by the bib of white fur that cloaks the chest and upper parts of the forelegs. The female on the far right is "spitting," an expulsion of air that expresses irritation or hostility toward another animal. In the distance to the left can be seen a nonterritorial herd of bachelor males. Such groups form and

break up casually while wandering from place to place in search of the best forage, and their members are always ready to take over the territory of a resident male if he weakens or disappears. Some of the plants of the harsh Andean environment are shown in the foreground. They include the grasses *Calamagrostis vicunarum* (far left) and *Festuca rigescens* (center), the lettucelike malvaceous *Nototriche transandica* in the lower left-hand corner, the composites *Baccharis microphylla* and *Lepidophyllum quadrangulare* just behind and to the right of the *Nototriche*, and the legume *Astragulus peruvianus* in the lower right-hand corner. All but the *Lepidophyllum* are eaten by the vicuñas. (Drawing by Sarah Landry; based on Koford, 1957, and Franklin, 1973.)

sition within the Ruminantia, the ungulate suborder containing the largest number of species and greatest diversity of social systems. The five living species are secretive, forest-dwelling animals seldom observed in the wild, and information on their behavior is unfortunately fragmentary.

Tragulids resemble nothing so much as large mice, being in many respects convergent to the acouchis and other large forest-dwelling caviimorph rodents of South America. Their movements are swift and agile. R. A. Sterndale (1884) said that "they trip about most daintily on the tips of their toes, and look as if a puff of wind would blow them away." Males are outwardly very similar to females, except for the possession of a pair of small tusks. The social organization appears to be simple in nature. Dorst (1970) reports that water chevrotains (*Hyemoschus aquaticus*), the only African tragulids, occur alone or in pairs. Males of the Asiatic *Tragulus* evidently maintain territories or at least aggressively protect females within their domain. Katherine Ralls (personal communication) has observed captive *T. napu* males mark their enclosures with scent from the intermandibular glands. They also smear the secretion over the backs of females. Strange males have been observed to fight when placed in the same enclosures, slashing at each other with their tusks (see Figure 23-1). However, males forced to live together in groups are seldom antagonistic, a condition that Ralls believes may be due to inbreeding over several generations.

The Vicuña (*Vicugna vicugna*)

High in the central Andes of western South America, above the limit of cultivated crops, lies a treeless pastoral zone, the puna. While scanning the bleak rolling grasslands of the puna a traveler may be startled by a prolonged screech. The cry attracts his gaze to a racing troop of fifty gazelle-like mammals, bright cinnamon in color—vicuñas! As they gallop up a barren slope he sees that a single large vicuña pursues them closely. The pursuer charges at one straggler, then another, as if to nip its heels. But suddenly the aggressor halts, stands tall with slender neck and stout tail erect, stares at a line of llamas in the distance, and whistles a high trill. Then it gallops away to join a band of several vicuñas, some obviously young, which graze close by.

Thus begins Carl B. Koford's classic account (1957) of the vicuña, one of the first of the studies of a vertebrate species to integrate social behavior and ecology in the modern way. The individual described in the passage is a male that is driving a herd of bachelors out of his territory and away from his harem of females and young. *Vicugna vicugna*, a member of the camel family, is notable for the fact that its males are the most strictly territorial mammals known. It is also one of the few ungulate species in which year-round harems are the norm.

The basic social unit of these ruminants of the high Andes is the territorial family group, consisting of the male and his harem (see Figure 23-2). At Huaylarco in the Peruvian Andes Koford found such bands to contain an average of 1 male, 4 females, and 2 juveniles, with the maximum numbers of females and juveniles ranging upward to 18 and 9, respectively. At the Pampa Galeras Reserve in Peru each group occupied both a feeding territory in which it fed and reproduced and a smaller sleeping territory in which it spent the nights. The holdings of six groups studied by Franklin (1973) varied from 7 to 30 hectares and averaged 17 hectares. Sometimes roads and streambeds serve as convenient physical barriers to separate the territories, but more often there is an invisible line recognized by the vicuñas alone. The males approach to within 2 or 3 meters of one another at this line and exchange threat displays. If one steps across he is promptly chased back.

The territories are dotted with large piles of dung that are treated in a ritual manner. All members of the family visit the piles regularly to sniff them, to knead them with the forefeet, and to add feces and urine. Although the piles' function is not clear, it is likely that they are used primarily not to keep intruders out, but to keep residents in, by serving as guideposts to territorial boundaries. However the territory is identified, its ultimate function is probably defense of the food supply, which in the barren puna environment appears to be the limiting resource through most or all of the year. The food-limitation hypothesis is strengthened by the fact that the size of the territory is greatest in areas with the least density of edible plants. Indeed, the very strictness of this limiting factor may have been responsible for the evolution of the unusual territorial system of the vicuña.

The male vicuña watches his little band at all times and leads it from one point in the territory to another. In times of danger he emits the screechlike alarm trill, consisting of several descending whistles delivered over about 4 seconds, and interposes himself between the source of the threat and the group. A nonterritorial male acquires a territory by taking over an unoccupied site or land abandoned by another male. At first he grazes and rests quietly, maintaining, as it were, a low profile. Then after a few days he begins to test neighboring males with aggressive encounters. By this means he appears to learn the precise limits of the land that can be safely occupied. Having thus consolidated his position, he sets about acquiring females to build a family group.

A few females are available throughout the year in the form of solitary yearlings, as well as unattached groups and older individuals somehow deprived of mates.

A second principal social unit is the nonterritorial male herd. This bachelor group usually contains from 15 to 25 members, but the total range is from 2 to 100, and solitary wanderers are common. The males aggregate loosely, with individuals coming and going in an evidently casual manner. The all-male groups wander widely along the fringes of the family territories, pausing to rest and feed. Individuals frequently test the defenses of the territorial males by deliberate intrusions and challenges—always ready to take over on a minute's notice if the resident weakens or leaves.

The Blue Wildebeest (*Connochaetes taurinus*)

The blue wildebeest, or brindled gnu, symbolizes the almost vanished glory of African wildlife. Regarded by zoologists as an aberrant form of antelope, it has been the most abundant ungulate of the African short grasslands. Its great migratory herds, containing thousands of individuals, once stretched to the horizon. Even today as many as a million wildebeest populate the Serengeti Plains.

Under consistently favorable grazing conditions, wildebeest populations are organized into resident herds of either females and calves (nursery herds) or bachelor males. In Tanzania's Ngorongoro Crater, the nursery herds contain an average of ten members and apparently occupy a consistent home range of at most several hundred hectares. They also appear to be relatively stable in composition and closed to outsiders, since strange cows attempting to join them are often harassed. During the dry season this arrangement is altered. The nursery herds begin to aggregate in the moister, low-lying areas to which the suitable forage becomes increasingly restricted. At first the herds return at night to their home ranges, but eventually they come to spend all of their time in the new feeding areas. Simultaneously, their numbers are swelled by an influx of bachelor herds and some of the territorial males. In regions with permanently drier conditions, wildebeest exist year-round in large aggregations that migrate from one site of suitable forage to another. In fact, permanent sedentary and migratory populations are the two poles of wildebeest social organization adapted respectively to very stable and very fluctuating environments. All intermediate stages are conceivable and do in fact occur. It is also possible for sedentary populations to bud off from migratory ones when local conditions become favorable, as reported for example in Rhodesia's Wankie National Park and in southern Botswana.

Blue wildebeest are well adapted to conduct mass migrations. They travel in single file along traditional game paths, leaving behind a scent from the interdigital glands of the hooves so strong that even a human being can follow them by smell alone. Their tolerated individual distance is less than in most other ungulates, permitting them to crowd closely together when occasion demands.

Superimposed upon the mostly female-centered herd system is the territorial organization of the solitary males. Where the vicuña male defends a territory to protect his harem and its food supply, the wildebeest bull defends it solely for purposes of courtship. The defense and the sexual advertisement associated with it are conducted throughout the year and are greatly intensified during the brief rutting season. In sedentary populations territories are moderate in size, averaging about 100–150 meters in diameter. But in the midst of migratory herds, where the males must shift their location at frequent intervals, the territories are often compressed to a diameter of 20 meters or less. During the most stringent period of the dry season, when the population is constantly on the move, territorial behavior is sometimes attenuated or lost altogether for brief periods. Only about half of the adult bulls are able to maintain territories in any season; the remainder are relegated to the bachelor herds.

The territorial advertisement displays of male wildebeest are among the most elaborate and spectacular to be found within the vertebrates (see Figure 23-3). Every day wildebeest bulls engage in their unique "challenge ritual" (Estes, 1969). Each male makes the round of all his territorial neighbors, performing the ceremony with each in turn for an average of 7 minutes. At least 45 minutes of the day is required to communicate with all of them. The apparent function of the challenge ritual is to reaffirm the male's property rights while testing those of his neighbors. The territorial owner seems to recognize his neighbors personally, and the exchanges are marked by what can be reasonably called mutual restraint. Fighting is extremely rare; real combat and injury usually occur at another time—when a male is first establishing its domain, in other words when it is still a stranger. About 30 distinct behavior patterns are employed in the ritual. They are used in almost every conceivable permutation, by either partner and at any moment in the ceremony. In addition to the basic repertory of the alcelaphine antelopes—head-up posture, pawing and ritual defecating, kneeling, and horning—the displays include lateral posturing; ritualized

Figure 23-3 The social organization of the blue wildebeest, or gnu, is depicted in this scene from the Serengeti Plains of Tanzania. In the foreground two males perform the challenge ritual, a daily exchange by which each reasserts his territorial rights and challenges those of his neighbors. The male on the left cavorts in front of his rival, who has just finished digging at the ground with his horns in another display of the ritual. The bulls appear to know each other, with the result that the challenge ritual lasts an average of only 7 minutes and almost never results in real fighting and injury. The exchange may take place on any site within the territory of either bull, including the stamping ground, an example of which is seen in the center foreground. To the right, a nursery herd feeds and rests while in transit

through the territory of one of the males. Any female in estrus is likely to be mated by the resident bull at this time. Two male calves play in a manner anticipating the elaborate aggressive rituals and combat that will consume so much of their adult lives. Other solitary bulls are seen standing on their territories, one beyond each of the two displaying individuals and another pair along the territorial boundary at the acacia tree in the right background. Other nursery herds graze to the left; in the center background is a loose herd of nonterritorial bachelor males. The dominant ground vegetation is bermuda grass, a tough colonial species that thrives under heavy grazing and manuring by the wildebeest. (Drawing by Sarah Landry; based on Estes, 1969 and personal communication.)

Figure 23-4 The two basic social groups of the African elephant are illustrated in this drawing. In the left foreground a family unit faces the observer in a tightly grouped defensive formation. Alertness and mild hostility are indicated by the erect stance of the animals, the forward position of their ears, and the extension of their trunks. The family unit consists entirely of cows and young elephants in various stages of growth. The matriarch is the second individual from the left; her greater age is revealed by her more wrinkled skin and tattered ears. Several infants and young juveniles belonging to both sexes have shifted to protected positions in the rear of the group. To the far right can be seen a cow about three-quarters grown, and next to her one about half grown. The larger individuals are all adult cows. If the group is forced to retreat, the matriarch will cover the rear, continuing to face the enemy and perhaps making

mock or real charges. When she moves away she will move no faster than the smallest, slowest calf. The family units constitute the central social grouping of elephants. These associations of individual cows, which are strongly dependent on the matriarchs, often last for decades. To the rear right is a loosely organized herd of bull elephants, two of whom are contending for dominance. The ranking males become the temporary consorts of estrous females in the cow herd. In the right foreground is an acacia tree recently broken down by a feeding elephant. This form of damage thins the vegetation. In regions supporting dense elephant populations dry forests are often converted into parklands of the kind shown in this illustration. (Drawing by Sarah Landry; based on Douglas-Hamilton, 1972 and personal communication, together with photographs by Peter Haas.)

grazing and grooming; cavorting, which includes head shaking, bucking, leaping, running about, and spinning; "pretended" alarm signals, in which one or both of the animals raise their heads, look away from each other, and stamp; and urine testing.

Although the histories of individual herd members have not been worked out in detail, the general life cycle is known. Before the calving season starts, young males are excluded from the nursery herds and begin to band together. By the time of the rutting season, four months later, all but a very few of the yearling males have joined the bachelor herds. While rejection by the mother and other females is a factor, the main force causing separation is the aggression of the territorial males, who treat the yearlings as rivals. Young females are treated more tolerantly, and it is possible that membership in the nursery herds is based at least to some extent on kinship through the female lines.

The African Elephant (*Loxodonta africana*)

The largest of land mammals is also distinguished by one of the most advanced social organizations. The African elephant is remarkable in the closeness and intimacy of the ties formed between the females, the power of the matriarch who rules over the family group, and the length of time these individual associations endure. This conception of elephant sociobiology is of recent vintage. The essential facts were inferred by Laws and Parker (1968) from demographic data and confirmed in direct behavioral observations by Hubert and Ursula Hendrichs (1971), who devoted two years to studying a population on the Serengeti Plains. More recently, Iain Douglas-Hamilton (1972, 1973) conducted a four-and-a-half-year study at Lake Manyara National Park, Tanzania, during which he came to recognize 414 of the approximately 500 elephants present and recorded an impressive amount of detail on their individual relationships and the histories of family groups.

Each population of the African elephant is organized into a two- or three-tiered hierarchy of social groupings. The most important grouping directly above the individual is the *family unit,* a tightly knit herd of 10–20 females and their offspring led by a powerful matriarch. At Manyara each unit contained an average of 3.4 female-offspring groups. Members appear never to wander from their unit for distances greater than a kilometer during intervals longer than a day. The matriarch is generally the oldest individual—and hence the largest and strongest, since elephants continue growing past maturity. Because of her age, the adult females around her are likely to include not only her daughters but also her granddaughters, and the female-female bonds can be assumed to last as long as 50 years. The matriarch rallies the others and leads them from one place to another. She takes the forward position when confronting danger and the rear position during retreats. When she grows old and feeble, a younger cow gradually takes her place. But in cases where the matriarch dies suddenly the effect is traumatic. The survivors mill around her body in panic, disorganized and seemingly unable to retreat or to mount a proper defense. Hunters have long known that when the leader is shot, the rest of the herd can easily be brought down in rapid succession. For this reason, Laws and Parker recommended that when culling is made necessary by population pressure, entire family units should be removed and not just individuals picked at random.

The second level in social organization is the *kinship group,* an ensemble of family units that remain near one another and whose members show some degree of personal familiarity. It is probable that such groups originate when family units divide by fission. That the units do split is indicated by the fact that few contain more than 20 individuals, even though most are constantly growing. Douglas-Hamilton witnessed the process of division in the largest unit at Manyara, which contained 22 members. Over a period of a year 2 young cows, an adolescent female, and 2 calves moved increasing distances from the remainder of the unit. After the adolescent female calved for the first time, the two subgroups remained apart for varying periods. Then one day the matriarch led the original family unit southward for a distance of 15 kilometers, producing the first major spatial separation of the two groups. When the parental unit returned to the original site, the derivative group rejoined it and continued to stay nearby. If this case history proves to be typical, the description of such complexes as kinship groups will be justified.

It is possible that population growth, expanding the assemblages of ultra-stable female groups, produces even larger social complexes which are coextensive with the local populations themselves. Such "clans" contain perhaps 100–250 individuals. During migrations as many as a thousand elephants form mobile aggregations that are evidently unorganized above the level of the kinship group. At Manyara, family units occupied home ranges 14 to 52 square kilometers in extent, through which they wandered in irregular patterns. The ranges overlapped greatly and there was no overt territorial behavior, possibly as a result of the kinship ties of adjacent groups.

The degree of cooperation and altruism displayed

within the family group is extraordinary. Young calves of both sexes are treated equally, and each is permitted to suckle from any nursing mother in the group. Adolescent cows serve as "aunts," restraining the calves from running ahead and nudging them awake from naps. When Douglas-Hamilton felled a young bull with an anesthetic dart, the adult cows rushed to his aid and tried to raise him to his feet. Similar behavior has been observed frequently by elephant hunters. In its adaptive value the response is basically similar to the raising of injured dolphins by their fellow school members. Because of the great bulk of the animal, a fallen elephant will soon suffocate from its own weight or overheat from lying still in the sun. Finally, the matriarch is exceptionally altruistic. She is ready to expose herself to danger while protecting her herd, and she is the most courageous individual when the group assembles in the characteristic circular defense formation (see Figure 23-4).

While still in the company of their mothers, young bulls begin rushing at one another in mock charges and play-fighting. In adolescence they begin to be pushed away by the cows, and at the age of 13 years, when almost grown, they are repeatedly chased away until they leave altogether. Adult males live alone or in loose bands and disperse more widely than the females. When in groups they compete for position in a dominance hierarchy, with the outcome usually being settled on the basis of size. The struggles become most strenuous in the presence of estrous females, but even then they seldom result in serious injury. In general the males appear to be sociobiologically less complex than the females, in a way that reflects their simpler, more selfish reproductive roles.

CHAPTER 24

The Carnivores

Among the mammalian orders the carnivores are surpassed only by the primates in the intricacy and variety of their social behavior. A majority of the 253 living species, which include dogs, cats, bears, raccoons, mongooses, and related forms, are wholly "solitary." This means that a society is comprised exclusively of the mother and her unweaned young, and adult males and females associate only during the breeding season. From this base several forms of more complex organization have evolved. One grade commonly encountered, for example in the jackals, raccoon dogs, foxes, and some mongooses, is characterized by pair bonding: the male remains with the female for extended periods of time and assists in some manner with the care and protection of the young. The coati *Nasua narica* represents another grade, distinguished by bands of females and offspring that are accompanied by males during the mating season. Many mongoose species possess a still higher form of organization, in which families headed by bonded male-female pairs cooperate during hunting. Sea otters display still another kind of organization. True to their marine environment, they gather like seals at safe places to live in loosely organized herds. There the males fight among themselves, and courtship and mating take place. Lions, the only cats with an advanced form of social organization, form prides of females to which one or two dominant males attach themselves in a nearly parasitic existence. Finally, at what might be called the summit of carnivore social evolution, packs of wolves and African hunting dogs display degrees of coordination and altruism attained elsewhere only by social insects and a few of the Old World monkeys and apes.

Social behavior is diversified not only within the Carnivora as a whole but also within single families and genera. The high evolutionary lability of individual social traits is comparable to that seen elsewhere in the mammals, making it difficult to represent trends by conventional phylogenetic diagrams. The carnivores are more social as a whole than the great majority of other mammalian orders. Not only are a higher percentage of the species organized above the elementary female-offspring unit, but more are in or near the highest evolutionary grades. But an even greater interest lies in the fact that most of carnivore social behavior serves to increase the efficiency of predation. This peculiarity has two consequences. First, in accordance with the ecological efficiency rule, carnivores live in far less dense populations than herbivores and their home ranges are correspondingly larger. Consequently, territories are spatiotemporal and in some cases consist of little more than broadly overlapping networks of traplines marked by scent posts. Second, being at the top of the energy pyramid, the largest carnivores are not themselves subject to significant predation. Lions, tigers, and wolves are the premier "top carnivores" usually cited by ecologists to illustrate this category. They present the results of a significant evolutionary experiment. Their social adaptations are almost certain to be keyed primarily or exclusively to hunting prey, and as such they can be contrasted profitably with the adaptations of rodents, antelopes, and other herbivores whose social systems represent to some extent devices for avoiding these same predators.

The species to be described in the following sections represent the best-studied paradigms of several of the carnivore social grades. Because some of the species concerned are also "big game" and popular zoo animals, interest in them has been more intense and field studies more careful than usual. Zoologists are consequently in a better position to consider the ecological basis of their social evolution.

The Black Bear (*Ursus americanus*)

Bears have long been considered to be exclusively solitary. In an admirable field study conducted in northern Minnesota, L. L. Rogers (1974) showed that although this is approximately the case in the American black bear, individual relationships are far more intimate and prolonged than had been suspected. In brief, females depend on the exclusive occupancy of feeding territories to breed, and in this sense they are solitary. But they also permit their female offspring to share subdivisions of the territories and bequeath their rights to these offspring when they move away or die. In order to learn these facts, Rogers trapped and tagged 94 individuals over a four-year period. With the aid of radio-telemetry he was able to trace the histories of 7 female cubs from birth to maturity.

During the mating season, from mid-May to late July, adult females defend exclusive territories, which in Minnesota average 15 square kilometers and range from 10 to 25 square kilometers in extent. There appears to be a clear cut-off point below which reproduction becomes difficult. Two females possessing territories of only 7 square kilometers did not produce litters, while a third left the area after having a single cub. As the end of the summer approached, aggressiveness toward intruders waned, even though most of the females remained within their territories.

Nine families monitored by Rogers broke up during the first three weeks of June, when the cubs were 16 to 17 months old. Each of the female yearlings then remained in the mother's territory, utilizing a subdivision of her own for a period of at least two years. In one case four young females lived close to older females that were probably siblings from previous litters. The ranges of both the mother and the young females tended to remain separate, despite the fact that the entire ensemble represented the mother's original mating territory. When a mother bear was killed, one of her daughters took sole possession of a 15-square-kilometer sector of the territory. She gave birth to a litter in the following winter and raised it in the inherited area. In another case a three-year-old female became the exclusive occupant of the eastern portion of her mother's territory when the latter shifted her site 2.4 kilometers to the west. Her sister, who acquired the smaller western portion, grew more slowly and failed to produce a litter. The mother made the move in the first place to occupy the former territory of a deceased neighbor. Her presence caused the neighbor's three-year-old daughter to move into the western half of the dead bear's former territory. The displaced daughter shared this portion with a five-year-old, who was probably a sibling from a previous litter. She was dominated by the older bear and did not reproduce the following winter.

Male black bears take no part in this inheritance system. They disperse from the maternal territories as subadults. During the mating season the fully mature males enter the male territories and displace one another by aggressive interactions, especially when they meet in the immediate vicinity of the females. Later, as their testosterone levels drop, they withdraw from the females and assemble in peaceful feeding aggregations wherever the richest food supplies are to be found. In the late fall they return to the female territories to den.

The Lion (*Panthera leo*)

To the zoocentric human mind the lion has long enjoyed an exalted status: king of beasts, symbol of the sun, even animal god. The Egyptian pharaoh Rameses II took lions with him into battle, and kings from Amenhotep II to Saint Louis have traditionally hunted them for sport. But only within the last ten years has *Panthera leo* been made the subject of intensive zoological studies. For three years, from 1966 into 1969, George Schaller followed lion prides over the grasslands of Tanzania's Serengeti Park, "a boundless region with horizons so wide that one can see clouds between the legs of an ostrich," where heat waves at noon transform "distant granite boulders into visions of castles and zebra into lean Giacometti sculptures." Schaller logged 149,000 kilometers of travel while keeping the lions under observation for a total of 2,900 hours. Subsequently Brian Bertram followed the same prides for an additional four years, confirming Schaller's results and acquiring valuable new insights into the ecological basis of their social behavior. Few animal populations have been studied for so long in the wild. As in Lynn Rogers' black bears, Iain Douglas-Hamilton's elephants, and Jane van Lawick-Goodall's chimpanzees, a new level of resolution has been attained, in which free-ranging individuals were tracked from birth through socialization, parturition, and death, and their idiosyncrasies and personal alliances recorded in clinical detail.

The core of a lion pride is a closed sisterhood of several adult females, related to one another at least as closely as cousins and associated for most or all of their lives within fixed territories passed from one generation to the next. In the prides most closely monitored by Schaller the average number of individuals per pride was 15 with a variation of 4 to 37. The degree of cooperation that the female members

Figure 24-1 In the Serengeti Park, a pride of lions devours a newly killed buffalo. The two males, who are brothers, have already eaten their fill and wandered away, permitting the remainder of the pride to approach and feed. The latter group consists of the lionesses, two 3-year-old males, a juvenile about 18 months old, and two cubs 5 months in age. In the background two black-backed jackals and a group of vultures wait for a chance to share in the remains. A herd of wildebeest can also be seen. The adult male to the rear displays a relaxed open-mouthed face, while his com-

panion stares at an unidentified object past the observer. Two of the lionesses snarl at each other during one of the frequent low-keyed aggressive exchanges that occur between pride members at the kills. One of the young males, temporarily displaced during the jostling, crouches behind the kill. In the dominance hierarchy of the pride, cubs are at the bottom, and they suffer a high mortality rate from malnutrition due to an inability to eat fully before the prey is consumed. (Drawing by Sarah Landry; based on Schaller, 1972, in consultation with Brian Bertram.)

display is one of the most extreme recorded for mammal species other than man. The lionesses often stalk prey by fanning out and then rushing simultaneously from different directions. Their young, like calves of the African elephant, are maintained in something approaching a crèche: each lactating female prefers to nurse her own cubs but will permit those of other pride members to suckle. A single cub may wander to three, four, or five nursing females in succession in order to obtain a full meal. The adult males, in contrast, exist as partial parasites on the females. Young males almost invariably leave the prides in which they were born, wandering either singly or in groups. (A minority of the young females also become nomads.) When the opportunity arises, these males attach themselves to a new pride, sometimes by aggressively displacing the resident males. Male bands both inside and outside the prides typically consist of brothers, or at least of individuals who have been associated through much of their lives. The pride males permit the females to lead them from one place to another, and they depend on them to hunt and kill most of the prey. Once the animal is downed, the males move in and use their superior size to push the lionesses and cubs aside and to eat their fill. Only after they have finished do the others gain full access to the prey (see Figure 24-1). Males also respond more aggressively to strangers, especially to other males who attempt to intrude into the pride domain. The larger the size of the brotherhood, the longer its members are able to maintain possession of a pride before being driven out by rivals.

What is the significance of this peculiar social structure, in a group of mammals (the cat family Felidae) otherwise celebrated for its solitary habits? Schaller convincingly argues that the prides evolved primarily because group hunting is a superior means of catching large herbivorous mammals in open terrain. His data show that several lions stalking together are generally twice as successful at catching prey as are solitary lions. They are also capable of bringing down exceptionally large and dangerous prey, particularly giraffes and adult male buffalos, which are virtually inaccessible to single individuals. Schaller further found that cubs are better protected from leopards and nomadic male lions when their mother belongs to a group. For both these reasons, prides are far more successful at rearing litters than mothers living alone.

A loose dominance order exists among the lions and lionesses, based entirely on strength. Each lion seems to know the fighting potential of every other. The result is a tense peace broken only by noisy sporadic clashes that are intimidating but ordinarily do little damage. However, real fighting occurs, especially as an outgrowth of quarreling over the spoils, and the big cats show little restraint when they start to slash and bite. The best strategy for a pride member is to anticipate the attacks and to stay out of harm's way. Sometimes lionesses are able to force male lions to back off by launching concerted attacks. Occasionally lions even kill each other. Schaller recorded several fights between males that resulted in death. He also witnessed a case of the killing and cannibalism of cubs after one of the resident males died and the territory was invaded by other prides.

Wolves and African Wild Dogs (Canidae)

Three species of the dog family Canidae hunt in packs: the wolf (with its derivative the domestic dog), the African wild dog, and the dhole of Asia. Mass predation requires the highest degree of cooperation and coordinated movement, which redound in all other aspects of social life. Pack hunting permits relatively small animals to exploit large, difficult prey. Bourlière (1963) and other zoologists have noted that predatory mammals hunt mostly animals their own size and smaller. By weight of numbers alone, the pack-hunting canids have been able to break this restriction. Their counterparts among the marine mammals are the killer whales, which attack much larger whales in coordinated groups. Among the insects, the socioecological analogs are the army ants, which employ group foraging and mass assaults to subdue colonies of other social insects, including those of ants. And according to prevailing theory, primitive man was the analog among the primates (see Chapter 26).

Two behavioral traits basic to the Canidae seem to have made it easy for pack hunting to evolve on multiple occasions (Kleiman and Eisenberg, 1973). There is first the unique form of the pair bond, in which the male provisions both the female and her young, so that large litters can be reared whenever sufficient prey are available. Packs have formed in the most social species by an extension of this economic system to hold groups of related families together. Second, canids, unlike the majority of cats and other carnivorous mammals, pursue their prey in the open instead of relying on stealth and ambush. It is easier for cooperative pack hunting to evolve from such an initial hunting strategy.

The wolf, *Canis lupus,* is the northern representative of the pack hunters. Before being largely exterminated by man it ranged throughout North America south to the highlands of Mexico, and from Eurasia to Arabia, India, and southern China. It is larger in size than all but the most massive breeds of domestic dogs. Adults weigh 35–45 kilograms on the average and in extreme instances reach 80 kilograms, with males being slightly heavier than females. In other words, wolves are as large as small human adults. They also occupy the top of the food web. Over 50 percent of their food items consist of mammals the size of beavers or larger. Typical prey in North America include beavers, deer, moose, caribou, elk, mountain sheep, and in the vicinity of settled areas, cattle, sheep, cats, and dogs (but seldom if ever human beings). Smaller prey, from mice to ptarmigans, add variety to the diet at all seasons but undoubtedly become more important in times of hardship. When packs sight an animal, they stalk and chase it as a coordinated unit. Smaller prey can be secured and disabled by the canine teeth of a single wolf. Larger prey must be literally torn down by concerted slashing and pulling on the part of the pack. Even group efforts frequently fail. The fleetest prey, such as deer and mountain sheep, often outrun the wolves, while an adult moose can fend off a large pack indefinitely if it stands its ground. Among 131 moose detected by wolves on Isle Royale as David Mech watched, only 6 were finally killed and eaten. Most of the remainder fled before the pack could close in, while the rest either stood at bay until the pack gave up or simply outran the wolves in straight pursuit. The literature contains many accounts of successful hunts that were made possible only because of concerted action. Usually the prey was either cornered or else flushed from impregnable positions by an onslaught from several directions. Several observers have witnessed wolves driving caribou toward other members of the pack lying in wait. Kelsall (1968), for example, saw a pack of five wolves wait quietly as a minor band of caribou moved into a small clump of stunted spruce. When the caribou were out of sight an adult wolf walked just uphill from the spruce and concealed itself directly in the path being followed by the caribou. The other four wolves simultaneously circled the spruce, spread out along its downhill side, and began a stealthy "drive" through it. The goal was evidently to move the caribou toward the wolf waiting uphill.

The details of wolf social behavior have been reviewed by Mech (1970), one of the principal observers of free-living packs, and Fox (1971), who has studied the socialization process in captive animals. Mech's account is the more detailed and has the added advantage of being collated with current knowledge of the ecology of the species. A new pack is formed when a mated pair leaves its parental group to produce a litter of its own. As the family grows, separate linear dominance orders form among the males and females, respectively, with the founding pair occupying the alpha positions for at least a time. Dominance is expressed in priority of access to food, favored resting places, and mates. It is not absolute, however. An "ownership zone" exists within about half a meter of any wolf's mouth, and food in the zone is not disputed by higher-ranking animals. Rank begins to be established early in life, when puppies play-fight. It is reinforced in maturity by repeated exchanges of hostile and submissive displays. Fights usually end quickly by the submission of one of the contenders. But occasionally, especially during the breeding season, all-out battles erupt that result in serious injury. Cliques of wolves have been seen to gang up on individuals during these disputes. The alpha male is the center of constant attention, in every sense the lord and master of the pack. He is the leader in most chases and reacts first and most strongly to intruders. Other members normally defer to him during the greeting ceremony, during which one wolf tenderly nips, licks, and smells the mouth of another. The ceremony appears to be a ritualized version of food-begging movements by puppies. Although conducted most commonly following a separation, it is on many occasions directed spontaneously at the alpha male. Sometimes whole groups crowd around the leader in this act of friendly obeisance.

The social behavior so well marked in the wolf is carried to further heights in the African wild dog *Lycaon pictus*, appropriately called by Hediger "the super beast of prey." The species is one of the scarcest yet most wide-ranging mammals of Africa. It occurs in most habitats other than extreme desert and dense forest. One of the strictest of carnivores, the wild dog usually hunts prey approximately its own size, such as Grant's and Thomson's gazelles, impala, and the calves of wildebeest. But it also attacks and consumes much larger animals, including adult wildebeest and zebras. The hunts are almost always conducted in a tight group. Lasting an average of only 30 minutes and usually ending in success, they are scenes of unparalleled ferocity. The pack leader selects the target while still at a distance and leads the others toward it in a determined sprint. Gazelles flee when the dogs approach to within 200–300 meters. The predators rely on a combination of speed, endurance, and numbers to capture even the fleetest animals. Running at 55 kilometers per hour and in bursts at 65 kilometers per hour, the dogs overtake most quarries within the first 3 kilometers. Occasionally they hold a 50 kilometer per hour pace for 5 kilometers or more. One dog, usually a member of the leadership "cadre," holds the lead throughout, while the others string out behind it for as much as a

Figure 24-2 The "super beasts of prey" and most highly social canids: a pack of wild dogs on the Serengeti Plains of Tanzania. Most of the adults are just returning from a successful hunt. In the foreground an adult prepares to regurgitate some of the fresh meat to pups who tumble out of the den. On the left the mother dog performs the greeting ceremony to the dominant male. In a moment she, too, will be fed by regurgitation. In the distance can be seen herds of zebras and wildebeest, which are among the largest animals attacked by the dogs. The exceptionally populous litter is another trait of this species. Only one or two females produce a litter in a given year, and the remaining adults

participate fully in the care and upbringing of the young animals. The exceptional altruism and cooperativeness of the species is associated with the habit of hunting in packs, a technique that increases the efficiency of capturing prey during daylight chases and makes possible the killing of animals much larger than the individual dogs. (Drawing by Sarah Landry; based on Estes and Goddard, 1967, and Hugo van Lawick-Goodall in van Lawick-Goodall and van Lawick-Goodall, 1971, in consultation with Richard D. Estes.)

kilometer or more. The advantages of group chasing are twofold. Some of the prey run in wide circles or in zigzag patterns in attempts to throw off pursuers at their heels. Other members of the pack running behind are able to cut across the curve and close the distance. Once the prey is seized, all members of the pack rush in to immobilize it, quickly tearing it to pieces by yanking in all directions. Gazelles can be killed and eaten within 10 minutes following capture. A bull wildebeest or zebra might require more than an hour, but it is still remarkable that a creature the size of a German shepherd can take such oversized prey at all.

During hundreds of hours of observations of wild dogs, zoologists have discovered a degree of cooperation and altruism unmatched by any other animals except elephants and chimpanzees. As soon as the pack has eaten its fill, it returns to the den to regurgitate to the pups, their mother, and any other adults who remained behind (see Figure 24-2). Even when the prey is not large enough to feed all of the dogs to repletion, the hunters still share their booty. Sick and crippled adults are thus cared for indefinitely. At the kill juveniles are given precedence by the adults, a complete reversal of the procedure in lions and wolves. Communal behavior is developed to such a degree that when a litter of nine pups watched by Estes and Goddard was orphaned at the age of five weeks, they were reared by the eight remaining members of the pack, all of which happened to be males.

Despite the savagery displayed in the hunts, wild dogs are relaxed and egalitarian in relations with one another. No individual distance is observed, and the pack members sometimes lie in heaps to keep warm. Females vie with one another for the privilege of nursing the pups, although the mother normally retains first rights. Separate dominance orders exist among the males and females, but they are so subtle in expression as to be easily overlooked by human observers. Threats are especially difficult to recognize. Instead of snarling and bristling like a wolf, the wild dog assumes a posture resembling that taken during stalking. The head is lowered to the level of the shoulder or below, the tail hangs quietly, and the dog either stands rigidly while facing its opponent or walks stiffly toward it. Submission, in contrast, is an elaborate and conspicuous performance. It grades insensibly into the greeting ceremony, by which the animals reestablish contact and on other occasions initiate pack chases. In potentially tense situations, the dogs seem to compete with one another in making submissive displays. Their lips draw back in a rictuslike grin, the forepart of the body is lowered, and the tail is lifted over the back. The animals excitedly twitter back and forth as each tries to burrow beneath the other. When ritualized begging behavior in the form of face licking and mouth snuffling is added, the performance turns into the full-fledged greeting ceremony.

Perhaps the most intriguing aspect of wild dogs is the timing of their reproduction. In any given year only one or two of the females produce a litter. Parturition, or at least success in bringing a litter to weaning, may depend on the position of the female in the dominance hierarchy. But whether or not this is so, it is indisputably true that the pack as a whole invests in only one or at most two litters at a time. These litters are relatively enormous in size, averaging about 10 pups in the wild and ranging to as many as 16. Most are born during the rainy season, when most herbivores are also born. The significance of the trait can be inferred, I believe, by comparing wild dogs with army ants. Both are extreme carnivores that use mass forays to conquer prey too large or otherwise too formidable for single predators. Probably as an ultimate consequence of this specialization, both the dogs and the ants are nomadic, shifting from site to site on an almost daily basis. Not to do so would be to reduce the food supply within striking range of the core area to below the maintenance level. Army ants are notable among social insects for the high degree of synchronization in their brood development, which is made possible by extraordinary bursts of oviposition over short periods of time and at regularly spaced intervals. These insects are nomadic only when the brood is in the larval stage. Thus synchronization of brood development means that the colony can remain safely in one well-entrenched home site for long stretches of time, when all of the young are in the egg and pupal stages. The wild dogs also benefit from synchronization but in a different way. When a litter is born the pack is tied down to one spot until the pups are large and strong enough to join the nomadic marches. If each bitch had a litter of the usual canid size and independently of the others, the pack would be forced to spend much longer periods of time in one place. Therefore it can be reasonably suggested that large litters by single females has as its raison d'être the synchronization of development, which permits the pack to be nomadic on a maximum number of days in each year.

CHAPTER 25

The Nonhuman Primates

The living primate species can be profitably viewed as a kind of *scala naturae* that proceeds from near the phylogenetic base of placental mammals upward in small steps through increasing anatomical specialization, behavioral complexity, and social organization. It embraces the following taxonomic sequence: the tree shrews, the tarsiers, the lemuroids, the New World monkeys, the Old World monkeys, the anthropoid apes, and finally man. As T. H. Huxley said in 1876, "Perhaps no order of mammals presents us with so extraordinary a series of gradations as this —leading us insensibly from the crown and summit of the animal creation down to creatures from which there is but a step, as it seems, to the lowest, smallest and least intelligent of the placental mammals." In modern terms the *scala* must be interpreted as a series of evolutionary grades transecting a branching phylogenetic tree rather than literal steps leading from ancestors to descendants among the living forms (see W. C. O. Hill, 1972). But the precise definition of the grades remains one of the key problems of current primate social studies, and special attention will be devoted to it here.

The Distinctive Social Traits of Primates

The scheme represented in Figure 25-1 postulates certain basic primate qualities to be evolutionary prime movers. Following the method outlined in Chapter 3, I have classified them as stemming either from phylogenetic inertia or from the major adaptive shift of primates to arboreal life. Both of these influences, the inertial and newly adaptive, triggered chains of other adaptations which together constitute the diagnostic social qualities of the primates.

The basic systems of mammalian reproduction and heredity are ultraconservative. An evolving mammalian population cannot easily alter the pituitary-gonadal endocrine system, substitute haplodiploidy for the XY sex-determination mechanism, or dispense with maternal care based on lactation. Consequently, the reproductive and genetic systems are inertial in their effects. Because of them certain ancient mammalian traits continue to prevail throughout the primates. There is a tendency for males to be polygynous and aggressive toward one another, although pair bonding and pacific associations are permissible minority strategies (Washburn et al., 1968). Where long-term sexual alliances are not the rule, the strongest and most enduring bonds are between the mother and her offspring, to an extent that matrilines can be said to be the heart of the society. Mothers are the principal socializing force in early life. In at least some of the aggressively organized species they exert an influence on the identity of the peers and social rank of their sons and daughters. Their influence may even extend to later generations (Kawamura, 1967; Marsden, 1968; Missakian, 1972).

The second class of ultimate determinants of primate social behavior consists of the basic postadaptive traits, shown as the right-hand side of Figure 25-1. The vast majority of arboreal animals, from insects to squirrels, are small and have no difficulty moving through the canopies of trees. The surfaces of trunks, limbs, and even leaves are broad enough in proportion to their bodies to be navigated as though they were uneven extensions of the ground. However, most primates, particularly the phylogenetically more advanced prosimians, monkeys, and apes, are unusual in being *large* arboreal animals. The ultimate reason why they filled the large-size categories is unknown, but the immediate physiological consequences of this adaptive shift are clear. For animals that must judge distances and the strength of supports with precision, vision is the paramount sense. Visual acuity in primates has been enhanced by moving the eyes to the front part of the head, making stereoscopic vision possible, and adding color vision, which increases the power of discrimination

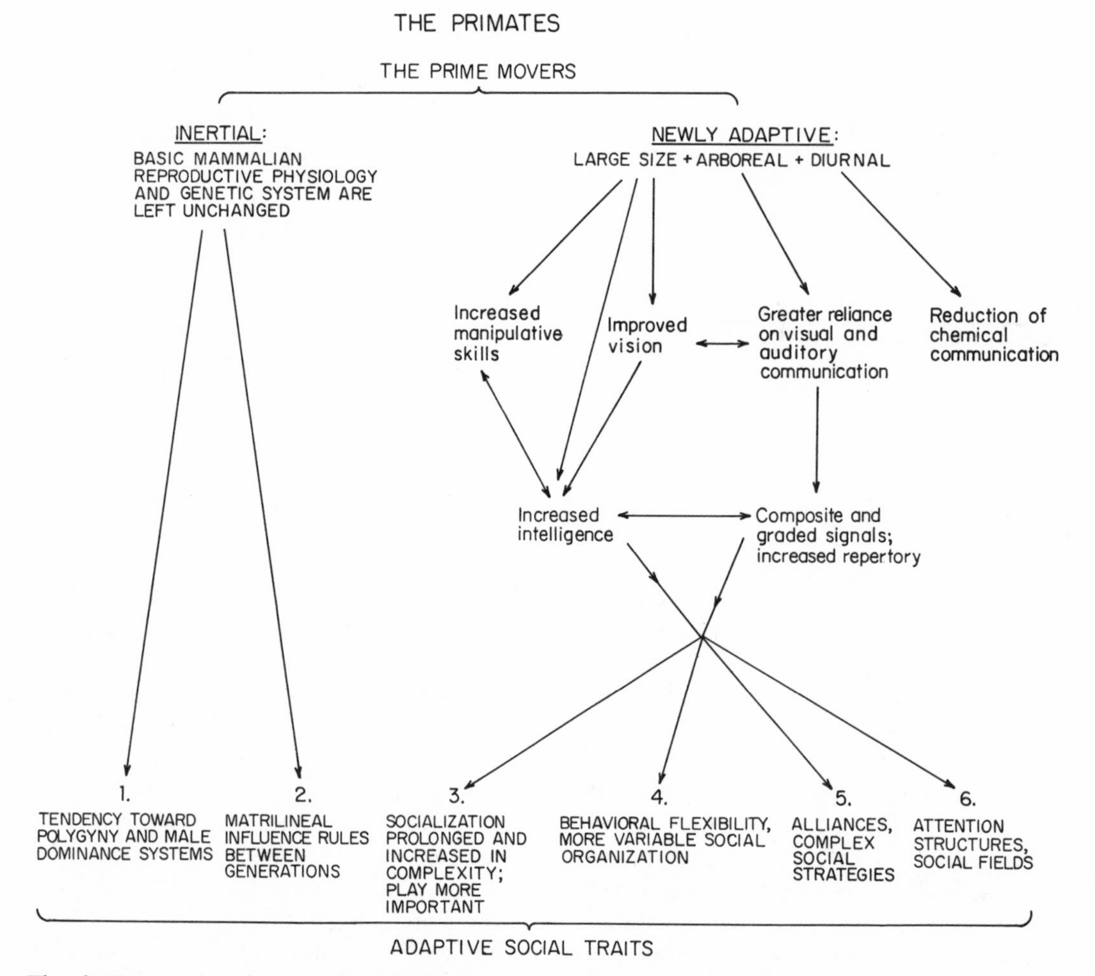

Figure 25-1 The distinctive social traits of the higher primates are viewed as the outcome of conservative mammalian qualities ("inertial" forces) and adaptation to arboreal life. Even phyletic lines that are now terrestrial have retained the evolutionary advances made by their arboreal ancestors.

of objects within the variegated foliage. Cartmill (1974) has suggested that the tendency to prey on small insects made these changes even more advantageous. Sound has taken on added significance as the only means of detecting other animals through dense foliage. At the same time the sense of smell has declined in importance. A large animal can depend less on the tracking of odors in the irregular air currents of the canopy. It moves too rapidly and must follow pathways through the branches too irregular to permit exact orientation along the active odor spaces emitted by other animals. As a consequence the primates have come to depend heavily on visual and auditory signals in their communication systems. The trend has been carried much further in the generally larger Old World monkeys and apes than in the prosimians and New World monkeys.

As Bernhard Rensch (1956, 1960) has argued on various occasions, large body size is crudely correlated in mammals with greater intelligence, seemingly as an inevitable result of the increase in absolute brain size. Thus the higher primates gained some component of their intelligence in the simple process of becoming large. Their mental capacity has been enhanced still more by the method of using the hands and feet to grasp branches during locomotion and rest. Both the New World monkeys and the Old World monkeys and apes have gone further by developing a "precision grip" as distinct from the more primitive "power grip." Instead of merely closing the hand around the object, whether for support or feeding, they invest some amount of separate control in the index finger and thumb, permitting the fine manipulation of food particles and the grooming of fur. In general, the larger the primate the more dexterous the manipulation. Chimpanzees are more skilled than macaques and baboons, which in turn are superior to langurs and guenons. Man represents the culmination of this evolutionary trend.

Intelligence is the prerequisite for the most complex societies in the vertebrate style. Individual relationships are personalized, finely graduated, and rapidly changing. There is a premium on the precise expression of mood. Higher primates have extended the basic mammalian tendency away from the use of elementary sign stimuli and toward the perception of gestalt, that is, toward the simultaneous summation of complex sets of signals. In vision, for example, a bird or fish may respond to a single patch of color or the correct performance of one movement of the head—and to virtually nothing else. The monkey or ape more consistently tends to act on the appearance of the entire body, the posture, and the history of previous encounters with the individual confronted. There is also a tendency to utilize information from more than one sensory modality. At close range visual and auditory signals are compatible and can be blended with tactile cues to form composite signals that convey messages redundantly and with greater exactitude (Marler, 1965). R. J. Andrew (1963a) has pointed out that the deep grunts commonly used by Old World monkeys and apes during close social encounters are particularly well suited to this purpose. The sounds are rich in overtones and therefore highly personalized, allowing the identification of individuals by voice alone. They are generated by the upper part of the respiratory tract, so that in addition to unique messages they carry redundant information concerning visual signals based on the shape of the mouth, position of the tongue, and other muscular, postures that determine the expression of the face. Increasing sophistication in the employment of such composite signals among primitive hominids could have set the stage for the origin of human speech. Another probable consequence was the use of the face for personal recognition. Jane Goodall, Schaller, and others have documented the striking, humanlike variation in the facial features of chimpanzees and gorillas. It is easy for human observers to recognize individuals at a glance and even to guess their parentage with a high level of accuracy.

In addition to monitoring multiple signals, higher primates evaluate the behavior of many individuals within the society simultaneously. The animal lives in a *social field* in which it responds to multiple individuals simultaneously, in ways that take differing relationships into account and often entail compromise. Observers of free-living societies of Old World monkeys and apes have noted the use of behavioral strategies that manipulate the social field. Kummer (1967), for example, described the "protected threat" tactic of the hamadryas baboon. A female competing with a rival moves next to the overlord male, where she is in a better position to intimidate and to resist attack. If she is threatened, the male is much more likely to drive her rival away than to punish her. As a result she is more likely to advance in social rank. Alliances are also commonplace, especially between mothers and their grown offspring. Alloparental care leads to coalitions between adults as well as to the more rapid extension of social contacts by developing young. In troops of macaques and baboons, adult males, not necessarily related, back one another up during aggressive encounters. The rank of an individual depends not only on his personal prowess but on the strength and dependability of his allies (Altmann, 1962a; Hall and DeVore, 1965). The dominant female of a troop of bonnet macaques studied by Simonds (1965) relied on the assistance of the dominant male to win encounters with every other member of the troop. But when her protector fell in status following the loss of a canine tooth and defeat in a major fight, the female was no longer able to dominate the other males.

Chance (1967) and Chance and Jolly (1970) have conceptualized the organization of individual social fields in terms of the *attention structures* of whole societies. Among the species of Old World monkeys and apes two categories of attention structure can be roughly distinguished. Centripetal societies, possessed by macaques, baboons, and most other cercopithecoid species, are organized around a dominant male. The members watch the male predominantly, shift their positions according to his approach or departure, and adjust their aggressive behavior toward others according to his responses. When the group is attacked from the outside, the dominant male and his allies lead the defense or retreat. The more pronounced the dominance structure, the stronger the centripetal orientation. When aggression breaks out inside the group, the members tend to move toward the cohort of dominant males, and this is sometimes true even when some of the males are the aggressors. Acentric societies, the second type, are exemplified by the patas, langurs, and gibbons. Although the attention structure varies in detail among species, all acentric societies are characterized by the tendency of the females and young to separate from the males during aggressive episodes. In other words, the society fragments in the face of tension. During peaceful moments the patas male lives mostly on the fringe of his little troop, serving principally as a watchdog. When a predator threatens, he runs to a low tree or other prominent position and threatens, while the females and young take refuge in another direction. Chance and Jolly view attention structure as basic and its analysis as the key to the understanding of primate societies. But in fact, attention structure is just one more parameter, compounded of a multiplicity of behaviors and evolving as an adaptation to special features of the environment. As such it can be fed into certain models of social organization along with other parameters such as age structure, group size, and signal transmission rates.

All of the distinctive primate traits just cited tighten the moment-to-moment adjustment by individuals to fluctuations in the environment—this in the most general terms is the key primate behavioral adaptation. To shift quickly and precisely from one response to another according to subtle changes in the social field requires that the structure of the society itself be malleable. The primate literature is filled with accounts of primate social malleability; Hans Kummer, Thelma Rowell, and others have stressed that it is one of the most distinctive phenomena seen in free-ranging societies. The anubis baboon is a particularly instructive species. On the savanna of Kenya's Nairobi Park, DeVore observed a definite marching order among the members of troops moving from one location to another. The dominant males accompanied females with small infants near the center, juveniles flanked these individuals near the center, and other adult males and females formed the van and rear. When a potential predator appeared, the dominant males at once moved to the front to meet it. Rowell (1966) discovered a different organization in anubis baboon troops in the forests of Uganda. Here troops progressed and communicated more like arboreal primate species than other baboons. The movements were less regular, with no consistent marching order. While moving through thicker vegetation, the baboons communicated more with grunts and showed a much greater concern for stragglers than did baboons on the savannas. Aggressive interactions among the males were also less frequent, and Rowell could find no evidence of the dominance hierarchies that are the hallmark of the savanna troops. Forest troops are casual about their sleeping places and generally avoid other troops. But on the open savanna of the Amboseli Reserve, where groves of sleeping trees are scarce, the anubis baboons tolerate the presence of other groups, and large sleeping aggregations sometimes form similar to those of the hamadryas baboon. At the Awash Falls of Ethiopia a troop of anubis baboons has penetrated into terrain otherwise occupied by hamadryas. By studying this transferred group in detail, Nagel (1973) was able to differentiate to some extent the genetic and learned components of the baboons' social behavior, in the special sense of determining which differences between the species persist when the two forms are placed in a similar environment. The anubis troop gathered in a common sleeping place and separated for foraging, the hamadryas-like pattern Rowell had seen in Uganda. They also resembled the hamadryas in the length of the foraging routes they followed and in the amount of time spent foraging in wooded areas. But they retained the one-level social organization characteristic of anubis elsewhere instead of shifting to the hamadryas two-level harem system.

The Ecology of Social Behavior in Primates

The principal organizing concept in the study of primate societies has been the theory that social parameters are fixed in each species as an adaptation to the particular environment in which the species lives. The parameters include size, demographic structure, home-range size and stability, and attention structure.

With the idea in mind of social behavior as a direct ecological adaptation, primatologists have proceeded to undertake a careful comparison of species occupying different habitats. Eisenberg and his coworkers (1972) have presented the most recent and cohesive analysis of the data available from field studies. As shown in Table 25-1, the key trait selected by these authors is the degree of male involvement in social life. This variable not only is satisfying by itself but also is reasonably well correlated with other social traits such as group size, the nature of the dominance system, and territoriality. Working with more data than had been available to previous authors, Eisenberg et al. recognized an intermediate social category, the age-graded-male troop. Some species that appear to be organized into multimale societies do not in practice adhere strictly to that pattern. Younger, weaker males may be tolerated but only in a subordinate status. After a time they take over the dominant position or else leave the troop altogether. Societies in this evolutionary grade do not contain ranking males of approximately the same age. Consequently, there are no alliances and cliques of the kind that form the central hierarchies in baboon and macaque troops.

The matrix of Table 25-1 provides the most efficient and heuristic system yet devised, but the correlations are nevertheless relatively weak. Insectivores remain in the bottom grade. Terrestrial and semiterrestrial species are still characterized by the most advanced social organization, and the same is true of omnivores. Little more can be extracted. Within single evolutionary grades it is possible to define subgrades based on additional social characteristics and to correlate them with certain aspects of the preferred niche. Thus folivores (leaf eaters) have smaller home ranges than frugivores (fruit eaters), and they are

Table 25-1 The arrangement of primate societies into evolutionary grades and their ecological correlates by Eisenberg et al. (1972 and personal communication). The grades are based on the degree of male involvement, given as the column headings. (Copyright © 1972 by the American Association for the Advancement of Science.)

Solitary species	Parental family	Minimal adult ♂ tolerance[a] (unimale troop)[b]	Intermediate ♂ tolerance[c] (age-graded-male troop)[b]	Highest ♂ tolerance[d] (multimale troop)[b]
A. Insectivore-frugivore	A. Frugivore-insectivore	A. Arboreal folivore	A. Arboreal folivore	A. Arboreal frugivore
Tupaiidae	Callithricidae (Hapalidae)	Colobinae	Colobinae	Indriidae
Tupaia glis	*Saguinus oedipus*	*Colobus guereza*	*Presbytis cristatus*	*Propithecus verreauxi*
Lemuridae	*Cebuella pygmaea*	*Presbytis senex*	*Presbytis entellus*	Lemuridae
Microcebus murinus	*Callithrix jacchus*	*Presbytis johnii*	Cebidae	*Lemur catta*
Cheirogaleus major	Cebidae	*Presbytis entellus*	*Alouatta villosa*	B. Semiterrestrial frugivore-omnivore
Daubentoniidae	*Callicebus moloch*	B. Arboreal frugivore	B. Arboreal frugivore	Cercopithecidae
Daubentonia madagascariensis	*Aotus trivirgatus*	Cebidae	Cebidae	*Cercopithecus aethiops*
Lorisidae	B. Folivore-frugivore	*Cebus capucinus*	*Ateles geoffroyi*	*Macaca fuscata*
Loris tardigradus	Indriidae	Cercopithecidae	*Saimiri sciureus*	*Macaca mulatta*
Perodictus potto	*Indri indri*	*Cercopithecus mitis*	Cercopithecidae	*Macaca radiata*
B. Folivore	Hylobatidae	*Cercopithecus cambelli*	*Cercopithecus talapoin*	*Papio cynocephalus*
Lemuridae	*Hylobates lar*	*Cercocebus albigena*	C. Semiterrestrial frugivore-omnivore	*Papio ursinus*
Lepilemur mustelinus	*Symphalangus syndactylus*	C. Semiterrestrial frugivore	Cercopithecidae	*Papio anubis*
		Cercopithecidae	*Cercopithecus aethiops*	*Macaca sinica*
		Erythrocebus patas	*Cercocebus torquatus*	Pongidae
		Theropithecus gelada	*Macaca sinica*	*Pan troglodytes*
		Mandrillus leucophaeus	D. Terrestrial folivore-frugivore	
		Papio hamadryas	Pongidae	
			Gorilla gorilla	

[a] Troop with one adult male and strong intolerance to maturing males.
[b] "Troop" refers to the basic social grouping of adult females and their dependent or semidependent offspring.
[c] Troop typically showing age-graded-male series.
[d] Troop with several mature, adult males and age-graded series of males.

more likely to employ individual calling or troop chorusing to maintain spacing between adjacent groups.

A very simple but promising start in model building using available field data has been made by Denham (1971), who stresses the crucial parameter of food distribution. His approach accords with current ecological theory and can be extended as follows. Earlier in this book (Chapter 3) it was argued that greater predictability of food in space and time promotes the evolution of territoriality. When the resources are dense and easily defensible, and when food is the limiting resource, the optimum strategy is double defense—by means of the monogamous pair bond. If the quality of the environment is not only predictable but also uniform from locality to locality, so that its variation is kept under the polygyny threshold, the tendency toward monogamy will be reinforced. This latter factor could explain one ecological difference between the "solitary" species of the first grade shown in Table 25-1, which include a high proportion of insectivores, and the pair-bonding species of the second grade, most or all of which are primarily vegetarian. The explanation is the hypothesis, both reasonable and testable, that the plant items vary in quantity and quality less from territory to territory than do the insects. The same hypothesis is consistent with the fact that leaf-eating species defend smaller territories and use more conspicuous vocal displays than do otherwise similar fruit-eating species. Higher social grades can be expected to evolve in accordance with the Horn principle, which states that when food becomes sufficiently patchy in space and unpredictable in time, the optimum strategy is to abandon feeding territories and to join groupings larger than the family. As Denham and others have pointed out, this may be the ultimate causation of greater group size in open-country species of Old World monkeys and apes. These primates live in an exceptionally patchy and unpredictable environment. The same principle can be extended to the forest-dwelling species in higher social grades, if further data reveal their food items to be similarly distributed. Tropical forests, contrary to the popular conception, are seasonal and usually strongly so. Many potential food sources within the forests, including the buds, flowers, and fruit of *particular* plant species, are not only seasonal but also patchy and unpredictable through time. Finally, predation plays an unquestioned auxiliary role in evolution, forcing species into one defensive strategy or another and thereby helping to shape group size and organization.

In the remainder of this chapter, we will view the full range of primate sociality by considering individual species that represent several of the evolutionary grades. Since the sequence will proceed upward through the grades rather than through phylogenetic groupings, the reader will note some curious taxonomic juxtapositions. For example, the anthropoid apes are distributed from one end of the array to the other. The mostly solitary orang-utan must be sociobiologically classed with the primitive prosimians, while gibbons are grouped with marmosets, titis, and the New World monkeys. The gorilla possesses age-graded-male societies with a reasonably complex organization, but it is still well behind the chimpanzee, which by all reasonable standards occupies the pinnacle of nonhuman primate social evolution. The apes are extreme in this display of diversity, but each of the remaining major phylogenetic groups spans several evolutionary grades as well.

The Lesser Mouse Lemur (*Microcebus murinus*)

The galagos, pottos, mouse lemurs, and other nocturnal prosimians are among the most primitively social primates. Because of the difficulty of studying these animals in the field, data on most of the species are still fragmentary and inconclusive. Thanks largely to the work of Petter (1962a,b) and R. D. Martin (1973), the population structure and behavior of one species, the lesser mouse lemur, has become sufficiently clear to serve as a paradigm for the lowest evolutionary grade. *Microcebus murinus* is the smallest and most widespread of all the prosimians of Madagascar, inhabiting nearly all forested areas along the coast. It is wholly nocturnal, spending the day in nests constructed from dry leaves in bushes and tree holes. Although primarily arboreal, it descends readily to the ground to cross gaps in the foliage and to forage through the leaf litter. The lesser mouse lemur is the most nearly omnivorous of all the known primates. Its diet includes fruits, flowers, and leaves of a variety of trees, bushes, and vines. It also eats insects, spiders, probably tree frogs and chameleons, and possibly mollusks. Individuals collect sap from holes that they score in live bark with rotating movements of their front teeth. In this last habit the species is convergent to the smallest of the New World primates, the pygmy marmoset.

Possibly as a result of its catholic diet, the lesser mouse lemur has a small home range, evidently 50 meters or less in diameter. The ranges tend to be exclusive, at least within sexes, and it is reasonable to hypothesize some form of territorial defense. Individuals placed by Petter into the same small enclosure all at the same time were compatible. But when one was allowed to occupy the space first, it subsequently attacked all newcomers. Even within the compatible groups males fought with each other whenever females came into estrus.

Martin's data show that the species is dispersed into localized population nuclei. Each nucleus contains a core of high density characterized by a proportion of 4 adult females to each adult male. At Mandena, Madagascar, the cores were located at sites containing a high frequency of the two preferred species of food trees. Since the sex ratio at birth is 1:1, it follows that males either emigrate or suffer higher mortality early in life. In fact, the surplus males are found concentrated in nests along the periphery of the core area. Females often nest in groups, evidently mating and rearing young compatibly. The number of females per nest in 1968 ranged from 1 to 15, with a median of 4. When females were in estrus, they were often accompanied by a single male. When the females passed out of estrus, the males evidently became more tolerant toward one another, and two and even three sometimes collected in the same female nest. Martin suggests that groups of females are often mothers and daughters. Sons, however, tend to be displaced to the periphery of the favored habitats, where they await an opportunity to join the dominant, breeding males. Communal nesting might be a consequence of the limited number of nest sites, perhaps abetted by kin selection. In any case the mouse lemur must still be regarded as an essentially solitary animal. No evidence exists of organized social life within the nests. Of equal importance, the lemurs forage wholly on their own.

The Orang-utan (*Pongo pygmaeus*)

Until recently the orang-utan was the least known of the great apes, the mysterious Old Man of the rain forests of Sumatra and Borneo, seldom more than glimpsed in the wild. Careful studies have now been completed by David Horr, P. S. Rodman (1973), and J. R. MacKinnon (1974). Rodman and his assistants logged the quite respectable observation time of 1,639 hours, in the course of which they came to know 11 orangs individually. They were able to follow their subjects continuously for hours or days through the nearly trackless forests of Borneo's Kutai Reserve.

As the orangs' unusual body form testifies, they are exclusively arboreal, relying extensively on brachiation to move through various levels in the rain forests from the canopy to near the ground. Although primarily frugivores, they also consume some leaves, bits of bark, and bird eggs. Their natural population densities had been previously judged to be 0.4 individuals per square kilometer or less. At the Kutai Reserve, which contains some of the least disturbed lowland habitats left in Asia, Rodman found the density to be closer to 3 individuals per square kilometer. Nuclear groups consist of females and their offspring, and are sometimes accompanied by an adult male. Solitary males are common, but juveniles or adult females are encountered alone only on rare occasions. Orang group size seldom if ever exceeds 4 individuals. At the Kutai Reserve, the following composition of seven such groups was recorded: adult female + infant male; adult female + infant male; adult female + infant, sex unknown; adult female + juvenile male; juvenile female alone; adult male alone; adult male alone. Occasionally these units met to form secondary groupings, the largest of which contained 6 individuals. Two of the temporary combinations were seen on multiple occasions and appeared to be based on kin ties. The others were passive aggregations brought together by the common attraction of a fruiting tree. The contacts were facilitated by a broad overlapping of the home ranges.

The orang society can be viewed as incorporating the loose fission-fusion structure so strikingly elaborated in the chimpanzee. But this is very elementary in form, and in most other respects the orang-utan is much closer to solitary prosimians such as the mouse lemurs. Specifically, females tend to aggregate, and males visit them only in order to copulate. As juvenile females mature, they disperse slowly from the mother's home range. Males disperse for great distances and wander a good deal before settling into home ranges of their own.

Social interactions among the orang-utans are few in kind and far simpler than in the other anthropoid apes. They are virtually limited to relations between mothers and their offspring and the brief, simple encounters between adult males and females. Aggression within the society is quite rare, and nothing resembling a dominance system has been established in studies to date. During their lengthy period of observation at the Kutai Reserve, Rodman and his coworkers recorded only one clear instance of open hostility—when one adult female drove another from a fruit tree.

But the adult males, wandering mostly in solitude, probably do repel one another in the vicinity of the females. Although direct confrontations have not yet been observed, a few pieces of indirect evidence suggest that such intrasexual conflict does exist. Sexual dimorphism is strongly developed, with the males averaging twice the size of the females and possessing large extensible vocal pouches. The males use their pouches to deliver the "long call," a loud, throaty scream that can be heard by human beings for as much as a kilometer away. They sound the call most frequently when they have separated from their temporary female consorts for a short period of time. The function seems to be to reestablish contact. But the males also call on occasion when they are with the females. Since the display is evidently designed for long-distance communication, its second function may be to threaten away rivals. Finally, it is surely significant that no more than one adult male is ever seen in the company of a receptive female.

The White-Handed Gibbon (*Hylobates lar*)

The six species of gibbons and their close relative the siamang (*Symphalangus syndactylus*) are the smallest of the great apes. As exemplified by the white-handed gibbon, the commonest and best-studied member of the group, they show a remarkable convergence in social behavior to the dusky titi and other monogamous New World primates. The white-handed gibbon, *Hylobates lar*, ranges from Indochina west to the Mekong River and south to Malaya and Sumatra. It is intensely arboreal in habit, depending on brachiation through the branches of trees for approximately 90 percent of its locomotion. It prefers the closed canopy of dense forest, where it can travel quickly from tree to tree. The gibbon occasionally descends to clumps of low bushes during feeding and all the way to the ground to drink from streams, although the great bulk of its liquid is obtained from eating fruit and licking bark and leaves after rain. In keeping with their monogamy, the sexes are similar in appearance and size, both ranging between 4 and 8 kilograms in total body weight. Gibbon troops defend territories 100–120 hectares in extent (Carpenter, 1940; Ellefson, 1968).

Carpenter and later researchers have found that the *Hylobates lar* society is identical to the family. There are two to six members, the mated pair plus up to four offspring. Occasionally an aging male is also retained in the group. Solitary individuals are sometimes encountered in the forests; they are evidently either aged animals or young adults still in search of mates and territories. The family stays close together, and dominance is weak or altogether absent. The female plays an equal role in territorial defense and in precoital sexual behavior (see Bernstein and Schusterman, 1964). The mother takes care of the infant, allowing it to cling to her belly when very young, nursing and playing with it, and leading it about when the youngster begins to travel on its own. The male's relation to the infant is also close. He frequently inspects, manipulates, and grooms it. Play sessions are frequent, during which the youngster is permitted to be the mock aggressor. When a young gibbon calls in alarm, the male quickly swings to its aid. He sometimes breaks up play between infants and juveniles that has become too rough. In a captive group of the dark-handed gibbon *H. agilis* assembled by Carpenter, a lone male allowed a small juvenile to adopt him. Thereafter he carried the smaller animal in the maternal position during much of the day. This observation suggests not only that paternal care is close under normal circumstances, but that the male is prepared to assume the role of the mother when she falls ill or dies.

The origin of new gibbon groups has never been observed in nature, but its course can be safely inferred from circumstantial evidence. As Berkson et al. (1971) have noted, young gibbons become aggressive at puberty, and adults placed close together are very hostile. Young adults tend to be excluded, especially at feeding sessions. It is probable that as relations between the parents and young adults become more abrasive, the offspring scatter to form families of their own. Carpenter observed one such pair that might have been in the process of forming an incestuous union, although their sexes and origin could not be ascertained. They remained close together at all times and often stayed well apart from the rest of the family. Berkson and his coworkers observed the formation of one pair from among a group of adults assembled as strangers in an outdoor enclosure.

The Ring-Tailed Lemur (*Lemur catta*)

The true lemurs, comprised of five species in the Madagascan genus *Lemur*, represent the pinnacle of social evolution within the Prosimii. As such they provide a separate natural experiment that can be compared with the attainment of higher evolutionary grades in the ceboid and cercopithecoid monkeys and apes.

The ring-tailed lemur (*Lemur catta*) inhabits the dry gallery and mixed deciduous forests of southern and western Madagascar. As shown by Alison Jolly (1966), it is the most terrestrial member of the genus, spending up to 20 percent of its time on the ground, four times more than the otherwise ecologically similar sifaka (*Propithecus verreauxi*) and almost as much as the "terrestrial" baboons. But it never strays far from the trees, to which it sprints at the slightest alarm. The lemur is exclusively vegetarian, feeding on the leaves, fruit, and seeds of a variety of tree species and a few ground plants. It obeys a strict cycle of diurnal activity. The troop begins to stir before dawn. No later than 8:30 A.M., the exact time depending on the temperature and weather conditions, it enters a period of sunning, feeding, and travel. Commonly two lengthy progressions occur during the morning, the first leading to feeding grounds in lower vegetation strata and the second to the place of the noon siesta. After further wandering and feeding in the afternoon, the troop returns to the feeding trees. There is a tendency to cover the same routes for three or four days in succession, then shift to a different part of the home range.

The population of lemurs at the Berenty Reserve that Jolly observed changed markedly in group composition and territorial occupancy over a period of several years. In 1963–64 there were two troops, consisting of 21 and 24 individuals, respectively. Adult males and females were equally numerous, and their total population was approximately matched by the combined numbers of juveniles and infants. Two or more subordinate males formed the "Drones' Club." They trailed the main group during progressions and tended to feed and take siestas by themselves. The troops avoided one another consistently and occupied mostly exclusive ranges. Fighting was rare. In 1970 the same population had subdivided into four troops with an average total membership of 11 adults and young. Now the home ranges overlapped widely, and feeding and drinking sites were shared on a time plan. Contacts and fighting were much more frequent, while the subordinate drone males often lagged so far behind as to be out of sight.

The lemur society is aggressively organized. Exchanges range from simple visual threats and cuffs to full-scale "jump-fights" during which the animals sometimes rake one another with long downward slashes of their canines. Adult females are dominant over adult males, a reversal of an otherwise nearly universal primate pattern. The female hierarchy is loose and at least partly nontransitive, while that of the males is strictly linear. Aggression among the males reaches its maximum during the April breeding season. Yet oddly, male dominance seems to have no influence on access to estrous females. Jolly saw a female copulate with three males in succession, while one subordinate male accomplished three out of six observed matings. Perhaps dominance determines which of the males remain with the troop over long stretches of time, and which succeed in staying close to the troop during the short breeding season. Leadership is divorced from dominance. During a group progression first one and then another adult takes the van. Occasionally the troop splits into fractions that move in different directions until finally some begin to mew loudly, a signal that brings the lemurs back together.

Lemur catta has a relatively complex communication system which is strikingly unique in some features. In particular, chemical communication is strongly developed and is employed principally during aggressive encounters. Both females and males mark small, vertical branches with genital secretions. They stand on their hands, hold on to the branch with their feet as high as possible, and rub their genitalia up and down in short strokes. The males also employ palmar marking, smearing an odorous secretion on branches by rubbing the surfaces with their forearms and hands. Brachial glands, which occur high on the male's chest, and conspicuous antebrachial organs on the forearms also produce odorous substances (see Figure 25-2). The male places the forearm glands against the chest glands, appearing to mix their secretions. During aggressive encounters the tail is pulled repeatedly between the forearms and waved in the air in a way that wafts the scent toward the opponents. A full-scale encounter between two males entails a flurry of chemical, visual, and vocal signaling. It is usually initiated by transfer of secretions to the tail and sometimes leads to a spectacular "stink fight": an extended series of palmar marking, tail marking, and tail waving by each of the two opposing males.

The Hamadryas Baboon (*Papio hamadryas*)

The hamadryas baboon, also called the sacred baboon, is a large, diurnal, almost exclusively terrestrial cercopithecoid monkey. It ranges across the arid acacia savanna and grassland in the region surrounding the mouth of the Red Sea—eastern Abyssinia, southern Somalia, and southwestern Arabia. Because hamadryas baboons hybridize extensively with anubis baboons, there is some question as to whether they really have the status of a full species (*Papio hamadryas*) or merely constitute a local subspecies (*Papio papio hamadryas*) of one unified baboon species. The former designation still seems the more prudent, especially in view of the strong morphological traits that distinguish this animal. The face is fleshy pink instead of black as in all other baboons. The males are twice the size of the females, their appearance made still more striking by a large mane of wavy gray hair. This dimorphism is related to the feature of hamadryas behavior that makes the species uniquely interesting: the extreme dominance of the adult male over females, who are kept forcibly in a permanent harem. This relationship influences virtually every other aspect of the social organization.

The sociobiology of the hamadryas baboon has been painstakingly documented by Hans Kummer over a 15-year period, first with captive animals and then in the field in Ethiopia (see especially Kummer,

Figure 25-2 The encounter of two ring-tailed lemur troops at the Berenty Reserve in Madagascar. The habitat is a riverside gallery forest, dominated in the foreground by a large tamarind tree (*Tamarindus indica*). The arboreal troop on the left is stirring into activity after a noontime siesta. One male faces the observer with a threat stare, his antebrachial gland visible on the inside of the left forearm. A second male behind him has begun to move down the tree trunk in the direction of the other troop. Directly to his rear two adults engage in mutual grooming, while other members stay clumped together in rest or in the early moments of arousal. The troop on the ground has begun its afternoon

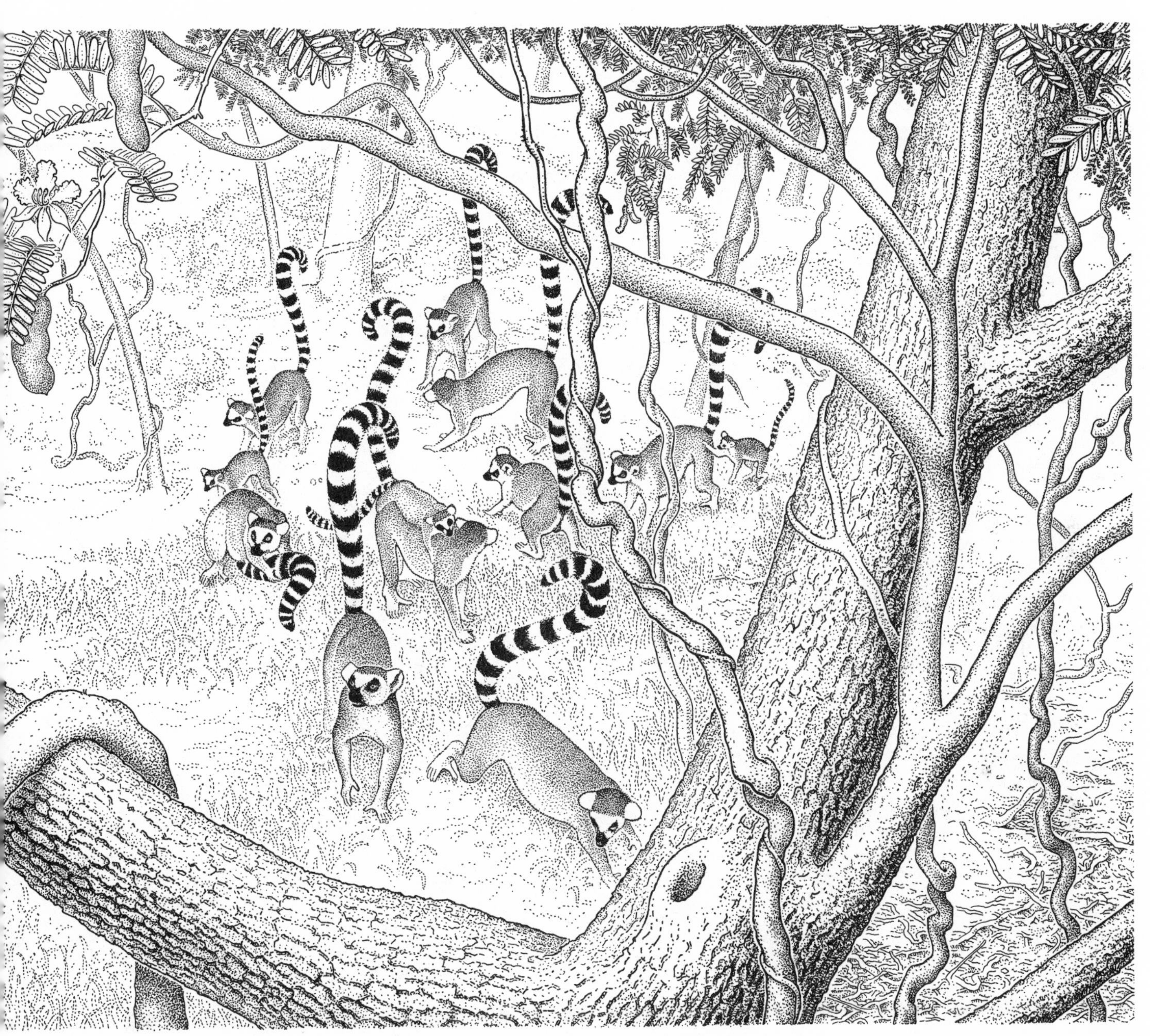

progression to a feeding site. Two adults to the left and front have spotted the group in the tree and are staring and barking in their direction. One of these, a male, draws its tail over the antebrachial glands in preparation for a hostile display. He is ready for a stink fight, during which the tail will be jerked back and forth to waft the scent toward the opponents. Well to the rear and in the center of this picture, two subordinate males of the "Drones' Club" trail the second troop. (Drawing by Sarah Landry; based on data from Alison Jolly, 1966, and personal communication.)

Figure 25-3 Social behavior in the hamadryas baboon. The scene is the arid grassland of the Danakil Plain near the low foothills of the Ahmar Mountains, seen along the horizon. In the early morning, a large group of baboons departs from the communal sleeping rock (left background) on the way to the feeding and watering sites. The procession is beginning to break up into the basic social units, which consist of single males and their harems of females and offspring. Aggressive interactions are frequent and animated. The two males in the foreground threaten each other, the one on the right using a hostile stare while his opponent responds with a more intense gaping display. This exchange might escalate into a ritual fight, with rapid boxing and mouth fencing. The females directly behind the two

males crouch, make fear faces, and scream; otherwise they stay out of the conflict. About 2 meters behind the right male, a younger follower male watches the exchange. Although he is teamed with the overlord and has been trying to acquire a harem of his own under the protection of this older partner, he is not likely to join the fight. Directly to his rear another overlord bites the neck of one of his females as punishment for straying too far. Her response will be to run closer to him. At the far left, to the rear of the mother carrying a young infant, two young bachelor males move along the procession in a social formation of their own. (Drawing by Sarah Landry; based on Kummer, 1968 and personal communication.)

1968, 1971). The species is totally social. Only one solitary individual, an adult male, was seen during months of observation over a substantial part of the species' range. The peculiarities of hamadryas organization are best understood by comparison with the more "conventional" system of other kinds of baboons. The basic unit of the *Papio anubis* society, as shown by DeVore, Hall, and others, is the *group*, an assembly of females, offspring, and multiple males. Aside from mothers and their dependents, no other distinct level of organization exists, at least not in the savanna population. The males are organized into dominance hierarchies. The group is ruled by a "central hierarchy" of dominant males who cooperate in defense and the control of subordinates. Access to females is determined to a large extent by rank, and possession is mostly limited to the time of estrus. In contrast, hamadryas males maintain permanent possession of females, and the societies are organized into three levels. The basic social element is the *one-male unit*, consisting of a mature male and the harem of females permanently associated with him. A limited number of one-male units combine in a *band*, the members of which stay together during part of the foraging expeditions and cooperate in the defense of food finds against other bands. The bands in turn collect at sleeping rocks to spend the night together more or less amicably. This sleeping unit, the *troop*, contains as many as 750 individuals in regions where suitable shelters are scarce, and as few as 12 where they are common. Finally, bachelor males, who constitute about 20 percent of the population, form little bands of their own.

The harems contain from one to as many as ten adult females. At their physical peak most males control from two to five of these adult consorts. The relationship is easily the most "sexist" known in all of the primates. The male herds the females, never letting them stray too far, associate with strangers, or quarrel too vigorously with one another. He employs forms of aggression that vary from a simple hostile stare or slap to a sharp bite on the neck (see Figure 25-3). The chastised female responds by running to the male.

Since the males sequester their harems with such jealousy, they are also responsible for most of the interactions with other hamadryas units. Young male leaders tend to initiate band movement by moving out with their families closely in tow. Older male leaders then either follow or remain seated, and their actions decide the issue for the band as a whole. When preparing to change position, the males notify one another with special gestures. Fighting between the bands is also conducted by the males. It consists almost entirely of spectacular bluffing during which the opponents fence at each other with open jaws and slap swiftly back and forth with their hands. Film analysis shows that in spite of appearances, physical contact seldom occurs. Only when one male turns and flees is he apt to receive a scratch on the anal region. Fights also end when one animal turns his head to expose the side of his neck. This surrender ritual stops the aggression of the winner instantly.

It is remarkable that in the face of all this jealousy and rage some overlord males tolerate the presence of a follower male. The attachment begins when a subadult male associates with estrous females in the harem. The overlord not only tolerates this intrusion, he allows the youngster to copulate with the females. Soon the subadult male accepts the older male as a leader, running to him like a female when threatened and following the unit out to the feeding sites. At this stage he shows the first, rudimentary tendency to form a harem of his own, by kidnapping infant males and females and holding on to them for periods of up to 30 minutes. As he matures, he ceases paying close attention to the overlord's females and begins to adopt and to mother juvenile females of his own. Thus the sexual bonds are formed, and reinforced by disciplinary aggression, long before copulation can be attempted. Now the two males, each with his own harem, constitute a team. As the older individual weakens with age, his mates stray and the harem grows smaller, but he can still count on the cooperation and support of his younger partner.

The Eastern Mountain Gorilla (*Gorilla gorilla beringei*)

The gorilla commands attention because it is the largest of the primates, its great males attaining a height of nearly 2 meters and a weight of 180 kilograms or more. But this "amiable vegetarian," as George Schaller called it, has sociobiological peculiarities that would make it worthy of attention even if it were a midget. The gorilla is the one anthropoid ape species organized into age-graded-male troops. Its social life is also one of the most muted in all the higher primates. Although the groups are cohesive and follow one another's movements closely, dominance behavior is very low-keyed and overt agression nearly nonexistent. Territorial spacing is either absent or extremely subtle and erratic, while sexual behavior is so rare that it has been observed in the wild only on a handful of occasions.

The species is comprised of isolated populations scattered across equatorial Africa. The easternmost fringe of the range is occupied by gorillas which are distinguished by longer hair and stronger development of the silverback trait in the males. They are collectively denoted as the subspecies *Gorilla g. beringei,* or, in the vernacular, the eastern mountain gorillas. Their range covers the Virunga Volcanoes and the Mt. Kahuzi district, which includes the mountains north and east of Lake Kivu and the surrounding lowlands. The gorillas are surprisingly adaptable, thriving in a diversity of habitats from lowland rain forest to the thick bamboo stands, *Hagenia* parkland, and *Lobelia-Senecio* groves of the high mountains. Total vegetarians, they eat the leaves, blossoms, shoots, fruit, and bark of many kinds of plants.

The key published work on the wild mountain gorilla is by Schaller (1963, 1965a,b). A newer, even more prolonged study by Dian Fossey has begun to add valuable supplementary information, but at the time of the present review had only been the subject of a preliminary report (Fossey, 1972). The social organization of at least this population of the species is now quite clear. The mountain gorillas live in groups of 2 to 30, which occupy home ranges that change slowly—over a period of weeks. In Schaller's overall census data silver-backed males, that is, those about ten years in age or older, constituted 13.1 percent of the group populations, black-backed (young adult) males 9.4 percent, adult females 34.1 percent, and infants and juveniles the remainder. A "typical" troop might consist of one silver-backed male, 0–2 black-backed males, about a half dozen females, and a comparable number of immature individuals. Lone males are relatively common, and Fossey observed one small troop composed entirely of bachelors. When these individuals are taken into account, the overall ratio in the population is approximately 1 male to 1.5 females. Some of the solitary males actively follow troops, giving the impression of being in leisurely transit from one group to another.

Mountain gorillas are organized into age-graded-male troops. The core of each troop is the silver-backed male, the adult females, and the young. Extra males, including both subordinate silverbacks and black-backed individuals, remain at the periphery. In spite of this form of dispersion, and the general slow tempo of gorilla social life, the groups are strongly cohesive. The cluster of individuals seldom exceeds 70 meters in diameter, and the dominant male is always within easy vocal range of the other troop members.

Dominance is well marked but subtle in expression. Rank is loosely correlated with size, so that the big silverback is usually at the top, with the somewhat smaller black-backed males dominating the females and young. If more than one silverback is present, the hierarchy is linear and influenced by age, with young and conspicuously aged males taking subordinate positions. Most dominance interactions consist of a mere acknowledgment of precedence. When two animals meet on a narrow trail the subordinate gives the right-of-way; subordinates also yield their sitting place if approached by superiors. Sometimes the dominant animal intimidates the subordinate by starting at it. At most it snaps its mouth or taps the body of the other animal with the back of its hand. Higher levels of aggression within the troop are quite rare, and even aggression directed at intruders is minimal, consisting mostly of bluffing on the part of the dominant male.

Gorillas communicate principally through the audiovisual channel. There are 16 or 17 distinct vocal displays, including the long-distance hooting of the silverbacks, and perhaps a somewhat smaller repertory of distinguishable facial expressions and postures (see Figure 25-4). It is a matter of considerable interest that this great ape, with its presumed higher level of intelligence, employs a communication system no richer than that of the majority of other social primates, or for that matter other social mammals and birds generally. It is only when we come to its nearest relative, the chimpanzee, that a new evolutionary grade in social behavior is approached. It has been stated that if the sociobiology of gorillas is not more advanced that that of other Old World monkeys and apes, it is at least qualitatively different in some important respects. The cumulative data of Schaller and Fossey now seem to contradict this notion. Gorilla life is certainly much quieter, slower in pace, and in some ways more subtle, but it does not seem to have diverged in any basic way from the mode of the great majority of other Old World species.

The Chimpanzee (*Pan troglodytes*)

By reference to most intuitive criteria chimpanzees are the socially most advanced of the nonhuman primates. They are organized into moderately large societies within which casual groups form, break up, and re-form with extraordinary fluidity. Although the societies are cohesive and occupy stable home ranges, their meetings are often amicable and they readily exchange adult females—not males as in other primate species. These two special qualities, great flexibility and openness, are enhanced by the intelligence and individuality in behavior of the troop members. The chimpanzee life cycle is characterized by a long period of socialization and loose but enduring ties between the mother and her adult offspring. Finally, the males are unique among nonhuman primates in the amount of cooperation they display

Figure 25-4 Gorillas have the most relaxed, amiable societies of any of the great apes and larger Old World monkeys. In the scene depicted here, a troop of mountain gorillas rests and feeds in a *Hypericum* forest 3,000 meters high in the Virunga Volcanoes of Uganda. The dominant silverback male stands in the left foreground. To his right are two adult females and a pair of two-year-old twins, who try to push each other off a branch in play resembling the human game "king of the castle." In the left rear three juveniles play "follow the leader." To the dominant male's left are another group of females; one cradles a nursing year-old infant, another grooms a three-year-old, while (on the far right) a third carries a two-year-old on her back while she feeds. Behind this group a black-backed male rests in a

hunched sitting position, while beyond him two black-backed males and a female forage and another silverback male rests in a prone position. In the upper right-hand corner can be seen a solitary male who watches the troop from afar. Notice the large amount of variation in facial features, which is believed to be used by the gorillas themselves in recognizing individual group members. The *Hypericum* forest is rich in wild celery and *Galium* vines, both of which are major foods of the gorillas. (Drawing by Sarah Landry; based on Schaller, 1965a,b, and personal communication, and Fossey, 1972.)

while hunting animals and in the subsequent begging and sharing of the meat.

The common chimpanzee *Pan troglodytes* ranges through equatorial Africa from Sierra Leone and Guinea on the Atlantic coast eastward to Lake Tanganyika and Lake Victoria. It occurs widely in many forested habitats, from rain forest to savanna-forest mosaics, at every elevation from sea level to 3,000 meters. It is semiterrestrial, spending 20–50 percent of its time on the ground on ordinary days. It forages during the day and builds sleeping nests in trees to spend the night. The chimpanzee is truly omnivorous, feeding to a large extent on fruit but also on the leaves, bark, and seeds of a wide variety of plant species. It collects termites and ants and at frequent intervals kills and eats small baboons and monkeys.

The results of extensive field studies by Jane Goodall and several Japanese investigators (see especially Izawa, 1970; Nishida and Kawanaka, 1972; and Sugiyama, 1968, 1973) indicate that the basic social unit of chimpanzees is a loose consociation of about 30–80 individuals that occupy a persistent and reasonably well-defined home range over a period of years. The home ranges are moderately large in size, 5–20 square kilometers at the Budongo Forest in Uganda and about 10 square kilometers near the Mahali Mountains in Tanzania, and they partially overlap one another. The duration of groups and their fidelity to the home ranges evidently persist over chimpanzee generations. Thus, contrary to earlier surmises, the societies are of demographic and not the casual form as defined in Chapter 6. When groups meet, for example at common feeding sites, they often travel together for short periods of time without evident antagonism. However, Sugiyama twice witnessed behavior at Budongo that resembles the territorial display of other kinds of primates. When two groups met, they mingled excitedly. They used exaggerated movements to eat leaves and fruits seldom touched under ordinary circumstances, ran over the ground, and clambered through the branches while shouting and barking. After about an hour of these noisy displays, each group withdrew toward the exclusive portions of its home range. At the Gombe Stream National Park in Tanzania, Goodall and her coworkers have witnessed fighting between the males of two adjacent groups, resulting in some injuries, fatalities, and territorial expansion. Even so, groups periodically exchange members during their brief encounters. Nishida and Kawanaka noted that the migrants at Budongo were mostly adult females, especially those that had become sexually receptive. Some females with children also transferred, but in each case they eventually returned to their home group.

In sum, chimpanzee populations appear to be organized along conventional lines. The temporary mingling of neighboring groups is unusual. But the apparent familiarity with neighbors as individuals is not at all out of the ordinary, having been documented in other species of mammals and birds. The transfer of females rather than males is peculiar, but its genetic consequences are the same as the all-male exchanges of other species.

The fluidity of chimpanzee internal organization is truly exceptional, however. The entire group is seldom seen together in the same place. The members gather during migrations from one part of the home range to another in search of special foods. For example, one group moved north in September at Budongo to find the juicy fruits of *Garcinia* plants. But most of the time much smaller parties form and reform within such groups in almost kaleidoscopic fashion. Except for the continued association of offspring with mothers, sometimes well past the time of weaning, the associations have no consistent demographic structure. They are in fact true casual groups of the sort common in human societies (see Chapter 6). Parties that discover fruit trees call in other chimpanzees by the carnival display, first described by the missionary Thomas S. Savage in 1844 as "hooting, screaming, and drumming with sticks on old logs." In fact, the chimpanzees spank tree trunks and buttresses with their hands, while running excitedly about, brachiating from branch to branch, and barking, whooping, and crying. The sounds can be heard more than a kilometer away, and parties within hearing distance often respond by running in the direction of the sound. The display is used in other contexts: when a party divides and one unit moves away, when a party starts to travel after resting or feeding, and sometimes with no apparent external stimulus. It serves to establish and strengthen ties within the group and perhaps, like the booming of howler monkeys and gibbons, to space the troops. When parties of the same group meet, it is usual for greeting ceremonies to be performed, especially between the adult males. A male newly arrived at a fruiting tree already occupied by a party may beat the buttresses of a tree while calling out. The male of the first party then approaches the newcomer, and the two mutually embrace and groom each other before settling down to eat. Sometimes the newcomer approaches the previous occupant directly and stretches out his hand. The other touches his hand, then the two embrace and groom.

Cooperation within chimpanzee parties is extraordinary in both kind and degree. Most of the time party members feed on fruit and other vegetable items in separate actions. But if the supply is limited

Figure 25-5 A temporary resting party of chimpanzees in the Gombe Stream National Park. Three adult males on the left are accompanied by two adult females, one with a female infant. (Photograph by Peter Marler and Richard Zigmond.)

—for example, if a human observer offers fruit and only the males are venturesome enough to pick it up—the chimps beg from one another and share the food. Cooperation of a different, more significant kind is displayed by chimpanzees while hunting animals. The cumulative observations of Suzuki (1971), Teleki (1973), and others have shown that predation on larger animals such as baboons is an infrequent but quite normal form of specialized behavior. The readiness to pursue, shown always by adult males, is conveyed by changes in posture, behavior, and facial expression. Other chimpanzees respond to these signals with alert, excited movements that often culminate in simultaneous pursuit. According to Teleki, predatory interest and intent are shown by a set or blank facial expression. The chimpanzee becomes unusually quiet and stares fixedly at the target prey. Its posture is tensed, and hair is partially erected all over its body. Ordinarily only mature males engage in the hunt, although on one occasion two females were seen to capture and kill a pair of young pigs. A notable aspect of the pursuit is the complete silence on the part of the chimpanzees until actual seizure is attempted. Such restraint on the part of one of the noisiest of all animals is most unusual.

Teleki distinguishes three modes of pursuit. In the first, the chimpanzees mingle with the prey and seize a victim with sudden, explosive movements. The second technique is a running pursuit. When the prey is a young baboon, capture can sometimes be achieved only after a battle with the adult males who rush out to defend it. The third mode, the most interesting of all, consists of stalking maneuvers in which the prey is helplessly treed or otherwise trapped. Most of the chimpanzees take positions to prevent the victim from escaping, while one of their companions moves in to try to seize it.

The distribution of the meat is a complicated procedure. As Jane Goodall and Teleki showed, various sequences of begging signals are used. The requesting animal may peer intently while placing its face close to the face of the meat eater or to the meat it is holding, or it may reach out and touch the meat itself or the chin and lips of the other animal. Alternatively, it extends an open hand with palm upward beneath the chin of the meat eater. Often a soft whimper or *hoo* accompanies these gestures. The individuals observed to beg belong to both sexes and all ages above two years. The meat eater sometimes rejects the request by pulling its booty away, moving to another position, or signaling refusal. Occasionally it acquiesces by allowing the other animal to chew directly on the meat or to remove small pieces with its hands. On four occasions during one year Teleki observed chimpanzees actually tear off pieces of meat and hand them over to supplicants.

Dominance behavior is well developed in the chimpanzee. A low-ranking individual gives way to a high-ranking one when they meet on a branch or when both approach the same piece of food. Its subordinate status is further signified when it detours around another animal or conciliates it by reaching out to touch it on the lips, thigh, or genital area. But these interactions are subtle. Overt threats and retreats are uncommon. The great majority of hostile acts involve adult males. Yet curiously in view of this fact, the dominance system appears to have no influence on access to females. Chimpanzee females are essentially promiscuous. They often copulate with more than one male in rapid succession, yet without provoking interference from nearby males. Once Jane Goodall saw seven males mount the same female, one after the other, with less than two minutes separating each of the first five copulations. On occasion the females themselves seek contact. An estrous female in Sugiyama's Budongo troop stopped grooming a dominant male, approached a young adult male on a nearby branch, copulated with him, and then resumed her ministrations to the first male. A second notable feature of chimpanzee dominance is that rank has little to do with allogrooming patterns. Chimpanzees groom one another regularly, seeming to use the behavior for mutual reassurance. Allogrooming occurs during a high percentage of those occasions, for example, when mothers and their offspring rejoin after a prolonged absence or when two parties of the same regional group meet during foraging excursions. Sometimes a dominant animal briefly grooms a subordinate that has approached it for reassurance, but in most cases it gives a mere token touch or pat.

Leadership, defined narrowly as the initiation of group movement, is well developed among chimpanzees. Ordinarily the dominant male of a party leads all the others. When the party is progressing rapidly from one food tree to another, the leader takes the front position. On other occasions it remains near the center or rear. Regardless of position it seldom loses control, because when it moves the rest move and when it halts, they halt also.

The rich communication system of chimpanzees has been described in detail by van Lawick-Goodall (1968b, 1971). It consists to a large extent of composite signals comprised of vocalizations, facial expressions, and body postures and movements. Touch, including allogrooming, is also frequently employed but is far poorer in signal diversity than the audiovisual system. Like human beings, the chimpanzee appears to make very little use of chemical signals. Yet it must be admitted that this subject has not been explicitly investigated with appropriate behavioral and chemical tests.

CHAPTER 26

Man: From Sociobiology to Sociology

Let us now consider man in the free spirit of natural history, as though we were zoologists from another planet completing a catalog of social species on Earth. In this macroscopic view the humanities and social sciences shrink to specialized branches of biology; history, biography, and fiction are the protocols of human ethology; and anthropology and sociology together constitute the sociobiology of a single primate species.

Homo sapiens is ecologically a very peculiar species. It occupies the widest geographical range and maintains the highest local densities of any of the primates. An astute ecologist from another planet would not be surprised to find that only one species of *Homo* exists. Modern man has preempted all the conceivable hominid niches. Two or more species of hominids did coexist in the past, when the *Australopithecus* man-apes in addition to an early *Homo* lived in Africa. But only one evolving line survived into late Pleistocene times to participate in the emergence of the most advanced human social traits.

Modern man is anatomically unique. His erect posture and wholly bipedal locomotion are not even approached in other primates that occasionally walk on their hind legs, including the gorilla and chimpanzee. The skeleton has been profoundly modified to accommodate the change: the spine is curved to distribute the weight of the trunk more evenly down its length; the chest is flattened to move the center of gravity back toward the spine; the pelvis is broadened to serve as an attachment for the powerful striding muscles of the upper legs and reshaped into a basin to hold the viscera; the tail is eliminated, its vertebrae (now called the coccyx) curved inward to form part of the floor of the pelvic basin; the occipital condyles have rotated far beneath the skull so that the weight of the head is balanced on them; the face is shortened to assist this shift in gravity; the thumb is enlarged to give power to the hand; the leg is lengthened; and the foot is drastically narrowed and lengthened to facilitate striding. Other changes have taken place. Hair has been lost from most of the body. It is still not known why modern man is a "naked ape." One plausible explanation is that nakedness served as a device to cool the body during the strenuous pursuit of prey in the heat of the African plains. It is associated with man's exceptional reliance on sweating to reduce body heat; the human body contains from two to five million sweat glands, far more than in any other primate species.

The reproductive physiology and behavior of *Homo sapiens* have also undergone extraordinary evolution. In particular, the estrous cycle of the female has changed in two ways that affect sexual and social behavior. Menstruation has been intensified. The females of some other primate species experience slight bleeding, but only in women is there a heavy sloughing of the wall of the "disappointed womb" with consequent heavy bleeding. The estrus, or period of female "heat," has been replaced by virtually continuous sexual activity. Copulation is initiated not by response to the conventional primate signals of estrus, such as changes in color of the skin around the female sexual organs and the release of pheromones, but by extended foreplay entailing mutual stimulation by the partners. The traits of physical attraction are. moreover, fixed in nature. They include the pubic hair of both sexes and the protuberant breasts and buttocks of women. The flattened sexual cycle and continuous female attractiveness cement the close marriage bonds that are basic to human social life.

At a distance a perceptive Martian zoologist would regard the globular head as a most significant clue to human biology. The cerebrum of *Homo* was expanded enormously during a relatively short span of evolutionary time (see Figure 26-1). Three million years ago *Australopithecus* had an adult cranial capacity of 400–500 cubic centimeters, comparable to that of the chimpanzee and gorilla. Two million years

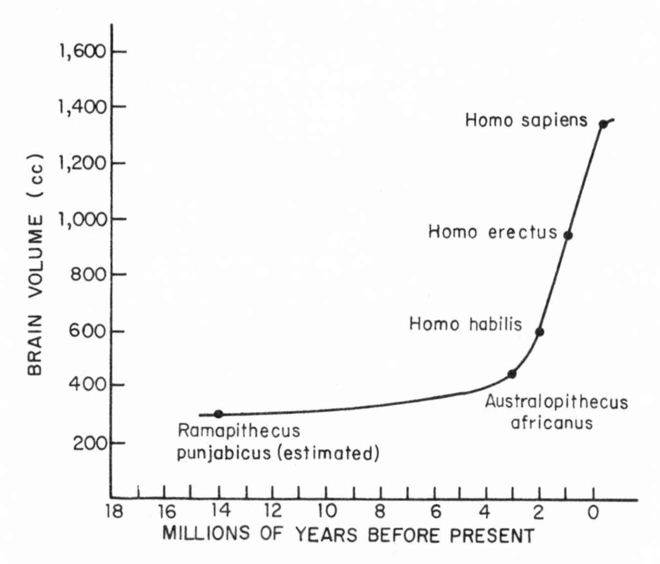

Figure 26-1 The increase in brain size during human evolution. (Redrawn from Pilbeam, 1972.)

later its presumptive descendant *Homo erectus* had a capacity of about 1,000 cubic centimeters. The next million years saw an increase to 1,400–1,700 cubic centimeters in Neanderthal man and 900–2,000 cubic centimeters in modern *Homo sapiens*. The growth in intelligence that accompanied this enlargement was so great that it cannot yet be measured in any meaningful way. Human beings can be compared among themselves in terms of a few of the basic components of intelligence and creativity. But no scale has been invented that can objectively compare man with chimpanzees and other living primates.

We have leaped forward in mental evolution in a way that continues to defy self-analysis. The mental hypertrophy has distorted even the most basic primate social qualities into nearly unrecognizable forms. Individual species of Old World monkeys and apes have notably plastic social organizations; man has extended the trend into a protean ethnicity. Monkeys and apes utilize behavioral scaling to adjust aggressive and sexual interactions; in man the scales have become multidimensional, culturally adjustable, and almost endlessly subtle. Bonding and the practices of reciprocal altruism are rudimentary in other primates; man has expanded them into great networks where individuals consciously alter roles from hour to hour as if changing masks.

It is the task of comparative sociobiology to trace these and other human qualities as closely as possible back through time. Besides adding perspective and perhaps offering some sense of philosophical ease, the exercise will help to identify the behaviors and rules by which individual human beings increase their Darwinian fitness through the manipulation of society. In a phrase, we are searching for the human biogram (Count, 1958; Tiger and Fox, 1971). One of the key questions, never far from the thinking of anthropologists and biologists who pursue real theory, is to what extent the biogram represents an adaptation to modern cultural life and to what extent it is a phylogenetic vestige. Our civilizations were jerry-built around the biogram. How have they been influenced by it? Conversely, how much flexibility is there in the biogram, and in which parameters particularly? Experience with other animals indicates that when organs are hypertrophied, phylogeny is hard to reconstruct. This is the crux of the problem of the evolutionary analysis of human behavior. In the remainder of the chapter, human qualities will be discussed insofar as they appear to be general traits of the species. Then current knowledge of the evolution of the biogram will be reviewed, and finally some implications for the planning of future societies will be considered.

Plasticity of Social Organization

The first and most easily verifiable diagnostic trait is statistical in nature. The parameters of social organization, including group size, properties of hierarchies, and rates of gene exchange, vary far more among human populations than among those of any other primate species. The variation exceeds even that occurring between the remaining primate species. Some increase in plasticity is to be expected. It represents the extrapolation of a trend toward variability already apparent in the baboons, chimpanzees, and other cercopithecoids. What is truly surprising, however, is the extreme to which it has been carried.

Why are human societies this flexible? Part of the reason is that the members themselves vary so much in behavior and achievement. Even in the simplest societies individuals differ greatly. Within a small tribe of !Kung Bushman can be found individuals who are acknowledged as the "best people"—the leaders and outstanding specialists among the hunters and healers. Even with an emphasis on sharing goods, some are exceptionally able entrepreneurs and unostentatiously acquire a certain amount of wealth. !Kung men, no less than men in advanced industrial societies, generally establish themselves by their mid-thirties or else accept a lesser status for life. There are some who never try to make it, live in run-down huts, and show little pride in themselves or their work (Pfeiffer, 1969). The ability to slip into such roles, shaping one's personality to fit, may itself be adaptive. Human societies are organized by high intelligence, and each member is faced by a mixture of social challenges that taxes all of his ingenuity. This baseline variation is amplified at the group level by other qualities exceptionally pronounced in human societies: the long, close period of socialization; the loose connectedness of the communication networks; the multiplicity of bonds; the capacity, especially within literate cultures, to communicate over long distances and periods of history; and from all these traits, the capacity to dissemble, to manipulate, and to exploit. Each parameter can be altered easily, and each has a marked effect on the final social structure. The result could be the observed variation among societies.

The hypothesis to consider, then, is that genes promoting flexibility in social behavior are strongly selected at the individual level. But note that variation in social organization is only a possible, not a necessary consequence of this process. In order to generate the amount of variation actually observed to occur, it is necessary for there to be multiple adaptive peaks. In other words, different forms of society within the same species must be nearly enough alike in survival ability for many to enjoy long tenure. The result would be a statistical ensemble of kinds of societies that, if not equilibrial, is at least not shifting rapidly toward one particular mode or another.

The alternative, found in some social insects, is flexibility in individual behavior and caste development, which nevertheless results in an approach toward uniformity in the statistical distribution of the kinds of individuals when all individuals within a colony are taken together. In honeybees and in ants of the genera *Formica* and *Pogonomyrmex*, "personality" differences are strongly marked even within single castes. Some individuals, referred to by entomologists as the elites, are unusually active, perform more than their share of lifetime work, and incite others to work through facilitation. Other colony members are consistently sluggish. Although they are seemingly healthy and live long lives, their per-individual output is only a small fraction of that of the elites. Specialization also occurs. Certain individuals remain with the brood as nurses far longer than the average, while others concentrate on nest building or foraging. Yet somehow the total pattern of behavior in the colony converges on the species average. When one colony with its hundreds or thousands of members is compared with another of the same species, the statistical patterns of activity are about the same. We know that some of this consistency is due to negative feedback. As one requirement such as brood care or nest repair intensifies, workers shift their activities to compensate until the need is met, then change back again. Experiments have shown that disruption of the feedback loops, and thence deviation by the colony from the statistical norms, can be disastrous. It is therefore not surprising to find that the loops are both precise and powerful (Wilson, 1971a).

The controls governing human societies are not nearly so strong, and the effects of deviation are not so dangerous. The anthropological literature abounds with examples of societies that contain obvious inefficiencies and even pathological flaws—yet endure. The slave society of Jamaica, compellingly described by Orlando Patterson (1967), was unquestionably pathological by the moral canons of civilized life. "What marks it out is the astonishing neglect and distortion of almost every one of the basic prerequisites of normal human living. This was a society in which clergymen were the 'most finished debauchees' in the land; in which the institution of marriage was officially condemned among both masters and slaves; in which the family was unthinkable to the vast majority of the population and promiscuity the norm; in which education was seen as an absolute waste of time and teachers shunned like the plague; in which the legal system was quite deliberately a travesty of anything that could be called justice; and in which all forms of refinements, of art, of folkways, were either absent or in a state of total disintegration. Only a small proportion of whites, who monopolized almost all of the fertile land in the island, benefited from the system. And these, no sooner had they secured their fortunes, abandoned the land which the production of their own wealth had made unbearable to live in, for the comforts of the mother country." Yet this Hobbesian world lasted for nearly two centuries. The people multiplied while the economy flourished.

How can such variation in social structure persist? The explanation may be lack of competition from other species, resulting in what biologists call ecological release. During the past ten thousand years or longer, man as a whole has been so successful in dominating his environment that almost any kind of culture can succeed for a while, so long as it has a modest degree of internal consistency and does not shut off reproduction altogether. No species of ant or termite enjoys this freedom. The slightest inefficiency in constructing nests, in establishing odor trails, or in conducting nuptial flights could result in the quick extinction of the species by predation and competition from other social insects. To a scarcely lesser extent the same is true for social carnivores and primates. In short, animal species tend to be tightly packed in the ecosystem with little room for experimentation or play. Man has temporarily escaped the constraint of interspecific competition. Although cultures replace one another, the process is much less effective than interspecific competition in reducing variance.

It is part of the conventional wisdom that virtually all cultural variation is phenotypic rather than genetic in origin. This view has gained support from the ease with which certain aspects of culture can be altered in the space of a single generation, too quickly to be evolutionary in nature. The drastic alteration in Irish society in the first two years of the potato blight (1846–1848) is a case in point. Another is the shift in the Japanese authority structure during the American occupation following World War II. Such examples can be multiplied endlessly—they are the substance of history.

The extreme orthodox view of the environmentalism goes further, holding that in effect there is no genetic variance in the transmission of culture. In other words, the capacity for culture is transmitted by a single human genotype. Dobzhansky (1963) stated this hypothesis as follows: "Culture is not inherited through genes, it is acquired by learning from other human beings . . . In a sense, human genes have surrendered their primacy in human evolution to an entirely new, nonbiological or superorganic agent, culture. However, it should not be forgotten that this agent is entirely dependent on the human genotype." Although the genes have given away most of their sovereignty, they maintain a certain amount of influence in at least the behavioral qualities that underlie variations between cultures. Moderately high heritability has been documented in introversion-extroversion measures, personal tempo, psychomotor and sports activities, neuroticism, dominance, depression, age of first sexual activity, timing of major cognitive development, and the tendency toward certain forms of mental illness such as schizophrenia (Parsons, 1967; Lerner, 1968; Martin et al., 1977; R. Wilson, 1979). Even a small portion of this variance invested in population differences might predispose societies toward cultural differences. At the very least, we should try to measure this amount. It is not valid to point to the absence of a behavioral trait in one or a few societies as conclusive evidence that the trait is environmentally induced and has no genetic disposition in man. The very opposite could be true.

In fact, Freedman (1974, 1979) and his associates have demonstrated marked racial differences in locomotion, posture, muscular tone, and emotional response of newborn infants that cannot reasonably be explained as the result of training or even conditioning within the womb. Chinese-American newborns, for example, tend to be less changeable, less easily perturbed by noise and movement, better able to adjust to new stimuli and discomfort, and quicker to calm themselves than Caucasian-American infants. To use a more precise phrasing, it can be said that a random sample of infants whose ancestors originated in certain parts of China differ in these behavioral traits from a comparable sample of European ancestry.

Clearly, we need a discipline of anthropological genetics. In the interval before we acquire it, it should be possible to characterize the human biogram by two indirect methods. First, models can be constructed from the most elementary rules of human behavior. Insofar as they can be tested, the rules will characterize the biogram in much the same way that ethograms drawn by zoologists identify the "typical" behavioral repertories of animal species. The rules can be legitimately compared with the ethograms of other primate species. Variation in the rules among human cultures, however slight, might provide clues to underlying genetic differences, particularly when it is correlated with variation in behavioral traits known to be heritable.

Social scientists have in fact begun to take this first approach, although in a different context from the one suggested here. Abraham Maslow (1954, 1972) postulated that human beings respond to a hierarchy of needs, such that the lower levels must be satisfied before much attention is devoted to the higher ones. The most basic needs are hunger and sleep. When these are met, safety becomes the primary consideration, then the need to belong to a group and receive love, next self-esteem, and finally self-actualization and creativity. The ideal society in Maslow's dream is one which "fosters the fullest development of human potentials, of the fullest degree of humanness." When the biogram is freely expressed, its center of gravity should come to rest in the higher levels. A second social scientist, George C. Homans (1961), has adopted a Skinnerian approach in an attempt to reduce human behavior to the basic processes of associative learning. The rules he postulates are the following:

1. If in the past the occurrence of a particular stimulus-situation has been the occasion on which a man's activity has been rewarded, then the more similar the present stimulus-situation is to the past one, the more likely the man is at the present time to emit this activity or one similar to it.

2. The more often within a given period of time a man's activity rewards the behavior of another, the more often the other will perform the behavior.

3. The more valuable to a man a unit of the activity another gives him, the more often he behaves in the manner rewarded by the activity of the other.

4. The more often a man has in the recent past received a rewarding activity from another, the less

valuable any further unit of that activity becomes to him.

Maslow the ethologist and visionary seems a world apart from Homans the behaviorist and reductionist. Yet their approaches are reconcilable. Homans' rules can be viewed as comprising some of the enabling devices by which the human biogram is expressed. His operational word is *reward*, which is in fact the set of all interactions defined by the emotive centers of the brain as desirable. According to evolutionary theory, desirability is measured in units of genetic fitness, and the emotive centers have been programmed accordingly. Maslow's hierarchy is simply the order of priority in the goals toward which the rules are directed.

The other indirect approach to anthropological genetics is through phylogenetic analysis. By comparing man with other primate species, it might be possible to identify basic primate traits that lie beneath the surface and help to determine the configuration of man's higher social behavior. This approach has been taken with great style and vigor in a series of popular books by Konrad Lorenz (*On Aggression*), Robert Ardrey (*The Social Contract*), Desmond Morris (*The Naked Ape*), and Lionel Tiger and Robin Fox (*The Imperial Animal*). Their efforts were salutary in calling attention to man's status as a biological species adapted to particular environments. The wide attention they received broke the stifling grip of the extreme behaviorists, whose view of the mind of man as a virtually equipotent response machine was neither correct nor heuristic. But their particular handling of the problem tended to be inefficient and misleading. They selected one plausible hypothesis or another based on a review of a small sample of animal species, then advocated the explanation to the limit.

The correct approach using comparative ethology is to base a rigorous phylogeny of closely related species on many biological traits. Then social behavior is treated as the dependent variable and its evolution deduced from it. When this cannot be done with confidence (and it cannot in man), the next best procedure is to establish the lowest taxonomic level at which each character shows significant intertaxon variation. Characters that shift from species to species or genus to genus are the most labile. We cannot safely extrapolate them from the cercopithecoid monkeys and apes to man. In the primates these labile qualities include group size, group cohesiveness, openness of the group to others, involvement of the male in parental care, attention structure, and the intensity and form of territorial defense. Characters are considered conservative if they remain constant at the level of the taxonomic family or throughout the order Primates, and they are the ones most likely to have persisted in relatively unaltered form into the evolution of *Homo*. These conservative traits include aggressive dominance systems, with males generally dominant over females; scaling in the intensity of responses, especially during aggressive interactions; intensive and prolonged maternal care, with a pronounced degree of socialization in the young; and matrilineal social organization. This classification of behavioral traits offers an appropriate basis for hypothesis formation. It allows a qualitative assessment of the probabilities that various behavioral traits have persisted into modern *Homo sapiens*. The possibility of course remains that some labile traits are homologous between man and, say, the chimpanzee. And conversely, some traits conservative throughout the rest of the primates might nevertheless have changed during the origin of man. Furthermore, the assessment is not meant to imply that conservative traits are more genetic—that is, have higher heritability—than labile ones. Lability can be based wholly on genetic differences between species or populations within species. Returning finally to the matter of cultural evolution, we can heuristically conjecture that the traits proven to be labile are also the ones most likely to differ from one human society to another on the basis of genetic differences. The evidence, reviewed in Table 26-1, is not inconsistent with this basic conception. Finally, it is worth special note that the comparative ethological approach does not in any way predict man's unique traits. It is a general rule of evolutionary studies that the direction of quantum jumps is not easily read by phylogenetic extrapolation.

Barter and Reciprocal Altruism

Sharing is rare among the nonhuman primates. It occurs in rudimentary form only in the chimpanzee and perhaps a few other Old World monkeys and apes. But in man it is one of the strongest social traits, reaching levels that match the intense trophallactic exchanges of termites and ants. As a result only man has an economy. His high intelligence and symbolizing ability make true barter possible. Intelligence also permits the exchanges to be stretched out in time, converting them into acts of reciprocal altruism (Trivers, 1971). The conventions of this mode of behavior are expressed in the familiar utterances of everyday life:

> "Give me some now; I'll repay you later."
> "Come to my aid this time, and I'll be your friend when you need me."
> "I really didn't think of the rescue as heroism; it was only what I would expect others to do for me or my family in the same situation."

Table 26-1 General social traits in human beings, classified according to whether they are unique, belong to a class of behaviors that are variable at the level of the species or genus in the remainder of the primates (labile), or belong to a class of behaviors that are uniform through the remainder of the primates (conservative).

Evolutionarily labile primate traits	Evolutionarily conservative primate traits	Human traits
		SHARED WITH SOME OTHER PRIMATES
Group size		Highly variable
Group cohesiveness		Highly variable
Openness of group to others		Highly variable
Involvement of male in parental care		Strong
Attention structure		Centripetal on leading males
Intensity and form of territorial defense		Highly variable, but territoriality is general
		SHARED WITH ALL OR ALMOST ALL OTHER PRIMATES
	Aggressive dominance systems, with males dominant over females	Consistent with other primates, although variable
	Scaling of responses, especially in aggressive interactions	Consistent with other primates
	Prolonged maternal care, pronounced socialization of young	Consistent with other primates
	Matrilineal organization	Mostly consistent with other primates
		UNIQUE
		True language, elaborate culture
		Sexual activity continuous through menstrual cycle
		Formalized incest taboos and marriage exchange rules with recognition of kinship networks
		Cooperative division of labor between adult males and females

Money, as Talcott Parsons has been fond of pointing out, has no value in itself. It consists only of bits of metal and scraps of paper by which men pledge to surrender varying amounts of property and services upon demand; in other words it is a quantification of reciprocal altruism.

Perhaps the earliest form of barter in early human societies was the exchange of meat captured by the males for plant food gathered by the females. If living hunter-gatherer societies reflect the primitive state, this exchange formed an important element in a distinctive kind of sexual bond.

Fox (1972), following Lévi-Strauss (1949), has argued from ethnographic evidence that a key early step in human social evolution was the use of women in barter. As males acquired status through the control of females, they used them as objects of exchange to cement alliances and bolster kinship networks. Preliterate societies are characterized by complex rules of marriage that can often be interpreted directly as power brokerage. This is particularly the case where the elementary negative marriage rules, proscribing certain types of unions, are supplemented by positive rules that direct which exchanges must be made. Within individual Australian aboriginal societies two moieties exist between which marriages are permitted. The men of each moiety trade nieces, or more specifically their sisters' daughters. Power accumulates with age, because a man can control the descendants of nieces as remote as the daughter of his sister's daughter. Combined with polygyny, the system insures both political and genetic advantage to the old men of the tribe.

For all its intricacy, the formalization of marital exchanges between tribes has the same approximate genetic effect as the haphazard wandering of male monkeys from one troop to another or the exchange of young mature females between chimpanzee populations. Approximately 7.5 percent of marriages contracted among Australian aborigines prior to European influence were intertribal, and similar rates have been reported in Brazilian Indians and other preliterate societies (Morton, 1969). The elementary theory of population genetics predicts that gene flow of the order of 10 percent per generation is more than enough to counteract fairly intensive natural pressures that tend to differentiate populations. Thus intertribal marital exchanges are a major factor in creating the observed high degree of genetic similarity among populations. The ultimate adaptive basis of exogamy is not gene flow per se but rather the avoidance of inbreeding. Again, a 10 percent gene flow is adequate for the purpose.

The microstructure of human social organization is based on sophisticated mutual assessments that lead to the making of contracts. As Erving Goffman correctly perceived, a stranger is rapidly but politely explored to determine his socioeconomic status, intelligence and education, self-perception, social attitudes, competence, trustworthiness, and emotional stability. The information, much of it subconsciously given and absorbed, has an eminently practical value. The probe must be deep, for the individual tries to create the impression that will gain him the maximum advantage. At the very least he maneuvers to avoid revealing information that will imperil his status. The presentation of self can be expected to contain deceptive elements:

> Many crucial facts lie beyond the time and place of interaction or lie concealed within it. For example, the "true" or "real" attitudes, beliefs, and emotions of the individual can be ascertained only indirectly, through his avowals or through what appears to be involuntary expressive behavior. Similarly, if the individual offers the others a product or service, they will often find that during the interaction there will be no time or place immediately available for eating the pudding that the proof can be found in. They will be forced to accept some events as conventional or natural signs of something not directly available to the senses. (Goffman, 1959)

Deception and hypocrisy are neither absolute evils that virtuous men suppress to a minimum level nor residual animal traits waiting to be erased by further social evolution. They are very human devices for conducting the complex daily business of social life. The level in each particular society may represent a compromise that reflects the size and complexity of the society. If the level is too low, others will seize the advantage and win. If it is too high, ostracism is the result. Complete honesty on all sides is not the answer. The old primate frankness would destroy the delicate fabric of social life that has built up in human populations beyond the limits of the immediate clan. As Louis J. Halle correctly observed, good manners have become a substitute for love.

Bonding, Sex, and Division of Labor

The building block of nearly all human societies is the nuclear family (Reynolds, 1968; Leibowitz, 1968). The populace of an American industrial city, no less than a band of hunter-gatherers in the Australian desert, is organized around this unit. In both cases the family moves between regional communities, maintaining complex ties with primary kin by means of visits (or telephone calls and letters) and the exchange of gifts. During the day the women and children remain in the residential area while the men forage for game or its symbolic equivalent in the form of barter and money. The males cooperate in bands to hunt or deal with neighboring groups. If not actually blood relations, they tend at least to act as "bands of brothers." Sexual bonds are carefully contracted in observance with tribal customs and are intended to be permanent. Polygamy, either covert or explicitly sanctioned by custom, is practiced predominantly by the males. Sexual behavior is nearly continuous through the menstrual cycle and marked by extended foreplay. Morris (1967a), drawing on the data of Masters and Johnson (1966) and others, has enumerated the unique features of human sexuality that he considers to be associated with the loss of body hair: the rounded and protuberant breasts of the young woman, the flushing of areas of skin during coition, the vasodilation and increased erogenous sensitivity of the lips, soft portions of the nose, ear, nipples, areolae, and genitals, and the large size of the male penis, especially during erection. As Darwin himself noted in 1871, even the naked skin of the woman is used as a sexual releaser. All of these alterations serve to cement the permanent bonds, which are unrelated in time to the moment of ovulation. Estrus has been reduced to a vestige, to the consternation of those who attempt to practice birth control by the rhythm method. Sexual behavior has been largely dissociated from the act of fertilization. It is ironic that religionists who forbid sexual activity except for purposes of procreation should do so on the basis of "natural law." Theirs is a misguided effort in comparative ethology, based on the incorrect assumption that in reproduction man is essentially like other animals.

The extent and formalization of kinship ties prevailing in almost all human societies are also unique features of the biology of our species. Kinship systems provide at least three distinct advantages. First, they bind alliances between tribes and subtribal units and provide a conduit for the conflict-free emigration of young members. Second, they are an important part of the bartering system by which certain males achieve dominance and leadership. Finally, they serve as a homeostatic device for seeing groups through hard times. When food grows scarce, tribal units can call on their allies for altruistic assistance in a way unknown in other social primates. The Athapaskan Dogrib Indians, a hunter-gatherer people of the northwestern Canadian arctic, provide one example. The Athapaskans are organized loosely by the bilateral primary linkage principle (June Helm, 1968). Local bands wander through a common territory, making intermittent contacts and exchanging members by intermarriage. When famine strikes, the endangered bands can coalesce with those temporarily better off. A second example is the Yanomamö of South America, who rely on kin when their crops are destroyed by enemies (Chagnon, 1968).

As societies evolved from bands through tribes into chiefdoms and states, some of the modes of bonding were extended beyond kinship networks to include other kinds of alliances and economic agreements. Because the networks were then larger, the lines of communication longer, and the interactions more diverse, the total systems became vastly more complex. But the moralistic rules underlying these arrangements appear not to have been altered a great deal. The average individaul still operates under a formalized code no more elaborate than that governing the members of hunter-gatherer societies.

Role Playing and Polyethism

The superman, like the super-ant or super-wolf, can never be an individual; it is the society, whose members diversify and cooperate to create a composite well beyond the capacity of any conceivable organism. Human societies have effloresced to levels of extreme complexity because their members have the intelligence and flexibility to play roles of virtually any degree of specification, and to switch them as the occasion demands. Modern man is an actor of many parts who may well be stretched to his limit by the constantly shifting demands of his environment. As Goffman (1961) observed, "Perhaps there are times when an individual does march up and down like a wooden soldier, tightly rolled up in a particular role. It is true that here and there we can pounce on a moment when an individual sits fully astride a single role, head erect, eyes front, but the next moment the picture is shattered into many pieces and the individual divides into different persons holding the ties of different spheres of life by his hands, by his teeth, and by his grimaces. When seen up close, the individual, bringing together in various ways all the connections he has in life, becomes a blur." Little wonder that the most acute inner problem of modern man is identity.

Roles in human societies are fundamentally different from the castes of social insects. The members of human societies sometimes cooperate closely in insectan fashion, but more frequently they compete for the limited resources allocated to their role-sector. The best and most entrepreneurial of the role-actors usually gain a disproportionate share of the rewards, while the least successful are displaced to other, less desirable positions. In addition, individuals attempt to move to higher socioeconomic positions by changing roles. Competition between classes also occurs, and in great moments of history it has proved to be a determinant of societal change.

A key question of human biology is whether there exists a genetic predisposition to enter certain classes and to play certain roles. Circumstances can be easily conceived in which such genetic differentiation might occur. The heritability of at least some parameters of intelligence and emotive traits is sufficient to respond to a moderate amount of disruptive selection. Dahlberg (1947) showed that if a single gene appears that is responsible for success and an upward shift in status, it can be rapidly concentrated in the uppermost socioeconomic classes. Suppose, for example, there are two classes, each beginning with only a 1 percent frequency of the homozygotes of the upwardly-mobile gene. Suppose further that 50 percent of the homozygotes in the lower class are transferred upward in each generation. Then in only ten generations, depending on the relative sizes of the groups, the upper class will be comprised of as many as 20 percent homozygotes or more and the lower class of as few as 0.5 percent or less. Using a similar argument, Herrnstein (1971b) proposed that as environmental opportunities become more nearly equal within societies, socioeconomic groups will be defined increasingly by genetically based differences in intelligence.

A strong initial bias toward such stratification is created when one human population conquers and subjugates another, a common enough event in human history. Genetic differences in mental traits, however slight, tend to be preserved by the raising of class barriers, racial and cultural discrimination, and physical ghettos. The geneticist C. D. Darlington (1969), among others, postulated this process to be a prime source of genetic diversity within human societies.

Yet despite the plausibility of the general argument, there is little evidence of any hereditary solidification of status. The castes of India have been in existence for 2,000 years, more than enough time for evolutionary divergence, but they differ only slighty in blood type and other measurable anatomical and physiological traits. Powerful forces can be identified that work against the genetic fixation of caste differences. First, cultural evolution is too fluid. Over a period of decades or at most centuries ghettos are replaced, races and subject people are liberated, the conquerors are conquered. Even within relatively stable societies the pathways of upward mobility are numerous. The daughters of lower classes tend to marry upward. Success in commerce or political life can launch a family from virtually any socioeconomic group into the ruling class in a single generation. Furthermore, there are many Dahlberg genes, not just the one postulated for argument in the simplest model. The hereditary factors of human success are strongly polygenic and form a long list, only a few of which have been measured. IQ constitutes only one subset of the components of intelligence. Less tangible but equally important qualities are creativity, entrepreneurship, drive, and mental stamina. Let us assume that the genes contributing to these qualities are scattered over many chromosomes. Assume further that some of the traits are uncorrelated or even negatively correlated. Under these circumstances only the most intense forms of disruptive selection could result in the formation of stable ensembles of genes. A much more likely circumstance is the one that apparently prevails: the maintenance of a large amount of genetic diversity within societies and the loose correlation of some of the genetically determined traits with success. This scrambling process is accelerated by the continuous shift in the fortunes of individual families from one generation to the next.

Even so, the influence of genetic factors in the adoption of certain *broad* roles cannot be discounted. Consider male homosexuality. The surveys of Kinsey and his coworkers showed that in the 1940's approximately 10 percent of the sexually mature males in the United States were mainly or exclusively homosexual for at least three years prior to being interviewed. Homosexuality is also exhibited by comparably high fractions of the male populations in many if not most other cultures. Kallmann's twin data indicate the probable existence of a genetic predisposition toward the condition. Accordingly, Hutchinson (1959) suggested that the homosexual genes may possess superior fitness in heterozygous conditions. His reasoning followed lines now standard in the thinking of population genetics. The homosexual state itself results in inferior genetic fitness, because of course homosexual men marry much less frequently and have far fewer children than their unambiguously heterosexual counterparts. The simplest way genes producing such a condition can be maintained in evolution is if they are superior in the heterozygous state, that is, if heterozygotes survive into maturity better, produce more offspring, or both. An interesting alternative hypothesis has been suggested to me by Herman T. Spieth (personal communication). The homosexual members of primitive societies may have functioned as helpers, either while hunting in company with other men or in more domestic occupations at the dwelling sites. Freed from the special obligations of parental duties, they could have operated wih special efficiency in assisting close relatives. Genes favoring homosexuality could then be sustained at a high equilibrium level by kin selection alone. It remains to be said that if such genes really exist, they are almost certainly incomplete in penetrance and variable in expressivity, meaning that which bearers of the genes develop the behavioral trait and to what degree they develop it depend on the presence or absence of modifier genes and the influence of the environment. A recent analysis of ethnographic data by Weinrich (1976) indicates that homosexuals in recent hunter-gatherer societies did indeed have a beneficent effect on relatives when they assumed their frequent roles of berdache and shaman.

Other basic types might exist, and perhaps the clues lie in full sight. In his study of British nursery children Blurton Jones (1969) distinguished two apparently basic behavioral types. "Verbalists," a small minority, often remained alone, seldom moved about, and almost never joined in rough-and-tumble play. They talked a great deal and spent much of their time looking at books. The other children were "doers." They joined groups, moved around a great deal, and spent much of their time painting and making objects instead of talking. Blurton Jones speculated that the dichotomy results from an early divergence in behavioral development persisting into maturity. Should it prove general, it might contribute fundamentally to diversity within cultures. There is no way of knowing whether the divergence is ultimately genetic in origin or triggered entirely by experiential events at an early age.

Communication

All of man's unique social behavior pivots on his use of language, which is itself unique. In any languge words are given arbitrary definitions within each culture and ordered according to a grammar that imparts new meaning above and beyond the definitions. The fully symbolic quality of the words and the sophistication of the grammar permit the creation of messages that are potentially infinite in number. Even communication about the system itself is made possible. This is the essential nature of human language. The basic attributes can be broken down, and other features of the transmission process itself can be added, to make a total of 16 design features (C. F. Hockett, reviewed by Thorpe, 1972a). Most of the features are found in at least rudimentary form in some other animal species. But the productivity and richness of human languages cannot be remotely approached even by chimpanzees taught to employ signs in simple sentences. The development of human speech represents a quantum jump in evolution comparable to the assembly of the eucaryotic cell.

Even without words human communication would be the richest known. The study of nonverbal communication has become a flourishing branch of the social sciences. Its codification is made difficult by the auxiliary role so many of the signals play to verbal communication. Categories of these signals are often defined inconsistently, and classifications are rarely congruent (see, for example, Renský, 1966; Crystal, 1969; Lyons, 1972). In Table 26-2 a composite arrangement is presented that I hope is both free of internal contradiction and consistent with current usage. The number of nonvocal signals, including all facial expressions, body postures and movement, and touch, probably number somewhat in excess of 100. Brannigan and Humphries (1972) have made a list of 136, which they believe is close to exhaustive. This number is consistent with the wholly independent estimate of Birdwhistle (1970), who believes that although the human face is capable of as many as 250,000 expressions, less than 100 sets of the expressions constitute distinct, meaningful symbols. Vocal paralanguage, insofar as it can be separated from the prosodic modifications of true speech, has not been cataloged so painstakingly. Grant (1969) recognized 6 distinct sounds, but several times this number would probably be distinguished by a zoologist accustomed to preparing ethograms of other primate species. In summary, all paralinguistic signals taken together almost certainly exceed 150 and may be close to 200. This repertory is larger than that of the majority of other mammals and birds by a factor of three or more, and it exceeds slightly the total repertories of both the rhesus monkey and the chimpanzee.

Table 26-2 The modes of human communication.

I. Verbal Communication (Language): the utterance of words and sentences
II. Nonverbal Communication
   A. Prosody: tone, tempo, rhythm, loudness, pacing, and other qualities of voice that modify the meaning of verbal utterances
   B. Paralanguage: signals separate from words used to supplement or to modify language
      1. Vocal paralanguage: grunts, giggles, laughs, sobs, cries, and other nonverbal sounds
      2. Nonvocal paralanguage: body posture, motion, and touch (kinesic communication); possibly also chemical communication

Another useful distinction in the analysis of human paralanguage can be made between signals that are prelinguistic, defined as having been in service before the evolutionary origin of true language, and those that are postlinguistic. The postlinguistic signals are most likely to have originated as pure auxiliaries to speech. One approach to the problem is through the phylogenetic analysis of the relevant properties of primate communications. Van Hooff (1972), for example, has established the homologues of smiling and laughing in facial expressions of the cercopithecoid monkeys and apes, thus classifying these human behaviors among our most primitive and universal signals.

Human language, as Marler (1965) argued, probably stemmed from richly graded vocal signals not unlike those employed by the rhesus monkey and chimpanzee, as opposed to the more discrete sounds characterizing the repertories of some of the lower primates. Human infants can utter a wide variety of vocalizations resembling those of macaques, baboons, and chimpanzees. But very early in their development they convert to the peculiar sounds of human speech. Multiple plosives, fricatives, nasals, vowels, and other sounds are combined to create the 40 or so basic phonemes. The human mouth and upper respiratory tract have been strongly modified to permit this vocal competence (see Figure 26-2). The crucial changes are associated with man's upright posture, which may have provided the initial

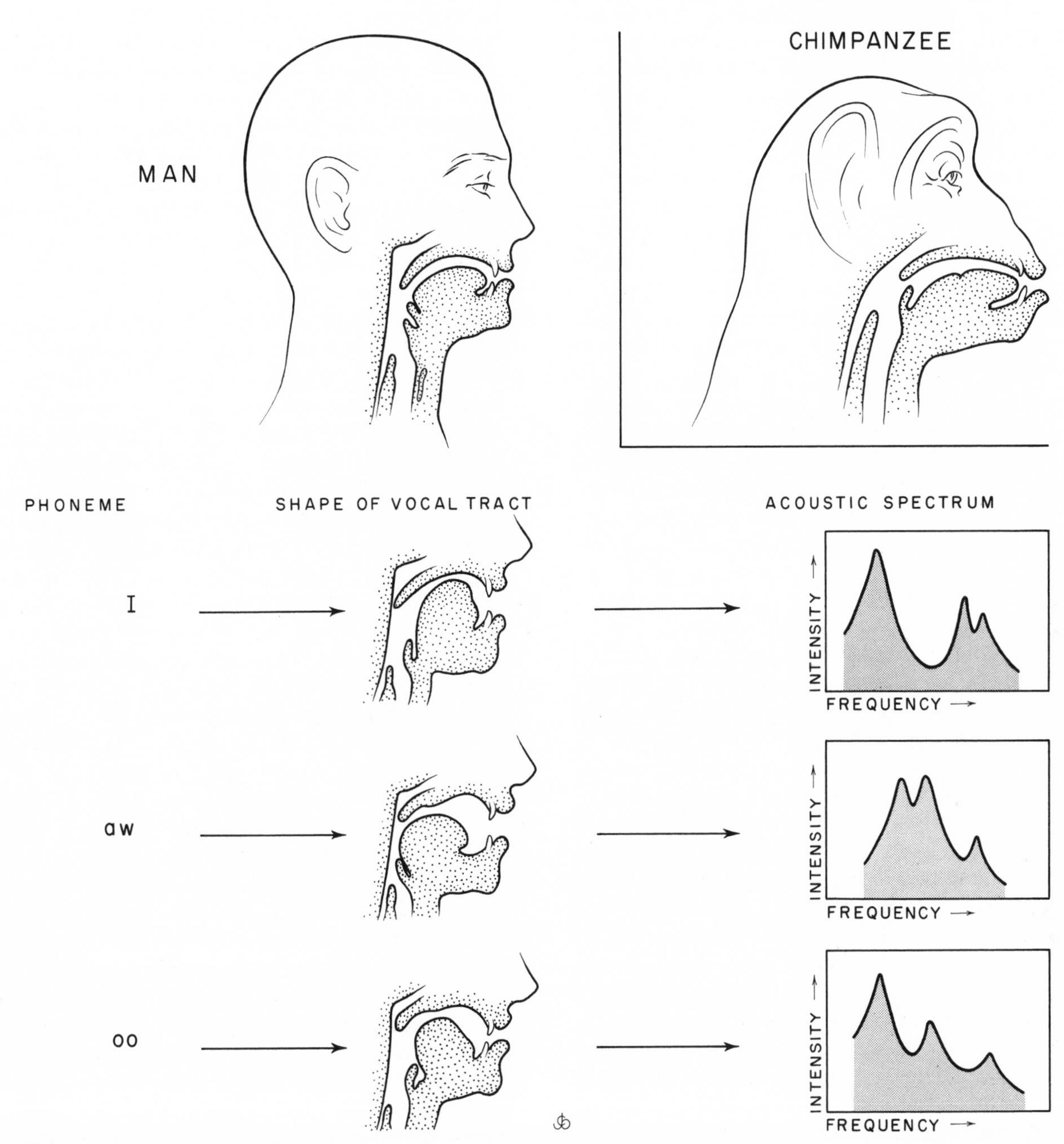

Figure 26-2 The human vocal apparatus has been modified in a way that greatly increases the variety of sounds that can be produced. The versatility was an essential accompaniment of the evolution of human speech. The upper diagrams show the ways in which man differs from the chimpanzee and other nonhuman primates: the angulation between the mouth and the upper respiratory tract is increased, the pharyngeal space is lengthened, and the back half of the tongue has come to form the front wall of the long tract above the vocal cords. The lower diagrams illustrate how movement of the tongue changes the shape of the air space to generate different sounds. (Modified from Howells, 1973, and Denes and Pinson, 1973.)

but still incomplete impetus toward the present modification. With the face directed fully forward, the mouth gave way to the upper pharyngeal space at a 90-degree angle. This configuration helped to push the rear of the tongue back until it formed part of the forward wall of the upper pharyngeal tract. Simultaneously the pharyngeal space and the epiglottis were both considerably lengthened.

These two principal changes, the shift in tongue position and lengthening of the pharyngeal tract, were responsible for the versatility in sound production. When air is forced upward through the vocal cords, it generates a buzzing noise that can be varied in intensity and duration but not in the all-important qualities of tone that produce phoneme differentiation. The latter effect is achieved as the air passes up through the pharyngeal tract and mouth cavity and out through the mouth. These structures together form an air tube which, like any cylinder, serves as a resonator. When its position and shape are altered, the tube emphasizes different combinations of frequencies emanating from the vocal cords. The result, illustrated in Figure 26-2, is the sounds we distinguish as phonemes (see also Lenneberg, 1967, and Denes and Pinson, 1973).

However, the great advance in language acquisition did not come from the ability to form many sounds. After all, it is theoretically possible for a highly intelligent being to speak only a *single* word and still communicate rapidly. It need only be programmed like a digital computer. Variation in loudness, duration, and pacing could be added to increase the transmission rate still more. Indeed, a single chemical substance, if modulated perfectly under ideal conditions, can generate up to 10,000 bits per second, far in excess of the capacity of human speech. Human languages gain their power instead from syntax, the dependence of meaning on the linear ordering of words. Each language possesses a grammar, the set of rules governing syntax. To truly understand the nature and origin of grammar would be to understand a great deal about the construction of the human mind. It is possible to distinguish three competing models that attempt to describe the known rules:

First hypothesis: *Probabilistic left-to-right model.* The explanation favored by extreme behavioristic psychologists is that the occurrence of a word is Markovian, meaning that its probability is determined by the immediately preceding word or string of words. The developing child learns which words to link together in each appropriate circumstance.

Second hypothesis: *Learned deep-structure model.* There exist a limited number of formal principles by which phrases of words are combined and juxtaposed to create various meanings. The child more or less unconsciously learns the deep structure of his own culture. Although the principles are finite in number, the sentences that can be generated from them are infinite in number. Animals cannot speak simply because they lack the necessary level of cognitive or intellectual ability, not because of the absence of any special "language faculty."

Third hypothesis: *Innate deep-structure model.* The formal principles exist as suggested in hypothesis number two, but they are partially or wholly genetic. In others words, at least some of the principles emerge by maturation in an invariant manner. A corollary of this proposition is that much of the deep structure of grammar is widespread if not universal in mankind, notwithstanding the profound differences in surface structure and word meaning that exist between languages. A second corollary is that animals cannot speak because they lack this inborn language faculty, which is a qualitatively unique human property and not simply an outcome of man's quantitatively superior intelligence. The innate deep-structure model is the one that has come to be associated most prominently with the name of Noam Chomsky, and appears to be currently favored by most psycholinguists.

The probabilistic left-to-right model has already been eliminated, at least in its extreme version. The number of transitional probabilities a child would have to learn in order to compute in a language such as English is enormous, and there is simply not enough time in childhood to master them all (Miller, Galanter, and Pribram, 1960). Grammatical rules are actually learned very rapidly and in a predictable sequence, with the child passing through forms of construction that anticipate the adult form while differing significantly from it (Brown, 1973). This kind of ontogeny is typical of the maturation of innate components of animal behavior. Nevertheless, the similarity cannot be taken as conclusive evidence of a genetic program general to humanity.

The ultimate resolution of the problem, as Roger Brown and other developmental psycholinguists have stressed, cannot be achieved until deep grammar itself has been securely characterized. This is a relatively new area of investigation, scarcely dating beyond Chomsky's *Syntactic Structures* (1957). From the beginning it has been marked by a complicated, rapidly shifting argumentation. The basic ideas have been presented in reviews by Slobin (1971) and Chomsky (1972). Here it will suffice to define the main processes recognized by the new linguistic analysis. *Phrase structure grammar,* which is exemplified in Figure 26-3, consists of the rules by which sentences are built up in a hierarchical manner. Phrases can be thought of as modules that are substituted for other, equivalent modules or added *de novo* into sentences to change meanings. These elements cannot be split and the parts interchanged without creating serious difficulties. In the example "The boy hit the ball," "the ball" is intuitively such a unit. It can be easily taken out and replaced with some other phrase such as "the shuttlecock" or simply the word "it." The combination "hit the" is not such a unit. Despite the fact that the two words are juxtaposed, they cannot be easily replaced without creating difficulties for the construction of the entire remainder of

RULES OF PHRASE STRUCTURE GRAMMAR

1. SENTENCE ⟶ NOUN PHRASE + VERB PHRASE
2. NOUN PHRASE ⟶ ARTICLE + NOUN
3. VERB PHRASE ⟶ VERB + NOUN PHRASE
4. ARTICLE ⟶ the, a
5. NOUN ⟶ boy, girl, ball
6. VERB ⟶ hit

TREE OF PHRASE STRUCTURES

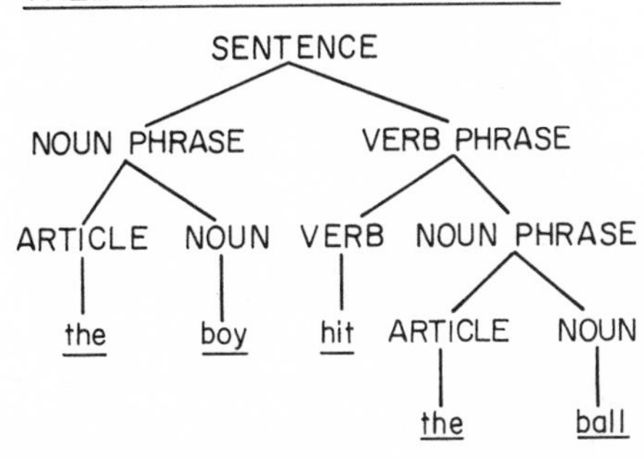

Figure 26-3 An example of the rules of phrase structure grammar in the English language. The simple sentence "The boy hit the ball" is seen to consist of a hierarchy of phrases. At each level one phrase can be substituted for another of equivalent composition, but the phrases cannot be split and their elements interchanged. (Based on Slobin, 1971.)

the sentence. By observing the rules we all know subconsciously, the sentence can be expanded by the insertion of appropriately selected phrases: *After taking his position,* the *little* boy *swung twice and finally* hit the ball *and ran to first base.*

In short, phrase structure grammar decrees the ways in which phrases can be formed. It generates what has been called the deep structure of the word strings as opposed to the surface structure, or the mere order in which the individual words appear. But of course the sequences in which phrases and terminal words appear are crucial to the meaning of the sentence. "The boy hit the ball" is very different from "What did the boy hit?" even though the deep (phrase) structure is similar. The rules by which the deep structures are converted into surface structures by the assembling of phrases are called *transformational grammar.* A transformation is an operation that converts one phrase structure into another. Among the most basic operations are substitutions ("what" for "the ball"), displacement (placing "what" before the verb), and permutation (switching the positions of related words).

The psycholinguists have described, for English, both phrase structure and transformational grammar. The evidence does not appear to be adequate, however, to choose between hypotheses two and three, in other words to decide whether the grammars are innately programmed or whether they are learned. The basic operations of transformation occur in all known human languages. However, this observation by itself does not establish that the precise rules of transformation are the same.

Is there a universal grammar? This question is difficult to answer because most attempts to generalize the rules of deep grammar have been based on the semantic content of one particular language. Students of the subject seldom confront the problem as if it were genuinely scientific, in a way that would reveal how concrete and soluble it might be. In fact, natural scientists are easily frustrated by the diffuse, oblique quality of much of the psycholinguistic literature, which often seems unconcerned with the usual canons of proposition and evidence. The reason is that many of the writers, including Chomsky, are structuralists in the tradition of Lévi-Strauss and Piaget. They approach the subject with the implicit world view that the processes of the human mind are indeed structured, and also discrete, enumerable, and evolutionarily unique with no great need to be referred to the formulations of other scientific disciplines. The analysis is nontheoretical in the sense that it fails to argue from postulates that can be tested and extended empirically. Some psychologists, including Roger Brown and his associates and Fodor and Garrett (1966), have adduced testable propositions and pursued them with mixed results, but the trail of speculation on deep grammar has not been easy to follow even for these skillful experimentalists.

Like poet-naturalists, the structuralists celebrate idiosyncratic personal visions. They argue from hidden premises, relying largely on metaphor and exemplification, and with little regard for the method of multiple competing hypotheses. Clearly, this discipline, one of the most important in all of science, is ripe for the application of rigorous theory and properly meshed experimental investigation.

A key question that the new linguistics may never answer is when human language originated. Did speech appear with the first use of stone tools and the construction of shelters by the *Australopithecus* man-apes, over two million years ago? Or did it await the emergence of fully modern *Homo sapiens,* perhaps even the development of religious rites in the past 100,000 years? Lieberman (1968) believes that the date was relatively recent. He interprets the Makapan *Australopithecus* restored by Dart to fall close to the chimpanzee in the form of its palate and pharyngeal tract. If he is right, this early hominid might not have been able to articulate the sounds of human speech. The same conclusion has been drawn with respect to the anatomy and vocal capacity of the Neanderthal man (Lieberman et al., 1972), which if true places the origin of language in the latest stages of speciation in the genus *Homo*. Other theoretical aspects of the evolutionary origin of human speech have been discussed by Jane Hill (1972) and I. G. Mattingly (1972). Lenneberg (1971) has hypothesized that the capacity for mathematical reasoning originated as a slight modification of linguistic ability.

Culture, Ritual, and Religion

The rudiments of culture are possessed by higher primates other than man, including the Japanese monkey and chimpanzee, but only in man has culture thoroughly infiltrated virtually every aspect of life. Ethnographic detail is genetically underprescribed, resulting in great amounts of diversity among societies. Underprescription does not mean that culture has been freed from the genes. What has evolved is the capacity for culture, indeed the overwhelming tendency to develop one culture or another. Robin Fox (1971) put the argument in the following form. If the proverbial experiments of the pharaoh Psammetichos and James IV of Scotland had worked, and children reared in isolation somehow survived in good health,

> I do not doubt that they *could* speak and that, theoretically, given time, they or their offspring would invent and develop a language despite their never having been taught one. Furthermore, this language, although totally different from any known to us, would be analyzable by linguists on the same basis as other languages and translatable into all known languages. But I would push this further. If our new Adam and Eve could survive and breed—still in total isolation from any cultural influences—then eventually they would produce a society which would have laws about property, rules about incest and marriage, customs of taboo and avoidance, methods of settling disputes with a minimum of bloodshed, beliefs about the supernatural and practices relating to it, a system of social status and methods of indicating it, initiation ceremonies for young men, courtship practices including the adornment of females, systems of symbolic body adornment generally, certain activities and associations set aside for men from which women were excluded, gambling of some kind, a tool- and weapon-making industry, myths and legends, dancing, adultery, and various doses of homicide, suicide, homosexuality, schizophrenia, psychosis and neuroses, and various practitioners to take advantage of or cure these, depending on how they are viewed.

Culture, including the more resplendent manifestations of ritual and religion, can be interpreted as a hierarchical system of environmental tracking devices. In Chapter 7 the totality of biological responses, from millisecond-quick biochemical reactions to gene substitutions requiring generations, was described as such a system. At that time culture was placed within the scheme at the slow end of the time scale. Now this conception can be extended. To the extent that the specific details of culture are nongenetic, they can be decoupled from the biological system and arrayed beside it as an auxiliary system. The span of the purely cultural tracking system parallels much of the slower segment of the biological tracking system, ranging from days to generations. Among the fastest cultural responses in industrial civilizations are fashions in dress and speech. Somewhat slower are political ideology and social attitudes toward other nations, while the slowest of all include incest taboos and the belief or disbelief in particular high gods. It is useful to hypothesize that cultural details are for the most part adaptive in a Darwinian sense, even though some may operate indirectly through enhanced group survival (Washburn and Howell, 1960; Masters, 1970). A second proposition worth considering, to make the biological analogy complete, is that the rate of change in a particular set of cultural behaviors reflects the rate of change in the environmental features to which the behaviors are keyed.

Slowly changing forms of culture tend to be encapsulated in ritual. Some social scientists have drawn an analogy between human ceremonies and the displays of animal communication. This is not correct. Most animal displays are discrete signals conveying limited meaning. They are commensurate with the postures, facial expressions, and elementary sounds of human paralanguage. A few animal displays, such as the most complex forms of sexual advertisement and nest changing in birds, are so impressively elaborate that they have occasionally been termed ceremonies by zoologists. But even here the comparison is misleading. Most human rituals have more than just an immediate signal value. As Durkheim stressed, they not only label but reaffirm and rejuvenate the moral values of the community.

The sacred rituals are the most distinctively human. Their most elementary forms are concerned with magic, the active attempt to manipulate nature and the gods. Upper Paleolithic art from the caves of Western Europe shows a preoccupation with game animals. There are many scenes showing spears and arrows embedded in the bodies of the prey. Other drawings depict men dancing in animal disguises or standing with heads bowed in front of animals. Probably the function of the drawings was sympathetic magic, based on the quite logical notion that what is done with an image will come to pass with the real thing. This anticipatory action is comparable to the intention movements of animals, which in the course of evolution have often been ritualized into communicative signals. The waggle dance of the honeybee, it will be recalled, is a miniaturized rehearsal of the flight from the nest to the food. Primitive man might have understood the meaning of such complex animal behavior easily. Magic was, and still is in some societies, practiced by special people variously called shamans, sorcerers, or medicine men. They alone were believed to have the secret knowledge and power to deal effectively with the supernatural, and their influence sometimes exceeded that of the tribal headmen.

Formal religion *sensu stricto* has many elements of magic but is focused on deeper, more tribally oriented beliefs. Its rites celebrate the creation myths, propitiate the gods, and resanctify the tribal moral codes. Instead of a shaman controlling physical power, there is a priest who communes with the gods and curries their favor through obeisance, sacrifice, and the proffered evidences of tribal good behavior. In more complex societies, polity and religion have always blended naturally. Power belonged to kings by divine right, but high priests often ruled over kings by virtue of the higher rank of the gods.

It is a reasonable hypothesis that magic and totemism constituted direct adaptations to the environment and preceded formal religion in social evolution. Sacred traditions occur almost universally in human societies. So do myths that explain the origin of man or at the very least the relation of the tribe to the rest of the world. But belief in high gods is not universal. Among 81 hunter-gatherer societies surveyed by Whiting (1968), only 28, or 35 percent, included high gods in their sacred traditions. The concept of an active, moral God who created the world is even less widespread. Furthermore, this concept most commonly arises with a pastoral way of life. The greater the dependence on herding, the more likely the belief in a shepherd god of the Judaeo-Christian model (see Table 26-3). In other kinds of societies the belief occurs in 10 percent or less of the cases. Also, the God of monotheistic religions is always male. This strong patriarchal tendency has several cultural sources (Lenski and Lenski, 1970). Pastoral societies are highly mobile, tightly organized, and often militant, all features that tip the balance toward male authority. It is also significant that herding, the main economic base, is primarily the responsibility of men. Because the Hebrews were originally a herding people, the Bible describes God as a shepherd and the chosen people as his sheep. Islam, one of the strictest of all monotheistic faiths, grew to early power among the herding people of the Arabian peninsula. The intimate relation of the shepherd to his flock apparently provides a microcosm which stimulates deeper questioning about the relation of man to the powers that control him.

Table 26-3 The religious beliefs of 66 agrarian societies, partitioned according to the percentage of subsistence derived from herding. (From *Human Societies*, by G. and Jean Lenski. Copyright © 1970 by McGraw-Hill Book Company. Used with permission.)

Percentage of subsistence from herding	Percentage of societies believing in an active, moral creator God	Number of societies
36–45	92	13
26–35	82	28
16–25	40	20
6–15	20	5

An increasingly sophisticated anthropology has not given reason to doubt Max Weber's conclusion that more elementary religions seek the supernatural for the purely mundane rewards of long life, abundant land and food, the avoidance of physical catastrophes, and the defeat of enemies. A form of group selection also operates in the competition between sects. Those that gain adherents survive; those that cannot, fail. Consequently, religions, like other human institutions, evolve so as to further the welfare of their practitioners. Because this demographic benefit applies to the group as a whole, it can be gained in part by altruism and exploitation, with certain segments profiting at the expense of others. Alternatively, it can arise as the sum of generally increased individual fitnesses. The resulting distinction in social terms is between the more oppressive and the more beneficent religions. All religions are probably oppressive to some degree, especially when they are promoted by chiefdoms and states. The tendency is intensified when societies compete, since religion can be effectively harnessed to the purposes of warfare and economic exploitation.

The enduring paradox of religion is that so much of its substance is demonstrably false, yet it remains a driving force in all societies. Men would rather believe than know, have the void as purpose, as Nietzsche said, than be void of purpose. At the turn of the century Durkheim rejected the notion that such force could really be extracted from "a tissue of illusions." And since that time social scientists have sought the psychological Rosetta stone that might clarify the deeper truths of religious reasoning. In a penetrating analysis of this subject, Rappaport (1971) proposed that virtually all forms of sacred rites serve the purposes of communication. In addition to institutionalizing the moral values of the community, the ceremonies can offer information on the strength and wealth of tribes and families. Among the Maring of New Guinea there are no chiefs or other leaders who command allegiance in war. A group gives a ritual dance, and individual men indicate their willingness to give military support by whether they attend the dance or not. The strength of the consortium can then be precisely determined by a head count. In more advanced societies military parades, embellished by the paraphernalia and rituals of the state religion, serve the same purpose. The famous potlatch ceremonies of the Northwest Coast Indians enable individuals to advertise their wealth by the amount of goods they give away. Rituals also regularize relationships in which there would otherwise be ambiguity and wasteful imprecision. The best examples of this mode of communication are the *rites de passage*. As a boy matures, his transition from child to man is very gradual in a biological and psychological sense. There will be times when he behaves like a child when an adult response would have been more appropriate, and vice versa. The society has difficulty in classifying him one way or the other. The *rite de passage* eliminates this ambiguity by arbitrarily changing the classification from a continuous gradient into a dichotomy. It also serves to cement the ties of the young person to the adult group that accepts him.

To sanctify a procedure or a statement is to certify it as beyond question and imply punishment for anyone who dares to contradict it. So removed is the sacred from the profane in everyday life that simply to repeat it in the wrong circumstance is a transgression. This extreme form of certification, the heart of all religions, is granted to the practices and dogmas that serve the most vital interests of the group. The individual is prepared by the sacred rituals for supreme effort and self-sacrifice. Overwhelmed by shibboleths, special costumes, and the sacred dancing and music so accurately keyed to his emotive centers, he has a "religious experience." He is ready to reassert allegiance to his tribe and family, perform charities, consecrate his life, leave for the hunt, join the battle, die for God and country. *Deus vult* was the rallying cry of the First Crusade. God wills it, but the summed Darwinian fitness of the tribe was the ultimate if unrecognized beneficiary.

It was Henri Bergson who first identified a second force leading to the formalization of morality and religion. The extreme plasticity of human social behavior is both a great strength and a real danger. If each family worked out rules of behavior on its own, the result would be an intolerable amount of tradition drift and growing chaos. To counteract selfish behavior and the "dissolving power" of high intelligence, each society must codify itself. Within broad limits virtually any set of conventions works better than none at all. Because arbitrary codes work, organizations tend to be inefficient and marred by unnecessary inequities. As Rappaport succinctly expressed it, "Sanctification transforms the arbitrary into the necessary, and regulatory mechanisms which are arbitrary are likely to be sanctified." The process engenders criticism, and in the more literate and self-conscious societies visionaries and revolutionaries set out to change the system. Reform meets repression, because to the extent that the rules have been sanctified and mythologized, the majority of the people regard them as beyond question, and disagreement is defined as blasphemy.

This leads us to the essentially biological question of the evolution of indoctrinability (Campbell, 1972). Human beings are absurdly easy to indoctrinate—they *seek* it. If we assume for argument that indoctrinability evolves, at what level does natural selection take place? One extreme possibility is that the group is the unit of selection. When conformity becomes too weak, groups become extinct. In this version selfish, individualistic members gain the upper hand and multiply at the expense of others. But their rising prevalence accelerates the vulnerability of the society and hastens its extinction. Societies containing higher frequencies of conformer genes replace those that disappear, thus raising the overall frequency of the genes in the metapopulation of societies. The spread of the genes will occur more rapidly if the metapopulation (for example, a tribal complex) is simultaneously enlarging its range. Formal models of the process show that if the rate of societal extinction is high enough relative to the intensity of the counteracting individual selection, the altruistic genes can rise to moderately high levels. The genes might be of the kind that favors indoctrinability even at the expense of the individuals who submit. For example, the willingness to risk death in battle can favor group survival at the expense of the genes that permitted the fatal military discipline. The group-selection hypothesis is sufficient to account for the evolution of indoctrinability.

The competing, individual-level hypothesis is equally sufficient. It states that the ability of individuals to conform permits them to enjoy the benefits of membership with a minimum of energy expenditure and risk. Although their selfish rivals may gain a momentary advantage, it is lost in the long run through ostracism and repression. The conformists perform altruistic acts, perhaps even to the extent of risking their lives, not because of self-denying genes selected at the group level but because the group is occasionally able to take advantage of the indoctrinability which on other occasions is favorable to the individual.

The two hypotheses are not mutually exclusive.

Group and individual selection can be reinforcing. If war requires spartan virtues and eliminates some of the warriors, victory can more than adequately compensate the survivors in land, power, and the opportunity to reproduce. The average individual will win the inclusive fitness game, making the gamble profitable, because the summed efforts of the participants give the average member a more than compensatory edge. (For a fuller discussion of the sociobiology of religion, see Wilson, 1978).

Ethics

Scientists and humanists should consider together the possibility that the time has come for ethics to be removed temporarily from the hands of the philosophers and biologized. The subject at present consists of several oddly disjunct conceptualizations. The first is *ethical intuitionism,* the belief that the mind has a direct awareness of true right and wrong that it can formalize by logic and translate into rules of social action. The purest guiding precept of secular Western thought has been the theory of the social contract as formulated by Locke, Rousseau, and Kant. In our time the precept has been rewoven into a solid philosophical system by John Rawls (1971). His imperative is that justice should be not merely integral to a system of government but rather the object of the original contract. The principles called by Rawls "justice as fairness" are those which free and rational persons would choose if they were beginning an association from a position of equal advantage and wished to define the fundamental rules of the association. In judging the appropriateness of subsequent laws and behavior, it would be necessary to test their conformity to the unchallengeable starting position.

The Achilles heel of the intuitionist position is that it relies on the emotive judgment of the brain as though that organ must be treated as a black box. While few will disagree that justice as fairness is an ideal state for disembodied spirits, the conception is in no way explanatory or predictive with reference to human beings. Consequently, it does not consider the ultimate ecological or genetic consequences of the rigorous prosecution of its conclusions. Perhaps explanation and prediction will not be needed for the millenium. But this is unlikely—the human genotype and the ecosystem in which it evolved were fashioned out of extreme unfairness. In either case the full exploration of the neural machinery of ethical judgment is desirable and already in progress. One such effort, constituting the second mode of conceptualization, can be called *ethical behaviorism.* Its basic proposition, which has been expanded most fully by J. F. Scott (1971), holds that moral commitment is entirely learned, with operant conditioning being the dominant mechanism. In other words, children simply internalize the behavioral norms of the society. Opposing this theory is the *developmental-genetic conception* of ethical behavior. The best-documented version has been provided by Lawrence Kohlberg (1969). Kohlberg's viewpoint is structuralist and specifically Piagetian, and therefore not yet related to the remainder of biology. Piaget has used the expression "genetic epistemology" and Kohlberg "cognitive-developmental" to label the general concept. However, the results will eventually become incorporated into a broadened developmental biology and genetics. Kohlberg's method is to record and classify the verbal responses of children to moral problems. He has delineated six sequential stages of ethical reasoning through which an individual may progress as part of his mental maturation. The child moves from a primary dependence on external controls and sanctions to an increasingly sophisticated set of internalized standards (see Table 26-4). The analysis has not yet been directed to the question of plasticity in the basic rules. Intracultural variance has not been measured, and heritability therefore not assessed. The difference between ethical behaviorism and the current version of developmental-genetic analysis is that the former postulates a mechanism (operant conditioning) without evidence and the latter presents evidence without postulating a mechanism. No great conceptual difficulty underlies this disparity. The study of moral development is only a more complicated and less tractable version of the genetic variance problem (see Chapter 2). With the accretion of data the two approaches can be expected to merge to form a recognizable exercise in behavioral genetics.

Even if the problem were solved tomorrow, however, an important piece would still be missing. This is the *genetic evolution of ethics.* In the first chapter of this book I argued that ethical philosophers intuit the deontological canons of morality by consulting the emotive centers of their own hypothalamic-limbic system. This is also true of the developmentalists, even when they are being their most severely objective. Only by interpreting the activity of the emotive centers as a biological adaptation can the meaning of the canons be deciphered. Some of the activity is likely to be outdated, a relic of adjustment to the most primitive form of tribal organization. Some of it may prove to be *in statu nascendi,* constituting new and quickly changing adaptations to agrarian and urban life. The resulting confusion will be reinforced by other factors. To the extent that unilaterally altruistic genes have been established in the population by group selection, they will be opposed by allelomorphs favored by individual selec-

Table 26-4 The classification of moral judgment into levels and stages of development. (Based on Kohlberg, 1969.)

Level	Basis of moral judgment	Stage of development
I	Moral value is defined by punishment and reward	1. Obedience to rules and authority to avoid punishment 2. Conformity to obtain rewards and to exchange favors
II	Moral value resides in filling the correct roles, in maintaining order and meeting the expectations of others	3. Good-boy orientation: conformity to avoid dislike and rejection by others 4. Duty orientation: conformity to avoid censure by authority, disruption of order, and resulting guilt
III	Moral value resides in conformity to shared standards, rights, and duties	5. Legalistic orientation: recognition of the value of contracts, some arbitrariness in rule formation to maintain the common good 6. Conscience or principle orientation: primary allegiance to principles of choice, which can overrule law in cases where the law is judged to do more harm than good

tion. The conflict of impulses under their various controls is likely to be widespread in the population, since current theory predicts that the genes will be at best maintained in a state of balanced polymorphism (Chapter 5). Moral ambivalency will be further intensified by the circumstance that a schedule of sex- and age-dependent ethics can impart higher genetic fitness than a single moral code which is applied uniformly to all sex-age groups. Some of the differences in the Kohlberg stages could be explained in this manner. For example, it should be of selective advantage for young children to be self-centered and relatively disinclined to perform altruistic acts based on personal principle. Similarly, adolescents should be more tightly bound by age-peer bonds within their own sex and hence unusually sensitive to peer approval. The reason is that at this time greater advantage accrues to the formation of alliances and rise in status than later, when sexual and parental morality become the paramount determinants of fitness. Genetically programmed sexual and parent-offspring conflict of the kind predicted by the Trivers models (Chapters 15 and 16) are also likely to promote age differences in the kinds and degrees of moral commitment. Finally, the moral standards of individuals during early phases of colony growth should differ in many details from those of individuals at demographic equilibrium or during episodes of overpopulation. Metapopulations subject to high levels of r extinction will tend to diverge genetically from other kinds of populations in ethical behavior (see Chapter 5).

If there is any truth to this theory of innate moral pluralism, the requirement for an evolutionary approach to ethics is self-evident. It should also be clear that no single set of moral standards can be applied to all human populations, let alone all sex-age classes within each population. To impose a uniform code is therefore to create complex, intractable moral dilemmas—these, of course, are the current condition of mankind. (For a fuller discussion of the implications of sociobiology for ethics and philosophy, see Wilson, 1978, and Stent, ed., 1978).

Esthetics

Artistic impulses are by no means limited to man. In 1962, when Desmond Morris reviewed the subject in *The Biology of Art*, 32 individual nonhuman primates had produced drawings and paintings in captivity. Twenty-three were chimpanzees, 2 were gorillas, 3 were orang-utans, and 4 were capuchin monkeys. None received special training or anything more than access to the necessary equipment. In fact, attempts to guide the efforts of the animals by inducing imitation were always unsuccessful. The drive to use the painting and drawing equipment was powerful, requiring no reinforcement from the human observers. Both young and old animals became so engrossed with the activity that they preferred it to being fed and sometimes threw temper tantrums when stopped. Two of the chimpanzees studied extensively were highly productive. "Alpha" produced over 200 pictures, while the famous "Congo," who deserves to be called the Picasso of the great apes, was responsible for nearly 400. Although most of the efforts consisted of scribbling, the patterns were far from random. Lines and smudges were spread over a blank page outward from a centrally located figure. When a drawing was started on one side of a blank page, the chimpanzee usually shifted to the opposite side to offset it. With time the calligraphy became bolder, starting with simple lines and progressing to more complicated multiple scribbles. Congo's patterns progressed along approximately the same developmental path as those of very young human children, yielding fan-shaped diagrams and even complete circles. Other chimpanzees drew crosses.

The artistic activity of chimpanzees may well be a special manifestation of their tool-using behavior. Members of the species display a total of about ten techniques, all of which require manual skill. Probably all are improved through practice, while at least a few are passed as traditions from one generation to the next. The chimpanzees have a considerable facility for inventing new techniques, such as the use of sticks to pull objects through cage bars and to pry open boxes. Thus the tendency to manipulate objects and to explore their uses appears to have an adaptive advantage for chimpanzees.

The same reasoning applies a fortiori to the origin of art in man. As Washburn (1970) pointed out, human beings have been hunter-gatherers for over 99 percent of their history, during which time each man made his own tools. The appraisal of form and skill in execution were necessary for survival, and they probably brought social approval as well. Both forms of success paid off in greater genetic fitness. If the chimpanzee Congo could reach the stage of elementary diagrams, it is not too hard to imagine primitive man progressing to representational figures. Once that stage was reached, the transition to the use of art in sympathetic magic and ritual must have followed quickly. Art might then have played a reciprocally reinforcing role in the development of culture and mental capacity. In the end, writing emerged as the idiographic representation of language.

Music of a kind is also produced by some animals. Human beings consider the elaborate courtship and territorial songs of birds to be beautiful, and probably ultimately for the same reasons they are of use to the birds. With clarity and precision they identify the species, the physiological condition, and the mental set of the singer. Richness of information and precise transmission of mood are no less the standards of excellence in human music. Singing and dancing serve to draw groups together, direct the emotions of the people, and prepare them for joint action. The carnival displays of chimpanzees described in earlier chapters are remarkably like human celebrations in this respect. The apes run, leap, pound the trunks of trees in drumming motions, and call loudly back and forth. These actions serve at least in part to assemble groups at common feeding grounds. They may resemble the ceremonies of earliest man. Nevertheless, fundamental differences appeared in subsequent human evolution. Human music has been liberated from iconic representation in the same way that true language has departed from the elementary ritualization characterizing the communication of animals. Music has the capacity for unlimited and arbitrary symbolization, and it employs rules of phrasing and order that serve the same function as syntax.

Territoriality and Tribalism

Anthropologists often discount territorial behavior as a general human attribute. This happens when the narrowest concept of the phenomenon is borrowed from zoology—the "stickleback model," in which residents meet along fixed boundaries to threaten and drive one another back. But earlier, in Chapter 12, I showed why it is necessary to define territory more broadly, as any area occupied more or less exclusively by an animal or group of animals through overt defense or advertisement. The techniques of repulsion can be as explicit as a precipitous all-out attack or as subtle as the deposit of a chemical secretion at a scent post. Of equal importance, animals respond to their neighbors in a highly variable manner. Each species is characterized by its own particular behavioral scale. In extreme cases the scale may run from open hostility, say, during the breeding season or when the population density is high, to oblique forms of advertisement or no territorial behavior at all. In less extreme forms, the scale may vary only from moderate aggressive displays at close range to calling at greater distances, or through several degrees of scent deposit, and so forth. One seeks to characterize the behavioral scale of the species and to identify the parameters that move individual animals up and down it.

If these qualifications are accepted, it is reasonable to conclude that territoriality is a general trait of hunter-gatherer societies. In a perceptive review of the evidence, Edwin Wilmsen (1973) found that these relatively primitive societies do not differ basically in their strategy of land tenure from many mammalian species. Systematic overt agression has been reported in a minority of hunter-gatherer peoples, for example the Chippewa, Sioux, and Washo of North America and the Murngin and Tiwi of Australia. Spacing and demographic balance were implemented by raiding parties, murder, and threats of witchcraft. The Washo of Nevada actively defended nuclear portions of their home ranges, within which they maintained their winter residences. Subtler and less direct forms of interaction can have the same result. The !Kung Bushman of the Nyae Nyae area refer to themselves as "perfect" or "clean" and other !Kung people as "strange" murderers who use deadly poisons.

Human territorial behavior is sometimes particularized in ways that are obviously functional. As recently as 1930 Bushman of the Dobe area in southwestern Africa recognized the principle of exclusive family land-holdings during the wet season. The rights extended only to the gathering of vegetable foods; other bands were allowed to hunt animals through the area (R. B. Lee in Wilmsen, 1973). Other hunter-gatherer peoples appear to have followed the same dual principle: more or less exclusive use by tribes or families of the richest sources of vegetable foods, opposed to broadly overlapping hunting ranges. Thus the original suggestion of Bartholomew and Birdsell (1953) that *Australopithecus* and the primitive *Homo* were territorial remains a viable hypothesis. Moreover, in obedience to the rule of ecological efficiency, the home ranges and territories were probably large and population density correspondingly low. This rule, it will be recalled, states that when a diet consists of animal food, roughly ten times as much area is needed to gain the same amount of energy yield as when the diet consists of plant food. Modern hunter-gatherer bands containing about 25 individuals commonly occupy between 1,000 and 3,000 square kilometers. This area is comparable to the home range of a wolf pack but as much as a hundred times greater than that of a troop of gorillas, which are exclusively vegetarian.

Hans Kummer (1971), reasoning from an assumption of territoriality, provided an important additional insight about human behavior. Spacing between groups is elementary in nature and can be achieved by a relatively small number of simple aggressive techniques. Spacing and dominance within groups is vastly more complex, being tied to all the remainder of the social repertory. Part of man's problem is that his intergroup responses are still crude and primitive, and inadequate for the extended extraterritorial relationships that civilization has thrust upon him. The unhappy result is what Garrett Hardin (1972) has defined as tribalism in the modern sense:

> Any group of people that perceives itself as a distinct group, and which is so perceived by the outside world, may be called a tribe. The group might be a race, as ordinarily defined, but it need not be; it can just as well be a religious sect, a political group, or an occupational group. The essential characteristic of a tribe is that it should follow a double standard of morality—one kind of behavior for in-group relations, another for out-group.
>
> It is one of the unfortunate and inescapable characteristics of tribalism that it eventually evokes counter-tribalism (or, to use a different figure of speech, it "polarizes" society).

Fearful of the hostile groups around them, the "tribe" refuses to concede to the common good. It is less likely to voluntarily curb its own population growth. Like the Sinhalese and Tamils of Sri Lanka, competitors may even race to outbreed each other. Resources are sequestered. Justice and liberty decline. Increases in real and imagined threats congeal the sense of group identity and mobilize the tribal members. Xenophobia becomes a political virtue. The treatment of nonconformists within the group grows harsher. History is replete with the escalation of this process to the point that the society breaks down or goes to war. No nation has been completely immune.

Early Social Evolution

Modern man can be said to have been launched by a two-stage acceleration in mental evolution. The first occurred during the transition from a larger arboreal primate to the first man-apes (*Australopithecus*). If the primitive hominid *Ramapithecus* is in the direct line of ancestry, as current opinion holds, the change may have required as much as ten million years. *Australopithecus* was present five million years ago, and by three million years B.P. it had speciated into several forms, including possibly the first primitive *Homo* (Tobias, 1973). As shown in Figure 26-1, the evolution of these intermediate hominids was marked by an accelerating increase in brain capacity. Simultaneously, erect posture and a striding, bipedal locomotion were perfected, and the hands were molded to acquire the precision grip. These early men undoubtedly used tools to a much greater extent than do modern chimpanzees. Crude stone implements were made by chipping, and rocks were pulled together to form what appear to be the foundations of shelters.

The second, much more rapid phase of acceleration began about 100,000 years ago. It consisted primarily of cultural evolution and must have been mostly phenotypic in nature, building upon the genetic potential in the brain that had accumulated over the previous millions of years. The brain had reached a threshold, and a wholly new, enormously more rapid form of mental evolution took over. This second phase was in no sense planned, and its potential is only now being revealed.

The study of man's origins can be referred to two questions that correspond to the dual stages of mental evolution:

——What features of the environment caused the hominids to adapt differently from other primates and started them along their unique evolutionary path?

——Once started, why did the hominids go so far?

The search for the prime movers of early human evolution has extended over more than 25 years. Participants in the search have included Dart (1949, 1956), Bartholomew and Birdsell (1953), Etkin (1954), Washburn and Avis (1958), Washburn et al. (1961), Rabb et al. (1967), Reynolds (1968), Schaller and Lowther (1969), C. J. Jolly (1970), and Kortlandt (1972). These writers have concentrated on two indisputably important facts concerning the biology of *Australopithecus* and early *Homo*. First, the evidence is strong that *Australopithecus africanus*, the species most likely to have been the direct ancestor of *Homo*, lived on the open savanna. The wear pattern of sand grains taken from the Sterkfontein fossils suggests a dry climate, while the pigs, antelopes, and other mammals found in association with the hominids are of the kind usually specialized for existence in grasslands. The australopithecine way of life came as the result of a major habitat shift. The ancestral *Ramapithecus* or an even more antecedent form lived in forests and was adapted for progression through trees by arm swinging. Only a very few other large-bodied primates have been able to join man in leaving the forest to spend most of their lives on the ground in open habitats (Figure 26-4). This is not to say that

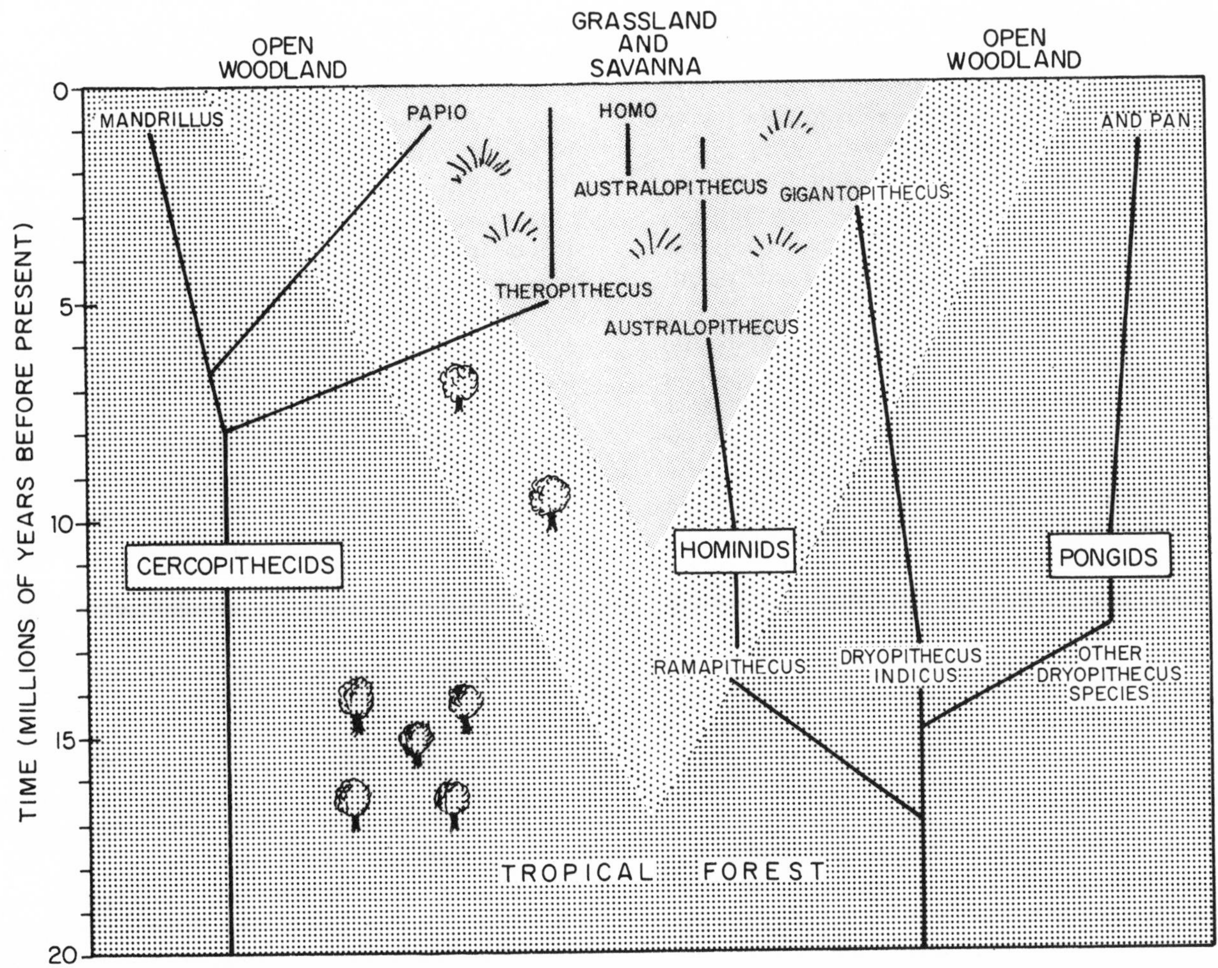

Figure 26-4 This simplified phylogeny of the Old World higher primates shows that only three existing groups have shifted from the forest to the savanna. They are the baboons (*Papio*), the gelada monkey (*Theropithecus gelada*), and man. (Based on Napier and Napier, 1967, and Simons and Ettel, 1970.)

bands of *Australopithecus africanus* spent all of their lives running about in the open. Some of them might have carried their game into caves and even lived there in permanent residence, although the evidence pointing to this often quoted trait is still far from conclusive (Kurtén, 1972). Other bands could have retreated at night to the protection of groves of trees, in the manner of modern baboons. The important point is that much or all of the foraging was conducted on the savanna.

The second peculiar feature of the ecology of early men was the degree of their dependence on animal food, evidently far greater than in any of the living monkeys and apes. The *Australopithecus* were catholic in their choice of small animals. Their sites contain the remains of tortoises, lizards, snakes, mice, rabbits, porcupines, and other small, vulnerable prey that must have abounded on the savanna. The man-apes also hunted baboons with clubs. From analysis of 58 baboon skulls, Dart estimated that all had been brought down by blows to the head, 50 from the front and the remainder from behind. The *Australopithecus* also appear to have butchered larger animals, including the giant sivatheres, or horned giraffes, and dinotheres, elephantlike forms with tusks that curved downward from the lower jaws. In early Acheulean times, when *Homo erectus* began employing stone axes, some of the species of large African mammals became extinct. It is reasonable to suppose that this impoverishment was due to excessive predation by the increasingly competent bands of men (Martin, 1966).

What can we deduce from these facts about the life of early man? Before an answer is attempted, it should be noted that very little can be inferred directly from comparisons with other living primates. Geladas and baboons, the only open-country forms, are primarily vegetarian. They represent a sample of at most six species, which differ too much from one another in social organization to provide a baseline for comparison. The chimpanzees, the most intelligent and socially sophisticated of the nonhuman primates, are forest-dwelling and mostly vegetarian. Only during their occasional ventures into predation do they display behavior that can be directly correlated with ecology in a way that has meaning for human evolution. Other notable features of chimpanzee social organization, including the rapidly shifting composition of subgroups, the exchange of females between groups, and the intricate and lengthy process of socialization, may or may not have been shared by primitive man. We cannot argue either way on the basis of ecological correlation. It is often stated in the popular literature that the life of chimpanzees reveals a great deal about the origin of man. This is not necessarily true. The manlike traits of chimpanzees could be due to evolutionary convergence, in which case their use in evolutionary reconstructions would be misleading.

The best procedure to follow, and one which I believe is relied on implicitly by most students of the subject, is to extrapolate backward from living hunter-gatherer societies. In Table 26-5 this technique is made explicit. Utilizing the synthesis edited by Lee and DeVore (1968; see especially J. W. M. Whiting, pp. 336–339), I have listed the most general traits of hunter-gatherer peoples. Then I have evaluated the lability of each behavioral category by noting the amount of variation in the category that occurs among the nonhuman primate species. The less labile the category, the more likely that the trait displayed by the living hunter-gatherers was also displayed by early man.

What we can conclude with some degree of confidence is that primitive men lived in small territorial groups, within which males were dominant over females. The intensity of aggressive behavior and the nature of its scaling remain unknown. Maternal care was prolonged, and the relationships were at least to some extent matrilineal. Speculation on remaining aspects of social life is not supported either way by the lability data and is therefore more tenuous. It is likely that the early hominids foraged in groups. To judge from the behavior of baboons and geladas, such behavior would have conferred some protection from large predators. By the time *Australopithecus* and early *Homo* had begun to feed on large mammals, group hunting almost certainly had become advantageous and even necessary, as in the African wild dog. But there is no compelling reason to conclude that men did the hunting while women stayed at home. This occurs today in hunter-gatherer societies, but comparisons with other primates offer no clue as to *when* the trait appeared. It is certainly not essential to conclude a priori that males must be a specialized hunter class. In chimpanzees males do the hunting, which may be suggestive. But in lions, it will be recalled, the females are the providers, often working in groups and with cubs in tow, while the males usually hold back. In the African wild dog both sexes participate. This is not to suggest that male group hunting was not an early trait of hominids, but only that there is no strong independent evidence to support the hypothesis.

This brings us to the prevailing theory of the origin of human sociality. It consists of a series of interlocking models that have been fashioned from bits of fossil evidence, extrapolations back from extant hunter-gatherer societies, and comparisons with other living primate species. The core of the theory

Table 26-5 Social traits of living hunter-gatherer groups and the likelihood that they were also possessed by early man.

Traits that occur generally in living hunter-gatherer societies	Variability of trait category among nonhuman primates	Reliability of concluding early man had the same trait through homology
Local group size:		
Mostly 100 or less	Highly variable but within range of 3–100	Very probably 100 or less but otherwise not reliable
Family as the nuclear unit	Highly variable	Not reliable
Sexual division of labor:		
Women gather, men hunt	Limited to man among living primates	Not reliable
Males dominant over females	Widespread although not universal	Reliable
Long-term sexual bonding (marriage) nearly universal; polygyny general	Highly variable	Not reliable
Exogamy universal, abetted by marriage rules	Limited to man among living primates	Not reliable
Subgroup composition changes often (fission-fusion principle)	Highly variable	Not reliable
Territoriality general, especially marked in rich gathering areas	Occurs widely, but variable in pattern	Probably occurred; pattern unknown
Game playing, especially games that entail physical skill but not strategy	Occurs generally, at least in elementary form	Very reliable
Prolonged maternal care; pronounced socialization of young; extended relationships between mother and children, especially mothers and daughters	Occurs generally in higher cercopithecoids	Very reliable

can be appropriately termed the *autocatalysis model.* It holds that when the earliest hominids became bipedal as part of their terrestrial adaptation, their hands were freed, the manufacture and handling of artifacts were made easier, and intelligence grew as part of the improvement of the tool-using habit. With mental capacity and the tendency to use artifacts increasing through mutual reinforcement, the entire materials-based culture expanded. Cooperation during hunting was perfected, providing a new impetus for the evolution of intelligence, which in turn permitted still more sophistication in tool using, and so on through cycles of causation. At some point, probably during the late *Australopithecus* period or the transition from *Australopithecus* to *Homo*, this autocatalysis carried the evolving populations to a certain threshold of competence, at which the hominids were able to exploit the antelopes, elephants, and other large herbivorous mammals teeming around them on the African plains. Quite possibly the process began when the hominids learned to drive big cats, hyenas, and other carnivores from their kills (see Figure 26-5). In time they became the primary hunters themselves and were forced to protect their prey from other predators and scavengers. The autocatalysis model usually includes the proposition that the shift to big game accelerated the process of mental evolution. The shift could even have been the impetus that led to the origin of early *Homo* from their australopithecine ancestors over two million years ago. Another proposition is that males became specialized for hunting. Child care was facilitated by close social bonding between the males, who left the domiciles to hunt, and the females, who kept the children and conducted most of the foraging for vegetable food. Many of the peculiar details of human sexual behavior and domestic life flow easily from this basic division of labor. But these details are not essential to the autocatalysis model. They are added because they are displayed by modern hunter-gatherer societies.

Although internally consistent, the autocatalysis model contains a curious omission—the triggering device. Once the process started, it is easy to see how it could be self-sustaining. But what started it? Why did the earliest hominids become bipedal instead of running on all fours like baboons and geladas? Clifford Jolly (1970) has proposed that the prime impetus was a specialization on grass seeds. Because the early pre-men, perhaps as far back as *Ramapithecus,* were the largest primates depending on grain, a premium was set on the ability to manipulate objects of very

Figure 26-5 At the threshold of autocatalytic social evolution two million years ago, a band of early men (*Homo habilis*) forages for food on the African savanna. In this speculative reconstruction the group is in the act of driving rival predators from a newly fallen dinothere. The great elephantlike creature had succumbed from exhaustion or disease, its end perhaps hastened by attacks from the animals closing in on it. The men have just entered the scene. Some drive away the predators by variously shouting, waving their arms, brandishing sticks, and throwing rocks, while a few stragglers, entering from the left, prepare to join the fray. To the right a female sabertooth cat (*Homotherium*) and her two grown cubs have been at least temporarily intimidated and are backing away. Their threat faces reveal the extraordinary gape of their jaws. In the left foreground, a pack of spotted hyenas (*Crocuta*) has also retreated but is ready to rush back the moment an opening is provided. The men are quite small, less than 1.5 meters in height, and

individually no match for the large carnivores. According to prevailing theory, a high degree of cooperation was therefore required to exploit such prey; and it evolved in conjunction with higher intelligence and the superior ability to use tools. In the background can be seen the environment of the Olduvai region of Tanzania as it may have looked at this time. The area was covered by rolling parkland and rimmed to the east by volcanic highlands. The herbivore populations were dense and varied, as they are today. In the left background are seen three-toed horses (*Hipparion*), while to the right are herds of wildebeest and giant horned giraffelike creatures called sivatheres. (Drawing by Sarah Landry; prepared in consultation with F. Clark Howell. The reconstruction of *Homotherium* was based in part on an Aurignacian sculpture; see Rousseau, 1971.)

small size relative to the hands. Man, in short, became bipedal in order to pick seeds. This hypothesis is by no means unsupported fantasy. Jolly points to a number of convergent features in skull and dental structure between man and the gelada, which feeds on seeds, insects, and other small objects. Moreover, the gelada is peculiar among the Old World monkeys and apes in sharing the following epigamic anatomical traits with man: growth of hair around the face and neck of the male and conspicuous fleshy adornments on the chest of the female. According to Jolly's model, the freeing of the hands of the early hominids was a preadaptation that permitted the increase in tool use and the autocatalytic concomitants of mental evolution and predatory behavior.

Later Social Evolution

Autocatalytic reactions in living systems never expand to infinity. Biological parameters normally change in a rate-dependent manner to slow growth and eventually bring it to a halt. But almost miraculously, this has not yet happened in human evolution. The increase in brain size and the refinement of stone artifacts indicate a gradual improvement in mental capacity throughout the Pleistocene. With the appearance of the Mousterian tool culture of *Homo sapiens neanderthalensis* some 75,000 years ago, the trend gathered momentum, giving way in Europe to the Upper Paleolithic culture of *Homo s. sapiens* about 40,000 years B.P. Starting about 10,000 years ago, agriculture was invented and spread, populations increased enormously in density, and the primitive hunter-gatherer bands gave way locally to the relentless growth of tribes, chiefdoms, and states. Finally, after A.D. 1400 European-based civilization shifted gears again, and knowledge and technology grew not just exponentially but superexponentially (see Figures 26-6, 26-7).

There is no reason to believe that during this final sprint there has been a cessation in the evolution of either mental capacity or the predilection toward special social behaviors. The theory of population genetics and experiments on other organisms show that substantial changes can occur in the span of less than 100 generations, which for man reaches back only to the time of the Roman Empire. Two thousand generations, roughly the period since typical *Homo sapiens* invaded Europe, is enough time to create new species and to mold them in major ways. Although we do not know how much mental evolution has actually occurred, it would be false to assume that modern civilizations have been built entirely on cap-

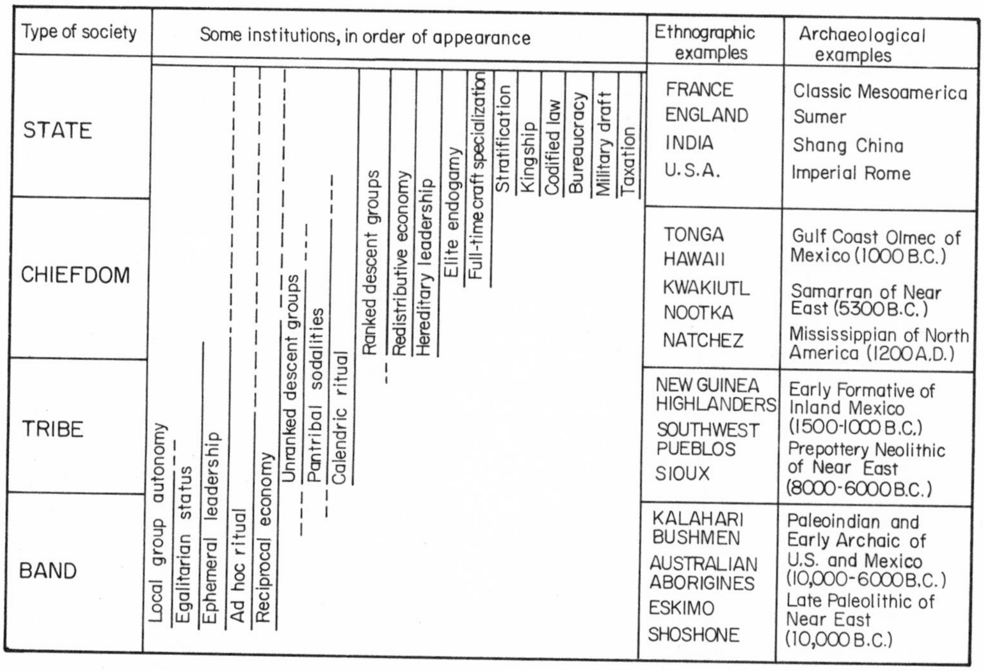

Type of society	Some institutions, in order of appearance	Ethnographic examples	Archaeological examples
STATE	Stratification, Kingship, Codified law, Bureaucracy, Military draft, Taxation	FRANCE ENGLAND INDIA U.S.A.	Classic Mesoamerica Sumer Shang China Imperial Rome
CHIEFDOM	Ranked descent groups, Redistributive economy, Hereditary leadership, Elite endogamy, Full-time craft specialization	TONGA HAWAII KWAKIUTL NOOTKA NATCHEZ	Gulf Coast Olmec of Mexico (1000 B.C.) Samarran of Near East (5300 B.C.) Mississippian of North America (1200 A.D.)
TRIBE	Unranked descent groups, Pantribal sodalities, Calendric ritual	NEW GUINEA HIGHLANDERS SOUTHWEST PUEBLOS SIOUX	Early Formative of Inland Mexico (1500-1000 B.C.) Prepottery Neolithic of Near East (8000-6000 B.C.)
BAND	Local group autonomy, Egalitarian status, Ephemeral leadership, Ad hoc ritual, Reciprocal economy	KALAHARI BUSHMEN AUSTRALIAN ABORIGINES ESKIMO SHOSHONE	Paleoindian and Early Archaic of U.S. and Mexico (10,000-6000 B.C.) Late Paleolithic of Near East (10,000 B.C.)

Figure 26-6 The four principal types of societies in ascending order of sociopolitical complexity, with living and extinct examples of each. A few of the sociopolitical institutions are shown, in the approximate order in which they are interpreted to have arisen. (From Flannery, 1972. Reproduced, with permission, from "The Cultural Evolution of Civilizations," *Annual Review of Ecology and Systematics*, Vol. 3, p. 401.)

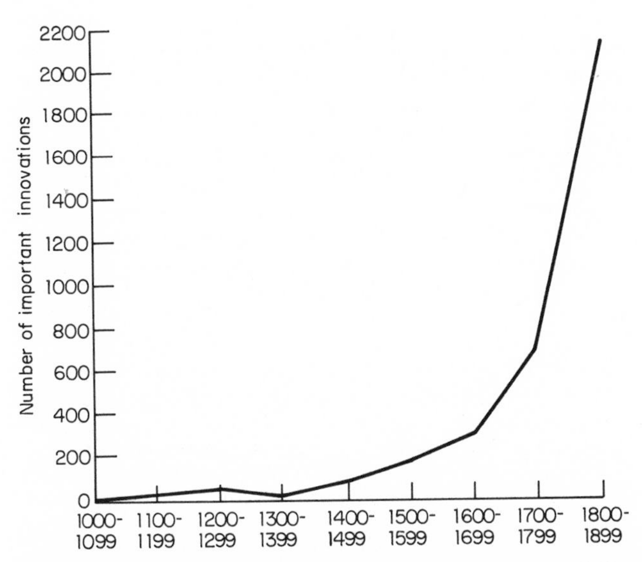

Figure 26-7 The number of important inventions and discoveries, by century, from A.D. 1000 to A.D. 1900. (From Lenski and Lenski, 1970; after Ogburn and Nimkoff, *Sociology*, 1958, published by Houghton Mifflin Company. Compiled from L. Darmstaedter and R. DuBois Reymond, *4000 Jahre-Pionier-Arbeit in den Exacten Wissenschaften*, Berlin, J. A. Stargart, 1904.)

ital accumulated during the long haul of the Pleistocene.

Since genetic and cultural tracking systems operate on parallel tracks, we can bypass their distinction for the moment and return to the question of the prime movers in later human social evolution in its broadest sense. Seed eating is a plausible explanation to account for the movement of hominids onto the savanna, and the shift to big-game hunting might account for their advance to the *Homo erectus* grade. But was the adaptation to group predation enough to carry evolution all the way to the *Homo sapiens* grade and farther, to agriculture and civilization? Anthropologists and biologists do not consider the impetus to have been sufficient. They have advocated the following series of additional factors, which can act singly or in combination.

Sexual Selection

Fox (1972), following a suggestion by Chance (1962), has argued that sexual selection was the auxiliary motor that drove human evolution all the way to the *Homo* grade. His reasoning proceeds as follows. Polygyny is a general trait in hunter-gatherer bands and may also have been the rule in the early hominid societies. If so, a premium would have been placed on sexual selection involving both epigamic display toward the females and intrasexual competition among the males. The selection would be enhanced by the constant mating provocation that arises from the female's nearly continuous sexual receptivity. Because of the existence of a high level of cooperation within the band, a legacy of the original *Australopithecus* adaptation, sexual selection would tend to be linked with hunting prowess, leadership, skill at tool making, and other visible attributes that contribute to the success of the family and the male band. Aggressiveness was constrained and the old forms of overt primate dominance replaced by complex social skills. Young males found it profitable to fit into the group, controlling their sexuality and aggression and awaiting their turn at leadership. As a result the dominant male in hominid societies was most likely to possess a mosaic of qualities that reflect the necessities of compromise; as Robin Fox expressed it, the male was "controlled, cunning, cooperative, attractive to the ladies, good with the children, relaxed, tough, eloquent, skillful, knowledgeable and proficient in self-defense and hunting." Since positive feedback occurs between these more sophisticated social traits and breeding success, social evolution can proceed indefinitely without additional selective pressures from the environment.

Multiplier Effects in Cultural Innovation and in Network Expansion

Whatever its prime mover, evolution in cultural capacity was implemented by a growing power and readiness to learn. The network of contacts among individuals and bands must also have grown. We can postulate a critical mass of cultural capacity and network size in which it became advantageous for bands actively to enlarge both. In other words, the feedback became positive. This mechanism, like sexual selection, requires no additional input beyond the limits of social behavior itself. But unlike sexual selection, it probably reached the autocatalytic threshold level very late in human prehistory.

Increased Population Density and Agriculture

The conventional view of the development of civilization used to be that innovations in farming led to population growth, the securing of leisure time, the rise of a leisure class, and the contrivance of civilized, less immediately functional pursuits. The hypothesis has been considerably weakened by the discovery that !Kung and other hunter-gatherer peoples work less and enjoy more leisure time than most farmers. Primitive agricultural people generally do not produce surpluses unless compelled to do so by political or religious authorities (Carneiro, 1970). Ester Boserup (1965) has gone so far as to suggest the reverse causation: population growth induces societies to deepen their involvement and expertise in agriculture. However, this explanation does not account for the population growth in the first place. Hunter-gatherer societies remained in approximate demographic equilibrium for hundreds of thousands of years. Something else tipped a few of them into becoming the first farmers. Quite possibly the crucial events were nothing more than the attainment of a certain level of intelligence and lucky encounters with wild-growing food plants. Once launched, agricultural economies permitted higher population densities, which in turn encouraged wider networks of social contact, technological advance, and further dependence on farming. A few innovations, such as irrigation and the wheel, intensified the process to the point of no return.

Warfare

Throughout recorded history the conduct of war has been common among tribes and nearly universal among chiefdoms and states. When Sorokin analyzed the histories of 11 European countries over periods of 275 to 1,025 years, he found that on the average they were engaged in some kind of military action 47 percent of the time, or about one year out of every two. The range was from 28 percent of the years in the case of Germany to 67 percent in the case of Spain. The early chiefdoms and states of Europe and the Middle East turned over with great rapidity, and much of the conquest was genocidal in nature. The spread of genes has always been of paramount importance. For example, after the conquest of the Midianites, Moses gave instructions identical in result to the aggression and genetic usurpation by male langur monkeys:

> Now kill every male dependent, and kill every woman who has had intercourse with a man, but spare for yourselves every woman among them who has not had intercourse. (Numbers 31)

And centuries later, von Clausewitz conveyed to his pupil the Prussian crown prince a sense of the true, biological joy of warfare:

> Be audacious and cunning in your plans, firm and persevering in their execution, determined to find a glorious end, and fate will crown your youthful brow with a shining glory, which is the ornament of princes, and engrave your image in the hearts of your last descendants.

The possibility that endemic warfare and genetic usurpation could be an effective force in group selection was clearly recognized by Charles Darwin. In *The Descent of Man* he proposed a remarkable model that foreshadowed many of the elements of modern group-selection theory:

> Now, if some one man in a tribe, more sagacious than the others, invented a new snare or weapon, or other means of attack or defence, the plainest self-interest, without the assistance of much reasoning power, would prompt the other members to imitate him; and all would thus profit. The habitual practice of each new art must likewise in some slight degree strengthen the intellect. If the invention were an important one, the tribe would increase in number, spread, and supplant other tribes. In a tribe thus rendered more numerous there would always be a rather greater chance of the birth of other superior and inventive members. If such men left children to inherit their mental superiority, the chance of the birth of still more ingenious members would be somewhat better, and in a very small tribe decidedly better. Even if they left no children, the tribe would still include their blood-relations, and it has been ascertained by agriculturists that by preserving and breeding from the family of an animal, which when slaughtered was found to be valuable, the desired character has been obtained.

Darwin saw that not only can group selection reinforce individual selection, but it can oppose it—and sometimes prevail, especially if the size of the breeding unit is small and average kinship correspondingly close. Essentially the same theme was later developed in increasing depth by Keith (1949), Bigelow (1969), and Alexander (1971). These authors envision some of the "noblest" traits of mankind, including team play, altruism, patriotism, bravery on the field of battle, and so forth, as the genetic product of warfare.

By adding the additional postulate of a threshold effect, it is possible to explain why the process has operated exclusively in human evolution (Wilson, 1972a). If any social predatory mammal attains a certain level of intelligence, as the early hominids, being large primates, were especially predisposed to do, one band would have the capacity to consciously ponder the significance of adjacent social groups and to deal with them in an intelligent, organized fashion. A band might then dispose of a neighboring band, appropriate its territory, and increase its own genetic representation in the metapopulation, retaining the tribal memory of this successful episode, repeating it, increasing the geographic range of its occurrence, and quickly spreading its influence still further in the metapopulation. Such primitive cultural capacity would be permitted by the possession of certain genes. Reciprocally, the cultural capacity might propel the spread of the genes through the genetic constitution of the metapopulation. Once begun, such a mutual reinforcement could be irreversible. The only combinations of genes able to confer superior fitness in contention with genocidal aggressors would be those that produce either a more effective technique of aggression or else the capacity to preempt genocide by some form of pacific maneuvering. Either probably entails mental and cultural advance. In addition to being autocatalytic, such evolution has the interesting property of requiring a selection episode only very occasionally in order to proceed as swiftly as individual-level selection. By current theory, genocide or genosorption strongly favoring the aggressor need take place only once every few generations to direct evolution. This alone could push truly altruistic genes to a high frequency within the bands (see Chapter 5). The turnover of tribes and chiefdoms estimated from atlases of early European and Mideastern history (for example, the atlas by McEvedy, 1967) suggests a sufficient magnitude of differential group fitness to have achieved this effect. Furthermore, it is to be expected that some isolated cultures will escape the process for generations at a time, in effect reverting temporarily to what ethnographers classify as a pacific state.

Multifactorial Systems

Each of the foregoing mechanisms could conceivably stand alone as a sufficient prime mover of social evolution. But it is much more likely that they contributed jointly, in different strengths and with complex interaction effects. Hence the most realistic model may be fully cybernetic, with cause and effect reciprocating through subcycles that possess high degrees of connectivity with one another. One such scheme, proposed by Adams (1966) for the rise of states and urban societies, is presented in Figure 26-8. Needless to say, the equations needed to translate this and similar models have not been written, and the magnitudes of the coefficients cannot even be guessed at the present time.

In both the unifactorial and multifactorial models of social evolution, an increasing internalization of the controls is postulated. This shift is considered to be the basis of the two-stage acceleration cited earlier. At the beginning of hominid evolution, the prime movers were external environmental pressures no different from those that have guided the social evolution of other animal species. For the moment, it seems reasonable to suppose that the hominids underwent two adaptive shifts in succession: first, to open-country living and seed eating, and second, after being preadapted by the anatomical and mental changes associated with seed eating, to the capture of large mammals. Big-game hunting induced further growth in mentality and social organization that brought the hominids across the threshold into the autocatalytic, more nearly internalized phase of evolution. This second stage is the one in which the most distinctive human qualities emerged. In stressing this distinction, however, I do not wish to imply that social evolution became independent of the environment. The iron laws of demography still clamped down on the spreading hominid populations, and the most spectacular cultural advances were impelled by the invention of new ways to control the environment. What happened was that mental and social change came to depend more on internal reorganization and less on direct responses to features in the surrounding environment. Social evolution, in short, had acquired its own motor.

The Future

When mankind has achieved an ecological steady state, probably by the end of the twenty-first century, the internalization of social evolution will be nearly complete. About this time biology should be at its peak, with the social sciences maturing rapidly. Some historians of science will take issue with this projection, arguing that the accelerating pace of discoveries in these fields implies a more rapid development. But historical precedents have misled us before: the subjects we are talking about are more difficult than physics or chemistry by at least two orders of magnitude.

Consider the prospects for sociology. This science is now in the natural history stage of its development. There have been attempts at system building

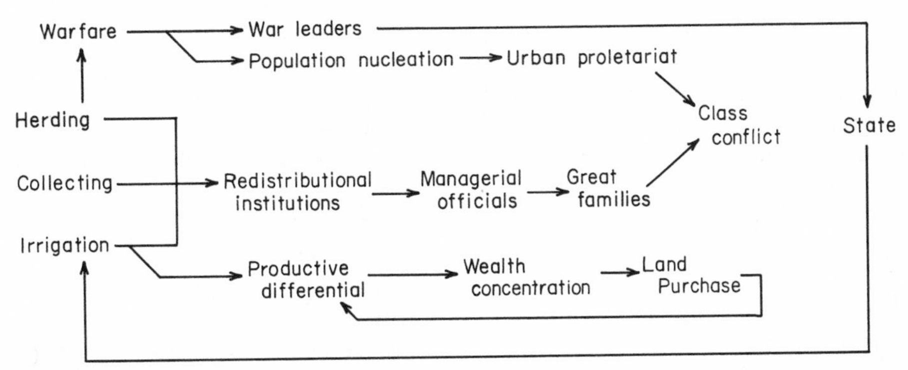

Figure 26-8 A multifactorial model of the origin of the state and urban society. (From Flannery, 1972; based on Adams, 1966. Reproduced, with permission, from "The Cultural Evolution of Civilizations," *Annual Review of Ecology and Systematics*, Vol. 3, p. 408.)

but, just as in psychology, they were premature and came to little. Much of what passes for theory in sociology today is really labeling of phenomena and concepts, in the expected manner of natural history. Process is difficult to analyze because the fundamental units are elusive, perhaps nonexistent. Syntheses commonly consist of the tedious cross-referencing of differing sets of definitions and metaphors erected by the more imaginative thinkers (see for example Inkeles, 1964, and Friedrichs, 1970). That, too, is typical of the natural history phase.

With an increase in the richness of descriptions and experiments, sociology is drawing closer each day to cultural anthropology, social psychology, and economics, and will soon merge with them. These disciplines are fundamental to sociology *sensu lato* and are most likely to yield its first phenomenological laws. In fact, some viable qualitative laws probably already exist. They include tested statements about the following relationships: the effects of hostility and stress upon ethnocentrism and xenophobia (LeVine and Campbell, 1972); the positive correlation between and within cultures of war and combative sports, resulting in the elimination of the hydraulic model of aggressive drive (Sipes, 1973); precise but still specialized models of promotion and opportunity within professional guilds (White, 1970); and, far from least, the most general models of economics.

The transition from purely phenomenological to fundamental theory in sociology must await a full, neuronal explanation of the human brain. Only when the machinery can be torn down on paper at the level of the cell and put together again will the properties of emotion and ethical judgment come clear. Simulations can then be employed to estimate the full range of behavioral responses and the precision of their homeostatic controls. Stress will be evaluated in terms of the neurophysiological perturbations and their relaxation times. Cognition will be translated into circuitry. Learning and creativeness will be defined as the alteration of specific portions of the cognitive machinery regulated by input from the emotive centers. Having cannibalized psychology, the new neurobiology will yield an enduring set of first principles for sociology.

The role of evolutionary sociobiology in this enterprise will be twofold. It will attempt to reconstruct the history of the machinery and to identify the adaptive significance of each of its functions. Some of the functions are almost certainly obsolete, being directed toward such Pleistocene exigencies as hunting and gathering and intertribal warfare. Others may prove currently adaptive at the level of the individual and family but maladaptive at the level of the group—or the reverse. If the decision is taken to mold cultures to fit the requirements of the ecological steady state, some behaviors can be altered experientially without emotional damage or loss in creativity. Others cannot. Uncertainty in this matter means that Skinner's dream of a culture predesigned for happiness will surely have to wait for the new neurobiology. A genetically accurate and hence completely fair code of ethics must also wait.

The second contribution of evolutionary sociobiology will be to monitor the genetic basis of social behavior. Optimum socioeconomic systems can never be perfect, because of Arrow's impossibility theorem and probably also because ethical standards are innately pluralistic. Moreover, the genetic foundation on which any such normative system is built can be expected to shift continuously. Mankind has never stopped evolving, but in a sense his populations are drifting. The effects over a period of a few generations could change the identity of the socioeconomic optima. In particular, the rate of gene flow around the world has risen to dramatic levels and is accelerating, and the mean coefficients of relationship within local communities are correspondingly diminishing. The result could be an eventual lessening of altruistic behavior through the maladaption and loss of group-selected genes (Haldane, 1932; Eshel, 1972). It was shown earlier that behavioral traits tend to be selected out by the principle of metabolic conservation when they are suppressed or when their original function becomes neutral in adaptive value. Such traits can largely disappear from populations in as few as ten generations, only two or three centuries in the case of human beings. With our present inadequate understanding of the human brain, we do not know how many of the most valued qualities are linked genetically to more obsolete, destructive ones. Cooperativeness toward groupmates might be coupled with aggressivity toward strangers, creativeness with a desire to own and dominate, athletic zeal with a tendency to violent response, and so on. In extreme cases such pairings could stem from pleiotropism, the control of more than one phenotypic character by the same set of genes. If the planned society—the creation of which seems inevitable in the coming century—were to deliberately steer its members past those stresses and conflicts that once gave the destructive phenotypes their Darwinian edge, the other phenotypes might dwindle with them. In this, the ultimate genetic sense, social control would rob man of his humanity.

It seems that our autocatalytic social evolution has locked us onto a particular course which the early hominids still within us may not welcome. To maintain the species indefinitely we are compelled to drive toward total knowledge, right down to the levels of the neuron and gene. When we have progressed enough to explain ourselves in these mechanistic terms, and the social sciences come to full flower, the result might be hard to accept. It seems appropriate therefore to close this book as it began, with the foreboding insight of Albert Camus:

> A world that can be explained even with bad reasons is a familiar world. But, on the other hand, in a universe divested of illusions and lights, man feels an alien, a stranger. His exile is without remedy since he is deprived of the memory of a lost home or the hope of a promised land.

This, unfortunately, is true. But we still have another hundred years.

GLOSSARY

BIBLIOGRAPHY

INDEX

Glossary

For a rapid comprehension of sociobiology the equivalent of a college-level course in biology is desirable. Also, some training in elementary mathematics, particularly calculus and probability theory, is needed to make the most technical chapters readily understandable. However, *Sociobiology: The Abridged Edition* has been written with the broadest possible audience in mind, and most of it can be read with full understanding by any intelligent person whether or not he or she has had formal training in science. To this end the following glossary has been stocked with the elementary terms of biology and mathematics that are most frequently used in the book. The reader will also find that it contains more technical expressions limited to sociobiology, including a few that appear only sparingly in the present book but are encountered consistently in the literature cited.

Active space The space within which the concentration of a pheromone (or any other behaviorally active chemical substance) is at or above threshold concentration. The active space of a pheromone is, in fact, the chemical signal itself.

Aculeate Pertaining to the Aculeata, or stinging Hymenoptera, a group including the bees, ants, and many of the wasps.

Adaptation In evolutionary biology, any structure, physiological process, or behavioral pattern that makes an organism more fit to survive and to reproduce in comparison with other members of the same species. Also, the evolutionary process leading to the formation of such a trait.

Adaptive Pertaining to any trait, anatomical, physiological, or behavioral, that has arisen by the evolutionary process of adaptation (*q.v.*).

Adaptive radiation The process of evolution in which species multiply, diverge into different ecological niches (for example, species that are predators on different kinds of prey, occupants of different habitats, and so forth), and come to occupy the same or at least overlapping ranges.

Age polyethism The regular changing of labor roles by members of a society as they age.

Aggregation A group of individuals of the same species, comprised of more than just a mated pair or a family, gathered in the same place but not internally organized or engaged in cooperative behavior. To be distinguished from a true society, *q.v.*

Aggression A physical act or threat of action by one individual that reduces the freedom or genetic fitness of another.

Agonistic Referring to any activity related to fighting, whether aggression or conciliation and retreat.

Alarm pheromone A chemical substance exchanged among members of the same species that induces a state of alertness or alarm in the face of a common threat.

Alate Winged. Sometimes also used as a noun to refer to a reproductive social insect still bearing wings.

Allele A particular form of a gene, distinguishable from other forms or alleles of the same gene.

Allogrooming Grooming directed at another individual, as opposed to self-grooming, which is directed at one's own body.

Allometry Any size relation between two body parts that can be expressed by $y = bx^a$, where a and b are fitted constants. In the special case of isometry, $a = 1$, and the relative proportions of the body parts therefore remain constant with change in total body size. In all other cases ($a \neq 1$) the relative proportions change as the total body size is varied. Allometry is important in the differentiation of castes of the social insects, especially ants.

Alloparent An individual that assists the parents in care of the young.

Alloparental care The assistance by individuals other than the parents in the care of offspring. The behavior may be shown either by females (allomaternal care) or by males (allopaternal care).

Allopatric Referring to populations, particularly species, that occupy different geographical ranges. (Contrast with sympatric.)

Allozygous Referring to two genes on the same chromosome locus that are different or at least whose identity is not due to common descent. (Contrast with autozygous.)

Alpha Referring to the highest-ranking individual within a dominance hierarchy.

Altricial Pertaining to young animals that are helpless for a substantial period following birth; used especially with reference to birds. (Contrast with precocial.)

Altruism Self-destructive behavior performed for the benefit of others. (See discussion in Chapter 5.)

Amphibian Any member of the vertebrate class Amphibia, such as a salamander, frog, or toad.

Analog signal Same as graded signal (*q.v.*).

Analogue Referring to structures, physiological processes, or behaviors that are similar owing to convergent evolution as opposed to common ancestry; hence, displaying analogy. (Opposed to homologue.)

Analogy A resemblance in function, and often in appearance also, between two structures, physiological processes, or behaviors that is due to evolution rather than to common ancestry. (Contrast with homology.)

Anisogamy The condition in which the female sex cell (ovum) is larger than the male sex cell (sperm). (Contrast with isogamy.)

Annual Referring to a life cycle, or the species possessing the life cycle, that is completed in one growing season.

Antennation Touching with the antennae. The movement can serve as a sensory probe or as a tactile signal to another insect.

Antisocial factor Any selection pressure that tends to inhibit or to reverse social evolution.

Aposematism The advertisement by dangerous animals of their identity. Thus the most venomous wasps, coral fishes, and snakes are also often the most brightly colored.

Arachnid A member of the Class Arachnida, such as a spider, mite, or scorpion.

Arena An area used consistently for communal courtship displays. Same as lek.

Army ant A member of an ant species that shows both nomadic and group-predatory behavior. In other words, the nest site is changed at relatively frequent intervals, in some cases daily, and the workers forage in groups. (Same as legionary ant.)

Arthropod Any member of the phylum Arthropoda, such as a crustacean, spider, millipede, centipede, or insect.

Artiodactyl Any mammal belonging to the order Artiodactyla, hence an ungulate with an even number of toes in each hoof. The commonest artiodactyls include the pigs, deer, camels, and antelopes. (Contrast with perissodactyl.)

Asexual reproduction Any form of reproduction that does not involve the actual fusion of sex cells (syngamy), such as budding and parthenogenesis.

Assembly The calling together of the members of a society for any communal activity.

Assortative mating The nonrandom pairing of individuals who resemble each other in one or more traits. (Contrast with disassortative mating.)

Aunt Any female who assists a parent in caring for the young.

Australopithecine Pertaining to the "man-apes," primates belonging to the genus *Australopithecus*, which were primitive forms that lived during the

Pleistocene Epoch and were ancestral to modern men (genus *Homo*). Australopithecines possessed postures and dentition similar to those of modern men but brains not much larger than those of modern apes.

Australopithecus See australopithecine.

Autocatalysis Any process the rate of which is increased by its own products. Thus autocatalytic reactions, fed by positive feedback, tend to accelerate until the ingredients are exhausted or some external constraint is imposed.

Automimicry The imitation by one sex or life stage of communication in another sex or life stage of the same species. An example is the imitation by the males of some monkey species of female sexual signals, which they appear to employ in appeasement rituals.

Autozygous Referring to two or more alleles on the same locus that are identical by common descent.

Auxiliaries Female social insects, especially bees, wasps, and ants, that associate with other females of the same generation and become workers.

Band The term applied to groups of certain social mammals, including coatis and human beings.

Behavioral biology The scientific study of all aspects of behavior, including neurophysiology, ethology, comparative psychology, sociobiology, and behavioral ecology.

Behavioral scale See behavioral scaling.

Behavioral scaling The range of forms and intensities of a behavior that can be expressed in an adaptive fashion by the same society or individual organism. For example, a society may be organized into individual territories at low densities but shift to a dominance system at high densities. (See discussion in Chapter 2.)

Biomass The weight of a set of plants, animals, or both. The set is chosen for convenience; it can be, for example, a colony of insects, a population of wolves, or an entire forest.

Bit The basic quantitative unit of information; specifically, the amount of information required to control, without error, which of two equiprobable alternates is to be chosen by the receiver.

Bonding Any close relationship formed between two or more individuals.

Brood Any young animals that are being cared for by adults. In social insects in particular, the immature members of a colony collectively, including eggs, nymphs, larvae, and pupae. In the strict sense eggs and pupae are not members of the society, but they are nevertheless referred to as part of the brood.

Brood cell A special chamber or pocket built to house immature stages of insects.

Budding The reproduction of organisms by the direct growth of a new individual from the body of an old one. Also, the multiplication of insect colonies by fission. (See colony fission.)

Canid A member of the mammal family Canidae, such as a wolf, domestic dog, or jackal.

Carnivore An animal that eats fresh meat.

Carrying capacity Usually symbolized by K, the largest number of organisms of a particular species that can be maintained indefinitely in a given part of the environment.

Carton In entomology, the chewed vegetable fibers used by many kinds of ants, wasps, and other insects to construct nests.

Caste Broadly defined, as in ergonomic theory (Chapter 14), any set of individuals of a particular morphological type, or age group, or both, that performs specialized labor in the colony. More narrowly defined, any set of individuals in a given colony that are both morphologically distinct from other individuals and specialized in behavior.

Casual society (or group) A temporary group formed by individuals within a society. The casual society is unstable, being open to new members and losing old ones at a high rate. Examples include feeding groups of monkeys within a troop and groups of playing children. (Contrast with demographic society.)

Central nervous system Often abbreviated as the CNS, that part of the nervous system which is condensed and centrally located; for example, the brain and spinal cord of vertebrates, and the brain and ladderlike chain of ganglia in insects.

Cercopithecoid Pertaining to the Old World monkeys and apes (classified as the superfamily Cercopithecoidea by many authors.)

Ceremony A highly evolved and complex display used to conciliate others and to establish and maintain social bonds.

Character In taxonomy and a few other fields of biology, the word *character* is commonly used a a synonym for trait. A particular trait possessed by one individual and not another, or by one species and not another, is often called a character state.

Chorus A group of calling anurans (frogs or toads) or insects.

Chromosome A complex, often rodlike structure found in the nucleus of a cell, bearing part of the basic genetic units (genes) of the cell.

Circadian rhythm A rhythm in behavior, metabolism, or some other activity that recurs about every 24 hours. (The prefix *circa-* refers to the lack of precision in the timing.)

Clade A species or set of species representing a distinct branch in a phylogenetic tree. (Contrast with evolutionary grade.)

Cladogram A phylogenetic tree that depicts only the splitting of species and groups of species through evolutionary time.

Class In systems of classification of organisms, the category below the phylum and above the order; a group of related, similar orders. Examples of classes are the Insecta (all true, six-legged insects) and Aves (all birds).

Claustral colony founding The procedure during which queens of ants and other social hymenopterans (or royal pairs in the case of termites) seal themselves off in cells and rear the first generation of workers on nutrients obtained mostly or entirely from their own storage tissues, including fat bodies and histolyzed wing muscles.

Clone A population of individuals all derived asexually from the same single parent.

Clutch The number of eggs laid by a female at one time.

Coefficient of consanguinity Same as coefficient of kinship (*q.v.*).

Coefficient of kinship Symbolized by F_{IJ} or f_{IJ}, the probability that a pair of alleles drawn at random from the same locus on two individuals are identical by virtue of common descent. Also called the coefficient of consanguinity.

Coefficient of relationship Also known as the degree of relatedness, and symbolized by r, the coefficient of relationship is the fraction of genes identical by descent between two individuals.

Colony A society which is highly integrated, either by physical union of the members, division of the members into specialized zooids or castes, or both. In vernacular usage the term colony is on occasion applied to almost any group of organisms, especially a group of nesting birds or a cluster of rodents living in dens.

Colony fission The multiplication of colonies by the departure of one or more reproductive forms, accompanied by groups of workers, from the parental nest, leaving behind comparable units to perpetuate the "parental" colony. This mode of colony multiplication is referred to occasionally as hesmosis in ant literature and sociotomy in termite literature. Swarming in honeybees can be regarded as a special form of colony fission.

Colony odor The odor found on the bodies of social insects that is peculiar to a given colony. By smelling the colony odor of another member of the same species, an insect is able to determine whether it is a nestmate. (See nest odor and species odor.)

Comb (of cells or cocoons) A layer of brood cells or cocoons crowded together in a regular arrangement. Combs are a characteristic feature of the nests of many species of social wasps and bees.

Communal Applied to the condition or to the group showing it in which members of the same generation cooperate in nest building but not in brood care.

Communication Action on the part of one organism (or cell) that alters the probability pattern of behavior in another organism (or cell) in an adaptive fashion. (See discussion in Chapter 8.)

Compartmentalization The manner and extent to which subgroups of societies act as discrete units.

Competition The active demand by two or more organisms (or two or more species) for a common resource.

Composite signal A signal composed of two or more simpler signals.

Connectedness The number and direction of communication links within and between societies.

Conspecific Belonging to the same species.

Control According to strict sociobiological usage, particularly in primate studies, control is intervention by one or more individuals to reduce or halt aggression between other members of the group.

Conventional behavior According to the hypothesis proposed by V. C. Wynne-Edwards, any behavior by which members of a population reveal their presence and allow others to assess the density of the population. A more elaborate form of such behavior is referred to as an epideictic display.

Coordination Interaction among individuals or subgroups such that the overall effort of the group is divided among these units without leadership being assumed by any one of them.

Core area The area of heaviest regular use within the home range.

Cormidium A group of zooids (individual members) of a siphonophore colony that can separate from the remainder of the colony and live an independent existence. The cormidium is the unit of organization between the zooid and the complete colony.

Counteracting selection The operation of selection pressures on two or more levels of organization, for example on the individual, family, and population, in such a way that certain genes are favored at one level but disfavored at another. (Contrast with reinforcing selection.)

Court The area defended by individual males within a lek, or communal display area; especially in birds. Also, the group of workers in an insect colony that surrounds the queen, especially in honeybees; the composition of such a court, also called a retinue, changes constantly.

Darwinism The theory of evolution by natural selection, as originally propounded by Charles Darwin. The modern version of this theory still recognizes natural selection as the central process and for this reason is often called neo-Darwinism.

Dealate Referring to an individual that has shed its wings, usually after mating; used both as an adjective and a noun.

Dealation The removal of the wings by the queens (and also males in the termites) during or immediately following the nuptial flight and prior to colony foundation.

Dear enemy phenomenon The recognition of territorial neighbors as individuals, with the result that aggressive interactions are kept at a minimum. The more intense forms of aggression are reserved for strangers.

Deme A local population within which breeding is completely random. Hence the largest population unit that can be analyzed by the simpler models of population genetics.

Demographic society A society that is stable enough through time, usually owing to its being relatively closed to newcomers, for the demographic processes of birth and death to play a significant role in its composition. (Contrast with casual group.)

Demography The rate of growth and the age structure of populations, and the processes that determine these properties.

Dendrogram A diagram showing evolutionary change in a biological trait, including the branching of the trait into different forms due to the multiplication of the species possessing it.

Density dependence The increase or decrease of the influence of a physiological or environmental factor on population growth as the density of the population increases.

Deterministic In mathematics, referring to a fixed relationship between two or more variables, without taking into account the effect of chance on the outcome of particular cases. (Contrast with stochastic.)

Developmental cycle The period from the birth of the egg to the eclosion of the adult insect. (Applied to social wasp studies.)

Dialect In the study of animal behavior, local geographic variants of bird songs, honeybee waggle dances, and other displays used in communication.

Dimorphism In caste systems, the existence in the same colony of two different forms, including two size classes, not connected by intermediates.

Diploid With reference to a cell or to an organism, having a chromosome complement consisting of two copies (called homologues) of each chromosome. A diploid cell or organism usually arises as the result of the union of two sex cells, each bearing just one copy of each chromosome. Thus, the two homologues in each chromosome pair in a diploid cell are of separate origin, one derived from the mother and the other from the father. (Contrast with haploid.)

Direct role A behavior or set of behaviors displayed by a subgroup of the society that benefits other subgroups and therefore the society as a whole. (Contrast with indirect role; see discussion in Chapter 14.)

Directional selection Selection that operates against one end of the range of variation and hence tends to shift the entire population toward the opposite end. (Contrast with disruptive and stabilizing selection.)

Disassortative mating The nonrandom pairing of individuals that differ from each other in one or more traits.

Discrete signal A signal used in communication that is turned either on or off, without significant intermediate gradations. (Contrast with graded signal.)

Displacement activity The performance of a behavioral act, usually in conditions of frustration or indecision, that is not directly relevant to the situation at hand.

Display A behavior pattern that has been modified in the course of evolution to convey information. A display is a special kind of signal, which in turn is broadly defined as *any* behavior that conveys information regardless of whether it serves other functions.

Disruptive selection Selection that operates against the middle of the range of variation and hence tends to split populations. (Contrast with directional and stabilizing selection.)

Distraction display A performance by a parent that draws the attention of predators away from its offspring.

DNA (Deoxyribonucleic acid) The basic hereditary material of all kinds of organisms. In higher organisms, including animals, the great bulk of DNA is located within the chromosomes.

Dominance hierarchy The physical domination of some members of a group by other members, in relatively orderly and long-lasting patterns. Except for the highest- and lowest-ranking individuals, a given member dominates one or more of its companions and is dominated in turn by one or more of the others. The hierarchy is initiated and sustained by hostile behavior, albeit sometimes of a subtle and indirect nature. (See discussion in Chapter 13.)

Dominance order Same as dominance hierarchy (*q.v.*).

Dominance system Same as dominance hierarchy (*q.v.*).

Drone A male social bee, especially a male honeybee or bumblebee.

Duetting The rapid and precise exchange of notes between two individuals, especially mated birds.

Dynamic selection Same as directional selection.

Eclosion Emergence of the adult (imago) insect from the pupa; less commonly, the hatching of an egg.

Ecological pressure See under prime movers.

Ecology The scientific study of the interaction of organisms with their environment, including both the physical environment and the other organisms that live in it.

Ecosystem All of the organisms of a particular habitat, such as a lake or a forest, together with the physical environment in which they live.

Effective population number The number of individuals in an ideal, randomly breeding population with a 1/1 sex ratio that would have the same rate of heterozygosity decrease as the real population under consideration.

Emigration The movement of an individual or society from one nest site to another.

Empathic learning See observational learning.

Enculturation The transmission of a particular culture, especially to the young members of the society.

Endemic Referring to a species that is native to a particular place and found nowhere else.

Endocrine gland Any gland, such as the adrenal or pituitary gland of vertebrates, that secretes hormones into the body through the blood or lymph. (Opposed to exocrine gland.)

Endocrinology The scientific study of endocrine glands and hormones.

Entomology The scientific study of insects.

Environmentalism In biology, the form of analysis that stresses the role of environmental influences in the development of behavioral or other biological traits. Also, the point of view that such influences tend to be paramount in behavioral development.

Epideictic display In theory at least, a display by which members of a population reveal their presence and allow others to assess the density of the population. An extreme form of "conventional behavior" as postulated by V. C. Wynne-Edwards.

Epidermis The outer layer of living cells in the skin.

Epigamic Any trait related to courtship and sex other than the essential organs and behavior of copulation.

Epigamic selection See under sexual selection.

Epizootic The spread of a disease through a population of animals; the equivalent of an epidemic in human beings.

Ergonomics The quantitative study of work, performance, and efficiency. (See discussion in Chapter 14.)

Estrous cycle The repeated series of changes in reproductive physiology and behavior that culminates in the estrus, or time of heat.

Estrus The period of heat, or maximum sexual receptivity, in the female. Ordinarily the estrus is also the time of the release of eggs in the female.

Ethocline A series of different behaviors observed among related species and interpreted to represent stages in an evolutionary trend.

Ethology The study of whole patterns of animal behavior in natural environments, stressing the analysis of adaptation and the evolution of the patterns.

Eusocial Applied to the condition or to the group possessing it in which individuals display all of the following three traits: cooperation in caring for the young; reproductive division of labor, with more or less sterile individuals working on behalf of individuals engaged in reproduction; and overlap of at least two generations of life stages capable of contributing to colony labor. "Eusocial" is the formal equivalent of the expressions "truly social" or "higher social," which are commonly used with less exact meaning in the study of social insects.

Eutherian Pertaining to the placental mammals (*q.v.*).

Evolution Any gradual change. Organic evolution, often referred to as evolution for short, is any genetic change in organisms from generation to generation; or more strictly, a change in gene frequencies within populations from generation to generation. (See discussion in Chapter 4.)

Evolutionary biology The collective disciplines of biology that treat the evolutionary process and the characteristics of populations of organisms, as well as ecology, behavior, and systematics.

Evolutionary convergence The evolutionary acquisition of a particular trait or set of traits by two or more species independently.

Evolutionary grade The evolutionary level of development in a particular structure, physiological process, or behavior occupied by a species or group of species. The evolutionary grade is distinguished from the phylogeny of a group, which is the relationship of species by descent.

Exocrine gland Any gland, such as the salivary gland, that secretes to the outside of the body or into the alimentary tract. Exocrine glands are the most common sources of pheromones, the chemical substances used in communication by most kinds of animals. (Opposed to endocrine gland.)

Exoskeleton The hardened outer body layer of insects and other arthropods that functions both as a protective covering and as a skeletal attachment for muscles.

Exponential growth Growth, especially in the number of organisms of a population, that is a simple function of the size of the growing entity; the larger the entity, the faster it grows.

F, f Symbol of the inbreeding coefficient (*q.v.*).

F_{IJ}, f_{IJ} Symbol of the coefficient of kinship (*q.v.*).

Facilitation See social facilitation.

Family In sociobiology, the word is given the conventional meaning of parents and offspring, together with other kin who are closely associated with them. In taxonomy, the family is the category below the order and above the genus; a group of related, similar genera. Examples of taxonomic families include the Formicidae, including all of the ants; and the Felidae, including all of the cats.

Fitness See genetic fitness.

Fixation In population genetics, the complete prevalence of one gene form (allele), resulting in the complete exclusion of another.

Flagellate A member of the phylum Mastigophora; a unicellular organism that propels itself by flagella, which are whiplike motile organs.

Floaters Individuals unable to claim a territory and hence forced to wander through less suitable surrounding areas.

Folivore An animal that eats leaves.

Food chain A portion of a food web, most frequently a simple sequence of prey species and the predators that consume them.

Food web The complete set of food links between species in a community; a diagram indicating which ones are the eaters and which are consumed.

Founder effect The genetic differentiation of an isolated population due to the fact that by chance alone its founders contained a set of genes statistically different from those of other populations.

Frequency curve A curve plotted on a graph to display a particular frequency distribution (*q.v.*).

Frequency distribution The array of numbers of individuals showing differing values of some variable quantity; for example, the numbers of animals of different ages, or the numbers of nests containing different numbers of young.

Frugivore An animal that eats fruit.

Gamete The mature sexual reproductive cell: the egg or the sperm.

Gametogenesis The specialized series of cellular divisions that leads to the production of sex cells (gametes).

Gene The basic unit of heredity.

Gene flow The exchange of genes between different species (an extreme case referred to as hybridization) or between different populations of the same species.

Gene pool All the genes—hence, hereditary material—in a population.

Genetic drift Evolution (change in gene frequencies) by chance processes alone.

Genetic fitness The contribution to the next generation of one genotype in a population relative to the contributions of other genotypes. By definition, this process of natural selection leads eventually to the prevalence of the genotypes with the highest fitnesses.

Genetic load The average loss of genetic fitness (*q.v.*) in an entire population due to the presence of individuals less fit than others.

Genome The complete genetic constitution of an organism.

Genotype The genetic constitution of an individual organism, designated with reference either to a single trait or to a set of traits. (Contrast with phenotype.)

Genus (plural: genera) A group of related, similar species. Examples include *Apis* (the four species of honeybees) and *Canis* (wolves, domestic dogs, and their close relatives).

Geographic race See subspecies.

Gonad An organ that produces sex cells; either an ovary (female gonad) or testis (male gonad).

Grade See evolutionary grade.

Graded signal A signal that varies in intensity or frequency or both, thereby transmitting quantitative information about such variables as mood of the sender, distance of the target, and so forth.

Grooming The cleaning of the body surface by licking, nibbling, picking with the fingers, or other kinds of manipulation. When the action is directed toward one's own body, it is called self-grooming; when directed at another individual, it is referred to as allogrooming.

Group Any set of organisms, belonging to the same species, that remain together for a period of time while interacting with one another to a distinctly greater degree than with other conspecific organisms. The term is also frequently used in a loose taxonomic sense to refer to a set of related species; thus a genus, or a division of a genus, would be an example of a taxonomic "group."

Group effect An alteration in behavior or physiology within a species brought about by signals that are directed in neither space nor time. A simple example is social facilitation, in which there is an increase of an activity merely from the sight or sound (or other form of stimulation) coming from other individuals engaged in the same activity.

Group predation The hunting and retrieving of living prey by groups of cooperating animals. This behavior pattern is developed, for example, in army ants and wolves.

Group selection Selection that operates on two or more members of a lineage group as a unit. Defined broadly, group selection includes both kin selection and interdemic selection (*q.v.*).

Habitat The organisms and physical environment in a particular place.

Haplodiploidy The mode of sex determination in which males are derived from haploid (hence, unfertilized) eggs and females from diploid, usually fertilized eggs.

Haploid Having a chromosome complement consisting of just one copy of each chromosome. Sex cells are typically haploid. (Contrast with diploid.)

Harem A group of females guarded by a male, who prevents other males from mating with them.

Hemimetabolous Undergoing development that is gradual and lacks a sharp separation into larval, pupal, and adult stages. Termites, for example, are hemimetabolous. (Opposed to holometabolus.)

Heritability The fraction of variation of a trait within a population—more precisely, the fraction of its variance, which is the statistical measure—due to heredity as opposed to environmental influences. A heritability score of one means that all the variation is genetic in basis; a heritability score of zero means that all the variation is due to the environment.

Hermaphroditism The coexistence of both female and male sex organs in the same individual.

Heterozygous Referring to a diploid organism having different alleles of a given gene on the pair of homologous chromosomes carrying that gene. (See chromosome.)

Hierarchy In general, a system of two or more levels of units, the higher levels controlling at least to some extent the activities of the lower levels in order to integrate the group as a whole. In dominance systems within societies, a hierarchy is the sequence of dominant and dominated individuals.

Holometabolous Undergoing a complete metamorphosis during development, with distinct larval, pupal, and adult stages. The Hymenoptera, for example, are holometabolous. (Opposed to hemimetabolous.)

Home range The area that an animal learns thoroughly and patrols regularly. The home range may or may not be defended; those portions that are defended constitute the territory.

Homeostasis The maintenance of a steady state, especially a physiological or social steady state, by means of self-regulation through internal feedback responses.

Hominid Pertaining to man, including early man. A term derived from the family Hominidae, the taxonomic group that includes modern man and his immediate predecessors.

Homo The genus of true men, including several extinct forms (*H. habilis, H. erectus, H. neanderthalensis*) as well as modern man (*H. sapiens*), who are or were primates characterized by completely erect stature, bipedal locomotion, reduced dentition, and above all an enlarged brain size.

Homogamy Same as assortative mating (*q.v.*).

Homologue Referring to a structure, physiological process, or behavior that is similar to another owing to common ancestry; hence, displaying homology. In genetics a homologue is one chromosome belonging to a group of chromosomes having the same overall genetic composition. (See diploid.)

Homology A similarity between two structures that is due to inheritance from a common ancestor. The structures are said to be homologous. (Contrast with analogy.)

Homopteran A member of, or pertaining to, the insect order Homoptera, which includes the aphids, jumping plant lice, treehoppers, spittlebugs, whiteflies, and related groups.

Homozygous Referring to a diploid organism possessing identical alleles of a given gene on both homologous chromosomes. An organism can be a homozygote with respect to one gene and at the same time a heterozygote with respect to another gene.

Honeybee A member of the genus *Apis*. Unless qualified otherwise, a honeybee is more particularly a member of the domestic species *A. mellifera*, and the term is usually applied to the worker caste.

Hormone Any substance, secreted by an endocrine gland into the blood or lymph, that affects the physiological activity of other organs in the body; hormones can also influence the nervous system, and through it, the behavior of the organism.

Hymenopteran Pertaining to the insect order Hymenoptera; also, a member of the order, such as a wasp, bee, or ant.

Imago The adult insect. In termites, the term is usually applied only to adult primary reproductives.

Imitation The copying of a novel or otherwise improbable act.

Inbreeding The mating of kin. The degree of inbreeding is measured by the fraction of genes that will be identical owing to common descent. (See inbreeding coefficient; and contrast with outcrossing.)

Inbreeding coefficient Symbolized by f or F, the probability that both alleles (gene forms) on one locus on a pair of chromosomes are identical by virtue of common descent.

Inclusive fitness The sum of an individual's own fitness plus all its influence on fitness in its relatives other than direct descendants; hence the total effect of kin selection with reference to an individual.

Indirect role A behavior or set of behaviors that benefits only the subgroups that display it and is neutral or even destructive to other subgroups of the society. (Opposed to direct role; see discussion in Chapter 14.)

Individual distance The fixed minimal distance an animal attempts to keep between itself and other members of the species.

Insect society In the strict sense, a colony of eusocial insects (ants, termites, eusocial wasps, or eusocial bees). In the broad sense adopted in this book, any group of presocial or eusocial insects.

Instar Any period between molts during the course of development of an insect or other arthropod.

Instinct Behavior that is highly stereotyped, more complex than the simplest reflexes, and usually directed at particular objects in the environment. Learning may or may not be involved in the development of instinctive behavior; the important point is that the behavior develops toward a narrow, predictable end product. (See discussion in Chapter 2.)

Intention movement The preparatory motions that an animal goes through prior to a complete behavioral response; for example, the crouch before the leap, the snarl before the bite, and so forth.

Intercompensation The effect caused by the domination of some density-dependent factors over others in population control. If the leading factor, say, food shortage, is eliminated, a second factor, for example disease, takes over. This compensation follows a sequence that is peculiar to each species.

Interdemic selection The selection of entire breeding populations (demes) as the basic unit. One of the extreme forms of group selection, to be contrasted with kin selection (*q.v.*).

Intrasexual selection See under sexual selection.

Intrinsic rate of increase Symbolized by r, the fraction by which a population is growing in each instant of time.

Invertebrate zoology The scientific study of invertebrate animals.

Invertebrates All kinds of animals lacking a vertebral column, from protozoans to insects and starfish. (See vertebrates.)

Isogamy The condition in which the male and female sex cells are of the same size. (Contrast with anisogamy.)

Iteroparity The production of offspring by an organism in successive groups. (Contrast with semelparity.)

K Symbol for the carrying capacity of the environment (*q.v.*).

K extinction The regular extinction of populations when they are at or near the carrying capacity of the environment (when there are K individuals in the population). (Contrast with r extinction.)

K selection Selection favoring superiority in stable, predictable environments in which rapid population growth is unimportant. (Contrast with r selection.)

Kin selection The selection of genes due to one or more individuals favoring or disfavoring the survival and reproduction of relatives (other than offspring) who possess the same genes by common descent. One of the extreme forms of group selection. (Contrast with interdemic selection.)

King In sociobiology, the male who accompanies the queen (egg-laying female) in termite colonies and inseminates her from time to time.

Kinship Possession of a common ancestor in the not too distant past. Kinship is measured precisely by the coefficient of kinship and coefficient of relationship (*q.v.*).

Lability In this book, the term is used with reference to evolutionary lability: the ease and speed with which particular categories of traits evolve. Thus, territorial behavior is usually highly labile, and maternal behavior much less so.

Langur An Asiatic monkey belonging to the genus *Presbytis*.

Larva An immature stage that is radically different in form from the adult; characteristic of many aquatic and marine invertebrate animals and the holometabolous insects, including the Hymenoptera. In the termites, the term is used in a special sense to designate an immature individual without any external trace of wing buds or soldier characteristics.

Leadership As narrowly used in sociobiology, leadership means only the role of leading other members of the society when the group progresses from one place to another.

Lek An area used consistently for communal courtship displays.

Lestobiosis The relation in which colonies of a small species of social insect nest in the walls of the nests of a larger species and enter the chambers of the larger species to prey on brood or to rob the food stores.

Life cycle The entire span of the life of an organism (or of a society) from the moment it originates to the time that it reproduces.

Lineage group A group of species allied by common descent.

Locus The location of a gene on the chromosome.

Logistic growth Growth, especially in the number of organisms constituting a population, that slows steadily as the entity approaches its maximum size. (Compare with exponential growth.)

Macaque Any monkey belonging to the genus *Macaca*, such as the rhesus monkey (*Macaca mulatta*).

Major worker A member of the largest worker subcaste, especially in ants. In ants the subcaste is usually specialized for defense, so that an adult belonging to it is often also referred to as a soldier. (See media worker and minor worker.)

Mammal Any animal of the class Mammalia, characterized by the production of milk by the female mammary glands and the possession of hair for body covering.

Mammalogy The scientific study of mammals.

Marsupial A mammal belonging to the subclass Metatheria; most marsupials, such as opossums and kangaroos, have a pouch (the marsupium) that contains the milk glands and shelters the young.

Mass communication The transfer among groups of individuals of information of a kind that cannot be transmitted from a single individual to another. Examples include the spatial organization of army ant raids, the regulation of numbers of worker ants on odor trails, and certain aspects of the thermoregulation of nests.

Mass provisioning The act of storing all of the food required for the development of a larva at the time the egg is laid. (Opposed to progressive provisioning.)

Matrifocal Pertaining to a society in which most of the activities and personal relationships are centered on the mothers.

Matrilineal Passed from the mother to her offspring, as for example access to a territory or status within a dominance system.

Maturation The automatic development of a behavioral pattern, which becomes increasingly complex and precise as the animal matures. Unlike learning, maturation does not require experience to occur.

Mean The numerical average.

Media worker In polymorphic ant series involving three or more worker subcastes, an individual belonging to the medium-sized subcaste(s). (See minor worker and major worker.)

Meiosis The cellular processes that lead to the formation of sex cells (gametes). In particular, a diploid cell divides twice to form four daughter cells; but the chromosomes are replicated only once, so that the four products are haploid (with only one complement of chromosomes each).

Metacommunication Communication about communication. A metacommunicative signal imparts information about how other signals should be interpreted. Thus a play-invitation signal indicates that subsequent threat displays should be taken as play and not as serious hostility.

Metapopulation A set of populations of organisms belonging to the same species and existing at the same time; by definition each population occupies a different area.

Metazoan Referring to any or all of the multicellular animals with the exception of the sponges.

Microevolution A small amount of evolutionary change, consisting of minor alterations in gene proportions, chromosome structure, or chromosome numbers. (A larger amount of change would be referred to as macroevolution or simply as evolution.)

Migrant selection Selection based on the different abilities of individuals of different genetic constitution to migrate. For example, if new populations are founded more consistently by individuals with gene *A* as opposed to those bearing gene *a*, gene *A* is said to be favored by migrant selection.

Minima In ants, a minor worker.

Minor worker A member of the smallest worker subcaste, especially in ants. Same as minima. (See media worker and major worker.)

Mobbing The joint assault on a predator too formidable to be handled by a single individual in an attempt to disable it or at least to drive it from the vicinity.

Molt (moult) The casting off of the outgrown skin or exoskeleton in the process of growth of an insect or other arthropod. Also the cast-off skin itself. The word is further used as an intransitive verb to designate the performance of the behavior.

Monogamy The condition in which one male and one female join to rear at least a single brood.

Monogyny In animals generally, the tendency of each male to mate with only a single female. In social insects, the term also means the existence of only one functional queen in the colony. (Opposed to polygyny.)

Monomorphism In entomology, the existence within an insect species or colony of only a single worker subcaste. (Opposed to polymorphism.)

Morphogenetic Pertaining to the development of anatomical structures during the growth of an organism.

Multiplier effect In sociobiology, the amplification of the effects of evolutionary change in behavior when the behavior is incorporated into the mechanisms of social organization.

Mutation In the broad sense, any discontinuous change in the genetic constitution of an organism. In the narrow sense, the word refers usually to a "point mutation," a change along a very narrow portion of the nucleic acid sequence.

Mutation pressure Evolution (change in gene frequencies) by different mutation rates alone.

Myrmecioid complex One of the two major taxonomic groups of ants; the name is based on the subfamily Myrmeciinae, one of the constituent taxa. It should not be confused with the subfamily Myrmicinae, which belongs to the poneroid complex.

Myrmecology The scientific study of ants.

Natural selection The differential contribution of offspring to the next generation by individuals of different genetic types but belonging to the same population. This is the basic mechanism proposed by Charles Darwin and is generally regarded today as the main guiding force in evolution. (See discussion in Chapter 4.)

Nest odor The distinctive odor of a nest, by which its inhabitants are able to distinguish the nest from nests belonging to other societies or at least from the surrounding environment. In some cases the animals, e.g., honeybees and some ants, can orient toward the nest by means of the odor. It is possible that the nest odor is the same as the colony odor in some cases. The nest odor of honeybees is often referred to as the hive aura or hive odor.

Net reproductive rate Symbolized by R_o, the average number of female offspring produced by each female during her entire lifetime.

Neurophysiology The scientific study of the nervous system, especially the physiological processes by which it functions.

Niche The range of each environmental variable, such as temperature, humidity, and food items, within which a species can exist and reproduce. The preferred niche is the one in which the species performs best, and the realized niche is the one in which it actually comes to live in a particular environment.

Nuptial flight The mating flight of the winged queens and males of an insect society.

Nymph In general entomology, the young stage of any insect species with hemimetabolous development. In termites, the term is used in a slightly more restricted sense to designate immature individuals that possess external wing buds and enlarged gonads and that are capable of developing into functional reproductives by further molting.

Observational learning Unrewarded learning that occurs when one animal watches the activities of another. Same as empathic learning.

Odor trail A chemical trace laid down by one animal and followed by another. The odorous material is referred to either as the trail pheromone or as the trail substance.

Oligogyny The occurrence in a single colony of social insects of from two to several functional queens. A special case of polygyny.

Omnivore An animal that eats both animal and vegetable materials.

Ontogeny The development of a single organism through the course of its life history. (Contrast with phylogeny.)

Opportunistic species Species specialized to exploit newly opened habitats. Such species usually are able to disperse for long distances and to reproduce rapidly; in other words, they are *r* selected (*q.v.*).

Optimal yield The highest rate of increase that a population can sustain in a given environment. Theoretically, there exists a particular size, less than the carrying capacity, at which this yield is realized.

Order In taxonomy, the category below the class and above the family; a group of related, similar families. Examples of orders include the Hymenoptera, which includes the wasps, ants, and bees; and the Primates, which includes the monkeys, apes, man, and other primates.

Organism Any living creature.

Ornithology The scientific study of birds.

Outcrossing The pairing of unrelated individuals. (Contrast with inbreeding.)

Ovariole One of the egg tubes that, together, form the ovary in female insects.

Pair bonding A close and long-lasting association formed between a male and female; in animals at least, pair bonding serves primarily for the cooperative rearing of young.

Panmictic Referring to a population in which mating is completely random; a panmictic population is often referred to as a deme.

Parameter In strict, mathematical usage a parameter is a quantity that can be held constant in a model while other quantities are being varied to study their relationships, but changed in value in other particular versions of the same model. Thus r, the rate of population increase, is a parameter that can be held constant at a particular value when varying N (number of organisms) and t (time), but changed to some new value in other versions of the same population-growth model. The term parameter is also used loosely to designate any variable property that exerts an effect upon a system.

Parasocial See presocial.

Parental investment Any behavior toward offspring that increases the chances of the offspring's survival at the cost of the parent's ability to invest in other offspring.

Parthenogenesis The production of an organism from an unfertilized egg.

Partially claustral colony founding The procedure during which an ant queen founds the colony by isolating herself in a chamber but still occasionally leaves to forage for part of her food supply.

Path analysis A graphical mode of analysis used to determine the inbreeding coefficient.

Peck order A term sometimes applied to a dominance order, especially in birds.

Perissodactyl Any mammal belonging to the order Perissodactyla, hence an ungulate with an odd number of toes in its hooves, such as a tapir or rhinoceros. (Contrast with artiodactyl.)

Permeability The degree of openness of a society to new members.

Phenotype The observable properties of an organism as they have developed under the combined influences of the genetic constitution of the individual and the effects of environmental factors. (Contrast with genotype.)

Pheromone A chemical substance, usually a glandular secretion, that is used in communication within a species. One individual releases the material as a signal and another responds after tasting or smelling it.

Philopatry The tendency of animals to remain at certain places or at least to return to them for feeding and resting.

Phyletic group A group of species related to one another through common descent.

Phylogenetic group Same as phyletic group.

Phylogenetic inertia See under prime movers.

Phylogeny The evolutionary history of a particular group of organisms; also, the diagram of the "family tree" that shows which species (or groups of species) gave rise to others. (Contrast with ontogeny.)

Phylum In taxonomy, a high-level category just beneath the kingdom and above the class; a group of related, similar classes. Examples of phyla include the Arthropoda, or all the crustaceans, spiders, insects, and related forms; and the Chordata, which includes the vertebrates, tunicates, and related forms.

Physiology The scientific study of the functions of organisms and of the individual organs, tissues, and cells of which they are composed. In its broadest sense physiology also encompasses most of molecular biology and biochemistry.

Placenta The organ, found in most mammals, that provides for the nourishment of the fetus and elimination of the fetal waste products. It is formed by the union of membranes from the fetus and the mother.

Placental Pertaining to mammals belonging to the subclass Eutheria, a group that is characterized by the presence of a placenta in the female and that contains the great majority of living mammal species. (Contrast with marsupial.)

Pleiotropism The control of more than one phenotypic characteristic, for example eye color, courtship behavior, or size, by the same gene or set of genes.

Pod A school of fish in which the bodies of the individuals actually touch. Also, a group of whales.

Point mutation A mutation resulting from a small, localized alteration in the chemical structure of a gene.

Pollen storers Bumblebee species that store pollen in abandoned cocoons. From time to time the adult females remove the pollen from the cocoons and feed it into a larval cell in the form of a liquid mixture of pollen and honey. (Opposed to pouch makers.)

Polyandry The acquisition by a female of more than one male as a mate. In the narrower sense of zoology, polyandry usually means that the males also cooperate with the female in raising a brood.

Polydomous Pertaining to single colonies that occupy more than one nest.

Polyethism Division of labor among members of a society. In social insects a distinction can be made between caste polyethism, in which morphological castes are specialized to serve different functions, and age polyethism, in which the same individual passes through different forms of specialization as it grows older. These two kinds of castes are referred to as physical castes and temporal castes, respectively.

Polygamy The acquisition, as part of the normal life cycle, of more than one mate. Polygyny: more than one female to a male. Polyandry: more than one male to a female. In the narrower sense of zoology, polygamy usually also implies a relationship in which the partners cooperate to raise a brood.

Polygenes Genes affecting the same trait but located on two or more loci on the chromosomes.

Polygyny In animals generally, the tendency of each male to mate with two or more females. In strict usage, the male also cooperates to some extent in rearing the young. In social insects, the term also means the coexistence in the same colony of two or more egg-laying queens. When multiple queens found a colony together, the condition is referred to as primary polygyny. When supplementary queens are added after colony foundation, the condition is referred to as secondary polygyny. The coexistence of only two or several queens is sometimes called oligogyny. (Opposed to monogyny.)

Polymorphism In social insects, the coexistence of two or more functionally different castes within the same sex. In ants it is possible to define polymorphism somewhat more precisely as the occurrence of nonisometric relative growth occurring over a sufficient range of size variation within a normal mature colony to produce individuals of distinctly different proportions at the extremes of the size range. In genetics, polymorphism is the maintenance of two or more forms of gene on the same locus at higher frequencies than would be expected by mutation and immigration alone.

Poneroid complex One of the two major taxonomic groups of ants; the name is based on the subfamily Ponerinae, one of the constituent taxa.

Pongid Any anthropoid ape other than a gibbon or siamang; the larger living apes (chimpanzee, gorilla, and orang-utan) together with certain fossil forms constitute the family Pongidae.

Population A set of organisms belonging to the same species and occupying a clearly delimited space at the same time. A group of populations of the same species, each of which by definition occupies a different area, is sometimes called a metapopulation.

Postadaptation An adaptation in the strict sense, meaning an evolutionary change in some trait that occurred in response to a particular selection pressure from the environment and did not accidentally precede it. (Contrast with preadaptation.)

Pouch makers Bumblebee species that build special wax pouches adjacent to groups of larvae and fill them with pollen. (Opposed to pollen storers.)

Preadaptation Any previously existing anatomical structure, physiological process, or behavior pattern that makes new forms of evolutionary adaptation more likely. (Contrast with adaptation and postadaptation.)

Precocial Referring to young animals who are able to move about and forage at a very early age; especially in birds. (Contrast with altricial.)

Predator Any organism that kills and eats other organisms.

Preferred niche See under niche.

Presocial Especially in insects, applied to the condition or to the group possessing it in which individuals display some degree of social behavior short of eusociality. Presocial species are either subsocial, i.e., the parents care for their own nymphs and larvae; or else parasocial, i.e., one or two of the following three traits are shown: cooperation in care of young, reproductive division of labor, and overlap of generations of life stages that contribute to colony labor.

Primary reproductive In termites, the colony-founding type of queen or male derived from the winged adult.

Primate Any member of the order Primates, such as a lemur, monkey, ape, or man.

Prime movers The ultimate factors that determine the direction and velocity of evolutionary change. There are two kinds of prime movers: phylogenetic inertia, which includes basic genetic mechanisms and prior adaptations that make certain changes more likely or less likely; and ecological pressure, the set of all environmental influences that constitute the agents of natural selection. (See discussion in Chapter 3.)

Primer pheromone A pheromone (chemical signal) that acts to alter the physiology of the receiving organism in some way and eventually causes the organism to respond differently. (Contrast with releaser pheromone.)

Primitive Referring to a trait that appeared first in evolution and gave rise to other, more "advanced" traits later. Primitive traits are often but not always less complex than the advanced ones.

Progressive provisioning The feeding of a larva in repeated meals. (Opposed to mass provisioning.)

Prosimian Any primate, such as a lemur or tarsier, belonging to the primitive suborder Prosimii.

Protease An enzyme that catalyzes the digestion of proteins.

Protistan Referring to the kingdom Protista, which embraces most of what used to be included in the old phylum Protozoa, including the flagellates, amebas, ciliates, and a few other unicellular organisms.

Protozoa A group of single-celled organisms classified by some zoologists as a single phylum; it includes the flagellates, amebas, and ciliates.

Proximate causation The conditions of the environment or internal physiology that trigger the responses of an organism. They are to be distinguished from the environmental forces, referred to as the ultimate causation, that led to the evolution of the response in the first place.

Pseudergate A special caste found in the lower termites, comprised of individuals who either have regressed from nymphal stages by molts that reduced or eliminated the wing buds, or else were derived from larvae by undergoing nondifferentiating molts. Pseudergates serve as the principal elements of the worker caste, but remain capable of developing into other castes by further molting.

Pupa The inactive developmental stage of the holometabolous insects (including the Hymenoptera) during which development into the final adult form is completed.

Pupate In insects, to change from a larva into a pupa.

Quasisocial Applied to the condition or to the group showing it in which members of the same generation use the same composite nest and cooperate in brood care.

Queen A member of the reproductive caste in semisocial or eusocial insect species. The existence of a queen caste presupposes the existence also of a worker caste at some stage of the colony life cycle. Queens may or may not be morphologically different from workers.

Queen substance Originally, the set of pheromones by which the queen honeybee continuously attracts the workers and controls their reproductive activities. The term is commonly used in a narrower sense to designate *trans*-9-keto-2-decenoic acid, the most potent component of the pheromone mixture. But it may also be defined more broadly in line with the original usage as any pheromone or set of pheromones used by a queen to control the reproductive behavior of the workers or of other queens.

Queenright Referring to a colony, especially a honeybee colony, that contains a functional queen.

r The symbol used to designate either the intrinsic rate of increase of a population, or the degree of relationship of two individuals.

***r* extinction** The extinction of entire populations shortly after colonization and while they are in an early stage of growth and expansion. (Contrast with *K* extinction.)

***r* selection** Selection favoring rapid rates of population increase, especially prominent in species that specialize in colonizing short-lived environments or undergo large fluctuations in population size. (Contrast with *K* selection.)

Race See subspecies.

Ramapithecus A small primate that lived in the Old World approximately 15 million years ago; its dental characteristics make it a likely candidate as one of the direct ancestors of man-apes (*Australopithecus*), which in turn gave rise to true men (*Homo*).

Realized niche See under niche.

Recessive In genetics, referring to an allele the phenotype of which is suppressed when it occurs in combination with a dominant allele.

Reciprocal altruism The trading of altruistic acts by individuals at different times. For example, one person saves a drowning person in exchange for the promise (or at least the expectation) that his altruistic act will be repaid if the circumstances are reversed at some time in the future.

Recombination The repeated formation of new combinations of genes through the processes of meiosis and fertilization that occur in the typical sexual cycle of most kinds of organisms.

Recruitment A special form of assembly by which members of a society are directed to some point in space where work is required.

Redirected activity The direction of some behavior, such as an act of aggression, away from the primary target and toward another, less appropriate object.

Reinforcing selection The operation of selection pressures on two or more levels of organization, for example on the individual, family, and population, in such a way that certain genes are favored at all levels and their spread through the population is accelerated.

Releaser A sign stimulus used in communication. Often the term is used broadly to mean any sign stimulus.

Releaser pheromone A pheromone (chemical signal) that is quickly perceived and causes a more or less immediate response. (Contrast with primer pheromone.)

Reproductive effort The effort required to reproduce, measured in terms of the decrease in the ability of the organism to reproduce at later times.

Reproductive success The number of surviving offspring of an individual.

Reproductive value Symbolized by v_x, the relative number of female offspring remaining to be born to each female of age x.

Reproductivity effect In social insects, the relation in which the rate of production of new individuals per colony member drops as the colony size increases.

Retinue The group of workers in an insect colony that surrounds the queen, especially in honeybees; the composition of the retinue changes constantly. Also known as the court.

Ritualization The evolutionary modification of a behavior pattern that turns it into a signal used in communication or at least improves its efficiency as a signal.

Role A pattern of behavior displayed by certain members of a society that has an effect on other members. (See discussion in Chapter 14.)

Royal cell In honeybees, the large, pitted, waxen cell constructed by the workers to rear queen larvae. In some species of termites, the special cell in which the queen is housed.

Royal jelly A material, supplied by workers to female larvae in royal cells, that is necessary for the transformation of larvae into queens. Royal jelly is secreted primarily by the hypopharyngeal glands and consists of a rich mixture of nutrient substances, many of them possessing a complex chemical structure.

Scaling See behavioral scaling.

School A group of fish or fishlike animals such as squid that swim together in an organized fashion; all or most of the school members are typically in the same stage of the life cycle.

Selection pressure Any feature of the environment that results in natural selection; for example, food shortage, the activity of a predator, or competition from other members of the same sex for a mate can cause individuals of different genetic types to survive to different average ages, to reproduce at different rates, or both.

Self-grooming Grooming directed at one's own body. Opposed to allogrooming, or the grooming of another individual.

Selfishness In the strict usage of sociobiology, behavior that benefits the individual in terms of genetic fitness at the expense of the genetic fitness of other members of the same species. (Compare with altruism and spite.)

Semelparity The production of offspring by an organism in one group all at the same time. (Contrast with iteroparity.)

Semiotics The scientific study of communication.

Semisocial In social insects, applied to the condition or to the group showing it in which members of the same generation cooperate in brood care and there is also a reproductive division of labor, i.e., some individuals are primarily egg layers and some are primarily workers.

Sensory physiology The study of sensory organs and the ways in which they receive stimuli from the environment and transmit them to the nervous system.

Sex determination The process by which the sex of an individual is determined. For example, the presence of a Y chromosome in a human embryo causes the fetus to develop into a male; while the fertilization of wasp and ant eggs causes them to develop into females.

Sex ratio The ratio of males to females (for example, 3 males/1 female) in a population, a society, a family, or any other group chosen for convenience.

Sexual dimorphism Any consistent difference between males and females beyond the basic functional portions of the sex organs.

Sexual selection The differential ability of individuals of different genetic types to acquire mates. Sexual selection consists of epigamic selection, based on choices made between males and females, and intrasexual selection, based on competition between members of the same sex.

Sib A close kinsman, especially a brother or sister.

Sign stimulus The single stimulus, or one out of a very few such crucial stimuli, by which an animal distinguishes key objects such as enemies, potential mates, and suitable nesting places.

Signal In sociobiology, any behavior that conveys information from one individual to another, regardless of whether it serves other functions as well. A signal specially modified in the course of evolution to convey information is called a display.

Social drift Random divergence in the behavior and mode of organization of societies.

Social facilitation An ordinary pattern of behavior that is initiated or increased in pace or frequency by the presence or actions of another animal.

Social homeostasis The maintenance of steady states at the level of the society either by control of the nest microclimate or by the regulation of the population density, behavior, and physiology of the group members as a whole.

Social insect In the strict sense, a "true social insect" is one that belongs to a eusocial species: in other words, it is an ant, a termite, or one of the eusocial wasps or bees. In the broad sense, a "social insect" is one that belongs to either a presocial or eusocial species.

Social releaser See releaser.

Sociality The combined properties and processes of social existence.

Socialization The total modification of behavior in an individual due to its interaction with other members of the society, including its parents.

Society A group of individuals belonging to the same species and organized in a cooperative manner. The diagnostic criterion is reciprocal communication of a cooperative nature, extending beyond mere sexual activity. (See further discussion in Chapter 2.)

Sociobiology The systematic study of the biological basis of all social behavior. (See discussion in Chapter 1.)

Sociocline A series of different social organizations observed among related species and interpreted to represent stages in an evolutionary trend.

Sociogram The full description, taking the form of a catalog, of all the social behaviors of a species, including a specification of the forms and frequencies of interactions.

Sociology The study of human societies.

Soldier A member of a worker subcaste specialized for colony defense.

Song In the study of animal behavior, any elaborate vocal signal.

Speciation The processes of the genetic diversification of populations and the multiplication of species.

Species The basic lower unit of classification in biological taxonomy, consisting of a population or series of populations of closely related and similar organisms. The somewhat more narrowly defined "biological species" consists of individuals that are capable of interbreeding freely with one another but not with members of other species under natural conditions.

Species odor The odor found on the bodies of social insects that is peculiar to a given species. It is possible that the species odor is merely the less distinctive component of a larger mixture comprising the colony odor (*q.v.*).

Spermatheca The receptacle in a female insect in which the sperm are stored.

Spite In the strict terminology of evolutionary biology, behavior that lowers the genetic fitnesses of both the perpetrator and the individual toward which the behavior is directed.

Stable age distribution The condition in which the proportions of individuals belonging to different age groups remain constant for generation after generation.

Stabilizing selection Selection that operates against the extremes of variation in a population and hence tends to stabilize the population around the mean. (Contrast with directional selection and disruptive selection.)

Steady state An apparently unchanging condition due to a balance between the synthesis (or arrival) and degradation (or departure) of all relevant components of a system.

Stochastic Referring to the properties of mathematical probability. A stochastic model takes into account variations in outcome that are due to chance alone. (Contrast with deterministic.)

Straight run The middle run made by a honeybee worker during the waggle dance and the element that contains most of the symbolical information concerning the location of the target outside the hive. The dancing bee makes a straight run, then loops back to the left (or right), then makes another straight run, then loops back in the opposite direction, and so on—the three basic movements together form the characteristic figure-eight pattern of the waggle dance.

Stridulation The production of sound by rubbing one part of the body surface against another. A common form of communication in insects.

Subsocial In the study of social insects, applied to the condition or to the group showing it in which the adults care for their nymphs or larvae for some period of time. (See also presocial.)

Subspecies A subdivision of a species. Usually defined narrowly as a geographical race: a population or series of populations occupying a discrete range and differing genetically from other geographical races of the same species.

Superfamily In taxonomy, the category between the family and the order; thus an order consists of a set of one or more superfamilies. Examples of superfamilies include the Apoidea, or all of the bees; and the Formicoidea, or all of the ants.

Superorganism Any society, such as the colony of a eusocial insect species, possessing features of organization analogous to the physiological properties of a single organism. The insect colony, for example, is divided into reproductive castes (analogous to gonads) and worker castes (analogous to somatic tissue); it may exchange nutrients by trophallaxis (analogous to the circulatory system), and so forth.

Supplementary reproductive A queen or male termite that takes over as the functional reproductive after the removal of the primary reproductive of the same sex. Supplementary reproductives are adultoid, mymphoid, or workerlike in form.

Swarming In honeybees, the normal method of colony reproduction, in which the queen and a large number of workers depart suddenly from the parental nest and fly to some exposed site. There they cluster while scout workers fly in search of a suitable new nest cavity. In ants and termites, the term "swarming" is often applied to the mass exodus of reproductive forms from the nests at the beginning of the nuptial flight.

Sympatric Referring to populations, particularly species, the geographical ranges of which at least partially overlap. (Contrast with allopatric.)

Syngamy The final step in fertilization, in which the nuclei of the sex cells meet and fuse.

Taxis The movement of an organism in a fixed direction with reference to a single stimulus. Thus, a phototaxis is a movement toward or away from a light, a geotaxis is a movement up or down in response to gravity, and so forth.

Taxon (plural: taxa) Any group of organisms representing a particular unit of classification, such as all the members of a given subspecies or of a species, genus, and so on. Thus *Homo sapiens,* the species of man, is one taxon; so is the order Primates, embracing all the species of monkeys, apes, men, and other kinds of primates.

Taxonomy The science of classification, especially of organisms.

Temporal polyethism Same as age polyethism (*q.v.*).

Territory An area occupied more or less exclusively by an animal or group of animals by means of repulsion through overt defense or advertisement.

Time-energy budget The amounts of time and energy allotted by animals to various activities.

Total range The entire area covered by an individual in its lifetime.

Tradition A specific form of behavior, or a particular site used for breeding or some other function, passed from one generation to the next by learning.

Tradition drift Social drift (random divergence in social behavior) that is based purely on differences in experience and hence passed on as part of tradition.

Trail pheromone A substance laid down in the form of a trail by one animal and followed by another member of the same species.

Trail substance Same as trail pheromone.

Troop In sociobiology, a group of lemurs, monkeys, apes, or some other kind of primate.

Trophallaxis In social insects, the exchange of alimentary liquid among colony members and guest organisms, either mutually or unilaterally. In stomodeal (oral) trophallaxis the material originates from the mouth; in proctodeal (anal) trophallaxis it originates from the anus.

Trophic Pertaining to food.

Trophic egg An egg, usually degenerate in form and inviable, that is fed to other members of the colony.

Trophic level The position of a species in a food chain, determined by which species it consumes and which consume it.

Ultimate causation The conditions of the environment that render certain traits adaptive and others nonadaptive; hence the adaptive traits tend to be retained in the population and are "caused" in this ultimate sense. (Contrast with proximate causation.)

Umwelt A German expression (loosely translated, "the world around me") used to indicate the total sensory input of an animal. Each species, including man, has its own distinctive *Umwelt.*

Unicolonial Pertaining to a population of social insects in which there are no behavioral colony boundaries. Thus the entire local population consists of one colony. (Opposed to multicolonial.)

Variance The most commonly used statistical measure of variation (dispersion) of a trait within a population. It is the mean squared deviation of all individuals from the sample mean.

Vertebrate zoology The scientific study of vertebrate animals.

Vertebrates Animals having a vertebral column ("backbone"), including all of the fishes, amphibians, reptiles, birds, and mammals.

Viscosity In sociobiology and population genetics, the slowness of individual dispersal and hence the low rate of gene flow.

Waggle dance The dance whereby workers of various species of honeybees (genus *Apis*) communicate the location of food finds and new nest sites. The dance is basically a run through a figure-eight pattern, with the middle, transverse line of the eight containing the information about the direction and distance of the target. (See Chapter 8.)

Worker A member of the nonreproductive, laboring caste in semisocial and eusocial insect species. The existence of a worker caste presupposes the existence also of royal (reproductive) castes. In termites, the term is used in a more restricted sense to designate individuals in the family Termitidae that completely lack wings and have reduced pterothoraces, eyes, and genital apparatus.

Zoology The scientific study of animals.

Zoosemiotics The scientific study of animal communication.

Zygote The cell created by the union of two gametes (sex cells), in which the gamete nuclei are also fused. The earliest stage of the diploid generation.

Bibliography

Adams, R. McC. 1966. *The evolution of urban society: early Mesopotamia and prehispanic Mexico.* Aldine Publishing Co., Chicago. xii + 191 pp.

Alexander, R. D. 1961. Aggressiveness, territoriality, and sexual behavior in field crickets (Orthoptera: Gryllidae). *Behaviour,* 17(2,3): 130–223.

——— 1971. The search for an evolutionary philosophy of man. *Proceedings of the Royal Society of Victoria,* 84(1): 99–120.

——— 1974. The evolution of social behavior. *Annual Review of Ecology and Systematics,* 5: 325–383.

Alexander, R. D., and T. E. Moore. 1962. The evolutionary relationships of 17-year and 13-year cicadas, and three new species (Homoptera, Cicadidae, *Magicicada*). *Miscellaneous Publications, Museum of Zoology, University of Michigan, Ann Arbor,* 21: 1–57.

Allee, W. C. 1931. *Animal aggregations: a study in general sociology.* University of Chicago Press, Chicago. ix + 431 pp.

——— 1938. *The social life of animals.* W. W. Norton, New York. 293 pp.

——— 1942. Group organization among vertebrates. *Science,* 95: 289–293.

Allee, W. C., N. E. Collias, and Catherine Z. Lutherman. 1939. Modification of the social order in flocks of hens by the injection of testosterone propionate. *Physiological Zoology,* 12(4): 412–440.

Allee, W. C., and J. C. Dickinson, Jr. 1954. Dominance and subordination in the smooth dogfish *Mustelus canis* (Mitchill). *Physiological Zoology,* 27(4): 356–364.

Allee, W. C., A. E. Emerson, O. Park, T. Park, and K. P. Schmidt. 1949. *Principles of Animal Ecology.* W. B. Saunders Co., Philadelphia. xii + 837 pp.

Altmann, Margaret. 1956. Patterns of herd behavior in free-ranging elk of Wyoming, *Cervus canadensis nelsoni. Zoologica, New York,* 41(2): 65–71.

——— 1958. Social integration of the moose calf. *Animal Behaviour,* 6(3,4): 155–159.

Altmann, S. A. 1956. Avian mobbing behavior and predator recognition. *Condor,* 58(4): 241–253.

——— 1962a. A field study of the sociobiology of rhesus monkeys, *Macaca mulatta. Annals of the New York Academy of Sciences,* 102(2): 338–435.

——— 1962b. Social behavior of anthropoid primates: analysis of recent concepts. In E. L. Bliss, ed., *Roots of behavior,* pp. 277–285. Harper and Brothers, New York. xi + 339 pp.

——— 1965. Sociobiology of rhesus monkeys: II, stochastics of social communication. *Journal of Theoretical Biology,* 8(3): 490–522.

Altmann, S. A., and Jeanne Altmann. 1970. *Baboon ecology: African field research.* University of Chicago Press, Chicago. viii + 220 pp.

Amadon, D. 1964. The evolution of low reproductive rates in birds. *Evolution,* 18(1): 105–110.

Andrew, R. J. 1956. Intention movements of flight in certain passerines, and their uses in systematics. *Behaviour,* 10(1,2): 179–204.

——— 1962. Evolution of intelligence and vocal mimicking. *Science,* 137: 585–589.

——— 1963a. Trends apparent in the evolution of vocalization in the Old World monkeys and apes. *Symposia of the Zoological Society of London,* 10: 89–107.

——— 1963b. The origin and evolution of the calls and facial expressions of the primates. *Behaviour,* 20(1,2): 1–109.

——— 1972. The information potentially available in mammal displays. In R. A. Hinde, ed. (*q.v.*), *Non-verbal communication,* pp. 179–206.

Archer, J. 1970. Effects of population density on behaviour in rodents. In J. H. Crook, ed., *Social behaviour in birds and mammals: essays on the social ethology of animals and man,* pp. 169–210. Academic Press, New York. xl + 492 pp.

Armstrong, E. A. 1947. *Bird display and behaviour: an introduction to the study of bird psychology,* 2d ed. Lindsay Drummond, London. 431 pp. (Reprinted by Dover, New York, 1965, 431 pp.)

Ashmole, N. P. 1963. The regulation of numbers of tropical oceanic birds. *Ibis,* 103b(3): 458–473.

Assem, J. van den. 1971. Some experiments on sex ratio and sex regulation in the pteromalid *Lariophagus distin-*

guendus. Netherlands Journal of Zoology, 21(4): 373–402.

Ayala, F. J. 1968. Evolution of fitness: II, correlated effects of natural selection on the productivity and size of experimental populations of *Drosophila serrata. Evolution,* 22(1): 55–65.

Bakker, R. T. 1968. The superiority of dinosaurs. *Discovery,* 3(2): 11–22.

——— 1971. Ecology of the brontosaurs. *Nature, London,* 229: 172–174.

Banta, W. C. 1973. Evolution of avicularia in cheilostome Bryozoa. In R. S. Boardman, A. H. Cheetham, and W. A. Oliver, Jr., eds. (*q.v.*), *Animal colonies: development and function through time,* pp. 295–303.

Barnes, H. 1962. So-called anecdysis in *Balanus balanoides* and the effect of breeding upon the growth of the calcareous shell of some common barnacles. *Limnology and Oceanography,* 7(4): 462–473.

Barrai, I., L. L. Cavalli-Sforza, and M. Mainardi. 1964. Testing a model of dominant inheritance for metric traits in man. *Heredity,* 19(4): 651–668.

Barth, R. H., Jr. 1970. Pheromone-endocrine interactions in insects. In G. K. Benson and J. G. Phillips, eds., *Hormones and the environment,* pp. 373–404. Memoirs of the Society for Endocrinology no. 18. Cambridge University Press, Cambridge. xvi + 629 pp.

Bartholomew, G. A., and J. B. Birdsell. 1953. Ecology and the protohominids. *American Anthropologist,* 55: 481–498.

Bastock, Margaret. 1956. A gene mutation which changes a behavior pattern. *Evolution,* 10(4): 421–439.

Bateman, A. J. 1948. Intra-sexual selection in *Drosophila. Heredity,* 2(3): 349–368.

Bates, B. C. 1970. Territorial behavior in primates: a review of recent field studies. *Primates,* 11(3): 271–284.

Bateson, G. 1955. A theory of play and fantasy. *Psychiatric Research Reports* (American Psychiatric Association), 2: 39–51.

Bateson, P. P. G. 1966. The characteristics and context of imprinting. *Biological Reviews, Cambridge Philosophical Society,* 41: 177–220.

Batra, Suzanne W. T. 1966. The life cycle and behavior of the primitively social bee, *Lasioglossum zephyrum* (Halictidae). *Kansas University Science Bulletin* (Lawrence), 46(10): 359–422.

Bayer, F. M. 1973. Colonial organization in octocorals. In R. S. Boardman, A. H. Cheetham, and W. A. Oliver, Jr., eds. (*q.v*), *Animal colonies: development and function through time,* pp. 69–93.

Beach, F. A. 1940. Effects of cortical lesions upon the copulatory behavior of male rats. *Journal of Comparative Psychology,* 29(2): 193–244.

——— 1964. Biological bases for reproductive behavior. In W. Etkin, ed. (*q.v.*), *Social behavior and organization among vertebrates,* pp. 117–142.

Beebe, W. 1947. Notes on the hercules beetle, *Dynastes hercules* (Linn.), at Rancho Grande, Venezuela, with special reference to combat behavior. *Zoologica, New York,* 32(2): 109–116.

Beklemishev, W. N. 1969. *Principles of comparative anatomy of invertebrates,* vol. 1, *Promorphology,* trans. by J. M. MacLennan, ed. by Z. Kabata. University of Chicago Press, Chicago. xxx + 490 pp.

Bequaert, J. C. 1935. Presocial behavior among the Hemiptera. *Bulletin of the Brooklyn Entomological Society,* 30(5): 177–191.

Berkson, G., B. A. Ross, and S. Jatinandana. 1971. The social behavior of gibbons in relation to a conservation program. In L. A. Rosenblum, ed., *Primate behavior: developments in field and laboratory research,* vol. 2, pp. 225–255. Academic Press, New York. xii + 400 pp.

Berkson, G., and R. J. Schusterman. 1964. Reciprocal food sharing of gibbons. *Primates,* 5(1,2): 1–10.

Bernstein, I. S., and R. J. Schusterman. 1964. The activities of gibbons in a social group. *Folia Primatologica,* 2(3): 161–170.

Bernstein, I. S., and L. G. Sharpe. 1966. Social roles in a rhesus monkey group. *Behaviour,* 26(1,2): 91–104.

Bertram, B. C. R. 1970. The vocal behaviour of the Indian hill mynah, *Gracula religiosa. Animal Behaviour Monographs,* 3(2): 79–192.

Best, J. B., A. B. Goodman, and A. Pigon. 1969. Fissioning in planarians: control by the brain. *Science,* 164: 565–566.

Bigelow, R. 1969. *The dawn warriors: man's evolution toward peace.* Atlantic Monthly Press, Little, Brown, Boston. xi + 277 pp.

Birch, H. G., and G. Clark. 1946. Hormonal modification of social behavior: II, the effects of sex-hormone administration on the social dominance status of the female-castrate chimpanzee. *Psychosomatic Medicine,* 8(5): 320–321.

Birdwhistle, R. L. 1970. *Kinesics and context: essays on body motion and communication.* University of Pennsylvania Press, Philadelphia. xiv + 338 pp.

Bishop, J. W., and L. M. Bahr. 1973. Effects of colony size on feeding by *Lophopodella carteri* (Hyatt). In R. S. Boardman, A. H. Cheetham, and W. A. Oliver, Jr., eds. (*q.v.*), *Animal colonies: development and function through time,* pp. 433–437.

Black-Cleworth, Patricia. 1970. The role of electrical discharges in the non-reproductive social behaviour of *Gymnotus carapo* (Gymnotidae, Pisces). *Animal Behaviour Monographs,* 3(1): 1–77.

Blair, W. F., and W. E. Howard. 1944. Experimental evidence of sexual isolation between three forms of mice of the cenospecies *Peromyscus maniculatus. Contributions from the Laboratory of Vertebrate Biology, University of Michigan, Ann Arbor,* 26: 1–19.

Boardman, R. S., A. H. Cheetham, and W. A. Oliver, Jr., eds. 1973. *Animal colonies: development and function through time.* Dowden, Hutchinson, and Ross, Stroudsburg, Pa. xiii + 603 pp.

Bonner, J. T. 1970. The chemical ecology of cells in the soil. In E. Sondheimer and J. B. Simeone, eds., *Chemical ecology,* pp. 1–19. Academic Press, New York. xvi + 336 pp.

Boorman, S. A., and P. R. Levitt. 1972. Group selection on the boundary of a stable population. *Proceedings of the National Academy of Sciences, U.S.A.,* 69(9): 2711–2713.

——— 1973. Group selection on the boundary of a stable population. *Theoretical Population Biology*, 4(1): 85–128.

Booth, A. H. 1960. *Small mammals of West Africa*. Longmans, Green, London. 68 pp. [Cited by Bradbury, 1975 (*q.v.*).]

Boserup, Ester. 1965. *The conditions of agricultural growth*. Aldine Publishing Co., Chicago. 124 pp.

Bossert, W. H., and E. O. Wilson. 1963. The analysis of olfactory communication among animals. *Journal of Theoretical Biology*, 5(3): 443–469.

Bourlière, F. 1963. Specific feeding habits of African carnivores. *African Wildlife*, 17(1): 21–27.

Bovbjerg, R. V., and Sandra L. Stephen. 1971. Behavioral changes in crayfish with increased population density. *Bulletin of the Ecological Society of America*, 52(4): 37–38.

Boyd, H. 1953. On encounters between wild white-fronted geese in winter flocks. *Behaviour*, 5(1): 85–129.

Bradbury, J. 1975. Social organization and communication. In W. Wimsatt, ed., *Biology of bats*, vol. 3, pp. 1–72. Academic Press, New York.

Brannigan, C. R., and D. A. Humphries. 1972. Human non-verbal behaviour, a means of communication. In N. Blurton Jones, ed. (*q.v.*), *Ethological studies of child behaviour*, pp. 37–64.

Brauns, H. 1926. A contribution to the knowledge of the genus *Allodape*, St. Farg. & Serv. Order Hymenoptera; section Apidae (Anthophila). *Annals of the South African Museum*, 23(3): 417–434.

Breder, C. M., Jr. 1959. Studies on social groupings in fishes. *Bulletin of the American Museum of Natural History*, 117(6): 393–482.

Breder, C. M., Jr., and C. W. Coates. 1932. A preliminary study of population stability and sex ratio of *Lebistes*. *Copeia*. 1932(3): 147–155.

Brémond, J. C. 1968. Recherches sur la sémantique et les éléments vecteurs d'information dans les signaux acoustiques du Rouge-gorge (*Erithacus rubecula L.*). *La Terre et la Vie*, 115(2): 109–220.

Brereton, J. L. G. 1971. Inter-animal control of space. In A. H. Esser, ed., *Behavior and environment: the use of space by animals and men*. pp. 69–91. Plenum Press, New York. xvii + 411 pp.

Brian, M. V. 1955. Food collection by a Scottish ant community. *Journal of Animal Ecology*, 24(2): 336–351.

——— 1956a. The natural density of *Myrmica rubra* and associated ants in West Scotland. *Insectes Sociaux*, 3(4): 473–487.

——— 1956b. Segregation of species of the ant genus *Myrmica*. *Journal of Animal Ecology*, 25(2): 319–337.

Brian, M. V., G. Elmes, and A. F. Kelly. 1967. Populations of the ant *Tetramorium caespitum* Latreille. *Journal of Animal Ecology*, 36(2): 337–342.

Brien, P. 1953. Étude sur les Phylactolemates. *Annales de la Société Royale Zoologique de Belgique*, 84(2): 301–444.

Brock, V. E., and R. H. Riffenburgh. 1960. Fish schooling: a possible factor in reducing predation. *Journal du Conseil, Conseil Permanent International pour l'Exploration de la Mer*, 25: 307–317.

Bro Larsen, Ellinor. 1952. On subsocial beetles from the salt-marsh, their care of progeny and adaptation to salt and tide. *Transactions of the Ninth International Congress of Entomology, Amsterdam, 1951*, 1: 502–506.

Bronson, F. H. 1969. Pheromonal influences on mammalian reproduction. In M. Diamond, ed., *Perspectives in reproduction and sexual behavior*, pp. 341–361. Indiana University Press, Bloomington. x + 532 pp.

——— 1971. Rodent pheromones. *Biology of Reproduction*, 4(3): 344–357.

Brothers, D. J., and C. D. Michener. 1974. Interactions in colonies of primitively social bees: III, ethometry of division of labor in *Lasioglossum zephyrum* (Hymenoptera: Halictidae). *Journal of Comparative Physiology*, 90(2): 129–168.

Brower, L. P. 1969. Ecological chemistry. *Scientific American*, 220(2) (February): 22–29.

Brown, D. H., D. K. Caldwell, and Melba C. Caldwell. 1966. Observations on the behavior of wild and captive false killer whales, with notes on associated behavior of other genera of captive delphinids. *Contributions in Science, Los Angeles County Museum*, 95: 1–32.

Brown, J. L. 1964. The evolution of diversity in avian territorial systems. *Wilson Bulletin*, 76(2): 160–169.

——— 1966. Types of group selection. *Nature, London*, 211(5051): 870.

——— 1969. Territorial behavior and population regulation in birds: a review and re-evaluation. *Wilson Bulletin*, 81(3): 293–329.

——— 1972. Communal feeding of nestlings in the Mexican jay (*Aphelocoma ultramarina*): interflock comparisons. *Animal Behaviour*, 20(2): 395–403.

——— 1974. Alternate-routes to sociality in jays—with a theory for the evolution of altruism and communal breeding. *American Zoologist*, 14(1): 63–80.

Brown, R. 1973. *A first language: the early stages*. Harvard University Press, Cambridge. xxii + 437 pp.

Brown, R. G. B. 1962. The aggressive and distraction behaviour of the western sandpiper *Ereunetes mauri*. *Ibis*, 104(1): 1–12.

Brown, W. L. 1968. An hypothesis concerning the function of the metapleural glands in ants. *American Naturalist*, 102(924): 188–191.

——— 1973. A comparison of the Hylean and Congo-West African rain forest ant faunas. In Betty J. Meggers, E. S. Ayensu, and W. D. Duckworth, eds., *Tropical forest ecosystems in Africa and South America: a comparative review*, pp. 161–185. Smithsonian Institution Press, Washington, D.C. viii + 350 pp.

Brown, W. L., T. Eisner, and R. H. Whittaker. 1970. Allomones and kairomones: transspecific chemical messengers. *BioScience*, 20(1): 21–22.

Bruce, H. M. 1966. Smell as an exteroceptive factor. *Journal of Animal Science*, supplement 25: 83–89.

Bruner, J. S. 1968. *Processes of cognitive growth: infancy*. Clark University Press, with Barre Publishers, Barre, Mass. vii + 75 pp.

Buck, J. B. 1938. Synchronous rhythmic flashing of fireflies. *Quarterly Review of Biology*, 13(3): 301–314.

Buechner, H. K., and H. D. Roth. 1974. The lek system in

Uganda kob antelope. *American Zoologist,* 14(1): 145–162.

Buettner-Janusch, J., and R. J. Andrew. 1962. The use of the incisors by primates in grooming. *American Journal of Physical Anthropology,* 20(1): 127–129.

Bullock, T. H. 1973. Seeing the world through a new sense: electroreception in fish. *American Scientist,* 61(3): 316–325.

Burchard, J. E., Jr. 1965. Family structure in the dwarf cichlid *Apistogramma trifasciatum* Eigenmann and Kennedy. *Zeitschrift für Tierpsychologie,* 22(2): 150–162.

Burnet, F. M. 1971. "Self-recognition" in colonial marine forms and flowering plants in relation to the evolution of immunity. *Nature, London,* 232(5308): 230–235.

Burt, W. H. 1943. Territoriality and home range concepts as applied to mammals. *Journal of Mammalogy,* 24(3): 346–352.

Burton, Frances D. 1972. The integration of biology and behavior in the socialization of *Macaca sylvana* of Gibraltar. In F. E. Poirier, ed. (q.v.), *Primate socialization,* pp. 29–62.

Busnel, R.-G., and A. Dziedzic. 1966. Acoustic signals of the pilot whale *Globicephala melaena* and of the porpoises *Delphinus delphis* and *Phocoena phocoena.* In K. S. Norris, ed., *Whales, dolphins, and porpoises,* pp. 607–646. University of California Press, Berkeley. xvi + 789 pp.

Butler, C. G. 1954a. *The world of the honeybee.* Collins, London. xiv + 226 pp.

——— 1954b. The method and importance of the recognition by a colony of honeybees (*A. mellifera*) of the presence of its queen. *Transactions of the Royal Entomological Society of London,* 105(2): 11–29.

Butler, C. G., and D. H. Calam. 1969. Pheromones of the honey bee—the secretion of the Nassanoff gland of the worker. *Journal of Insect Physiology,* 15(2): 237–244.

Butler, C. G., and J. B. Free. 1952. The behaviour of worker honeybees at the hive entrance. *Behaviour,* 4(4): 262–292.

Caldwell, Melba C., and D. K. Caldwell. 1972. Behavior of marine mammals: sense and communication. In S. H. Ridgway, ed. (*q.v.*), *Mammals of the sea: biology and medicine,* pp. 419–502.

Calhoun, J. B. 1962a. *The ecology and sociology of the Norway rat.* U.S. Department of Health, Education, and Welfare, Public Health Service Document no. 1008. Superintendent of Documents, U.S. Government Printing Office, Washington, D.C. viii + 288 pp.

——— 1962b. Population density and social pathology. *Scientific American,* 206(2) (February): 139–148.

Campbell, D. T. 1972. On the genetics of altruism and the counterhedonic components in human culture. *Journal of Social Issues,* 28(3): 21–37.

Candland, D. K., and A. I. Leshner. 1971. Formation of squirrel monkey dominance order is correlated with endocrine output. *Bulletin of the Ecological Society of America,* 52(4): 54.

Carl, E. A. 1971. Population control in arctic ground squirrels. *Ecology,* 52(3): 395–413.

Carneiro, R. L. 1970. A theory of the origin of the state. *Science,* 169: 733–738.

——— 1940. A field study in Siam of the behavior and social relations of the gibbon (*Hylobates lar*). *Comparative Psychology Monographs,* 16(5): 1–212.

Cartmill, M. 1974. Rethinking primate origins. *Science,* 184: 436–443.

Cavalli-Sforza, L. L., and W. F. Bodmer. 1971. *The genetics of human populations.* W. H. Freeman, San Francisco. xvi + 965 pp.

Chagnon, N. A. 1968. *Yanomamö: the fierce people.* Holt, Rinehart and Winston, New York. xviii + 142 pp.

Chance, M. R. A. 1962. Social behaviour and primate evolution. In M. F. Ashley Montagu, ed., *Culture and the evolution of man,* pp. 84–130. Oxford University Press, New York. xiii + 376 pp.

——— 1967. Attention structure as the basis of primate rank orders. *Man,* 2(4): 503–518.

Chance, M. R. A., and C. J. Jolly. 1970. *Social groups of monkeys, apes and men.* E. P. Dutton, New York. 224 pp.

Cherry, C. 1957. *On human communication.* John Wiley & Sons, New York. xvi + 333 pp.

Chitty, D. 1967. What regulates bird populations? *Ecology,* 48(4): 698–701.

Chomsky, N. 1957. *Syntactic structures.* Mouton, The Hague. 118 pp.

——— 1972. *Language and mind,* enlarged ed. Harcourt, Brace, Jovanovich, New York. xii + 194 pp.

Christian, J. J. 1961. Phenomena associated with population density. *Proceedings of the National Academy of Sciences, U.S.A.,* 47(4): 428–449.

——— 1970. Social subordination, population density, and mammalian evolution. *Science,* 168: 84–90.

Clark, Eugenie. 1972. The Red Sea's garden of eels. *National Geographic,* 142(5) (November): 724–735.

Clausen, J. A., ed. 1968. *Socialization and society.* Little, Brown, Boston. xvi + 400 pp.

Cleveland, L. R., S. R. Hall, Elizabeth P. Sanders, and Jane Collier. 1934. The wood-feeding roach *Cryptocercus,* its Protozoa, and the symbiosis between Protozoa and roach. *Memoirs of the American Academy of Arts and Sciences,* 17(2): 185–342.

Cody, M. L. 1966. A general theory of clutch size. *Evolution,* 20(2): 174–184.

Cohen, J. E. 1969a. Grouping in a vervet monkey troop. *Proceedings of the Second International Congress of Primatology, Atlanta, Georgia (U.S.A.), 1968,* 1: 274–278.

——— 1969b. Natural primate troops and a stochastic population model. *American Naturalist,* 103(933): 455–477.

——— 1971. *Casual groups of monkeys and men: stochastic models of elemental social systems.* Harvard University Press, Cambridge. xiii + 175 pp.

Collias, N. E. 1943. Statistical analysis of factors which make for success in initial encounters between hens. *American Naturalist,* 77(773): 519–538.

——— 1950. Social life and the individual among vertebrate animals. *Annals of the New York Academy of Sciences,* 51(6): 1076–1092.

Conder, P. J. 1949. Individual distance. *Ibis,* 91(4): 649–655.

Cooper, K. W. 1957. Biology of eumenine wasps: V, digital communication in wasps. *Journal of Experimental Zoology,* 134(3): 469–509.

Coulson, J. C. 1966. The influence of the pair-bond and age

on the breeding biology of the kittiwake gull *Rissa tridactyla. Journal of Animal Ecology*, 35(2): 269–279.

Count, E. W. 1958. The biological basis of human sociality. *American Anthropologist*, 60(6): 1049–1085.

Cousteau, J.-Y., and P. Diolé. 1972. Killer whales have fearsome teeth and a strange gentleness to man. *Smithsonian*, 3(3) (June): 66–73. (Reprinted in modified form from J.-Y. Cousteau, *The whale: mighty monarch of the sea*, Doubleday, Garden City, N. Y., 1972.)

Craig, G. B. 1967. Mosquitoes: female monogamy induced by male accessory gland substance. *Science*, 156: 1499–1501.

Craig, J. V., and A. M. Guhl. 1969. Territorial behavior and social interactions of pullets kept in large flocks. *Poultry Science*, 48(5): 1622–1628.

Creighton, W. S. 1953. New data on the habits of *Camponotus (Myrmaphaenus) ulcerosus* Wheeler. *Psyche, Cambridge*, 60(2): 82–84.

Crook, J. H. 1964. The evolution of social organization and visual communication in the weaver birds (Ploceinae). *Behaviour*, supplement 10. 178 pp.

——— 1971. Sources of cooperation in animals and man. In J. F. Eisenberg and W. S. Dillon, eds. (*q.v.*), *Man and beast: comparative social behaviour*, pp. 237–272.

——— 1972. Sexual selection, dimorphism, and social organization in the primates. In B. G. Campbell, ed., *Sexual selection and the descent of man, 1871–1971*, pp. 231–281. Aldine Publishing Co., Chicago. x + 378 pp.

Crow, J. F., and M. Kimura. 1970. *An introduction to population genetics theory*. Harper & Row, New York. xiv + 591 pp.

Crystal, D. 1969. *Prosodic systems and intonation in English*. Cambridge University Press, London. viii + 381 pp.

Curtis, Helena. 1968. *Biology*. Worth Publishers, New York. 854 pp.

Curtis, H. J. 1971. Genetic factors in aging. *Advances in Genetics*, 16: 305–325.

Curtis, R. F., J. A. Ballantine, E. B. Keverne, R. W. Bonsall, and R. P. Michael. 1971. Identification of primate sexual pheromones and the properties of synthetic attractants. *Nature, London*, 232(5310): 396–398.

Daanje, A. 1950. On locomotory movements in birds and the intention movements derived from them. *Behaviour*, 3(1): 48–98.

Dahlberg, G. 1947. *Mathematical methods for population genetics*. S. Karger, New York. 182 pp.

Darling, F. F. 1937. *A herd of red deer*. Oxford University Press, London. x + 215 pp. (Reprinted as a paperback, Doubleday, Garden City, N. Y., 1964. xiv + 226 pp.)

——— 1938. *Bird flocks and the breeding cycle: a contribution to the study of avian sociality*. Cambridge University Press, Cambridge. x + 124 pp.

Darlington, C. D. 1969. *The evolution of man and society*. Simon and Schuster, New York. 753 pp.

Darlington, P. J. 1971. Interconnected patterns of biogeography and evolution. *Proceedings of the National Academy of Sciences, U. S. A.*, 68(6): 1254–1258.

Dart, R. A. 1949. The predatory implemental technique of *Australopithecus*. *American Journal of Physical Anthropology*, n.s. 7: 1–38.

——— 1953. The predatory transition from ape to man. *International Anthropological and Linguistic Review*, 1(4): 201–213.

——— 1956. Cultural status of the South African man-apes. *Report of the Smithsonian Institution, Washington, D. C., 1955*, pp. 317–338.

Darwin, C. 1871. *The descent of man, and selection in relation to sex*, 2 vols. Appleton, New York. Vol. 1: vi + 409 pp.; vol. 2: viii + 436 pp.

Davis, D. E. 1957. Aggressive behavior in castrated starlings. *Science*, 126: 253.

——— 1964. The physiological analysis of aggressive behavior. In W. Etkin, ed. (*q.v.*), *Social behavior and organization among vertebrates*, pp. 53–74. University of Chicago Press, Chicago. xii + 307 pp.

Davis, R. T., R. W. Leary, Mary D. C. Smith, and R. F. Thompson. 1968. Species differences in the gross behaviour of nonhuman primates. *Behaviour*, 31(3,4): 326–338.

DeFries, J. C., and G. E. McClearn. 1970. Social dominance and Darwinian fitness in the laboratory mouse. *American Naturalist*, 104(938): 408–411.

Deligne, J. 1965. Morphologie et fonctionnement des mandibules chez les soldats des termites. *Biologia Gabonica*, 1(2): 179–186.

Denes, P. B., and E. N. Pinson. 1973. *The speech chain: the physics and biology of spoken language*, rev. ed. Anchor Press, Doubleday, Garden City, N. Y. xviii + 217 pp.

Denham, W. W. 1971. Energy relations and some basic properties of primate social organization. *American Anthropologist*, 73: 77–95.

DeVore, B. I. 1972. Quest for the roots of society. In P. R. Marler, ed., *The marvels of animal behavior*, pp. 393–408. National Geographic Society, Washington, D. C. 422 pp.

Dingle, H., and R. L. Caldwell. 1969. The aggressive and territorial behaviour of the mantis shrimp *Gonodactylus bredini* Manning (Crustacea: Stomatopoda). *Behaviour*, 33(1,2): 115–136.

Dobzhansky, T. 1963. Anthropology and the natural sciences—the problem of human evolution. *Current Anthropology*, 4: 138, 146–148.

Donisthorpe, H. St. J. K. 1915. *British ants, their life-history and classification*. William Brendon and Son, Plymouth, England. xv + 379 pp.

Dorst, J. 1970. *A field guide to the larger mammals of Africa*. Houghton Mifflin Co., Boston. 287 pp.

Douglas-Hamilton, I. 1972. On the ecology and behaviour of the African elephant: the elephants of Lake Manyara. Ph.D. Thesis, Oriel College, Oxford University, Oxford. xiv + 268 pp.

——— 1973. On the ecology and behaviour of the Lake Manyara elephants. *East African Wildlife Journal*, 11(3,-4): 401–403.

Dreher, J. J., and W. E. Evans. 1964. Cetacean communication. In W. N. Tavolga, ed., *Marine bio-acoustics*, pp. 373–393. Pergamon, New York. xiv + 413 pp.

Ducke, A. 1910. Révision des guêpes sociales polygames d'Amérique. *Annales Historico-Naturales Musei Nationales Hungarici*, 8(2): 449–544.

——— 1914. Über Phylogenie and Klassifikation der sozialen Vespiden. *Zoologische Jahrbücher, Abteilungen Sys-*

tematik, Ökologie und Geographie der Tiere, 36(2,3): 303–330.

Duellman, W. E. 1966. Aggressive behavior in dendrobatid frogs. *Herpetologica*, 22(3): 217–221.

Dunbar, M. J. 1960. The evolution of stability in marine environments: natural selection at the level of the ecosystem. *American Naturalist*, 94(875): 129–136.

——— 1972. The ecosystem as a unit of natural selection. In E. S. Deevey, ed., *Growth by intussusception: ecological essays in honor of G. Evelyn Hutchinson*, pp. 114–130. Transactions of the Academy, vol. 44. Connecticut Academy of Arts and Sciences, New Haven. 442 pp.

Eberhard, Mary Jane West. 1969. The social biology of polistine wasps. *Miscellaneous Publications, Museum of Zoology, University of Michigan, Ann Arbor*, 140: 1–101.

Eibl-Eibesfeldt, I. 1966. Das Verteidigen der Eiablageplätze bei der Hood-Meerechse (*Amblyrhynchus cristatus venustissimus*). *Zeitschrift für Tierpsychologie*, 23(5): 627–631.

——— 1970. *Ethology: the biology of behavior*. Holt, Rinehart and Winston, New York. xiv + 530 pp.

Eimerl, S., and I. DeVore. 1965. *The primates*. Time-Life Books, Chicago. 200 pp.

Eisenberg, J. F. 1966. The social organization of mammals. *Handbuch der Zoologie*, 10(7): 1–92.

Eisenberg, J. F., and R. E. Kuehn. 1966. The behavior of *Ateles geoffroyi* and related species. *Smithsonian Miscellaneous Collections*, 151(8). iv + 63 pp.

Eisenberg, J. F., and M. Lockhart. 1972. An ecological reconnaissance of Wilpattu National Park, Ceylon. *Smithsonian Contributions to Zoology*, 101. vi + 118 pp.

Eisner, T. 1957. A comparative morphological study of the proventriculus of ants (Hymenoptera: Formicidae). *Bulletin of the Museum of Comparative Zoology, Harvard*, 116(8): 439–490.

Ellefson, J. O. 1968. Territorial behavior in the common white-handed gibbon, *Hylobates lar* Linn. In Phyllis C. Jay, ed., *Primates: studies in adaptation and variability*, pp. 180–199. Holt, Rinehart and Winston, New York. xiv + 529 pp.

Ellis, Peggy E. 1959. Learning and social aggregation in locust hoppers. *Animal Behaviour*, 7(1,2): 91–106.

Emerson, A. E. 1956. Regenerative behavior and social homeostasis in termites. *Ecology*, 37(2): 248–258.

Emlen, S. T. 1968. Territoriality in the bullfrog, *Rana catesbeiana*. *Copeia*, 1968, no. 2, pp. 240–243.

——— 1971. The role of song in individual recognition in the indigo bunting. *Zeitschrift für Tierpsychologie*, 28(3): 241–246.

——— 1972. An experimental analysis of the parameters of bird song eliciting species recognition. *Behaviour*, 41(1,2): 130–171.

Erickson, J. G. 1967. Social hierarchy, territoriality, and stress reactions in sunfish. *Physiological Zoology*, 40(1): 40–48.

Eshel, I. 1972. On the neighbor effect and the evolution of altruistic traits. *Theoretical Population Biology*, 3(3): 258–277.

Esser, A. H., ed. 1971. *Behavior and environment: the use of space by animals and men*. Proceedings of an international symposium held at the 1968 meeting of the American Association for the Advancement of Science, Dallas, Texas. Plenum Press, New York. xvii + 411 pp.

Estes, R. D. 1966. Behaviour and life history of the wildebeest (*Connochaetes taurinus* Burchell). *Nature, London*, 212(5066): 999–1000.

——— 1969. Territorial behavior of the wildebeest (*Connochaetes taurinus* Burchell, 1823). *Zeitschrift für Tierpsychologie*, 26(3): 284–370.

——— 1974. Social organization of the African Bovidae. In V. Geist and F. Walther, eds., *The behaviour of ungulates and its relation to management*, IUCN Publications, n. s., no. 24, vol. 1, pp. 166–205. International Union for the Conservation of Nature and Natural Resources, Morges, Switzerland. vol. 1, pp. 1–511; vol. 2, pp. 512–940.

Estes, R. D., and J. Goddard. 1967. Prey selection and hunting behavior of the African wild dog. *Journal of Wildlife Management*, 31(1): 52–70.

Etkin, W. 1954. Social behavior and the evolution of man's mental faculties. *American Naturalist*, 88(840): 129–142.

Evans, H. E. 1958. The evolution of social life in wasps. *Proceedings of the Tenth International Congress of Entomology, Montreal, 1956*, 2: 449–457.

Evans, H. E., and Mary Jane West Eberhard. 1970. *The wasps*. University of Michigan Press, Ann Arbor. vi + 265 pp.

Evans, S. M. 1973. A study of fighting reactions in some nereid polychaetes. *Animal Behaviour*, 21(1): 138–146.

Evans, W. E., and J. Bastian. 1969. Marine mammal communication: social and ecological factors. In H. T. Andersen, ed., *The biology of marine mammals*, pp. 425–475. Academic Press, New York. 511 pp.

Ewing, L. S. 1967. Fighting and death from stress in a cockroach. *Science*, 155: 1035–1036.

Fagen, R. M. 1972. An optimal life-history strategy in which reproductive effort decreases with age. *American Naturalist*, 106(948): 258–261.

——— 1974. Selective and evolutionary aspects of animal play. *American Naturalist*, 108(964): 850–858.

——— 1977. Selection for optimal age-dependent schedules of play behavior. *American Naturalist*, 111: 395–414.

Falls, J. B. 1969. Functions of territorial song in the white-throated sparrow. In R. A. Hinde, ed. (*q.v.*), *Bird vocalizations: their relations to current problems in biology and psychology: essays presented to W. H. Thorpe*, pp. 207–232.

Fisher, J. 1954. Evolution and bird sociality. In J. Huxley, A. C. Hardy, and E. B. Ford, eds., *Evolution as a process*, pp. 71–83. George Allen & Unwin, London. 376 pp. (Reprinted as a paperback, Collier Books, New York, 1963. 416 pp.)

Fisher, R. A. 1930. *The genetical theory of natural selection*. Clarendon Press, Oxford. xiv + 272 pp.

Flanders, S. E. 1956. The mechanisms of sex-ratio regulation in the (parasitic) Hymenoptera. *Insectes Sociaux*, 3(2): 325–334.

Flannery, K. V. 1972. The cultural evolution of civilizations. *Annual Review of Ecology and Systematics*, 3: 399–426.

Fodor, J., and M. Garrett. 1966. Some reflections on competence and performance. In J. Lyons and R. J. Wales, eds., *Psycholinguistic papers*, pp. 133–163. Edinburgh University Press, Edinburgh. 243 pp.

Fossey, Dian. 1972. Living with mountain gorillas. In P. R. Marler, ed., *The marvels of animal behavior*, pp. 209–229. National Geographic Society, Washington, D. C. 422 pp.

Fox, M. W. 1972. Socio-ecological implications of individual differences in wolf litters: a developmental and evolutionary perspective. *Behaviour*, 46(3,4): 298–313.

Fox, R. 1971. The cultural animal. In J. F. Eisenberg and W. S. Dillon, eds., *Man and beast: comparative social behavior*, pp. 263–296. Smithsonian Institution Press, Washington, D. C. 401 pp.

——— 1972. Alliance and constraint: sexual selection in the evolution of human kinship systems. In B. G. Campbell, ed., *Sexual selection and the descent of man 1871–1971*, pp. 282–331. Aldine Publishing Co., Chicago. x + 378 pp.

Franklin, W. L. 1973. High, wild world of the vicuña. *National Geographic*, 143(1) (January): 76–91.

Franzisket, L. 1960. Experimentelle Untersuchung über die optische Wirkung der Streifung beim Preussenfisch (*Dascyllus aruanus*). *Behaviour*, 15(1,2): 77–81.

Fraser, A. F. 1968. *Reproductive behaviour in ungulates*. Academic Press, New York. x + 202 pp.

Free, J. B. 1961a. The social organization of the bumble-bee colony. A lecture given to The Central Association of Bee-keepers on 18th January 1961. North Hants Printing and Publishing Co., Fleet, Hants, England. 11 pp.

——— 1961b. Hypopharyngeal gland development and division of labour in honey-bee (*Apis mellifera L.*) colonies. *Proceedings of the Royal Entomological Society of London*, ser. A, 36(1–3): 5–8.

Free, J. B., and C. G. Butler. 1959. *Bumblebees*. New Naturalist, Collins, London. xiv + 208 pp.

Freedman, D. G. 1974. *Human infancy: an evolutionary perspective*. Lawrence Erlbaum, Hillsdale, N. J. xii + 212 pp.

——— 1979. *Human sociobiology*. Free Press, New York. 188 pp.

Friedrichs, R. W. 1970. *A sociology of sociology*. Free Press, Collier-Macmillan, New York. xxxiv + 429 pp.

Frisch, K. von 1967. *The dance language and orientation of bees*, trans. by L. E. Chadwick. Belknap Press of Harvard University Press, Cambridge. xiv + 566 pp.

Frisch, K. von, and G. A. Rösch. 1926. Neue Versuche über die Bedeutung von Duftorgan und Pollenduft für die Verständigung im Bienenvolk. *Zeitschrift für Vergleichende Physiologie*, 4(1): 1–21.

Furuya, Y. 1963. On the Gagyusan troop of Japanese monkeys after the first separation. *Primates*, 4(1): 116–118.

Gadgil, M. 1975. Evolution of social behavior through interpopulation selection. *Proceedings of the National Academy of Sciences, U. S. A.*, 72(3): 1199–1201.

Gadgil, M., and W. H. Bossert. 1970. Life history consequences of natural selection. *American Naturalist*, 104(935): 1–24.

Garcia, J., B. K. McGowan, F. R. Ervin, and R. A. Koelling. 1968. Cues: their relative effectiveness as a function of the reinforcer. *Science*, 160: 794–795.

Gartlan, J. S. 1968. Structure and function in primate society. *Folia Primatologica*, 8(2): 89–120.

——— 1969. Sexual and maternal behavior of the vervet monkey, *Cercopithecus aethiops*. *Journal of Reproduction and Fertility*, supplement 6: 137–150.

Geist, V. 1971a. *Mountain sheep: a study in behavior and evolution*. University of Chicago Press, Chicago. xvi + 383 pp.

——— 1971b. The relation of social evolution and dispersal in ungulates during the Pleistocene, with emphasis on the Old World deer and the genus *Bison*. *Quarternary Research*, 1(3): 283–315.

Ghiselin, M. T. 1969. The evolution of hermaphroditism among animals. *Quarterly Review of Biology*, 44(2): 189–208.

Gilbert, J. J. 1966. Rotifer ecology and embryological induction. *Science*, 151: 1234–1237.

——— 1973. Induction and ecological significance of gigantism in the rotifer, *Asplancha sieboldi*. *Science*, 181: 63–66.

Gill, J. C., and W. Thomson. 1956. Observations on the behaviour of suckling pigs. *British Journal of Animal Behaviour*, 4(2): 46–51.

Ginsburg, B., and W. C. Allee. 1942. Some effects of conditioning on social dominance and subordination in inbred strains of mice. *Physiological Zoology*, 15(4): 485–506.

Glancey, B. M., C. E. Stringer, C. H. Craig, P. M. Bishop, and B. B. Martin. 1970. Pheromone may induce brood tending in the fire ant, *Solenopsis saevissima*. *Nature, London*, 226(5248): 863–864.

Goffman, E. 1959. *The presentation of self in everyday life*. Doubleday Anchor Books, Doubleday, Garden City, N. Y. xvi + 259 pp.

——— 1961. *Encounters: two studies in the sociology of interaction*. Bobbs-Merrill, Indianapolis. 152 pp.

Goin, C. J., and Olive B. Goin. 1962. *Introduction to herpetology*. W. H. Freeman, San Francisco. 341 pp.

Gottschalk, L. A., S. M. Kaplan, Goldine C. Gleser, and Carolyn Winget. 1961. Variations in the magnitude of anxiety and hostility with phases of the menstrual cycle. *Psychosomatic Medicine*, 23(5): 448.

Gramza, A. F. 1967. Responses of brooding nighthawks to a disturbance stimulus. *Auk*, 84(1): 72–86.

Grant, E. C. 1969. Human facial expression. *Man*, 4(4): 525–536.

Green, R. G., C. L. Larson, and J. F. Bell. 1939. Shock disease as the cause of the periodic decimation of the snowshoe hare. *American Journal of Hygiene*, ser. B, 30: 83–102.

Greer, A. E., Jr. 1971. Crocodilian nesting habits and evolution. *Fauna*, 2: 20–28.

Groot, A. P. de. 1953. Protein and amino acid requirements of the honey-bee (*Apis mellifica L.*). *Physiologia Comparata et Oecologia*, 3(2,3): 197–285.

Gwinner, E. 1966. Über einige Bewegungsspiele des Kolkraben (*Corvus corax L.*). *Zeitschrift für Tierpsychologie*, 23(1): 28–36.

Hailman, J. P. 1960. Hostile dancing and fall territory of a color-banded mockingbird. *Condor*, 62(6): 464–468.

Haldane, J. B. S. 1932. *The causes of evolution*. Longmans, Green, London. vii + 234 pp. (Reprinted as a paperback, Cornell University Press, Ithaca, N. Y., 1966. vi + 235 pp.)

Hall, K. R. L. 1960. Social vigilance behaviour of the chacma baboon, *Papio ursinus*. *Behaviour*, 16(3,4): 261–294.

Hall, K. R. L., and I. DeVore. 1965. Baboon social behavior. In I. DeVore, ed., *Primate behavior: field studies of monkeys and apes*, pp. 53–110. Holt, Rinehart and Winston, New York. xiv + 654 pp.

Hamilton, W. D. 1964. The genetical theory of social behaviour, I, II. *Journal of Theoretical Biology*, 7(1): 1–52.

——— 1966. The moulding of senescence by natural selection. *Journal of Theoretical Biology*, 12(1): 12–45.

——— 1967. Extraordinary sex ratios. *Science*, 156: 477–488.

——— 1970. Selfish and spiteful behaviour in an evolutionary model. *Nature, London*, 228(5277): 1218–1220.

——— 1971a. Geometry for the selfish herd. *Journal of Theoretical Biology*, 31(2): 295–311.

——— 1971b. Selection of selfish and altruistic behavior in some extreme models. In J. F. Eisenberg and W. S. Dillon, eds., *Man and beast: comparative social behavior*, pp. 57–91. Smithsonian Institution Press, Washington, D. C. 401 pp.

——— 1972. Altruism and related phenomena, mainly in social insects. *Annual Review of Ecology and Systematics*, 3: 193–232.

Hamilton, W. J., III, and W. M. Gilbert. 1969. Starling dispersal from a winter roost. *Ecology*, 50(5): 886–898.

Hangartner, W. 1969a. Structure and variability of the individual odor trail in *Solenopsis geminata* Fabr. (Hymenoptera, Formicidae). *Zeitschrift für Vergleichende Physiologie*, 62(1): 111–120.

——— 1969b. Carbon dioxide, a releaser for digging behavior in *Solenopsis geminata* (Hymenoptera: Formicidae). *Psyche, Cambridge*, 76(1): 58–67.

Hansen, E. W. 1966. The development of maternal and infant behavior in the rhesus monkey. *Behaviour*, 27(1,2): 107-149.

Hardin, G. 1972. Population skeletons in the environmental closet. *Bulletin of the Atomic Scientists*, 28(6) (June): 37–41.

Harlow, H. F., M. K. Harlow, R. O. Dodsworth, and G. L. Arling. 1966. Maternal behavior of rhesus monkeys deprived of mothering and peer associations in infancy. *Proceedings of the American Philosophical Society*, 110(1): 58–66.

Harris, G. W., and R. P. Michael. 1964. The activation of sexual behaviour by hypothalamic implants of oestrogen. *Journal of Physiology*, 171(2): 275–301.

Harris, V. T. 1952. An experimental study of habitat selection by prairie and forest races of the deermouse, *Peromyscus maniculatus*. *Contributions from the Laboratory of Vertebrate Biology, University of Michigan, Ann Arbor*, no. 56. 53 pp.

Harrison, C. J. O. 1965. Allopreening as agonistic behaviour. *Behaviour*, 34(3,4): 161–209.

Hartley, P. H. T. 1950. An experimental analysis of interspecific recognition. *Symposia of the Society for Experimental Biology*, 4: 313–336.

Haskins, C. P. 1970. Researches in the biology and social behavior of primitive ants. In L. R. Aronson, Ethel Tobach, D. S. Lehrman, and J. S. Rosenblatt, eds., *Development and evolution of behavior: essays in memory of T. C. Schneirla*, pp. 355–388. W. H. Freeman, San Francisco. xviii + 656 pp.

Haskins, C. P., and Edna F. Haskins. 1950. Notes on the biology and social behavior of the archaic ponerine ants of the genera *Myrmecia* and *Promyrmecia*. *Annals of the Entomological Society of America*, 43(4): 461–491.

Hay, D. A. 1972. Recognition by *Drosophila melanogaster* of individuals from other strains or cultures: support for the role of olfactory cues in selective mating. *Evolution*, 26(2): 171–176.

Haydak, M. H. 1935. Brood rearing by honeybees confined to a pure carbohydrate diet. *Journal of Economic Entomology*, 28(4): 657–660.

Hediger, H. 1941. Biologische Gesetzmässigkeiten im Verhalten von Wirbeltieren. *Mitteilungen der Naturforschenden Gesellschaft Bern, 1940*, pp. 37–55.

——— 1950. *Wildtiere in Gefangenschaft—ein Grundriss der Tiergartenbiologie*. Benno Schwabe, Basle. (Reprinted as *Wild animals in captivity: an outline of the biology of zoological gardens*, trans. by G. Sircom, Butterworth Scientific Publications, London. 207 pp.)

——— 1955. *Studies of the psychology and behaviour of captive animals in zoos and circuses*, trans. by G. Sircom. Criterion Books, New York. vii + 166 pp. (Reprinted as *The psychology and behaviour of animals in zoos and circuses*, Dover, New York, 1968. vii + 166 pp.)

Heimburger, N. 1959. Das Markierungsverhalten einiger Caniden. *Zeitschrift für Tierpsychologie*, 16(1): 104–113.

Heldmann, G. 1936a. Ueber die Entwicklung der polygynen Wabe von *Polistes gallica L. Arbeiten über Physiologische und Angewandte Entomologie aus Berlin-Dahlem*, 3: 257–259.

——— 1936b. Über das Leben auf Waben mit mehreren überwinterten Weibchen von *Polistes gallica L. Biologisches Zentralblatt*, 56(7,8): 389–401.

Heller, H. C. 1971. Altitudinal zonation of chipmunks (*Eutamias*): interspecific aggression. *Ecology*, 52(2): 312–329.

Helm, June. 1968. The nature of Dogrib socioterritorial groups. In R. B. Lee and I. DeVore, eds. (*q.v.*), *Man the hunter*, pp. 118–125.

Hendrichs, H., and Ursula Hendrichs. 1971. *Dikdik und Elefanten*. R. Piper, Munich. 173 pp.

Henry, C. S. 1972. Eggs and repagula of *Ululodes* and *Ascaloptynx* (Neuroptera: Ascalaphidae): a comparative study. *Psyche, Cambridge*, 79(1,2): 1–22.

Hensley, M. M., and J. B. Cope. 1951. Further data on removal and repopulation of the breeding birds in a spruce-fir forest community. *Auk*, 68(4): 483–493.

Herrnstein, R. J. 1971b. I. Q. *Atlantic Monthly*, 228(3) (September): 43–64.

Hill, C. 1946. Playtime at the zoo. *Zoo Life* (Zoological Society of London), 1(1): 24–26.

Hill, Jane H. 1972. On the evolutionary foundations of language. *American Anthropologist,* 74(3): 308–317.
Hill, W. C. O. 1972. *Evolutionary biology of the primates.* Academic Press, New York. x + 233 pp.
Hinde, R. A. 1952. The behaviour of the great tit (*Parus major*) and some other related species. *Behaviour,* supplement 2. x + 201 pp.
——— 1956. The biological significance of the territories of birds. *Ibis,* 98(3): 340–369.
——— ed. 1969. *Bird vocalizations: their relations to current problems in biology and psychology: essays presented to W. H. Thorpe.* Cambridge University Press, Cambridge. xvi + 394 pp.
——— 1970. *Animal behaviour: a synthesis of ethology and comparative psychology,* 2d ed. McGraw-Hill Book Co., New York. xvi + 876 pp.
——— ed. 1972. *Non-verbal communication.* Cambridge University Press, Cambridge. xiii + 423 pp.
——— 1974. *Biological bases of human social behaviour.* McGraw-Hill Book Co., New York. xvi + 462 pp.
Hinde, R. A., and Lynda M. Davies. 1972a. Changes in mother-infant relationship after separation in *Rhesus* monkeys. *Nature, London,* 239(5366): 41–42.
——— 1972b. Removing infant rhesus from mother for 13 days compared with removing mother from infant. *Journal of Child Psychology and Psychiatry,* 13: 227–237.
Hinde, R. A., and Yvette Spencer-Booth. 1971. Effects of brief separation from mother on rhesus monkeys. *Science,* 173: 111–118.
Hjorth, I. 1970. Reproductive behaviour in Tetraonidae with special references to males. *Viltrevy,* 7(4): 183–596.
Hochbaum, H. A. 1955. *Travels and traditions of waterfowl.* University of Minnesota Press, Minneapolis. xii + 301 pp.
Hoese, H. D. 1971. Dolphin feeding out of water in a salt marsh. *Journal of Mammalogy,* 52(1): 222–223.
Hogan-Warburg, A. J. 1966. Social behavior of the ruff, *Philomachus pugnax* (L.). *Ardea,* 54(3,4): 109–229.
Holgate, P. 1967. Population survival and life history phenomena. *Journal of Theoretical Biology,* 14(1): 1–10.
Holling, C. S. 1959. Some characteristics of simple types of predation and parasitism. *Canadian Entomologist,* 91(7): 385–398.
Homans, G. C. 1961. *Social behavior: its elementary forms.* Harcourt, Brace & World, New York. xii + 404 pp.
Hooff, J. A. R. A. M. van. 1972. A comparative approach to the phylogeny of laughter and smiling. In R. A. Hinde, ed., *Non-verbal communication,* pp. 209–241. Cambridge University Press, Cambridge. xiii + 423 pp.
Horn, H. S. 1968. The adaptive significance of colonial nesting in the Brewer's blackbird (*Euphagus cyanocephalus*). *Ecology,* 49(4): 682–694.
Howells, W. W. 1973. *Evolution of the genus Homo.* Addison-Wesley, Reading, Mass. 188 pp.
Hrdy, Sarah Blaffer. 1976. Care and exploitation of nonhuman primate infants by conspecifics other than the mother. *Advances in the Study of Behavior,* 6: 101–158.
——— 1977. *The langurs of Abu: female and male strategies of reproduction.* Harvard University Press, Cambridge. xx + 361 pp.
Hunsaker, D. 1962. Ethological isolating mechanisms in the *Sceloporus torquatus* group of lizards. *Evolution,* 16(1): 62–74.
Hunsaker, D., and T. C. Hahn. 1965. Vocalization of the South American tapir, *Tapirus terrestris. Animal Behaviour,* 13(1): 69–78.
Hunter, J. R. 1969. Communication of velocity changes in jack mackerel (*Trachurus symmetricus*) schools. *Animal Behaviour,* 17(3): 507–514.
Hutchinson, G. E. 1948. Circular causal systems in ecology. *Annals of the New York Academy of Sciences,* 50(4): 221–246.
——— 1951. Copepodology for the ornithologist. *Ecology,* 32(3): 571–577.
——— 1959. A speculative consideration of certain possible forms of sexual selection in man. *American Naturalist,* 93(869): 81–91.
——— 1961. The paradox of the plankton. *American Naturalist,* 95(882): 137–145.
Hutt, Corinne. 1966. Exploration and play in children. *Symposia of the Zoological Society of London,* 18: 61–81.
Huxley, J. S. 1934. A natural experiment on the territorial instinct. *British Birds,* 27(10): 270–277.
Ihering, H. von. 1896. Zur Biologie der socialen Wespen Brasiliens. *Zoologischer Anzeiger,* 19(516): 449–453.
Imanishi, K. 1958. Identification: a process of enculturation in the subhuman society of *Macaca fuscata. Primates,* 1(1): 1–29. (In Japanese with English introduction.)
——— 1963. Social behavior in Japanese monkeys, *Macaca fuscata.* In C. H. Southwick, ed., *Primate social behavior: an enduring problem,* pp. 68–81. Van Nostrand Co., Princeton, N. J. viii + 191 pp. (Originally published in Japanese in *Psychologia,* 1[1]: 47–54, 1957.)
Immelmann, K. 1966. Beobachtungen an Schwalbenstaren. *Journal für Ornithologie,* 107(1): 37–69.
——— 1972. Sexual and other long-term aspects of imprinting in birds and other species. *Advances in the Study of Behavior,* 4: 147–174.
Inkeles, A. 1964. *What is sociology? An introduction to the discipline and profession.* Prentice-Hall, Englewood Cliffs, N. J. viii + 120 pp.
Itani, J. 1959. Paternal care in the wild Japanese monkey, *Macaca fuscata fuscata. Primates,* 2(1): 61–93.
——— 1972. A preliminary essay on the relationship between social organization and incest avoidance in nonhuman primates. In F. E. Poirier, ed. (*q.v.*), *Primate socialization,* pp. 165–171.
Ivey, M. E., and Judith M. Bardwick. 1968. Patterns of affective fluctuation in the menstrual cycle. *Psychosomatic Medicine,* 30(3): 336–345.
Izawa, K. 1970. Unit groups of chimpanzees and their nomadism in the savanna woodland. *Primates,* 11(1): 1–46.
Jarman, P. J. 1974. The social organisation of antelope in relation to their ecology. *Behaviour,* 58(3,4): 215–267.
Jay, Phyllis C. 1965. The common langur of North India. In I. DeVore, ed., *Primate behavior: field studies of monkeys and apes,* pp. 197–249. Holt, Rinehart, and Winston, New York. xiv + 654 pp.
Jennrich, R. I., and F. B. Turner. 1969. Measurement of

non-circular home range. *Journal of Theoretical Biology*, 22(2): 227–237.

Johnsgard, P. A. 1967. Dawn rendezvous on the lek. *Natural History*, 76(3) (March): 16–21.

Johnston, Norah C., J. H. Law, and N. Weaver. 1965. Metabolism of 9-ketodec-2-enoic acid by worker honeybees (*Apis mellifera L.*). *Biochemistry*, 4: 1615–1621.

Jolicoeur, P. 1959. Multivariate geographical variation in the wolf *Canis lupus L. Evolution*, 13(3): 283–299.

Jolly, Alison. 1966. *Lemur behavior: a Madagascar field study*. University of Chicago Press, Chicago. xiv + 187 pp.

Jolly, C. J. 1970. The seed-eaters: a new model of hominid differentiation based on a baboon analogy. *Man*, 5(1): 5–26.

Jullien, J. 1885. Monographie des bryozoaires d'eau douce. *Bulletin de la Société Zoologique de France*, 10: 91–207.

Kaiser, P. 1954. Über die Funktion der Mandibeln bei den Soldaten von *Neocapritermes opacus* (Hagen). *Zoologischer Anzeiger*, 152(9,10): 228–234.

Kalela, O. 1954. Über den Revierbesitz bei Vögeln und Säugetieren als populationsökologischer Faktor. *Annales Zoologici Societatis Zoologicae Botanicae Fennicae "Vanamo"* (Helsinki), 16(2): 1–48.

——— 1957. Regulation of reproductive rate in subarctic populations of the vole *Clethrionomys rufocanus* (Sund.). *Annales Academiae Scientiarum Fennicae* (Suomalaisen Tiedeakatemian Toimituksia), ser. A (IV, Biologica), 34: 1–60.

Kalmijn, A. J. 1971. The electric sense of sharks and rays. *Journal of Experimental Biology*, 55(2): 371–383.

Karlson, P., and A. Butenandt. 1959. Pheromones (ectohormones) in insects. *Annual Review of Entomology*, 4: 39–58.

Kaufmann, J. H. 1962. Ecology and social behavior of the coati, *Nasua narica*, on Barro Colorado Island, Panama. *University of California Publications in Zoology*, 60(3): 95–222.

——— 1966. Behavior of infant rhesus monkeys and their mothers in a free-ranging band. *Zoologica, New York*, 51(1): 17–27.

——— 1967. Social relations of adult males in a free-ranging band of rhesus monkeys. In S. A. Altmann, ed., *Social communication among primates*, pp. 73–98. University of Chicago Press, Chicago. xiv + 392 pp.

——— 1974. Social ethology of the whiptail wallaby, *Macropus parryi*, in northeastern New South Wales. *Animal Behaviour*, 22(2): 281–369.

Kawai, M. 1958. On the system of social ranks in a natural troop of Japanese monkeys: I, basic rank and dependent rank. *Primates*, 1–2: 111–130. (In Japanese; translated in S. A. Altmann, ed., 1965. *Japanese monkeys, a collection of translations*, selected by K. Imanishi. The Editor, Edmonton, Alberta. v + 151 pp.)

——— 1965. Newly acquired pre-cultural behavior of the natural troop of Japanese monkeys on Koshima Islet. *Primates*, 6(1): 1–30.

Kawamura, S. 1954. A new type of action expressed in the feeding behavior of the Japanese monkey in its wild habitat. *Organic Evolution*, 2(1): 10–13. (In Japanese; cited by K. Imanishi, 1963 [*q.v.*].)

——— 1958. Matriarchal social ranks in the Minoo-B troop: a study of the rank system of Japanese monkeys. *Primates*, 1–2: 149–156. (In Japanese; translated in S. A. Altmann, ed., 1965, *Japanese monkeys, a collection of translations*, selected by K. Imanishi. The Editor, Edmonton. v + 151 pp.

——— 1963. The process of sub-culture propagation among Japanese macaques. In C. H. Southwick, ed., *Primate social behavior: an enduring problem*, pp. 82–90. Van Nostrand Co., Princeton, N. J. viii + 191 pp. (Originally published in Japanese in *Journal of Primatology*, 1959, 2[1]: 43–60.)

——— 1967. Aggression as studied in troops of Japanese monkeys. In Carmine D. Clemente and D. B. Lindsley, eds., *Brain function*, vol. 5, *Aggression and defense, neural mechanisms and social patterns*, pp. 195–223. University of California Press, Berkeley. xv + 361 pp.

Keith, A. 1949. *A new theory of human evolution*. Philosophical Library, New York. x + 451 pp.

Kendeigh, S. C. 1952. *Parental care and its evolution in birds*. Illinois Biological Monographs, 22(1–3). x + 356 pp.

King, C. E., and W. W. Anderson. 1971. Age-specific selection: II, the interaction between *r* and *K* during population growth. *American Naturalist*, 105(942): 137–156.

King, J. A. 1955. Social behavior, social organization, and population dynamics in a black-tailed prairiedog town in the Black Hills of South Dakota. *Contributions from the Laboratory of Vertebrate Biology, University of Michigan, Ann Arbor*, no. 67. 123 pp.

——— 1957. Relationships between early social experience and adult aggressive behavior in inbred mice. *Journal of Genetic Psychology*, 90: 151–166.

Kislak, J. W., and F. A. Beach. 1955. Inhibition of aggressiveness by ovarian hormones. *Endocrinology*, 56(6): 684–692.

Kleiman, Devra G., and J. F. Eisenberg. 1973. Comparisons of canid and felid social systems from an evolutionary perspective. *Animal Behaviour*, 21(4): 635–659.

Klopman, R. B. 1968. The agonistic behavior of the Canada goose (*Branta canadensis canadensis*): I, attack behavior. *Behaviour*, 30(4): 287–319.

Kluijver, H. N., and L. Tinbergen. 1953. Territory and the regulation of density in titmice. *Archives Néerlandaises de Zoologie, Leydig*, 10(3): 265–289.

Koenig, O. 1962. *Kif-Kif*. Wollzeilen-Verlag, Vienna. [Cited by W. Wickler, 1972 (*q.v.*).]

Koford, C. B. 1957. The vicuña and the puna. *Ecological Monographs*, 27(2): 153–219.

——— 1963. Rank of mothers and sons in bands of rhesus monkeys. *Science*, 141: 356–357.

Kohlberg, L. 1969. Stage and sequence: the cognitive-developmental approach to socialization. In D. A. Goslin, ed., *Handbook of socialization theory and research*, pp. 347–480. Rand McNally Co., Chicago. xiii + 1182 pp.

Kortlandt, A. 1940. Eine Übersicht der angeboren Verhaltungsweisen des Mittel-Europäischen Kormorans (*Phalocrocorax carbo sinensis* [Shaw & Nodd.]), ihre Funktion, ontogenetische Entwicklung und phylogenetische Herkunft. *Archives Néerlandaises de Zoologie, Leydig*, 4(4): 401–442.

——— 1972. *New perspectives on ape and human evolution*.

Stichting voor Psychobiologie, Universiteit van Amsterdam, The Netherlands. 100 pp.

Krames, L., W. J. Carr, and B. Bergman. 1969. A pheromone associated with social dominance among male rats. *Psychonomic Science,* 16(1): 11–12.

Krebs, C. J. 1972. *Ecology: the experimental analysis of distribution and abundance.* Harper & Row, New York. x + 694 pp.

Krebs, J. R. 1971. Territory and breeding density in the great tit, *Parus major L. Ecology,* 52(1): 2–22.

Kruuk, H. 1972. *The spotted hyena: a study of predation and social behavior.* University of Chicago Press, Chicago. xvi + 335 pp.

Kühlmann, D. H. H., and H. Karst. 1967. Freiwasserbeobachtungen zum Verhalten von Tobiasfischschwärmen (*Ammodytidae*) in der westlichen Ostsee. *Zeitschrift für Tierpsychologie,* 24(3): 282–297.

Kühne, W. 1965. Communal food distribution and division of labour in African hunting dogs. *Nature, London,* 205(4970): 443–444.

Kullenberg, B. 1956. Field experiments with chemical sexual attractants on aculeate Hymenoptera males. *Zoologiska Bidrag från Uppsala,* 31: 253–354.

Kummer, H. 1967. Tripartite relations in hamadryas baboons. In S. A. Altmann, ed., *Social communication among primates,* pp. 63–71. University of Chicago Press, Chicago. xiv + 392 pp.

——— 1968. *Social organization of hamadryas baboons: a field study.* University of Chicago Press, Chicago. viii + 189 pp.

——— 1971. *Primate societies: group techniques of ecological adaptation.* Aldine-Atherton, Chicago. 160 pp.

Kurtén, B. 1972. *Not from the apes.* Vintage Books, Random House, New York. viii + 183 pp.

Lack, D. 1954. *The natural regulation of animal numbers.* Oxford University Press, Oxford. viii + 343 pp.

——— 1966. *Population studies of birds.* Oxford University Press, Oxford. v + 341 pp.

——— 1968. *Ecological adaptations for breeding in birds.* Methuen, London. xii + 409 pp.

Lancaster, Jane B. 1971. Play-mothering: the relations between juvenile females and young infants among free-ranging vervet monkeys (*Cercopithecus aethiops*). *Folia Primatologica,* 15(3,4): 161–182.

Lawick, H. van, and Jane van Lawick-Goodall. 1971. *Innocent killers.* Houghton Mifflin Co., Boston. 222 pp.

Lawick-Goodall, Jane van. 1967. *My friends the wild chimpanzees.* National Geographic Society, Washington, D. C. 204 pp.

——— 1968a. The behaviour of free-living chimpanzees in the Gombe Stream Reserve. *Animal Behaviour Monographs,* 1(3): 161–311.

——— 1968b. A preliminary report on expressive movements and communication in the Gombe Stream chimpanzees. In Phyllis C. Jay, ed., *Primates: studies in adaptation and variability,* pp. 313–374. Holt, Rinehart and Winston, New York. xiv + 529 pp.

——— 1969. Mother-offspring relationships in free-ranging chimpanzees. In D. Morris, ed., *Primate ethology: essays on the socio-sexual behavior of apes and monkeys,* pp. 364–436. Anchor Books, Doubleday, Garden City, N. Y. vii + 471 pp.

——— 1971. *In the shadow of man.* Houghton Mifflin Co., Boston. xx + 297 pp.

Laws, R. M., and I. S. C. Parker. 1968. Recent studies on elephant populations in East Africa. *Symposia of the Zoological Society of London,* 21: 319–359.

Lee, R. B. 1968. What hunters do for a living, or how to make out on scarce resources. In R. B. Lee and I. DeVore, eds. (*q.v.*), *Man the hunter,* pp. 30–48.

Lee, R. B., and I. DeVore, eds. 1968. *Man the hunter.* Aldine Publishing Co., Chicago. xvi + 415 pp.

Lees, A. D. 1966. The control of polymorphism in aphids. *Advances in Insect Physiology,* 2: 207–277.

Lehrman, D. S. 1964. The reproductive behavior of ring doves. *Scientific American,* 211(5) (November): 48–54.

——— 1965. Interaction between internal and external environments in the regulation of the reproductive cycle of the ring dove. In F. A. Beach, ed., *Sex and behavior,* pp. 355–380. John Wiley & Sons, New York. xvi + 592 pp.

Leibowitz, Lila. 1968. Founding families. *Journal of Theoretical Biology,* 21(2): 153–169.

Lemon, R. E. 1967. The response of cardinals to songs of different dialects. *Animal Behaviour,* 15(4): 538–545.

Lenneberg, E. H. 1967. *Biological foundations of language.* John Wiley & Sons, New York. xviii + 489 pp.

——— 1971. Of language knowledge, apes, and brains. *Journal of Psycholinguistic Research,* 1(1): 1–29.

Lenski, G., and Jean Lenski. 1970. *Human societies: a macrolevel introduction to sociology.* McGraw-Hill Book Co., New York. xvi + 515 pp.

Lerner, I. M. 1954. *Genetic homeostasis.* Oliver and Boyd, London. vii + 134 pp.

——— 1968. *Heredity, evolution, and society.* W. H. Freeman, San Francisco. xviii + 307 pp.

Leuthold, W. 1966. Variations in territorial behavior of Uganda kob *Adenota kob thomasi* (Neumann 1896). *Behaviour,* 27(3,4): 215–258.

——— 1974. Observations on home range and social organization of lesser kudu, *Tragelaphus imberbis* (Blyth, 1869). In V. Geist and F. Walther, eds., *The behaviour of ungulates and its relation to management,* vol. 1, pp. 206–234. IUCN Publications, N. S., no. 24. International Union for the Conservation of Nature and Natural Resources, Morges, Switzerland.

LeVine, R. A., and D. T. Campbell. 1972. *Ethnocentrism: theories of conflict, ethnic attitudes, and group behavior.* John Wiley & Sons, New York. x + 310 pp.

Levins, R. 1970. Extinction. In M. Gerstenhaber, ed., *Some mathematical questions in biology,* pp. 77–107. Lectures on Mathematics in the Life Sciences, vol. 2. American Mathematical Society, Providence, R. I. vii + 156 pp.

Lévi-Strauss, C. 1949. *Les structures élémentaires de la parenté.* Presses Universitaires de France, Paris. xiv + 639 pp. (*The elementary structures of kinship,* rev. ed., trans. by J. H. Bell and J. R. von Sturmer and ed. by R. Needham, Beacon Press, Boston, 1969. xlii + 541 pp.)

Lewontin, R. C. 1972. Testing the theory of natural selection. (Review of R. Creed, ed., *Ecological genetics and evolution*, Blackwell Scientific Publications, Oxford, 1971.) *Nature, London*, 236(5343): 181–182.

Leyhausen, P. 1956. Verhaltensstudien an Katzen. *Zeitschrift für Tierpsychologie*, supplement 2. vi + 120 pp.

——— 1965. The communal organization of solitary mammals. *Symposia of the Zoological Society of London*, 14: 249–263.

——— 1971. Dominance and territoriality as complemented in mammalian social structure. In A. H. Esser, ed., *Behavior and environment: the use of space by animals and men*, pp. 22–33. Plenum Press, New York. xvii + 411 pp.

Lidicker, W. Z., Jr. 1962. Emigration as a possible mechanism permitting the regulation of population density below carrying capacity. *American Naturalist*, 96(886): 29–33.

Lieberman, P. 1968. Primate vocalizations and human linguistic ability. *Journal of the Acoustic Society of America*, 44: 1574–1584.

Lieberman, P., E. S. Crelin, and D. H. Klatt. 1972. Phonetic ability and related anatomy of the newborn and adult human, Neanderthal man, and the chimpanzee. *American Anthropologist*, 74(3): 287–307.

Lill, A. 1968. An analysis of sexual isolation in the domestic fowl: I, the basis of homogamy in males; II, the basis of homogamy in females. *Behaviour*, 30(2,3): 107–145.

Lilly, J. C. 1961. *Man and dolphin*. Doubleday, New York. (Reprinted as a paperback, Pyramid Books, New York, 1969. 191 pp.)

——— 1967. *The mind of the dolphin: a nonhuman intelligence*. Doubleday, New York. (Reprinted as a paperback, Avon Books, Hearst Corporation, New York, 1969. 286 pp.)

Lin, N., and C. D. Michener. 1972. Evolution of sociality in insects. *Quarterly Review of Biology*, 47(2): 131–159.

Lindauer, M. 1961. *Communication among social bees*. Harvard University Press, Cambridge. ix + 143 pp.

——— 1970. Lernen und Gedächtnis—Versuche an der Honigbiene. *Naturwissenschaften*, 57: 463–467.

Lindburg, D. G. 1971. The rhesus monkey in North India: an ecological and behavioral study. In L. A. Rosenblum, ed. (*q.v.*), *Primate behavior: developments in field and laboratory research*, vol. 2, pp. 1–106.

Linsenmair, K. E. 1967. Konstruktion und Signalfunktion der Sandpyramide der Reiterkrabbe *Ocypode saratan* Forsk. (Decapoda Brachyura Ocypodidae). *Zeitschrift für Tierpsychologie*, 24(4): 403–456.

——— 1972. Die Bedeutung familienspezifischer "Abzeichen" für den Familienzusammenhalt bei der sozialen Wüstenassel *Hemilepistus reaumuri* Audouin u. Savigny (Crustacea, Isopoda, Oniscoidea). *Zeitschrift für Tierpsychologie*, 31(2): 131–162.

Linsenmair, K. E., and Christa Linsenmair. 1971. Paarbildung and Paarzusammenhalt bei der monogamen Wüstenassel *Hemilepistus reaumuri* (Crustacea, Isopoda, Oniscoidea). *Zeitschrift für Tierpsychologie*, 29(2): 134–155.

Lissmann, H. W. 1958. On the function and evolution of electric organs in fish. *Journal of Experimental Biology*, 35(1): 156–191.

Lloyd, J. E. 1966. *Studies on the flash communication system in* Photinus *fireflies*. Miscellaneous Publications, Museum of Zoology, University of Michigan, Ann Arbor, 130. 95 pp.

——— 1973. Fireflies of Melanesia: bioluminescense, mating behavior, and synchronous flashing (Coleoptera: Lampyridae). *Annals of the Entomological Society of America*, 2(6): 991–1008.

Loizos, Caroline. 1966. Play in mammals. In P. A. Jewell and Caroline Loizos, ed., *Play, exploration and territory in mammals*, pp. 1–9. Symposia of the Zoological Society of London, no. 18. Academic Press, New York. xiii + 280 pp.

——— 1967. Play behaviour in higher primates: a review. In D. Morris, ed., *Primate ethology: essays on the socio-sexual behavior of apes and monkeys*, pp. 226–282. Aldine Publishing Co., Chicago. x + 374 pp.

Łomnicki, A., and L. B. Slobodkin. 1966. Floating in *Hydra littoralis*. *Ecology*, 47(6): 881–889.

Lorenz, K. Z. 1935. Der Kumpan in der Umwelt des Vogels. *Journal für Ornithologie*, 83(2): 137–213.

——— 1970. *Studies in animal and human behaviour*, vol. 1, trans. by R. Martin. Harvard University Press, Cambridge. xx + 403 pp.

——— 1971. *Studies in animal and human behaviour*, vol. 2, trans. by R. Martin. Harvard University Press, Cambridge. xxiv + 366 pp.

Low, R. M. 1971. Interspecific territoriality in a pomacentrid reef fish, *Pomacentrus flavicauda* Whitley. *Ecology*, 52(4): 648–654.

Lüscher, M., ed. 1977. *Phase and caste determination in insects: endocrine aspects*. Pergamon Press, Elmsford, N. Y. 130 pp.

Lyons, J. 1972. Human language. In R. A. Hinde, ed. (*q.v.*), *Non-verbal communication*, pp. 49–85.

MacArthur, R. H. 1972. *Geographical ecology: patterns in the distribution of species*. Harper & Row, New York. xviii + 269 pp.

MacArthur, R. H., and E. O. Wilson. 1967. *The theory of island biogeography*. Princeton University Press, Princeton, N. J. xi + 203 pp.

Mackie, G. O. 1964. Analysis of locomotion in a siphonophore colony. *Proceedings of the Royal Society*, ser. B, 159: 366–391.

——— 1973. Coordinated behavior in hydrozoan colonies. In R. S. Boardman, A. H. Cheetham, and W. A. Oliver, Jr., eds. (*q.v.*), *Animal colonies: development and function through time*, pp. 95–106.

MacKinnon, J. 1974. The behaviour and ecology of wild orang-utans (*Pongo pygmaeus*). *Animal Behaviour*, 22(1); 3–74.

Mann, T. 1964. *The biochemistry of semen and of the male reproductive tract*. Methuen, London. xxiii + 493 pp.

Marchal, P. 1897. La castration nutriciale chez les Hyménoptères sociaux. *Compte Rendu de la Société de Biologie, Paris*, pp. 556–557.

Markl, H. 1968. Die Verständigung durch Stridulationssignale bei Blattschneiderameisen: II, Erzeugung und Ei-

genschaften der Signale. *Zeitschrift für Vergleichende Physiologie*, 60(2): 103–150.
Marler, P. R. 1956. Behaviour of the chaffinch, *Fringilla coelebs. Behaviour*, supplement 5. vii + 184 pp.
——— 1957. Specific distinctiveness in the communication signals of birds. *Behaviour*, 11(1): 13–39.
——— 1959. Developments in the study of animal communication. In P. R. Bell, ed., *Darwin's biological work: some aspects reconsidered*, pp. 150–206. Cambridge University Press, Cambridge. xiii + 342 pp.
——— 1961. The logical analysis of animal communication. *Journal of Theoretical Biology*, 1(3): 295–317.
——— 1965. Communication in monkeys and apes. In I. DeVore, ed., *Primate behavior: field studies of monkeys and apes*, pp. 544–584. Holt, Rinehart and Winston, New York. xiv + 654 pp.
——— 1967. Animal communication signals. *Science*, 157: 769–774.
Marler, P. R., and W. J. Hamilton III. 1966. *Mechanisms of animal behavior*. John Wiley & Sons, New York. xi + 771 pp.
Marler, P. R., and P. Mundinger. 1971. Vocal learning in birds. In H. Moltz, ed., *The ontogeny of vertebrate behavior*, pp. 389–450. Academic Press, New York. xi + 500 pp.
Marsden, H. M. 1968. Agonistic behaviour of young rhesus monkeys after changes induced in social rank of their mothers. *Animal Behaviour*, 16(1): 38–44.
——— 1971. Intergroup relations in rhesus monkeys (*Macaca mulatta*). In A. H. Esser, ed. (*q.v.*), *Behavior and environment: the use of space by animals and men*, pp. 112–113.
Marshall, A. J. 1954. *Bower-birds, their displays and breeding cycles*. Clarendon Press of Oxford University Press, Oxford. x + 208 pp.
Martin, N. G., L. J. Eaves, and H. J. Eysenck. 1977. Genetical, environmental and personality factors in influencing the age of first sexual intercourse in twins. *Journal of Biosocial Science*, 9(1): 91–97.
Martin, P. S. 1966. Africa and Pleistocene overkill. *Nature, London*, 212(5060): 339–342.
Martin, R. D. 1968. Reproduction and ontogeny in tree shrews (*Tupaia belangeri*) with reference to their general behavior and taxonomic relationships. *Zeitschrift für Tierpsychologie*, 25(4): 409–495; 25(5): 505–532.
——— 1973. A review of the behaviour and ecology of the lesser mouse lemur (*Microcebus murinus* J. F. Miller 1777). In R. P. Michael and J. H. Crook, eds. (*q.v.*), *Comparative ecology and behaviour of primates*, pp. 1–68. Academic Press, New York. xvi + 847 pp.
Maslow, A. H. 1936. The role of dominance in the social and sexual behavior of infra-human primates: IV, the determination of hierarchy in pairs and in a group. *Journal of Genetic Psychology*, 49(1): 161–198.
——— 1954. *Motivation and personality*. Harper, New York. 411 pp.
——— 1972. *The farther reaches of human nature*. Viking Press, New York. xxii + 423 pp.
Masters, R. D. 1970. Genes, language, and evolution. *Semiotica*, 2(4): 295–320.
Masters, W. H., and Virginia E. Johnson. 1966. *Human sexual response*. Little, Brown, Boston. xiii + 366 pp.
Mathewson, Sue F. 1961. Gonadotrophic control of aggressive behavior in starlings. *Science*, 134: 1522–1523.
Matthews, L. H. 1971. *The life of mammals*, vol. 2. Universe Books, New York. 440 pp.
Matthews, R. W. 1968a. *Microstigmus comes:* sociality in a sphecid wasp. *Science*, 160: 787–788.
——— 1968b. Nesting biology of the social wasp *Microstigmus comes. Psyche, Cambridge*, 75(1): 23–45.
Mattingly, I. G. 1972. Speech cues and sign stimuli. *American Scientist*, 60(3): 327–337.
Mautz, D., R. Boch, and R. A. Morse. 1972. Queen finding by swarming honey bees. *Annals of the Entomological Society of America*, 65(2): 440–443.
May, R. M. 1973. *Stability and complexity in model ecosystems*. Princeton University Press, Princeton, N. J. x + 235 pp.
——— ed. 1976. *Theoretical ecology: Principles and applications*. Saunders, Philadelphia. viii + 317 pp.
Maynard Smith, J. 1964. Group selection and kin selection. *Nature, London*, 201(4924): 1145–1147.
——— 1965. The evolution of alarm calls. *American Naturalist*, 99(904): 59–63.
——— 1976. Evolution and the theory of games. *American Scientist*, 64: 41–45.
——— 1978. *The evolution of sex*. Cambridge University Press. Cambridge. viii + 222 pp.
Maynard Smith, J., and G. R. Price. 1973. The logic of animal conflict. *Nature, London*, 246(5427): 15–18.
Maynard Smith, J., and M. G. Ridpath. 1972. Wife sharing in the Tasmanian native hen, *Tribonyx mortierii:* a case of kin selection? *American Naturalist*, 106(950): 447–452.
Mayr, E. 1935. Bernard Altum and the territory theory. *Proceedings of the Linnaean Society of New York* (1933–34), nos. 45, 46, pp. 24–38.
——— 1969. *Principles of systematic zoology*. McGraw-Hill Book Co., New York. xi + 428 pp.
McBride, A. F., and D. O. Hebb. 1948. Behavior of the captive bottle-nose dolphin, *Tursiops truncatus. Journal of Comparative and Physiological Psychology*, 41: 111–123.
McBride, G. 1963. The "teat order" and communication in young pigs. *Animal Behaviour*, 11(1): 53–56.
McClearn, G. E. 1970. Behavioral genetics. *Annual Review of Genetics*, 4: 437–468.
McClintock, Martha. 1971. Menstrual synchrony and suppression. *Nature, London*, 229(5282): 244–245.
McCook, H. C. 1879. Combats and nidification of the pavement ant, *Tetramorium caespitum. Proceedings of the Academy of Natural Sciences of Philadelphia*, 31: 156–161.
McDonald, A. L., N. W. Heimstra, and D. K. Damkot. 1968. Social modification of agonistic behaviour in fish. *Animal Behaviour*, 16(4): 437–441.
McEvedy, C. 1967. *The Penguin atlas of ancient history*. Penguin Books, Baltimore Md. 96 pp.
Mead, Margaret, 1963. Socialization and enculturation. *Current Anthropology*, 4(1): 184–188.
Mech, L. D. 1970. *The wolf: the ecology and behavior of an*

endangered species. Natural History Press, Garden City, N. Y. xx + 384 pp.

Merrell, D. J. 1968. A comparison of the estimated size and the "effective size" of breeding populations of the leopard frog, *Rana pipiens*. *Evolution*, 22(2): 274–283.

Mesarović, M. D., D. Macko, and Y. Takahara. 1970. *Theory of hierarchical, multilevel systems*. Academic Press, New York. xiii + 294 pp.

Meyerriecks, A. J. 1960. *Comparative breeding behavior of four species of North American herons*. Publication no. 2. The Nuttall Ornithological Club, Cambridge, Mass. viii + 158 pp.

Michael, R. P. 1966. Action of hormones on the cat brain. In R. A. Gorski and R. E. Whalen, eds., *Brain and behavior*, vol. 3, *The brain and gonadal function*, pp. 81–98. University of California Press, Berkeley. xv + 289 pp.

Michener, C. D. 1958. The evolution of social behavior in bees. *Proceedings of the Tenth International Congress of Entomology, Montreal, 1956*, 2: 441–447.

——— 1964a. Reproductive efficiency in relation to colony size in hymenopterous societies. *Insectes Sociaux*, 11(4): 317–341.

——— 1964b. The bionomics of *Exoneurella*, a solitary relative of *Exoneura* (Hymenoptera: Apoidea: Ceratinini). *Pacific Insects*, 6(3): 411–426.

——— 1965. The life cycle and social organization of bees of the genus *Exoneura* and their parasite, *Inquilina* (Hymenoptera: Xylocopinae). *Kansas University Science Bulletin*, 46(9): 317–358.

——— 1969. Comparative social behavior of bees. *Annual Review of Entomology*, 14: 299–342.

Michener, C. D., and D. J. Brothers. 1974. Were workers of eusocial Hymenoptera initially altruistic or oppressed? *Proceedings of the National Academy of Sciences, U.S.A.*, 71(3): 671–674.

Milkman, R. D. 1970. The genetic basis of natural variation in *Drosophila melanogaster*. *Advances in Genetics*, 15: 55–114.

Miller, G. A., E. Galanter, and K. H. Pribram. 1960. *Plans and the structure of behavior*. Henry Holt, New York. xii + 226 pp.

Miller, N. E. 1948. Theory and experiment relating psychoanalytic displacement to stimulus-response generalization. *Journal of Abnormal and Social Psychology*, 43(2): 155–178.

Missakian, Elizabeth A. 1972. Genealogical and cross-genealogical dominance relations in a group of free-ranging rhesus monkeys (*Macaca mulatta*) on Cayo Santiago. *Primates*, 13(2): 169–180.

Mizuhara, H. 1964. Social changes of Japanese monkey troops in the Takasakiyama. *Primates*, 5(1,2): 27–52.

Montagu, M. F. Ashley. 1968a. The new litany of "innate depravity," or original sin revisited. In M. F. Ashley Montagu, ed. (*q.v.*), *Man and aggression*, pp. 3–17.

——— ed. 1968b. *Man and aggression*. Oxford University Press, Oxford. xiv + 178 pp. (2d ed., 1973.)

Moore, B. P. 1964. Volatile terpenes from *Nasutitermes* soldiers (Isoptera, Termitidae). *Journal of Insect Physiology*, 10(2): 371–375.

Moreau, R. E. 1960. Conspectus and classification of the ploceine weaver-birds. *Ibis*, 102(2): 298–321; 102(3): 443–471.

Morris, C. 1946. *Signs, language, and behavior*. Prentice-Hall, Englewood Cliffs, N. J. xiv + 365 pp.

Morris, D. 1957. "Typical intensity" and its relation to the problem of ritualization. *Behaviour*, 11(1): 1–12.

——— 1962. *The biology of art*. Alfred Knopf, New York. 176 pp.

——— 1967. *The naked ape: a zoologist's study of the human animal*. McGraw-Hill Book Co., New York. 252 pp.

Morton, N. E. 1969. Human population structure. *Annual Review of Genetics*, 3: 53–74.

Mosebach-Pukowski, Erna. 1937. Uber die Raupengesellschaften von *Vanessa io* und *Vanessa urticae*. *Zeitschrift für Morphologie und Okologie der Tiere*, 33(3): 358–380.

Moynihan, M. H. 1969. Comparative aspects of communication in New World primates. In D. Morris, ed., *Primate ethology: essays on the socio-sexual behavior of apes and monkeys*, pp. 306–342. Anchor Books, Doubleday, Garden City, N. Y. vii + 471 pp.

Müller-Schwarze, D. 1971. Pheromones in black-tailed deer (*Odocoileus hemionus columbianus*). *Animal Behaviour*, 19(1): 141–152.

Murchison, C. 1935. The experimental measurement of a social hierarchy in *Gallus domesticus*: IV, loss of body weight under conditions of mild starvation as a function of social dominance. *Journal of General Psychology*, 12: 296–312.

Murdoch, W. W. 1966. Population stability and life history phenomena. *American Naturalist*, 100(910): 5–11.

Murphy, G. I. 1968. Patterns in life history. *American Naturalist*, 102(927): 391–403.

Murton, R. K., A. J. Isaacson, and N. J. Westwood. 1966. The relationships between wood-pigeons and their clover food supply and the mechanism of population control. *Journal of Applied Ecology*, 3(1): 55–96.

Mykytowycz, R. 1962. Territorial function of chin gland secretion in the rabbit, *Oryctolagus cuniculus* (L.). *Nature, London*, 193(4817): 799.

Nagel, U. 1973. A comparison of anubis baboons, hamadryas baboons and their hybrids at a species border in Ethiopia. *Folia Primatologica*, 19(2,3): 104–165.

Napier, J. R. 1960. Studies of the hands of living primates. *Proceedings of the Zoological Society of London*, 134(4): 647–657.

Napier, J. R., and P. H. Napier. 1967. *A handbook of living primates*. Academic Press, New York. xiv + 456 pp.

Neal, E. 1948. *The badger*. Collins, London. xvi + 158 pp.

Neel, J. V. 1970. Lessons from a "primitive" people. *Science*, 170: 815–822.

Nero, R. W. 1956. A behavior study of the red-winged blackbird: I, mating and nesting activities. *Wilson Bulletin*, 68(1): 5–37.

Neuweiler, G. 1969. Verhaltensbeobachtungen an einer indischen Flughundkolonie (*Pteropus g. giganteus* Brünn). *Zeitschrift für Tierpsychologie*, 26(2): 166–199.

Nice, Margaret M. 1941. The role of territory in bird life. *American Midland Naturalist*, 26(3): 441–487.

Nicholson, A. J. 1954. An outline of the dynamics of animal populations. *Australian Journal of Zoology*, 2(1): 9–65.

Nisbet, I. C. T. 1973. Courtship-feeding, egg-size and breeding success in common terns. *Nature, London,* 241(5385): 141–142.

Nishida, T., and K. Kawanaka. 1972. Inter-unit-group relationships among wild chimpanzees of the Mahali Mountains. *Kyoto University African Studies,* 7: 131–169.

Noble, G. K. 1939. The role of dominance in the social life of birds. *Auk,* 56(3): 263–273.

Ogburn, W. F. and M. Nimkoff. 1958. *Sociology,* 3d ed. Houghton Mifflin Co., Boston. x + 756 pp.

Oliver, J. A. 1956. Reproduction in the king cobra, *Ophiophagus hannah* Cantor. *Zoologica, New York,* 41(4): 145–152.

Orians, G. H. 1961. The ecology of blackbird (*Agelaius*) social systems. *Ecological Monographs,* 31(3): 285–312.

Orians, G. H., and G. M. Christman. 1968. A comparative study of the behavior of red-winged, tricolored, and yellow-headed blackbirds. *University of California Publications in Zoology,* 84. 81 pp.

Oster, G. F., and E. O. Wilson. 1978. *Caste and ecology in the social insects.* Princeton University Press, Princeton, N. J. xvi + 352 pp.

Ostrom, J. H. 1972. Were some dinosaurs gregarious? *Palaeogeography, Palaeoclimatology, Palaeoecology,* 11: 287–301.

Otte, D. 1970. *A comparative study of communicative behavior in grasshoppers.* Miscellaneous Publications, Museum of Zoology, University of Michigan, Ann Arbor, 141. 168 pp.

——— 1972. Simple versus elaborate behavior in grasshoppers: an analysis of communication in the genus *Syrbula. Behaviour,* 42(3,4): 291–322.

Pardi, L. 1940. Ricerche sui Polistini: I, poliginia vera ed apparente in *Polistes gallicus* (L.). *Processi Verbali della Società Toscana di Scienze Naturali in Pisa,* 49: 3–9.

——— 1948. Dominance order in *Polistes* wasps. *Physiological Zoology,* 21(1): 1–13.

Parker, G. A. 1970. Sperm competition and its evolutionary consequences in the insects. *Biological Reviews, Cambridge Philosophical Society,* 45: 525–568.

Parsons, P. A. 1967. *The genetic analysis of behaviour.* Methuen, London. x + 174 pp.

Patterson, I. J. 1965. Timing and spacing of broods in the black-headed gull *Larus ridibundus. Ibis,* 107(4): 433–459.

Patterson, O. 1967. *The sociology of slavery: an analysis of the origins, development and structure of Negro slave society in Jamaica.* Fairleigh Dickinson University Press, Cranbury, N. J. 310 pp.

Payne, R. S., and S. McVay. 1971. Songs of humpback whales. *Science,* 173: 585–597.

Peacock, A. D., and A. T. Baxter. 1950. Studies in Pharaoh's ant, *Monomorium pharaonis* (L.): 3, life history. *Entomologist's Monthly Magazine,* 86: 171–178.

Petter, J.-J. 1962a. Recherches sur l'écologie et l'éthologie des lémuriens malgaches. *Mémoires du Muséum National d'Histoire Naturelle, Paris,* ser. A. (Zoology), 27(1): 1–146.

——— 1962b. Ecological and behavioral studies of Madagascar lemurs in the field. *Annals of the New York Academy of Sciences,* 102(2): 267–281.

Pfeffer, P. 1967. Le mouflon de Corse (*Ovis ammon musimom* Schreber 1782); position systématique, écologie et éthologie comparées. *Mammalia,* 31, supplement. 262 pp.

Pfeiffer, J. E. 1969. *The emergence of man.* Harper & Row, New York. xxiv + 477 pp.

Pilbeam, D. 1972. *The ascent of man: an introduction to human evolution.* Macmillan Co., New York. x + 207 pp.

Pilleri, G., and J. Knuckey. 1969. Behaviour patterns of some Delphinidae observed in the western Mediterranean. *Zeitschrift für Tierpsychologie,* 26(1): 48–72.

Pitcher, T. J. 1973. The three-dimensional structure of schools in the minnow, *Phoxinus phoxinus* (L.). *Animal Behaviour,* 21(4): 673–686.

Pitelka, F. A. 1959. Numbers, breeding schedule, and territoriality in pectoral sandpipers of northern Alaska. *Condor,* 61(4): 233–264.

Poirier, F. E. 1968. The Nilgiri langur (*Presbytis johnii*) mother-infant dyad. *Primates,* 9(1,2): 45–68.

——— 1969. Behavioral flexibility and intergroup variation among Nilgiri langurs (*Presbytis johnii*) of South India. *Folia Primatologica,* 11(1,2): 119–133.

——— 1970a. The Nilgiri langur (*Presbytis johnii*) of South India. In *L. A. Rosenblum, ed., Primate behavior: developments in field and laboratory research,* vol. 1, pp. 251–383. Academic Press, New York. xii + 400 pp.

——— 1970b. Dominance structure of the Nilgiri langur (*Presbytis johnii*) of South India. *Folia Primatologica,* 12(3): 161–186.

——— ed. 1972a. *Primate socialization.* Random House, New York. x + 260 pp.

——— 1972b. Introduction. In F. E. Poirier, ed. (*q.v.*), *Primate socialization,* pp. 3–28.

Pontin, A. J. 1961. Population stabilization and competition between the ants *Lasius flavus* (F.) and *L. niger* (L.). *Journal of Animal Ecology,* 30(1): 47–54.

——— 1963. Further considerations of competition and the ecology of the ants *Lasius flavus* (F.) and *L. niger* (L.). *Journal of Animal Ecology,* 32(3): 565–574.

Pulliam, R., B. Gilbert, P. Klopfer, D. McDonald, Linda McDonald, and G. Millikan. 1972. On the evolution of sociality, with particular reference to *Tiaris olivacea. Wilson Bulletin,* 84(1): 77–89.

Rabb, G. B., J. H. Woolpy, and B. E. Ginsburg. 1967. Social relationships in a group of captive wolves. *American Zoologist,* 7(2): 305–311.

Radakov, D. V. 1973. *Schooling in the ecology of fish,* trans. by H. Mills. Halsted Press, Wiley, New York. viii + 173 pp.

Rand, A. S. 1967a. The adaptive significance of territoriality in iguanid lizards. In W. W. Milstead, ed., *Lizard ecology: a symposium,* pp. 106–115. University of Missouri Press, Columbia. xi + 300 pp.

——— 1967b. Ecology and social organization in the iguanid lizard *Anolis lineatopus. Proceedings of the United States National Museum, Smithsonian Institution,* 122: 1–79.

Rappaport, R. A. 1971. The sacred in human evolution. *Annual Review of Ecology and Systematics,* 2:23–44.

Rau, P. 1933. *The jungle bees and wasps of Barro Colorado Island (with notes on other insects).* Published by the author, Kirkwood, St. Louis County, Mo. 324 pp.

Rawls, J. 1971. *A theory of justice.* Belknap Press of Harvard University Press, Cambridge. xvi + 607 pp.

Reid, M. J., and J. W. Atz. 1958. Oral incubation in the cichlid fish *Geophagus jurupari* Heckel. *Zoologica, New York,* 43(5): 77–88.

Rensch, B. 1956. Increase of learning ability with increase of brain size. *American Naturalist,* 90(851): 81–95.

——— 1960. *Evolution above the species level.* Columbia University Press, New York. xvii + 419 pp.

Renský, M. 1966. The systematics of paralanguage. *Travaux linguistiques de Prague,* 2:97–102.

Reynolds, V. 1965. Some behavioural comparisons between the chimpanzee and the mountain gorilla in the wild. *American Anthropologist,* 67(3): 691–706.

——— 1968. Kinship and the family in monkeys, apes and man. *Man,* 3(2): 209–233.

Rheingold, Harriet L. 1963a. Maternal behavior in the dog. In Harriet Rheingold, ed. (*q.v.*), *Maternal behavior in mammals,* pp. 169–202.

——— ed. 1963b. *Maternal behavior in mammals.* John Wiley & Sons, New York. viii + 349 pp.

Rhijn, J. G. van. 1973. Behavioural dimorphism in male ruffs, *Philomachus pugnax* (L.). *Behaviour,* 47(3,4): 153–229.

Richards, O. W. 1965. Concluding remarks on the social organization of insect communities. *Symposia of the Zoological Society of London,* 14: 169–172.

——— 1971. The biology of the social wasps (Hymenoptera, Vespidae). *Biological Reviews, Cambridge Philosophical Society,* 46(4): 483–528.

Riemann, J. G., Donna J. Moen, and Barbara J. Thorson. 1967. Female monogamy and its control in houseflies. *Journal of Insect Physiology,* 13(3): 407–418.

Ripley, Suzanne. 1967. Intertroop encounters among Ceylon gray langurs (*Presbytis entellus*). In S. A. Altmann, ed., *Social communication among primates,* pp. 237–253. University of Chicago Press, Chicago. xiv + 392 pp.

Robins, C. R., C. Phillips, and Fanny Phillips. 1959. Some aspects of the behavior of the blennioid fish *Chaenopsis ocellata* Poey. *Zoologica, New York,* 44(2): 77–84.

Rodman, P. S. 1973. Population composition and adaptive organisation among orang-utans of the Kutai Reserve. In R. P. Michael and J. H. Crook, eds., *Comparative ecology and behaviour of primates,* pp. 171–209. Academic Press, New York. xvi + 847 pp.

Roelofs, W. L., and A. Comeau. 1969. Sex pheromone specificity: taxonomic and evolutionary aspects in Lepidoptera. *Science,* 165: 398–400.

Rogers, L. L. 1974. Movement patterns and social organization of black bears in Minnesota. Ph.D. thesis, University of Minnesota, Minneapolis.

Ropartz, P. 1966. Contribution à l'étude du déterminisme d'un effet de groupe chez les souris. *Comptes Rendus de l'Académie des Sciences, Paris,* 263: 2070–2072.

——— 1968. Olfaction et comportement social chez les rongeurs. *Mammalia,* 32(4): 550–569.

Rose, R. M., J. W. Holaday, and I. S. Bernstein. 1971. Plasma testosterone, dominance rank and aggressive behaviour in male rhesus monkeys. *Nature, London,* 231(5302): 366–368.

Rosenblatt, J. S. 1972. Learning in newborn kittens. *Scientific American,* 227(6) (December): 18–25.

Rosenblum, L. A. 1971. The ontogeny of mother-infant relations in macaques. In H. Moltz, ed., *The ontogeny of vertebrate behavior,* pp. 315–367. Academic Press, New York. xi + 500 pp.

Rothballer, A. B. 1967. Aggression, defense and neurohumors. In Carmine D. Clemente and D. B. Lindsley, eds., *Brain function,* vol. 5, *Aggression and defense, neural mechanisms and social patterns,* pp. 135–170. University of California Press, Berkeley. xv + 361 pp.

Roubaud, E. 1916. Recherches biologiques sur les guêpes solitaires et sociales d'Afrique: la genèse de la vie sociale et l'évolution de l'instinct maternel chez les vespides. *Annales des Sciences Naturelles,* 10th ser. (Zoologie), 1: 1–160.

Roughgarden, J. 1971. Density-dependent natural selection. *Ecology,* 52(3): 453–468.

Rousseau, M. 1971. Un machairodonte dans l'art Aurignacien? *Mammalia,* 35(4): 648–657.

Rowell, Thelma E. 1963. Behaviour and female reproductive cycles of rhesus macaques. *Journal of Reproduction and Fertility,* 6: 193–203.

——— 1966. Forest living baboons in Uganda. *Journal of Zoology, London,* 149(3): 344–364.

——— 1967. A quantitative comparison of the behaviour of a wild and a caged baboon troop. *Animal Behaviour,* 15(4): 499–509.

——— 1969. Long-term changes in a population of Ugandan baboons. *Folia Primatologica,* 11(4): 241–254.

——— 1972. *Social behaviour of monkeys.* Penguin Books, Harmondsworth, Middlesex. 203 pp.

Rowell, Thelma E., R. A. Hinde, and Yvette Spencer-Booth. 1964. "Aunt"-infant interaction in captive rhesus monkeys. *Animal Behaviour,* 12(2,3): 219–226.

Rowley, I. 1965. The life history of the superb blue wren, *Malurus cyaneus. Emu* 64(4): 251–297.

Rumbaugh, D. M. 1970. Learning skills of anthropoids. In L. A. Rosenblum, ed. (*q.v.*), *Primate behavior: developments in field and laboratory research,* vol. 1, pp. 1–70. Academic Press, New York. xii + 400 pp.

Ryland, J. S. 1970. *Bryozoans.* Hutchinson University Library, London. 175 pp.

Saayman, G. S. 1971. Behaviour of the adult males in a troop of free-ranging chacma baboons (*Papio ursinus*). *Folia Primatologica,* 15(1,2): 36–57.

Saayman, G. S., C. K. Tayler, and D. Bower. 1973. Diurnal activity cycles in captive and free-ranging Indian Ocean bottlenose dolphins (*Tursiops aduncus* Ehrenburg). *Behaviour,* 44(3,4): 212–233.

Sade, D. S. 1965. Some aspects of parent-offspring and sibling relations in a group of rhesus monkeys, with a discussion of grooming. *American Journal of Physical Anthropology,* 23(1): 1–17.

——— 1967. Determinants of dominance in a group of free-ranging rhesus monkeys. In S. A. Altmann, ed. (*q.v.*), *Social communication among primates,* pp. 99–114.

Sakagami, S. F., and Y. Akahira. 1960. Studies on the Japanese honeybee, *Apis cerana cerana* Fabricius: 8, two opposing adaptations in the post-stinging behavior of honeybees. *Evolution*, 14(1): 29–40.

Sakagami, S. F., and Y. Oniki. 1963. Behavior studies of the stingless bees, with special reference to the oviposition process: 1, *Melipona compressipes manaosensis* Schwarz. *Journal of the Faculty of Science, Hokkaido University*, 6th ser. (Zoology), 15(2): 300–318.

Sale, P. F. 1972. Effect of cover on agonistic behavior of a reef fish: a possible spacing mechanism. *Ecology*, 53(4): 753–758.

Sanders, C. J., and F. B. Knight, 1968. Natural regulation of the aphid *Pterocomma populifoliae* on bigtooth aspen in northern lower Michigan. *Ecology*, 49(2): 234–244.

Schaller, G. B. 1963. *The mountain gorilla: ecology and behavior*. University of Chicago Press, Chicago. xviii + 431 pp.

——— 1965a. The behavior of the mountain gorilla. In I. DeVore, ed., *Primate behavior: field studies of monkeys and apes*, pp. 324–367. Holt, Rinehart and Winston, New York. xiv + 654 pp.

——— 1965b. *The year of the gorilla*. Ballantine Books, New York. 285 pp.

——— 1972. *The Serengeti lion: a study of predator-prey relations*. University of Chicago Press, Chicago. xiii + 480 pp.

Schaller, G. B., and G. R. Lowther. 1969. The relevance of carnivore behavior to the study of early hominids. *Southwestern Journal of Anthropology*, 25(4): 307–341.

Schenkel, R. 1947. Ausdrucks-Studien an Wölfen. Gefangenschafts-Beobachtungen. *Behaviour*, 1(2): 81–129.

——— 1966. Play, exploration and territoriality in the wild lion. *Symposia of the Zoological Society of London*, 18: 11–22.

Schevill, W. E. 1964. Underwater sounds of cetaceans. In W. N. Tavolga, ed., *Marine bio-acoustics*, pp. 307–316. Pergamon, New York. xiv + 413 pp.

Schevill, W. E., and W. A. Watkins. 1962. *Whale and porpoise voices: a phonograph record*. Contribution no. 1320. Woods Hole Oceanographic Institution, Woods Hole, Mass. 24 pp.

Schjelderup-Ebbe, T. 1922. Beiträge zur Sozialpsychologie des Haushuhns. *Zeitschrift für Psychologie*, 88(3–5): 225–252.

Schneider, D. 1969. Insect olfaction: deciphering system for chemical messages. *Science*, 163: 1031–1037.

Schneirla, T. C., J. S. Rosenblatt, and Ethel Tobach. 1963. Maternal behavior in the cat. In Harriet L. Rheingold, ed. (*q.v.*), *Maternal behavior in mammals*, pp. 122–168.

Schoener, T. W. 1965. The evolution of bill size differences among sympatric congeneric species of birds. *Evolution*, 19(2): 189–213.

——— 1967. The ecological significance of sexual dimorphism in size in the lizard *Anolis conspersus*. *Science*, 155: 474–477.

——— 1968a. Sizes of feeding territories among birds. *Ecology*, 49(1): 123–141.

——— 1968b. The *Anolis* lizards of Bimini: resource partitioning in a complex fauna. *Ecology*, 49(4): 704–726.

——— 1971. Theory of feeding strategies. *Annual Review of Ecology and Systematics*, 2: 369–404.

——— 1973. Population growth regulated by intraspecific competition for energy or time: some simple representations. *Theoretical Population Biology*, 4(1): 56–84.

Schull, W. J., and J. V. Neel. 1965. *The effects of inbreeding on Japanese children*. Harper & Row, New York. xii + 419 pp.

Schultz, A. H. 1958. The occurrence and frequency of pathological and teratological conditions and of twinning among non-human primates. *Primatologia, Handbuch der Primatenkunde*, 1: 965–1014.

Schultze-Westrum, T. 1965. Innerartliche Verständigung durch Düfte beim Gleitbeutler *Petaurus breviceps papuanus* Thomas (Marsupialia, Phalangeridae). *Zeitschrift für Vergleichende Physiologie*, 50(2): 151–220.

Scott, J. F. 1971. *Internalization of norms: a sociological theory of moral commitment*. Prentice-Hall, Englewood Cliffs, N. J. xviii + 237 pp.

Scott, J. P., and E. Fredericson. 1951. The causes of fighting in mice and rats. *Physiological Zoology*, 24(4): 273–309.

Scott, J. P., and J. L. Fuller. 1965. *Genetics and the social behavior of the dog*. University of Chicago Press, Chicago. xviii + 468 pp.

Scott, J. W. 1950. A study of the phylogenetic or comparative behavior of three species of grouse. *Annals of the New York Academy of Sciences*, 51(6): 1062–1073.

Scudo, F. M. 1967. The adaptive value of sexual dimorphism: 1, anisogamy. *Evolution*, 21(2): 285–291.

Seay, B. 1966. Maternal behavior in primiparous and multiparous rhesus monkeys. *Folia Primatologica*, 4(2): 146–168.

Sebeok, T. A. 1962. Coding in the evolution of signalling behavior. *Behavioral Science*, 7(4): 430–442.

——— 1963. Communication among social bees; porpoises and sonar; man and dolphin. *Language*, 39(3): 448–466.

——— 1965. Animal communication. *Science*, 174: 1006–1014.

Seemanova, Eva. 1971. A study of children of incestuous matings. *Human Heredity*, 21: 108–128.

Selander, R. K. 1965. On mating systems and sexual selection. *American Naturalist*, 99(906): 129–141.

——— 1966. Sexual dimorphism and differential niche utilization in birds. *Condor*, 68(2): 113–151.

——— 1972. Sexual selection and dimorphism in birds. In B. Campbell, ed., *Sexual selection and the descent of man, 1871–1971*, pp. 180–230. Aldine Publishing Co., Chicago, x + 378 pp.

Seton, E. T. 1909. *Life-histories of northern animals: an account of the mammals of Manitoba*, 2 vols. Charles Scribner's Sons, New York. Vol. 1: xxx + 673 pp.; vol. 2: xii + 590 pp.

Shaw, Evelyn. 1962. The schooling of fishes. *Scientific American*, 206(6) (June): 128–138.

——— 1970. Schooling in fishes: critique and review. In L. R. Aronson, Ethel Tobach, D. S. Lehrman, and J. S. Rosenblatt, eds., *Development and evolution of behavior: essays in memory of T. C. Schneirla*, pp. 452–480. W. H. Freeman, San Francisco. xviii + 656 pp.

Shearer, D., and R. Boch. 1965. 2-Heptanone in the mandibular gland secretion of the honey-bee. *Nature, London,* 206(4983): 530.

Shepher, J. 1971. Mate selection among second-generation kibbutz adolescents and adults: incest avoidance and negative imprinting. *Archives of Sexual Behavior,* 1(4): 293–307.

Sherman, P. W. 1977. Nepotism and the evolution of alarm calls. *Science,* 197: 1246–1253.

Shettleworth, Sara J. 1972. Constraints on learning. *Advances in the Study of Behavior,* 4: 1–68.

Shoemaker, H. H. 1939. Social hierarchy in flocks of the canary. *Auk,* 56(4): 381–406.

Silberglied, R. E., and O. R. Taylor. 1973. Ultraviolet differences between the sulfur butterflies, *Colias eurytheme* and *C. philodice,* and a possible isolating mechanism. *Nature, London,* 241(5389): 406–408.

Silén, L. 1942. Origin and development of the cheilo-ctenostomatous stem of Bryozoa. *Zoologiska Bidrag* (Uppsala), 22: 1–59.

Simmons, K. E. L. 1970. Ecological determinants of breeding adaptations and social behaviour in two fish-eating birds. In J. H. Crook, ed., *Social behaviour in birds and mammals: essays on the social ethology of animals and men,* pp. 37–77. Academic Press, New York. xl + 492 pp.

Simonds, P. E. 1965. The bonnet macaque in South India. In I. DeVore, ed., *Primate behavior: field studies of monkeys and apes,* pp. 175–196. Holt, Rinehart and Winston, New York. xiv + 654 pp.

Simons, E. L., and P. C. Ettel. 1970. Gigantopithecus. *Scientific American,* 222(1) (January): 76–85.

Simpson, G. G. 1953. *The major features of evolution.* Columbia University Press, New York. xx + 434 pp.

——— 1961. *Principles of animal taxonomy.* Columbia University Press, New York. xii + 247 pp.

Sipes, R. G. 1973. War, sports and aggression: an empirical test of two rival theories. *American Anthropologist,* 75(1): 64–86.

Skaife, S. H. 1954a. The black-mound termite of the Cape, *Amitermes atlanticus* Fuller. *Transactions of the Royal Society of South Africa,* 34(1): 251–271.

——— 1954b. Caste differentiation among termites. *Transactions of the Royal Society of South Africa,* 34(2): 345–353.

——— 1955. *Dwellers in darkness.* Longmans, Green, London. x + 134 pp.

Skutch, A. F. 1961. Helpers among birds. *Condor,* 63(3): 198–226.

Sladen, F. W. L. 1912. *The humble-bee, its life-history and how to domesticate it, with descriptions of all the British species of* Bombus *and* Psithyrus. Macmillan Co., London. xiii + 283 pp.

Slobin, D. 1971. *Psycholinguistics.* Scott, Foresman, Glenview, Ill. xii + 148 pp.

Smith, C. C. 1968. The adaptive nature of social organization in the genus of tree squirrels *Tamiasciurus. Ecological Monographs,* 38(1): 31–63.

Smith, E. A. 1968. Adoptive suckling in the grey seal. *Nature, London,* 217(5130): 762–763.

Smith, W. J. 1969a. Messages of vertebrate communication. *Science,* 165: 145–150.

——— 1969b. Displays of *Sayornis phoebe* (Aves, Tryannidae). *Behaviour,* 33(3,4): 283–322.

Smythe, N. 1970. The adaptive value of the social organization of the coati (*Nasua narica*). *Journal of Mammalogy,* 51(4): 818–820.

Snow, D. W. 1958. *A study of blackbirds.* Allen and Unwin, London. 192 pp.

——— 1961. The natural history of the oilbird, *Steatornis caripensis,* in Trinidad, W.I.: 1, general behavior and breeding habits. *Zoologica, New York,* 46(1): 27–48.

Sorenson, M. W. 1970. Behavior of tree shrews. In L. A. Rosenblum, ed., *Primate behavior: developments in field and laboratory research,* vol. 1, pp. 141–193. Academic Press, New York. xii + 400 pp.

Southwick, C. H. 1967. An experimental study of intragroup agonistic behavior in rhesus monkeys (*Macaca mulatta*). *Behaviour,* 28(1,2): 182–209.

——— 1969. Aggressive behaviour of rhesus monkeys in natural and captive groups. In S. Garattini and E. B. Sigg, eds., *Aggressive behaviour,* Proceedings of the Symposium on the Biology of Aggressive Behaviour, Milan, May 1968, pp. 32–43. Excerpta Medica, Amsterdam. 369 pp.

Sparks, J. H. 1965. On the role of allopreening invitation behaviour in reducing aggression among red avadavats, with comments on its evolution in the Spermestidae. *Proceedings of the Zoological Society of London,* 145(3): 387–403.

——— 1969. Allogrooming in primates: a review. In D. Morris, ed., *Primate ethology: essays on the socio-sexual behavior of apes and monkeys,* pp. 190–225. Aldine Publishing Co., Chicago. x + 374 pp.

Spencer-Booth, Yvette. 1968. The behaviour of group companions towards rhesus monkey infants. *Animal Behaviour,* 16(4): 541–557.

Spradbery, J. P., 1973. *Wasps: an account of the biology and natural history of solitary and social wasps.* Sidgwick and Jackson, London, xvi + 408 pp.

Stamps, Judy A. 1973. Displays and social organization in female *Anolis aeneus. Copeia,* 1973, no. 2, pp. 264–272.

Stent, G. S., ed. 1978. *Morality as a biological phenomenon.* Dahlem Workshop. Abakon Verlagsgesellschaft, Berlin. 323 pp.

Sterndale, R. A. 1884. *Natural history of the Mammalia of India and Ceylon.* Calcutta. [Cited by L. H. Matthews, 1971 (*q.v.*).]

Stewart, R. E., and J. W. Aldrich. 1951. Removal and repopulation of breeding birds in a spruce-fir forest community. *Auk,* 68(4): 471–482.

Stiles, F. G. 1971. Time, energy, and territoriality of the Anna hummingbird (*Calypte anna*). *Science,* 173: 818–821.

Stimson, J. 1970. Territorial behavior of the owl limpet, *Lottia gigantea. Ecology,* 51(1): 113–118.

Struhsaker, T. T. 1967a. Behavior of vervet monkeys (*Cercopithecus aethiops*). *University of California Publications in Zoology,* 82. 64 pp.

——— 1967b. Social structure among vervet monkeys (*Cercopithecus aethiops*). *Behaviour,* 29(2–4): 83–121.

Sugiyama, Y. 1960. On the division of a natural troop of

Japanese monkeys at Takasakiyama. *Primates*, 2(2): 109–148.

——— 1967. Social organization of hanuman langurs. In S. A. Altmann, ed., *Social communication among primates*, pp. 221–236. University of Chicago Press, Chicago. xiv + 392 pp.

——— 1968. Social organization of chimpanzees in the Budongo Forest, Uganda. *Primates*, 9(3): 225–258.

——— 1972. Social characteristics and socialization of wild chimpanzees. In F. E. Poirier, ed. (*q.v.*), *Primate socialization*, pp. 145–163.

——— 1973. Social organization of wild chimpanzees. In C. R. Carpenter, ed., *Behavioral regulators of behavior in primates*, pp. 68–80. Bucknell University Press, Lewisburg, Pa. 303 pp.

Suzuki, A. 1971. Carnivory and cannibalism observed among forest-living chimpanzees. *Journal of the Anthropological Society of Nippon*, 79(1): 30–48.

Tavolga, Margaret C. 1966. Behavior of the bottlenose dolphin (*Tursiops truncatus*); social interactions in a captive colony. In K. S. Norris, ed., *Whales, dolphins and porpoises*, pp. 718–730. University of California Press, Berkeley. xvi + 789 pp.

Tavolga, Margaret C., and F. S. Essapian. 1957. The behavior of the bottle-nosed dolphin (*Tursiops truncatus*): mating, pregnancy, parturition, and mother-infant behavior. *Zoologica, New York*, 42(1): 11–31.

Tayler, C. K., and G. S. Saayman. 1973. Imitative behaviour by Indian Ocean bottlenose dolphins (*Tursiops aduncus*) in captivity. *Behaviour*, 44(3,4): 286–298.

Teleki, G. 1973. *The predatory behavior of wild chimpanzees.* Bucknell University Press, Lewisburg, Pa. 232 pp.

Thielcke, G. 1969. Geographic variation in bird vocalizations. In R. A. Hinde, ed. (*q.v.*), *Bird vocalizations: their relation to current problems in biology and psychology: essays presented to W. H. Thorpe*, pp. 311–339.

Thielcke, G., and Helga Thielcke. 1970. Die sozialen Funktionen verschiedener Gesangsformen des Sonnenvogels (*Leiothrix lutea*). *Zeitschrift für Tierpsychologie*, 27(2): 177–185.

Thiessen, D. D., K. Owen, and G. Lindzey. 1971. Mechanisms of territorial marking in the male and female Mongolian gerbils (*Meriones unguiculatus*). *Journal of Comparative and Physiological Psychology*, 77(1): 38–47.

Thoday, J. M. 1953. Components of fitness. *Symposia of the Society for Experimental Biology*, 7: 96–113.

Thorpe, W. H. 1972. The comparison of vocal communication in animals and man. In R. A. Hinde, ed. (*q.v.*), *Non-verbal communication*, pp. 27–47. Cambridge University Press, Cambridge. xiii + 423 pp.

Tiger, L. 1969. *Men in groups.* Random House, New York. xx + 254 pp.

Tiger, L., and R. Fox. 1971. *The imperial animal.* Holt, Rinehart and Winston, New York. xi + 308 pp.

Tinbergen, N. 1939. Field observations of East Greenland birds: II, the behavior of the snow bunting (*Plectrophenax nivalis subnivalis* [Brehm]) in spring. *Transactions of the Linnaean Society of New York*, 5: 1–94.

——— 1951. *The study of instinct.* Clarendon Press of Oxford University Press, Oxford. xii + 228 pp.

——— 1952. "Derived" activities; their causation, biological significance, origin, and emancipation during evolution. *Quarterly Review of Biology*, 27(1): 1–32.

——— 1960. The evolution of behavior in gulls. *Scientific American*, 203(6) (December): 118–130.

Tinkle, D. W. 1965. Population structure and effective size of a lizard population. *Evolution*, 19(4): 569–573.

——— 1969. The concept of reproductive effort and its relation to the evolution of life histories of lizards. *American Naturalist*, 103(933): 501–516.

Tobias, P. V. 1973. Implications of the new age estimates of the early South African hominids. *Nature, London*, 246(5428): 79–83.

Tordoff, H. B. 1954. Social organization and behavior in a flock of captive, nonbreeding red crossbills. *Condor*, 56(6): 346–358.

Tretzel, E. 1966. Artkennzeichnende und reaktionsauslösende Komponenten im Gesang der Heidelerche (*Lullula arborea*). *Verhandlungen der Deutschen Zoologischen Gesellschaft, Jena, 1965*, pp. 367–380.

Trivers, R. L. 1971. The evolution of reciprocal altruism. *Quarterly Review of Biology*, 46(4): 35–57.

——— 1972. Parental investment and sexual selection. In B. Campbell, ed., *Sexual selection and the descent of man, 1871–1971*, pp. 136–179. Aldine Publishing Co., Chicago. x + 378 pp.

——— 1974. Parent-offspring conflict. *American Zoologist*, 14(1): 249–264.

Trivers, R. L., and Hope Hare. 1976. Haplodiploidy and the evolution of the social insects. *Science*, 191: 249–263.

Trivers, R. L., and D. E. Willard. 1973. Natural selection of parental ability to vary the sex ratio of offspring. *Science*, 179: 90–92.

Truman, J. W., and Lynn M. Riddiford. 1974. Hormonal mechanisms underlying insect behaviour. *Advances in Insect Physiology*, 10: 297–352.

Trumler, E. 1959. Das "Rossigkeitsgesicht" und ähnliches Ausdrucksverhalten bei Einhufern. *Zeitschrift für Tierpsychologie*, 16(4): 478–488.

Tsumori, A. 1967. Newly acquired behavior and social interactions of Japanese monkeys. In S. A. Altmann, ed., *Social communication among primates*, pp. 207–219. University of Chicago Press, Chicago. xiv + 392 pp.

Tsumori, A., M. Kawai, and R. Motoyoshi. 1965. Delayed response of wild Japanese monkeys by the sand-digging method: 1, case of the Koshima troop. *Primates*, 6(2): 195–212.

Turnbull, C. M. 1968. The importance of flux in two hunting societies. In R. B. Lee and I. DeVore, eds. (*q.v.*), *Man the hunter*, pp. 132–137.

Uzzell, T. 1970. Meiotic mechanisms of naturally occurring unisexual vertebrates. *American Naturalist*, 104(939): 433–445.

Vandenbergh, J. G. 1971. The effects of gonadal hormones on the aggressive behaviour of adult golden hamsters (*Mesocricetus auratus*). *Animal Behaviour*, 19(3): 589–594.

Van Denburgh, J. 1914. The gigantic land tortoises of the Galapagos Archipelago. *Proceedings of the California Academy of Sciences, San Francisco*, 4th ser. 2(1): 203–374.

Velthuis, H. H. V., and J. van Es. 1964. Some functional aspects of the mandibular glands of the queen honeybee. *Journal of Apicultural Research*, 3(1): 11–16.

Verheyen, R. 1954. *Monographie éthologique de l'hippopotame* (Hippopotamus amphibius *Linné*). Institut des Parcs Nationaux du Congo Belge. Exploration du Parc National Albert, Brussels. 91 pp.

Verner, J., and Gay H. Engelsen. 1970. Territories, multiple nest building, and polygyny in the long-billed marsh wren. *Auk*, 87(3): 557–567.

Verron, H. 1963. Rôle des stimuli chimiques dans l'attraction sociale chez *Calotermes flavicollis* (Fabr.). *Insectes Sociaux*, 10(2): 167–184; 10(3): 185–296; 10(4): 297–335.

Verwey, J. 1930. Die Paarungsbiologie des Fischreihers. *Zoologische Jahrbücher, Abteilungen Physiologie*, 48: 1–120.

Vince, Margaret A. 1969. Embryonic communication, respiration and the synchronization of hatching. In R. A. Hinde, ed. (*q.v.*), *Bird vocalizations: their relations to current problems in biology and psychology*, pp. 233–260.

Wallace, B. 1958. The average effect of radiation-induced mutations on viability in *Drosophila melanogaster*. *Evolution*, 12(4): 532–556.

——— 1968. *Topics in population genetics*. W. W. Norton, New York. x + 481 pp.

——— 1973. Misinformation, fitness, and selection. *American Naturalist*, 107(953): 1–7.

Washburn, S. L., ed. 1961. *Social life of early man*. Viking Fund Publications in Anthropology no. 31. Aldine Publishing Co., Chicago. ix + 299 pp.

——— 1970. Comment on: "A possible evolutionary basis for aesthetic appreciation in men and apes." *Evolution*, 24(4): 824–825.

Washburn, S. L., and Virginia Avis. 1958. Evolution of human behavior. In Anne Roe and G. G. Simpson, eds., *Behaviour and evolution*, pp. 421–436. Yale University Press, New Haven, Conn. vii + 557 pp.

Washburn, S. L., and F. C. Howell. 1960. Human evolution and culture. In S. Tax, ed., *Evolution after Darwin*, vol. 2, *Evolution of man*, pp. 33–56. University of Chicago Press, Chicago. viii + 473 pp.

Washburn, S. L., Phyllis C. Jay, and Jane B. Lancaster. 1968. Field studies of Old World monkeys and apes. *Science*, 150: 1541–1547.

Watson, A., and R. Moss. 1971. Spacing as affected by territorial behavior, habitat and nutrition in red grouse (*Lagopus l. scoticus*). In A. H. Esser, ed., *Behavior and environment: the use of space by animals and men*, pp. 92–111. Plenum Press, New York. xvii + 411 pp.

Watts, C. R., and A. W. Stokes. 1971. The social order of turkeys. *Scientific American*, 224(6) (June): 112–118.

Weber, N. A. 1972. *Gardening ants: the attines*. Memoirs of the American Philosophical Society no. 92. American Philosophical Society, Philadelphia. xx + 146 pp.

Wecker, S. C. 1963. The role of early experience in habitat selection by the prairie deer mouse, *Peromyscus maniculatus bairdi*. *Ecological Monographs*, 33(4): 307–325.

Weeden, Judith Stenger. 1965. Territorial behavior of the tree sparrow. *Condor*, 67(3): 193–209.

Weeden, Judith Stenger, and J. B. Falls. 1959. Differential responses of male ovenbirds to recorded songs of neighboring and more distant individuals. *Auk*, 76(3): 343–351.

Weinrich, J. D. 1976. Human reproductive strategy: the importance of income unpredictability, and the evolution of non-reproduction. Ph.D. thesis, Harvard University, Cambridge. 231 pp.

Weir, J. S. 1959. Egg masses and early larval growth in *Myrmica*. *Insectes Sociaux*, 6(2): 187–201.

Weiss, R. F., W. Buchanan, Lynne Altstatt, and J. P. Lombardo. 1971. Altruism is rewarding. *Science*, 171: 1262–1263.

West, Mary Jane. 1967. Foundress associations in polistine wasps: dominance hierarchies and the evolution of social behavior. *Science*, 157: 1584–1585.

Weygoldt, P. 1972. Geisselskorpione und Geisselspinnen (*Uropygi* und *Amblypygi*). *Zeitschrift des Kölner Zoo*, 15(3): 95–107.

Wheeler, W. M. 1923. *Social life among the insects*. Harcourt, Brace, New York. vii + 375 pp.

——— 1927a. *Emergent evolution and the social*. Kegan Paul, Trench, Trubner, London. 57 pp.

——— 1927b. The physiognomy of insects. *Quarterly Review of Biology*, 2(1): 1–36.

——— 1928. *The social insects: their origin and evolution*. Harcourt, Brace, New York. xviii + 378 pp.

——— 1933. *Colony-founding among ants, with an account of some primitive Australian species*. Harvard University Press, Cambridge. x + 179 pp.

White, H. C. 1970. *Chains of opportunity: system models of mobility in organizations*. Harvard University Press, Cambridge. xvi + 418 pp.

Whiting, J. W. M. 1968. Discussion, "Are the hunter-gatherers a cultural type?" In R. B. Lee and I. DeVore, eds. (*q.v.*), *Man the hunter*, pp. 336–339.

Wickler, W. 1967a. Vergleichende Verhaltensforschung und Phylogenetik. In G. Heberer, ed., *Die Evolution der Organismen*, vol. 1, pp. 420–508. G. Fischer, Stuttgart. xvi + 754 pp.

——— 1967b. Specialization of organs having a signal function in some marine fish. *Studies in Tropical Oceanography, Miami*, 5: 539–548.

——— 1972. *The sexual code: the social behavior of animals and men*. Doubleday, Garden City, N. Y. xxxi + 301 pp. (Translated from *Sind Wir Sünder?*, Droemer Knaur, Munich, 1969.)

Wickler, W., and Uta Seibt. 1970. Das Verhalten von *Hymenocera picta* Dana, einer Seesterne fressenden Garnele (Decapoda, Natantia, Gnathophyllidae). *Zeitschrift für Tierpsychologie*, 27(3): 352–368.

Williams, C. B. 1964. *Patterns in the balance of nature and related problems in quantitative biology*. Academic Press, New York. vii + 324 pp.

Williams, E. E. 1972. The origin of faunas, evolution of lizard congeners in a complex island fauna: a trial analysis. *Evolutionary Biology*, 6: 47–89.

Williams, G. C. 1964. Measurement of consociation among fishes and comments on the evolution of schooling. *Publications of the Museum, Michigan State University, East Lansing, Biological Series*, 2(7): 351–383.

——— 1966. *Adaptation and natural selection: a critique of*

some current evolutionary thought. Princeton University Press, Princeton, N. J. x + 307 pp.

Williams, G. C., and Doris C. Williams. 1957. Natural selection of individually harmful social adaptations among sibs with special reference to social insects. *Evolution,* 11(1): 32–39.

Wilmsen, E. N. 1973. Interaction, spacing behavior, and the organization of hunting bands. *Journal of Anthropological Research,* 29(1): 1–31.

Wilson, D. S. 1975. A theory of group selection. *Proceedings of the National Academy of Sciences, U. S. A.,* 72: 143–146.

Wilson, E. O. 1958. A chemical releaser of alarm and digging behavior in the ant *Pogonomyrmex badius* (Latreille). *Psyche, Cambridge,* 65(2,3): 41–51.

——— 1962. Chemical communication among workers of the fire ant *Solenopsis saevissima* (Fr. Smith): 1, the organization of mass-foraging; 2, an information analysis of the odour trail; 3, the experimental induction of social responses. *Animal Behaviour,* 10(1,2): 134–164.

——— 1968a. The ergonomics of caste in the social insects. *American Naturalist,* 102(923): 41–66.

——— 1968b. Chemical systems. In T. A. Sebeok, ed., *Animal communication: techniques of study and results of research,* pp. 75–102. Indiana University Press, Bloomington. xviii + 686 pp.

——— 1970. Chemical communication within animal species. In E. Sondheimer and J. B. Simeone, eds., *Chemical ecology,* pp. 133–155. Academic Press, New York. xvi + 336 pp.

——— 1971a. *The insect societies.* Belknap Press of Harvard University Press, Cambridge. x + 548 pp.

——— 1971b. Competitive and aggressive behavior. In J. F. Eisenberg and W. Dillon, eds., *Man and beast: comparative social behavior,* pp. 183–217. Smithsonian Institution Press, Washington, D. C. 401 pp.

——— 1972a. On the queerness of social evolution. *Bulletin of the Entomological Society of America,* 19(1): 20–22.

——— 1972b. Animal communication. *Scientific American,* 227(3) (September): 52–60.

——— 1973. Group selection and its significance for ecology. *BioScience,* 23(11): 631–638.

——— 1974a. The soldier of the ant *Camponotus* (*Colobopsis*) *fraxinicola* as a trophic caste. *Psyche, Cambridge,* 81(1): 182–188.

——— 1974b. The population consequences of polygyny in the ant *Leptothorax curvispinosus* Mayr (Hymenoptera: Formicidae). *Annals of the Entomological Society of America,* 67(5): 781–786.

——— 1975. *Sociobiology: The new synthesis.* Belknap Press of Harvard University Press, Cambridge. x + 697 pp.

——— 1978. *On human nature.* Harvard University Press, Cambridge. xii + 260 pp.

Wilson, E. O., and W. H. Bossert. 1963. Chemical communication among animals. *Recent Progress in Hormone Research,* 19: 673–716.

——— 1971. *A primer of population biology.* Sinauer Associates, Sunderland, Mass. 192 pp.

Wilson, E. O., T. Eisner, W. R. Briggs, R. E. Dickerson, R. L. Metzenberg, R. D. O'Brien, M. Susman, and W. E. Boggs. 1973. *Life on earth.* Sinauer Associates, Sunderland, Mass. xiv + 1053 pp.

Wilson, E. O., and F. E. Regnier. 1971. The evolution of the alarm-defense system in the formicine ants. *American Naturalist,* 105(943): 279–289.

Wilson, R. S. 1978. Synchronies in mental development: an epigenetic perspective. *Science,* 202: 939–948.

Wood-Gush, D. G. M. 1955. The behaviour of the domestic chicken: a review of the literature. *British Journal of Animal Behaviour,* 3(3): 81–110.

Woolfenden, G. E. 1973. Nesting and survival in a population of Florida scrub jays. *Living Bird,* 12: 25–49.

——— 1974a. Florida scrub jay helpers at the nest. *Auk,* 92(1): 1–15.

——— 1974b. The effect and source of Florida scrub jay helpers. (Unpublished manuscript.)

Wright, S. 1943. Isolation by distance. *Genetics,* 28(2): 114–138.

——— 1945. Tempo and mode in evolution: a critical review. *Ecology,* 26(4): 415–419.

Wynne-Edwards, V. C. 1962. *Animal dispersion in relation to social behaviour.* Oliver and Boyd, Edinburgh. xi + 653 pp.

Zuckerman, S. 1932. *The social life of monkeys and apes.* Harcourt, Brace, New York. xii + 356 pp.

Zumpe, Doris. 1965. Laboratory observations on the aggressive behaviour of some butterfly fishes (*Chaetodontidae*). *Zeitschrift für Tierpsychologie,* 22(2): 226–236.

Index

A

B

C

D

E

F

G

H

I

J

K

L

M

N

O

P

Q

R

S

T

U

V

W

X

Y

Z